Telemetry Techniques:
A User Guide for Fisheries Research

Telemetry Techniques: A User Guide for Fisheries Research

Edited by

Noah S. Adams
United States Geological Survey, Western Fisheries Research Center
Columbia River Research Laboratory
5501 A Cook-Underwood Road, Cook, Washington 98605, USA

John W. Beeman
United States Geological Survey, Western Fisheries Research Center
Columbia River Research Laboratory
5501 A Cook-Underwood Road, Cook, Washington 98605, USA

John H. Eiler
National Marine Fisheries Service, Alaska Fisheries Science Center
Auke Bay Laboratories
TSMRI 17109 Point Lena Loop Road, Juneau, Alaska, 99801, USA

American Fisheries Society
Bethesda, Maryland
2012

A suggested citation format for this book follows.

Entire Book

Adams, N. S., J. W. Beeman, and J. H. Eiler, editors. 2012. Telemetry techniques: a user guide for fisheries research. American Fisheries Society, Bethesda, Maryland.

Chapter within the Book

Liedtke, T. L., and A. M. W. Rub. 2012. Techniques for telemetry transmitter attachment and evaluation of transmitter effects on fish performance. Pages 45–87 *in* N. S. Adams, J. W. Beeman, and J. H. Eiler, editors. Telemetry techniques: a user guide for fisheries research. American Fisheries Society, Bethesda, Maryland.

Printed in the United States of America on acid-free paper.

Library of Congress Control Number 2012942368
ISBN 978-1-934874-26-4

American Fisheries Society Web site address: *www.fisheries.org*

American Fisheries Society
5410 Grosvenor Lane, Suite 100
Bethesda, Maryland 20814
USA

Table of Contents

Preface

The idea for this book came about in the late 1990s when we realized that the popularity of using radio and acoustic telemetry in fisheries research was progressing faster than the guidelines for its implementation. Most fishery biologists have at best a brief exposure to this tool in their formal schooling, but few are aware of the underlying theory behind sound and radio wave propagation, the variety of equipment available, or how to apply telemetry effectively to collect the information needed to address research and management questions. Journal articles usually focus on the information obtained, and do not include many of the practical considerations related to field use. We suggest that understanding the basics is crucial to the efficient use of this tool. In response to this need, we convened a session at the 2007 Annual Meeting of the American Fisheries Society as a forum for discussing the basic methods and new advancements in the fields of acoustic and radio telemetry. Many of the participants in that session have authored Sections in this book.

The book is written primarily as a handbook for fishery biologists working in the field, but it may also be useful to those interested in studying other aquatic animals or managers looking for some background to help them interpret results of studies based on radio or acoustic telemetry. Our intent is to provide a basic understanding of the theory of signal transmission, propagation, and reception of radio and acoustic waves from the transmitter to the receiver, and to build on that foundation with practical considerations and guidelines from the engineers designing the equipment and the biologists using it in the field. Specific case studies are also presented as examples of how these methods can be applied to address a variety of topics. We strived to present information that is not specific to any particular manufacturer's equipment since information presented in this way provides a more functional foundation that will have a longer, more useful life in the ever-changing field of acoustic and radio telemetry.

Some of the topics stressed in this book will be vital in some applications and trivial in others. For example, overcoming radio signal loss from the receive antenna to the receiver is important when the transmitters are relatively weak, water conductivity is high, and distance between antenna and receiver requires long cable lengths. However, this is not as critical when using tags that emit a powerful signal and cable lengths are relatively short. The key is to understand enough about the technology to know the difference and enable useful results from either application. Only the reader can determine the applicability of this book to their work, but we hope most will find something useful within it.

The United States Geological Survey provided financial support to assist with the publishing of this book. The use of trade names does not imply endorsement by the U.S. Government.

Noah S. Adams
John W. Beeman
John H. Eiler
June 2012

Contributors

Noah S. Adams (Coeditor): United States Geological Survey, Western Fisheries Research Center, Columbia River Research Laboratory, 5501 A Cook-Underwood Road, Cook, Washington 98605, USA
nadams@usgs.gov

John W. Beeman (Coeditor, Sections 2, 6.5, and 9.5): United States Geological Survey, Western Fisheries Research Center, Columbia River Research Laboratory, 5501 A Cook-Underwood Road, Cook, Washington 98605, USA
jbeeman@usgs.gov

Sam J. Brenkman (Section 6.1): Department of Interior, National Park Service, Olympic National Park, 600 East Park Avenue, Port Angeles, Washington 98362, USA
sam_brenkman@nps.gov

Edd J. Brooks (Section 8.2): Cape Eleuthera Institute, Eleuthera, The Bahamas
eddbrooks@ceibahamas.org

Theodore Castro-Santos (Sections 9.1 and 9.2): United States Geological Survey, Leetown Science Center, Conte Anadromous Fish Research Center, One Migratory Way, P.O. Box 796, Turners Falls, Massachusetts 01376, USA
tcastrosantos@usgs.gov

Steven J. Cooke (Section 8.2): Fish Ecology and Conservation Physiology Laboratory, Department of Biology, Carleton University, 1125 Colonel By Drive, Ottawa, Ontario, K1S 5B6, Canada
Steven_cooke@carleton.ca

Stephen C. Corbett (Section 6.1): Pacific States Marine Fisheries Commission, Northwest Fisheries Science Center, Seattle, Washington 98112, USA
Steve.Corbett@noaa.gov

Timothy D. Counihan (Section 9.3): United States Geological Survey, Western Fisheries Research Center, Columbia River Research Laboratory, 5501 A Cook-Underwood Road, Cook, Washington 98605, USA
tcounihan@usgs.gov

Andy J. Danylchuk (Section 8.2): Department of Environmental Conservation, University of Massachusetts, 160 Holdsworth Way, Amherst, Massachusetts 01003, USA
adanylch@nrc.umass.edu

John H. Eiler (Coeditor, Section 5.3): National Marine Fisheries Service, Alaska Fisheries Science Center, Auke Bay Laboratories, TSMRI 17109 Point Lena Loop Road, Juneau, Alaska, 99801, USA
john.eiler@noaa.gov

Karl K. English (Section 9.4): LGL Limited, 9768 Second Street, Sidney, British Columbia, V8L 3Y8, Canada
kenglish@lgl.com

Scott D. Evans (Section 5.2): United States Geological Survey, Western Fisheries Research Center, Columbia River Research Laboratory, 5501 A Cook-Underwood Road, Cook, Washington 98605, USA
sdevans@usgs.gov

Chris R. Haak (Section 8.2): Department of Environmental Conservation, University of Massachusetts, 160 Holdsworth Way, Amherst, Massachusetts, 010035, USA
chaak@eco.umass.edu

Lisiane Hahn (Section 6.3): Neotropical Environmental Consulting and Advisory Services, Rua Cesário Rossetto, n. 265, Grove Lucas Araújo, RS, 990740-210, Passo Fundo, Brazil
lisiane@neotropical.com.br

Jill M. Hardiman (Section 9.3): United States Geological Survey, Western Fisheries Research Center, Columbia River Research Laboratory, 5501 A Cook-Underwood Road, Cook, Washington 98605, USA
jhardiman@usgs.gov

Clinton J. Hay (Section 6.2): Formerly from the Ministry of Fisheries and Marine Resources, Private Bag 2116, Mariental, Nambia
cjhay@telkomsa.net

Christopher M. Holbrook (Sections 8.1 and 9.2): United States Geological Survey, Great Lakes Science Center, Hammond Bay Biological Station, Western Basin Ecosystem Branch, 11188 Ray Road, Millersburg, Michigan 49759, USA
cholbrook@usgs.gov

Eric E. Hockersmith (Sections 2 and 6.5): United States Army Corps of Engineers, Walla Walla District, 201 North Third Ave, Walla Walla, Washington, 99362, USA
Eric.E.Hockersmith@usace.army.mil

Samuel V. Johnston (Section 7.1): Hydroacoustic Technology, Inc., 715 NE Northlake Way, Seattle, Washington 98105
sjohnston@htisonar.com

John D. Koehn (Section 3): Arthur Rylah Institute for Environmental Research, Department of Sustainability and Environment, 123 Brown Street, Heidelberg, Victoria 3079, Australia
john.koehn@dse.vic.gov.au

Valerian B. Kuechle (Section 5.1): Advanced Telemetry Systems, 470 First Avenue North, P.O. Box 398, Isanti, Minnesota 55040, USA
lkuechle@atstrack.com

Peter J. Kuechle (Section 5.1): Advanced Telemetry Systems, 470 First Avenue North, P.O. Box 398, Isanti, Minnesota 55040, USA
pkuechle@atstrack.com

Theresa L. Liedtke (Section 4): United States Geological Survey, Western Fisheries Research Center, 5501 A Cook-Underwood Road, Cook, Washington 98605, USA
tliedtke@usgs.gov

Michael C. Melnychuk (Section 7.2): Fisheries Centre and Department of Zoology, University of British Columbia, 2202 Main Hall, Vancouver, British Columbia, V6T 1Z4, Canada
mikem@zoology.ubc.ca

Karen J. Murchie (Section 8.2): Fish Ecology and Conservation Physiology Laboratory, Department of Biology, Carleton University, 1125 Colonel By Drive, Ottawa, Ontario K1S 5B6, Canada
kmurchie@connect.carleton.ca

Tor F. Næsje (Section 6.2): Norwegian Institute for Nature Research, P.O. Box 5685 Sluppen, NO-7485 Trondheim, Norway
tor.naesje@nina.no

Finn Økland (Section 6.2): Norwegian Institute for Nature Research, P.O. Box 5685 Sluppen, NO-7485 Trondheim, Norway
finn.okland@nina.no

Amanda C. O'Toole (Section 8.2): Fish Ecology and Conservation Physiology Laboratory, Department of Biology, Carleton University, 1125 Colonel By Drive, Ottawa, Ontario K1S 5B6, Canada
aotoole@connect.carleton.ca

Russell W. Perry (Sections 9.1, 9.2, and 9.5): United States Geological Survey, Western Fisheries Research Center, Columbia River Research Laboratory, 5501 A Cook-Underwood Road, Cook, Washington 98605, USA
rperry@usgs.gov

Douglas G. Pincock (Section 7.1): Vemco Division, Amirix Systems Incorporated, 211 Horseshoe Lake Drive, Halifax, Nova Scotia, B3S 0B9, Canada
pincock@amirix.com

Dave Robichaud (Section 9.4): LGL Limited, 9768 Second Street, Sidney, British Columbia V8L 3Y8, Canada
drobichaud@lgl.com

A. Michelle Wargo Rub (Section 4): Fish Ecology Division, Northwest Fisheries Science Center, National Marine Fisheries Service, National Oceanic and Atmospheric Administration, 2725 Montlake Boulevard East, Seattle, Washington 98112, USA
michelle.rub@noaa.gov

Benjamin P. Sandford (Section 9.2): National Oceanic and Atmospheric Administration, Fisheries, 3305 East Commerce Street, Pasco, Washington, 99361, USA
ben.sandford@noaa.gov

Aaron D. Shultz (Section 8.2): Cape Eleuthera Institute, Eleuthera, The Bahamas
aaronshultz@ceibahamas.org

Tracey W. Steig (Section 8.1): Hydroacoustic Technology Incorporated, 715 NE Northlake Way, Seattle, Washington 98105, USA
tsteig@htisonar.com

John R. Stevenson (Sections 5.2 and 6.5): BioAnalysts, Incorporated, 16541 Redmond Way, # 339, Redmond, Washington 98052, USA
john.stevenson@bioanalysts.net

Cory D. Suski (Section 8.2): Department of Natural Resources and Environmental Sciences, University of Illinois, 1102 S. Goodwin Avenue, Urbana, Illinois 61801, USA
suski@illinois.edu

Eva Bonsak Thorstad (Section 6.2): Norwegian Institute for Nature Research, P.O. Box 5685 Sluppen, NO-7485 Trondheim, Norway
eva.thorstad@nina.no

Peter Wainwright (Section 9.4): LGL Limited, 9768 Second Street, Sidney, British Columbia V8L 3Y8, Canada
pwainwright@lgl.com

Christopher E. Walker (Section 9.3): United States Army Corps of Engineers, Portland District, 333 Southwest First Street, Portland, Oregon 97208, USA
Christopher.E.Walker@usace.army.mil

Fred G. Whoriskey (Section 8.3): Dalhousie University, Ocean Tracking Network, 1355 Oxford Street, Halifax, Nova Scotia B3H 4J1, Canada
FWhoriskey@Dal.Ca

Jim D. Winter (Section 6.4): University of Arkansas, Little Rock Graduate Institute of Technology, 2801 South University Avenue, 215 FH, Little Rock, Arkansas 72204, USA
jdwinter@ualr.edu

Symbols and Abbreviations

The following symbols and abbreviations may be found in this book without definition. Also undefined are standard mathematical and statistical symbols given in most dictionaries.

A	ampere
AC	alternating current
Bq	becquerel
C	coulomb
°C	degrees Celsius
cal	calorie
cd	candela
cm	centimeter
Co.	Company
Corp.	Corporation
cov	covariance
DC	direct current; District of Columbia
D	dextro (as a prefix)
d	day
d	dextrorotatory
df	degrees of freedom
dL	deciliter
E	east
E	expected value
e	base of natural logarithm (2.71828…)
e.g.	(exempli gratia) for example
eq	equivalent
et al.	(et alii) and others
etc.	et cetera
eV	electron volt
F	filial generation; Farad
°F	degrees Fahrenheit
fc	footcandle (0.0929 lx)
ft	foot (30.5 cm)
ft^3/s	cubic feet per second ($0.0283\ m^3/s$)
g	gram
G	giga (10^9, as a prefix)
gal	gallon (3.79 L)
Gy	gray
h	hour
ha	hectare (2.47 acres)
hp	horsepower (746 W)
Hz	hertz
in	inch (2.54 cm)
Inc.	Incorporated
i.e.	(id est) that is
IU	international unit
J	joule
K	Kelvin (degrees above absolute zero)
k	kilo (10^3, as a prefix)
kg	kilogram
km	kilometer
l	levorotatory
L	levo (as a prefix)
L	liter (0.264 gal, 1.06 qt)
lb	pound (0.454 kg, 454g)
lm	lumen
log	logarithm
Ltd.	Limited
M	mega (10^6, as a prefix); molar (as a suffix or by itself)
m	meter (as a suffix or by itself); milli (10^{-3}, as a prefix)
mi	mile (1.61 km)
min	minute
mol	mole
N	normal (for chemistry); north (for geography); newton
N	sample size

NS not significant
n ploidy; nanno (10^{-9}, as a prefix)
o ortho (as a chemical prefix)
oz ounce (28.4 g)
P probability
p para (as a chemical prefix)
p pico (10^{-12}, as a prefix)
Pa pascal
pH negative log of hydrogen ion activity
ppm parts per million
qt quart (0.946 L)
R multiple correlation or regression coefficient
r simple correlation or regression coefficient
rad radian
S siemens (for electrical conductance); south (for geography)
SD standard deviation
SE standard error
s second
T tesla
tris tris(hydroxymethyl)-aminomethane (a buffer)
UK United Kingdom
U.S. United States (adjective)
USA United States of America (noun)
V volt
V, Var variance (population)
var variance (sample)
W watt (for power); west (for geography)
Wb weber
yd yard (0.914 m, 91.4 cm)
α probability of type I error (false rejection of null hypothesis)
β probability of type II error (false acceptance of null hypothesis)
Ω ohm
μ micro (10^{-6}, as a prefix)
′ minute (angular)
″ second (angular)
° degree (temperature as a prefix, angular as a suffix)
% per cent (per hundred)
‰ per mille (per thousand)

Introduction and Overview

Section 1

Introduction

Telemetry provides a powerful and flexible tool for studying aquatic animals, making it possible to repeatedly locate and identify individuals in remote or inaccessible settings—a task that would be difficult (if not impossible) to accomplish using other methods. The use of telemetry has increased dramatically in recent years, and its application is limited only by the capabilities of the equipment and the researcher's imagination. In spite of these advantages, telemetry is equipment-intensive and generally requires more specialized knowledge and training than many other field techniques. The electronic equipment associated with its use can often dazzle, intimidate, and confuse those just starting out. Even experienced users are often hard-pressed to keep up with the technological advances. Answers to such basic questions as "what equipment do I need?" or "how do I get started?" are not always evident or straightforward. These are valid concerns, since the equipment and methods used can affect the success of the study and the quality of information collected.

The purpose of this book is to provide a guide for using telemetry to study aquatic animals—call it a user's manual or Telemetry 101. Our principal intent is to provide the basic information to plan, implement, and conduct telemetry studies under field conditions. Considerations related to data collection and interpretations are also discussed. As with any scientific procedure, collecting usable information and accurately interpreting study results depends on an understanding of the underlying principles of the methods used. A wide range of telemetry equipment and field techniques are available. Clearly defined research objectives and knowledge of the various options, capabilities, and limitations of the equipment and methods is essential for developing projects that effectively address the research or management questions being asked. Telemetry is a tool, and like any tool it will only function effectively when used properly.

So what is telemetry? Telemetry is derived from the Greek words "tele" meaning remote and "metron" meaning to measure. Over the years this term has expanded to encompass a wide range of data collection techniques. In this book, we confine our discussion to the methods used to tag and track animals with active transmitters (i.e., electronic tags equipped with their own energy source) in either the radio or acoustic frequency range. Radio and acoustic telemetry has traditionally focused on locating tagged individuals, although the advent of transmitter sensors for collecting proximal data, such as water depth and temperature, has also made it possible to address questions related to the surrounding environment and why animals are exhibiting certain behaviors. There are several related topics that this book does not cover. Although passive integrated transponder (PIT) technology can also be viewed as telemetry, the energy needed to transmit the data from the tag is generated by the receiving system. The passive nature and technical considerations associated with this approach distinguishes it from the methods covered in this book, and have been addressed by other authors (Prentice et al. 1990). Similarly, tracking large marine animals (e.g., marine mammals and other surface-oriented species) with satellite-linked transmitters and the use of pop-up satellite archival tags for collecting movement, depth, and temperature data are relatively distinct telemetry applications and are discussed in other forums (Timko and Kolz 1982; Block et al. 1998; Hart and Hyrenbach 2009). Passive hydroacoustic systems including SONAR (sound navigation and ranging) are increasingly being used to monitor aquatic animals. However, this approach does not utilize tags, and instead detects untagged individuals in the water by sending out sound waves and measuring the time taken by the sound to travel back to the receiver after reflecting off objects in the water (Brandt 1996). This discipline is also beyond the scope of this book.

Telemetry is a complex and wide ranging field and this book has been organized with the goal of making it easy for you to access the information pertinent to your particular application. The first three Sections of the book (Sections 2–4) contains information on several general topics. Section 2 provides a historical account of how telemetry has developed and evolved since its inception in the 1950s. In particular, how the collaboration between biologists and engineers, combined with the surge of new technologies, has contributed to the enhanced equipment, field techniques, and data processing methods currently available. The next two Sections are devoted to topics common to all telemetry studies. Section 3 discusses study design and factors that should be considered during the initial planning process. Although telemetry can provide detailed and hard-to-get information, the costs in terms of time, effort, and resources can be substantial. This Section examines the principal options and decisions that need to be considered in order to maximize study success. Telemetry involves the tagging of animals, and is based on the assumption that the handling and tagging methods used have minimal impact (i.e., behaviors of tagged animals represent behaviors of untagged animals). Choosing methods that are minimally invasive and are defendable in terms of animal care concerns is essential for collecting meaningful data. Section 4 describes the principal methods used to tag aquatic animals and reviews the basic factors that should be considered to minimize tagging and handling effects.

The remaining Sections discuss specific telemetry applications. Although many papers in the scientific literature report using telemetry, specific details which would benefit other users are often lacking. Sections 5.1–5.3 are devoted to the basics of radio telemetry and describe the key principles and theories, the basic equipment used, and practical methods for monitoring and tracking tagged individuals. Sections 6.1–6.5 contain case studies that describe how radio telemetry has been used to address research objectives in different environmental settings. Sections 7.1 and 7.2 treat acoustic telemetry in a similar manner, describing the underlying principles, theories, and basic equipment as well as means to describe and estimate the probability of detecting tagged animals. Specific case studies on acoustic telemetry are presented in Sections 8.1–8.3. As is the case with most of the information in the book, even though these sections focus on their particular perspective (radio or acoustic telemetry), many of the techniques discussed for one are applicable to the other. Sections 9.1–9.5 describe methods for managing telemetry information and discusses research questions less frequently confronted using telemetry. The topics include analyzing animal movements in relation to time (Section 9.1), using telemetry-based mark-recapture models to estimate fish survival (Section 9.2), aspects related to data management and quality control (Sections 9.3 and 9.4), and the effects of false-positive records, errant records likely present in all telemetry data, and methods to remove them prior to analysis (Section 9.5). Although these topics often receive less attention, their importance in making meaningful inferences from telemetry data is obvious.

Telemetry is a diverse field, and there are undoubtedly additional topics that could have been included or covered in more detail. Field work is a dynamic process due to the diverse and varying conditions often encountered. Although it is impossible to cover every contingency, this book will hopefully provide you with the basic information needed to help identify and address the problems and challenges that often arise and are an inevitably part of field research.

The material contained in this book is the product of many people from around the world, including biologists, technicians, and engineers from the private, academic, and government sectors. Their experience with the design, function, and use of telemetry equipment has made it possible to apply this technology to a wide range of research questions concerning the biology and management of aquatic resources. Telemetry is a rapidly expanding field, and some of the information provided will likely become outdated with future advances in technology. Such is the way of science, and these advances may provide new avenues of study and exciting insights into the aquatic environment. As technology changes, it is our hope this book will continue to be a useful guide to telemetry techniques and assist your efforts to address the complex environmental challenges and opportunities that face our world.

REFERENCES

Block, B. A., H. Dewar, C. Farwell, and E. D. Prince. 1998. Proceedings of the National Academy of Science 95:9384–9389.

Brandt, S. B., Acoustic assessment of fish abundance and distribution. 1996. Pages 385–419 *in* B. R. Murphy and D. W. Willis, editors. Fisheries Techniques, 2nd edition. American Fisheries Society, Bethesda, Maryland.

Hart, K. M., and K. D. Hyrenbach. 2009. Satellite telemetry of marine megavertebrates: the coming of age of an experimental science. Endangered Species Research 10:9–20.

Prentice, E. F., T. A. Flagg, and C. A. McCutcheon. 1990. Feasibility of using implantable passive integrated transponder (PIT) tags in salmonids. Pages 317–322 *in* N. C. Parker, A. E. Giorgi, R. C. Heidinger, D. B. Jester, Jr., E. D. Prince, and G. A. Winans, editors. American Fisheries Society, Symposium 7, Bethesda, Maryland.

Timko, R. E., and A. L. Kolz. 1982. Satellite sea turtle tracking. Marine Fisheries Review 44(4):19–24.

Section 2

A History of Telemetry in Fishery Research

Eric E. Hockersmith and John W. Beeman

Biotelemetry has been defined as "the instrumental technique for gaining and transmitting information from a living organism and its environment to a remote observer" (Slater 1965). Biotelemetry typically utilizes wireless transmission of either an audible signal or electronic data to determine location of a tagged animal. Fisheries researchers use location information to gain a variety of insights into migration, habitat use, behavior, productivity, or survival of fish. Biotelemetry can be divided into two basic categories, acoustic or radio, based on mode of transmission, mechanical or electromagnetic energy, and operating frequency. Most acoustic systems in use today transmit at low frequency, between 30 and 300 kHz, while most radio systems transmit at very high frequency, between 30 and 300 MHz (Sisak and Lotimer 1998).

Acoustic telemetry is based on the principals of sonar (sound navigation and ranging), which was developed to detect submarines during World War I. The properties of acoustic systems favor their use in deep waters with high conductivity and low turbulence (Winter 1996). Radio telemetry is based on the principals of wireless radio communication, which were first demonstrated by Nikola Tesla in 1893. Radio systems are best suited in shallow waters with relatively low conductivity but have the added benefit of improved signal detection in turbulent conditions and with aerial antennas. Advances in both technologies have resulted in highly efficient transmitter and receiving systems.

Advancements in products used for animal telemetry over the past 50 years have generally followed those in the electronics field (Figure 1). Bell Laboratories[1] ushered in the age of digital electronics with the invention of the transistor in 1947 (Mann 2000). Today transistors are common in everyday items such as radios, televisions, hearing aids, computers, cell phones and even MP3 players. Consumer demand for inexpensive small electronic devices with increased functionality has continually driven advancements in the field of electronics. These advancements have subsequently led to improvements in biotelemetry transmitters and receivers such as miniaturization of components, increased battery performance, and more powerful micro-processing.

The earliest use of biotelemetry to study fish and wildlife occurred in the mid- to late 1950s. Research covered a broad field, including examination of salmon migrations (Trefethen 1956), measurement of incubation temperatures for penguin eggs (Ecklund and Charltin 1959), and investigation of woodchuck movements (LeMunyan et al. 1959). Telemetry brought two new advantages to fish and wildlife research: the ability to identify individual

[1]Reference to trade names does not imply endorsement by the U.S. Government.

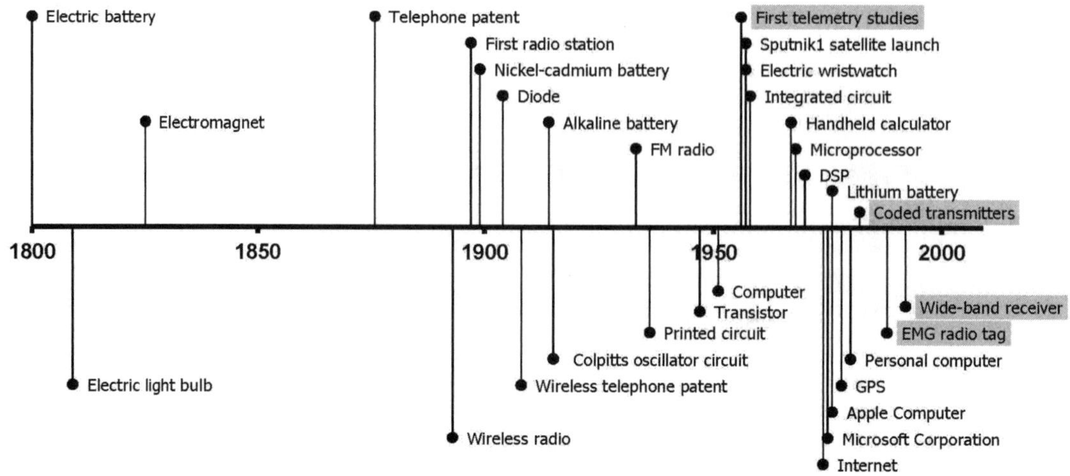

FIGURE 1. Time line of major developments in electronics that contributed to advances in fisheries telemetry over the past 200 years. Telemetry specific developments are highlighted.

animals, and the ability to locate individual animals without having to observe or recapture them, greatly enhancing behavior and migration studies. Further, the quantity and detail of data generated from studies using biotelemetry has often been greater than that generated from mark–recapture or direct observation methods (Adams and Davis 1967).

The first acoustic telemetry equipment for studying fish was developed in 1956 by the U.S. Bureau of Commercial Fisheries (BCF) and the Minneapolis-Honeywell Regulator Corporation (Figure 2; Trefethen 1956; Trefethen et al. 1957). Fish were externally tagged with a transmitter attached to the dorsal area of the fish (Figure 3). These studies examined adult salmon passage behavior at Bonneville Dam on the Columbia River (Trefethen 1956; Johnson 1960). Fish locations and movements were determined by mobile tracking from a boat (Figure 4) and using fixed-site acoustic receivers that recorded the data on paper and magnetic tape (Figure 5). In considering the advances in knowledge of fish movement and migration, Arnold (2000) concluded that the insight gained during the 30 years following application of biotelemetry for fisheries research were greater than those during the preceding 75 years.

Acoustic telemetry use in fisheries research continued to develop during the 1960s, following invention of the integrated circuit by Texas Instruments in 1958. Because of their size, early acoustic transmitters could be used only with large adult fish. However, within a few years of initial development, an acoustic tag was produced that was approximately 25% smaller and had twice the detection range and three times the tag life of earlier designs (Novotny and Esterberg 1962). At approximately the same time, University of Wisconsin researchers developed a much smaller acoustic transmitter for use with white bass. These fish had an average length of only 300 mm, a size considerably less than adult Chinook salmon *Oncorhynchus tshawytscha*, but due to battery limitations, the tag had a life of less than 24 h (Henderson et al. 1966).

In the early 1960s, the BCF used acoustic-tagged fish to determine whether the newly formed Brownlee Reservoir on the Snake River impacted the ability of adult Chinook salmon to migrate to traditional spawning grounds (Trefethen 1968). They used both mobile tracking and fixed-site monitoring to locate tagged fish. Similar equipment was used to study Chinook

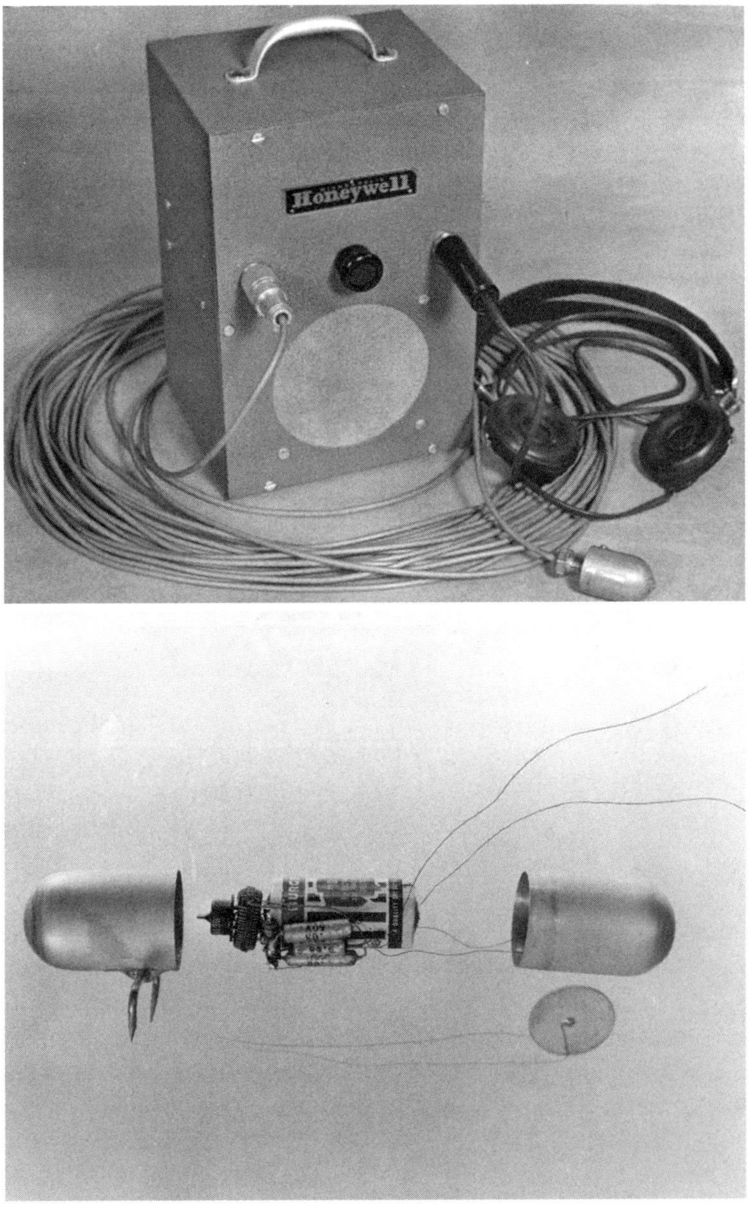

FIGURE 2. The first acoustic receiver and transmitter developed to study migration and behavior of adult salmon in the Columbia River (Trefethen 1956; Trefethen et al. 1957).

Figure 3. Early acoustic tag which used a hog ring clamped into the dorsal muscle of the fish for attachment.

Figure 4. Acoustic tracking system used for mobile tracking adult salmon from a boat during the late 1950s near Bonneville Dam on the Columbia River.

FIGURE 5. Fixed-site acoustic receiver system used during the late 1950s near Bonneville Dam on the Columbia River. Data was recorded on both paper and magnetic tape.

salmon movements through the San Joaquin River delta in California in 1964 (Hallock et al. 1970). In the USSR, acoustic transmitters developed in 1964 were used to study sturgeon migrations in 1965 (Malinin and Svirskii 1973). By 1967, acoustic telemetry was used routinely, and applications were expanded to investigation of more fine-scale fish movement, such as a study of net avoidance by American shad in the Connecticut River (Leggett and Jones 1971).

During the late 1960s and early 1970s, research programs such as those at the Marine Biotelemetry Laboratory at the University of New Brunswick led to improvements in acoustic signal transmission efficiency in water from about 1–80%. The ability to automatically receive and decode transmitted information was also developed around this time (D. Pincock, Electrical Engineering Department, Technical University of Nova Scotia, personal communication), along with the ability to transmit pressure and temperature information from acoustic tags (Luke et al. 1973; Rochelle and Coutant 1973). Many transmitters used in these programs were still hand-built from analog parts soldered together under stereoscopes. The increasing availability of digital rather than analog equipment during the 1970s enabled more efficient construction techniques and products that were more reliable.

Radio telemetry was first used by wildlife biologists to study movement of terrestrial animals in 1956 (LeMunyan et al. 1959), and its use grew rapidly during the 1960s (Cochran and Lord 1963). However, this technology was not used to study fish until 1968 (Lonsdale and Baxter 1968). During the 1970s, a variety of researchers began modifying radio telemetry systems used in terrestrial habitats for use in freshwater.

As the use of acoustic telemetry increased, fisheries researchers found that it performed poorly in some environments, such as areas of high turbulence (Trefethen 1956; Johnson 1960) or large areas with limited access for monitoring. In 1971, National Marine Fisheries Service (formerly BCF) researchers switched from acoustic to radio tags to evaluate the upstream migration behavior of Chinook salmon. Radio tags were chosen because they performed well in the turbulent conditions below dams and had higher detection rates than acoustic transmitters (Monan et al. 1975). The ability to detect radio transmitters in air using highly portable aerial antennas also provided a major advantage for migration studies in large or inaccessible river systems (Koehn 1999). Furthermore, fisheries researchers benefitted significantly by using the methods, equipment, and analytical techniques developed from wildlife telemetry. By the mid-1970s, radio telemetry and its newly discovered advantages had eclipsed acoustic telemetry to the point where use of the latter was limited to systems with high conductivity where radio telemetry worked poorly (Stasko and Pincock 1977).

Prior to about 1970, most researchers using acoustic and radio telemetry made at least some of their own equipment (Lonsdale and Baxter 1968; Standora et al. 1972), or worked with private sector electronic engineers, as in the relationship between the BCF and Honeywell (Trefethen et al. 1957). It was common for fisheries researchers to describe circuit diagrams of telemetry transmitters when publishing their biological results (Cochran and Lord 1963; Luke et al. 1973). Radio telemetry transmitters and receivers were first made commercially available in 1967 by AVM Instrument Company (Figure 6). For acoustic telemetry, commercial equipment was first produced by Sonotronics, founded in 1971. During the early 1970s acoustic receivers were also developed by the Marine Biotelemetry Laboratory at the University of New Brunswick and Smith Root (Figure 7).

Both technological advances and biotelemetry companies have proliferated since the 1970s. Today more than 30 companies produce equipment for biotelemetry applications worldwide, with the majority located in North America. Roughly twice as many manufacturers produce equipment for radio telemetry compared to acoustic telemetry, and only a few produce both. The higher prevalence of radio telemetry equipment manufacturers is likely due to greater demand, resulting from the ability of radio telemetry to be used in both aquatic and

Figure 6. Early versions of commercially produced radio telemetry receivers. The receiver on the left was produced by AVM in the late 1960s and the receiver on the right was produced by Smith-Root in the early 1970s.

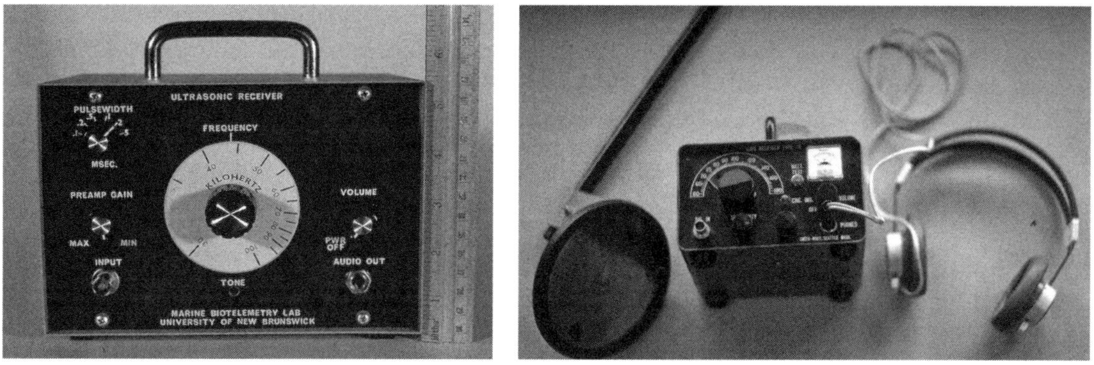

FIGURE 7. Acoustic telemetry receivers from the 1970s. The receiver on the left was developed by The University of New Brunswick Marine Biotelemetry Lab (photo courtesy of Doug Pincock) and the receiver on the right was produced by Smith-Root.

terrestrial settings. About one-third of all biotelemetry startup companies formed during the past 50 years began during the 1980s following the rapid advances in electronic components during this period (Arnold and Dewar 2001).

Commercialization of telemetry equipment allowed the technology to be used by researchers who did not have the capability of manufacturing their own equipment. Between 1970 and 1980, the same advances in electronics that led to the first digital cell phones, lithium batteries, and personal computers were used by biotelemetry firms to increase the capability of commercial transmitters and receivers. Data-logging radio and acoustic receivers became more refined and more available during this period, and transmitters continued to become smaller and more capable. Reduction in tag size, while either maintaining or improving detection range and tag life, continues to be a primary objective of innovation in transmitter development (Figure 8). Today very small acoustic and radio transmitters are readily available, with some as light as 0.2 g in air (Figure 9; Naef-Daenzer et al. 2005) and have provided opportunities to study species that previously could not be tagged due to their small size, as well as the juvenile life stages of many species.

There are two types of transmitters—noncoded (typically referred to as pinger tags) or coded. Further description and an explanation of transmitters and coding systems is provided in Section 5. The earliest studies used noncoded transmitters because existing receivers were simple in design and did not contain the circuitry to process the signals from coded transmitters. In 1982, the National Marine Fisheries Service developed the first pulse code modulation radio transmitter, with a code set that provided the ability to study up to 600 individuals at a time (Stuehrenberg et al. 1990). They used the new transmitters in the Columbia River Basin to study salmonid migration and dam passage, first of adults during 1982, and subsequently of juveniles during 1985. The coding of these transmitters provided the ability to greatly increase the number of animals tracked, as well as to decrease the scan time of receivers. By the early 2000s there were several commercial coding schemes available, with most transmitter manufacturers using their own.

Wide-band receivers that could simultaneously scan multiple radio frequencies were developed shortly after the advent of coded transmitters. One of the first wide-band receivers could decode and record up to 144 unique transmitters that simultaneously occupied a single detection field (Stuehrenberg et al. 1990). Prior to the 1990s, most multi-frequency receivers

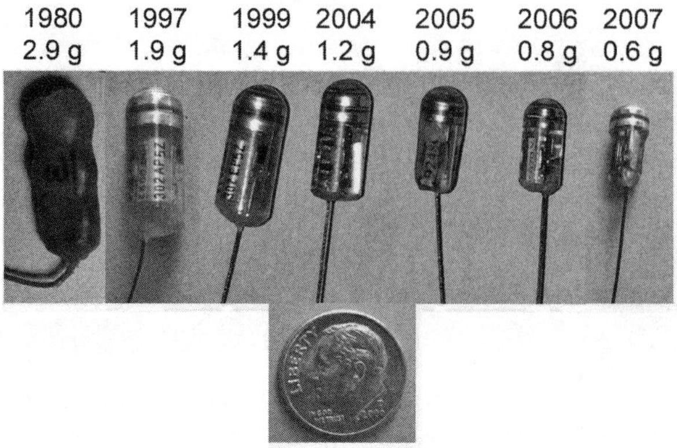

Figure 8. Relative size and weight in air of radio transmitters used by National Marine Fisheries Service to study juvenile salmon migration in the Columbia River over a 27-year period from 1980 through 2007.

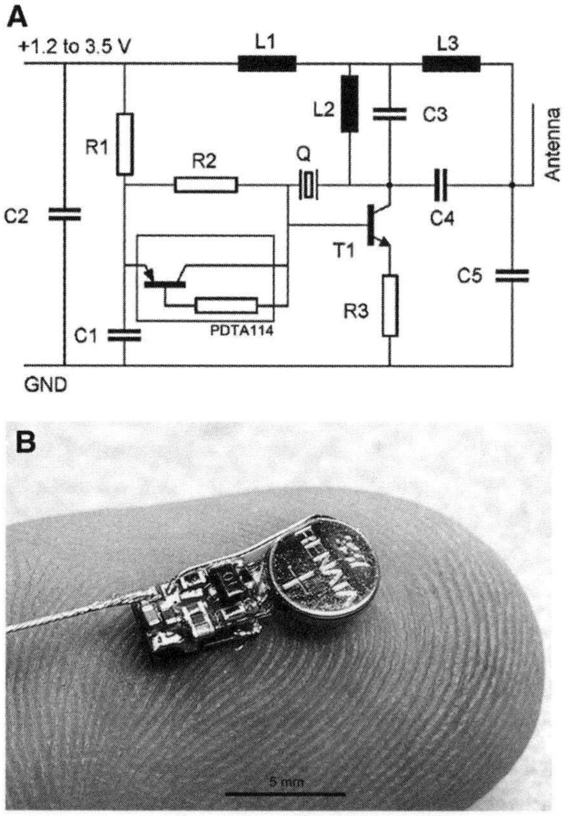

Figure 9. Circuit schematic (A) and photograph (B) of a 0.2 g radio transmitter (Naef-Daenzer, et al. 2005).

either mechanically or automatically stepped sequentially through each frequency or antenna, one at a time. Most wide-band receivers today are based on digital spectrum processors, which were first developed in the late 1970s. They virtually eliminate the scan time required by narrow-band receivers, but are less sensitive to weak transmissions. Wide-band receivers are available from several manufacturers today and are particularly advantageous for detecting tags in applications where fish could move through a detection field rapidly.

Although the first artificial satellite was launched in 1957, satellites were not used in conjunction with fish and wildlife telemetry until the early 1970s. The first integration of satellite technology and telemetry was used in a study tracking radio-tagged bears (Buechner et al. 1971). In 1978, a worldwide tracking and environmental monitoring system was started through a collaborative relationship between the French Space Agency, U.S. National Aeronautics and Space Administration, and the U.S. National Oceanic and Atmospheric Administration. This partnership led to the Advanced Research and Global Observation Satellite (ARGOS) system that used polar-orbiting satellites and up to 50 ground stations to collect data from more than 20,000 ocean platforms. ARGOS was originally developed for meteorological and oceanographic applications, but in the 1980s, transmitter miniaturization combined with an effective data logger, led to applications for tracking fish. Frequencies used to track fish are low compared to those used for satellite communications. The ground stations and floating platforms are used to collect data which is then converted to the frequency used by the satellites for data transmission.

The Global Positioning System (GPS) was developed in the 1970s to provide three-dimensional positioning for military applications using satellites (Musser 1992). Applications of GPS have been used with aquatic telemetry to describe locations of radio-tagged fish migrating through large river systems (Hockersmith and Peterson 1997) and for tasks as simple as synchronizing clocks in a series of telemetry receivers. Satellite systems have also been used for retrieval of data from receivers (Eiler 1995), or the tags themselves (Block et al. 1998), thus expanding the possible applications in remote areas.

In addition to the traditional methods, several telemetry applications using specialized tags have been developed for fisheries research. Some of these include data archival and geo-locating archival transmitters; the CART, or combined acoustic and radio transmitter (Solomon and Potter 1988); the PSATS, or pop-up satellite archival tag (Block et al. 1998); the SPOT, or smart position transmitting tag; and the CHAT, or communicating histogram acoustic transponder tag (Voegeli et al. 2001).

Researchers have either combined sensors with telemetry or used telemetry to relay data from sensors since 1957 (Eklund and Charlton 1959). The first description of sensor use and telemetry in the aquatic environment was the measurement of body temperature for an untethered dolphin (Mackay 1964). Sensor technologies have been combined with both acoustic and radio transmitters to measure either the environment occupied by or the physiology of the tagged animal, along with its location.

In fish telemetry studies that measure environmental conditions, typical sensors used include depth (Beeman et al. 1998), temperature (Haynes et al. 1986), and ambient light (Standora et al. 1972). Tilt sensors have been used to indicate activity and mortality in telemetry studies (Eiler 1990). Early versions of these used a mercury switch; however these have been replaced by switches that incorporate a metal rolling ball. Physiological studies of tagged fish have primarily used electromyogram (EMG) transmitters, which measure the electrical activity of muscles from electrodes inserted into the musculature of the fish (Kaseloo et al. 1991).

EMG has been used to measure swimming speed and to estimate energetic costs associated with physical activity such as movement through a fishway at a dam (Geist et al. 2000). EMG transmitters can also be used to measure mortality, since muscle activity ceases at death.

The field of specialized transmitters and their application continues to expand rapidly with growth in the electronics industry. This expansion is reflected in the prevalence of telemetry publications, which has increased greatly over the past 50 years. For example, in 1972, a review of underwater biotelemetry examined 52 papers (Malinin and Svirskii 1973). Five years later, a similar review was comprised of 147 publications (Stasko and Pincock 1977), and fourteen years later, a review by Baras (1991) included over 1,100 references. Other biotelemetry reviews include those completed by Ireland and Kanwisher (1978) and Mitson (1978).

Expansion of biotelemetry applications was also reflected in the appearance of several conference series specific to the field of biotelemetry. The longest running of these was organized by the International Society of Biotelemetry (ISOB; www.biotelmetery.org), which held 17 conferences between 1974 and 2003. The ISOB was founded in 1973 to encourage research in biotelemetry and related fields, and to promote its application in medicine and life sciences. More recently, ISOB requests for presenters have seen a poor response, and no conferences were held since 2003 (J. Eiler, National Marine Fisheries Service, personal communication). Another long-term series was the International Conference on Wildlife Biotelemetry, first held in Laramie, WY, in 1977. This conference was held every two years until the last meeting in 1985. The Conference on Fish Telemetry in Europe began meeting every other year in 1995 and is still well attended today.

These specialized meetings provided a vital mechanism for the development of technical aspects and applications in an emerging field prior to the startup of the commercial sector. They provided a venue for researchers to demonstrate advances in technology and commercial vendors to promote new products. When wildlife and fish telemetry was in its infancy, getting a diverse group of researchers together furthered the development of technical aspects and applications, particularly since those using the technique were relatively few. As the technology developed, it has become widely accepted by the main stream disciplines (e.g., American Fisheries Society, Wildlife Society, etc.). Today biotelemetry equipment is widely available and its applications are common, with frequent presentations of telemetry studies at fisheries conferences, and conferences in many other fields. The transition from an emerging tool that is continually improving and evolving to one that is commonly used, combined with an increasing shift towards holistic ecosystem studies, has likely shifted interest to specific fields rather than general methodology.

Advances in animal telemetry have opened the door for collaboration among researchers across the globe, and several ambitious projects were underway as of this writing. Below we enumerate some of these applications to show the level of creativity and diversity within the field. However, due to the broad scope and speed of expansion in this field, we do not attempt to provide a comprehensive list or to cover these topics in detail.

One large collaborative project is the Ocean Tracking Network, a multinational effort to understand marine life, ocean conditions, and how the changing earth affects them. It is based on acoustic tracking equipment installed in 14 ocean regions of all seven continents. Transmitters in marine animals transmit data to underwater hydrophones, to one another, and eventually to land-based system via satellite or cables (see OceanTrackingNetwork.org).

A second large marine system collaboration using telemetry is the Census of Marine Life. Telemetry projects associated with the census include the Pacific Ocean Shelf Tracking (www.postcomol.org) and Tagging of Pelagic Predators (www.topp.org) projects. The POST project uses a series of listening lines placed across the continental shelf of North America to monitor passing marine animals tagged with compatible technology. The TOPP project uses a variety of animal-locating tags, such as PSAT and SPOT, to study movements of large pelagic animals in the North Pacific Ocean.

Large studies examining extensive migration are also being conducted in freshwater. For example, international studies between the U.S. and Canada used radio telemetry to examine adult salmon migrations over thousands of miles of the Yukon River and provide information on basin-wide stock composition and timing (Eiler et al. 2004). Extensive systems of acoustic receivers have been deployed in freshwater environments such as the Sacramento and Columbia Rivers in the United States. These systems were installed to study factors affecting the migration and survival of juvenile salmonids during their seaward migration, such as passage of hydropower dams or water withdrawal projects.

With continued technological innovation, future applications of telemetry will likely provide insights that are currently unavailable. Improvements in battery technology, combined with continued miniaturization of transmitter components, will likely reduce transmitter size further, while increasing efficiency and extending either detection range or tag life. As transmitter circuitry becomes more efficient, surplus battery power can be used to power sensors, so that a common practice of future studies will be to study not only the movement and behavior of tagged fish but also to simultaneously collect information on the environment fish use. This will lead to more comprehensive multidiscipline studies addressing behavior, biology, and ecology. In the future, technological advances may eventually lead to a transmitter capable of following the movements and behavior of individuals throughout their life cycle.

References

Adams, L. and S. D. Davis. 1967. The internal anatomy of home range. Journal of Mammalogy 48(4) 529–536.

Arnold, G. P. 2000. Unraveling mechanisms of fish migrations: a 30-year perspective. Pages 7–20 *in* D. A. Hancock, D. C. Smith, and J. D. Koehn editors. Fish movement and migrations. Australian Society for Fish Biology Workshop, Bendigo, Victoria.

Arnold, G., and H. Dewar. 2001. Electronic tags in marine fisheries research: a 30-year perspective. Pages 7–64 *in* J. Sibert and J. Nielsen editors. Electronic tagging and tracking in marine fisheries. Reviews: methods and technologies in fish biology and fisheries, Vol 1. Kluwer Academic Press, Dordrecht, The Netherlands.

Baras, E. 1991. A bibliography on underwater telemetry, 1956–1990. Canadian Technical Report of Fisheries and Aquatic Sciences 1819.

Beeman, J. W., P. V. Haner, and A. G. Maule. 1998. Evaluation of a new miniature pressure-sensitive radio transmitter. North American Journal of Fisheries and Management 18:458–464.

Block, B. A., H. Dewar, C. Farwell, and E. D. Prince. 1998. A new satellite technology for tracking the movements of Atlantic bluefin tuna. Proceedings of the National Academy of Science, USA 95:9384–9389.

Buechner, H. K., F. C. Craighead Jr., J. J. Craighead, C. E. Cote. 1971. Satellites for research on free-roaming animals. Bioscience 21:1201–1205.

Cochran, W. W., and R. D. Lord. 1963. A radio-tracking system for wild animals. Journal of Wildlife Management 27:9–24.

Eiler, J. H. 1990. Radio transmitters used to study salmon in glacial rivers. Pages 370–374 *in* N. C. Parker, A. E. Giorgi, R. C. Heidinger, D. B. Jester, Jr., E. D. Prince, and G. A. Winans, editors. Fish-marking techniques. American Fisheries Society, Symposium 7, Bethesda, Maryland.

Eiler, J. H. 1995. A remote satellite-linked tracking system for studying Pacific salmon with radio telemetry. Transactions of the American Fisheries Society 124:184–193.

Eiler, J. H., T. R. Spencer, J. J. Pella, M. M. Masuda, and R.R. Holder. 2004. Distribution and movement patterns of Chinook salmon returning to the Yukon River Basin in 2000–2002. U.S. Department of Commerce, NOAA Technical Memorandum NMFS-AFSC-148.

Eklund, C. R., and F. E. Charlton. 1959. Measuring the temperature of incubating penguin eggs. American Scientist 47:80–86.

Geist, D. R., C. S. Abernathy, S. L. Blanton, and V. I. Cullinan. 2000. The use of electromyogram telemetry to estimate energy expenditure of adult fall Chinook salmon. Transactions of the American Fisheries Society 129:16–135.

Hallock, R. J., R. F. Elwell, and D. H. Fry Jr. 1970. Migrations of adult king salmon, *Onchorhynchus tshawytscha*, in the San Joaquin delta as demonstrated by the use of sonic tags. California Fish and Game Fishery Bulletin 151:1–92.

Haynes, J. M., D. C. Nettles, K. M. Parnell, M. P. Voiland, R. A. Olsen, and J. D. Winter. 1986. Movements of rainbow steelhead trout (*Salmon gairdneri*)in Lake Ontario and a hypothesis for the influence of spring thermal structure. Journal of Great Lakes Research 12:304–313.

Henderson, H. F., A. D. Hasler, and G. G. Chipman. 1966. An ultrasonic transmitter for use in studies of movements of fishes. Transactions of American Fisheries Society 95(4):350–356.

Hockersmith, E. E., and B. W. Peterson. 1997. Use of the global positioning system for locating radio-tagged fish from aircraft. North American Journal of Fisheries Management 17:457–460.

Ireland, L. C. and J. S. Kanwisher. 1978. Underwater acoustic biotelemetry: procedures for obtaining information on the behavior and physiology of free-swimming aquatic animals in their natural environments. Pages 341–379 *in* D. I. Mostofsky, editor. The behavior of fish and other aquatic animals. Academic Press, New York.

Johnson, J. H. 1960. Sonic tracking of adult salmon at Bonneville Dam, 1957. U.S. Fish Wildlife Service Fishery Bulletin 176(60):471–485.

Kaseloo, P. A., Weatherley, A. H., Lotimer, J. S., and Farina, M. D. 1991. A biotelemetry system for transmission of electromyograms in fish as indicators of physical activity. Journal of Fish Biology 48:664–674.

Koehn J. D. 1999. Why use radio tags to study freshwater fish? Pages 24–32 *in* Fish Movement and Migrations. D. A. Hancock, D. C. Smith and J. D. Koehn editors. Australian Society for Fish Biology Workshop, Bendigo, Victoria, September 1999.

LeMunyan, C. D., W. White, E. Nybert, and J. J. Christian. 1959. Design of a miniature radio transmitter for use in animal studies. Journal of Wildlife Management 23(1):107–110.

Leggett, W. C., and R. A. Jones. 1971. Net avoidance behavior in American shad (Alosa sapidissima) as observed by ultrasonic tracking techniques. Journal of Fisheries Research Board of Canada 28:1167–1171.

Lonsdale, E. M., and G. T. Baxter. 1968. Design and field test of a radio-wave transmitter for fish tagging. The Progressive Fish-Culturist 30:47–52.

Luke, D. McD., D. G. Pincock, and A. B. Stasko. 1973. Pressure sensing ultrasonic transmitter for tracking aquatic animals. Journal of the Fisheries Research Board of Canada 30:102–1404.

Mackay, R. S. 1964. Deep body temperature of untethered dolphin recorded by ingested radio transmitter. Science 144:864–866.

Malinin, L. K. and A. M. Svirskii. 1973. Application of biotelemetry to ichthyology. Translated by Fisheries Research Board of Canada 2707 (in Russian).

Mann, C. C. 2000. The end of Moore's Law. Technology Review May/June 2000.

Mitson, R. B. 1978. A review of biotelemetry techniques using acoustic tags. Pages 269–283 *in* J. E. Thorpe, editor. Rhythmic activities of fishes. Academic Press, New York.

Monan, G. E., J. J. Johnson, and G. F. Esterberg. 1975. Electronic tags and related tracking techniques aid in study of migrating salmon and steelhead trout in the Columbia River Basin. Marine Fisheries Review 37(2):9–15.

Musser, D. D. 1992. GPS/DGPS in offshore navigation, positioning. Sea Technology 33:61–66.

Naef-Daenzer, B., D. Früh, M. Stalder, P. Wetli, and E. Weise. 2005. Miniaturization (0.2 g) and evaluation of attachment techniques of telemetry transmitters. The Journal of Experimental Biology 208:4063–4068.

Novotny, A. J., and G. F. Esterberg. 1962. A 132-kilocycle sonic fish tag. The Progressive Fish-Culturist 24:139–141.

Rochelle, J. M., and C. C. Coutant. 1973. Temperature sensitive ultra-sonic fish tag, Q-5009. ORNL/TM-4438, Oak Ridge National Laboratory, Oak Ridge, Tennessee, USA.

Sisak, M. M., and J. S. Lotimer. 1998. Frequency choice for radio telemetry: the HF vs, VHF conundrum Hydrobiologia 371/ 372:53–59.

Slater, L. E. 1965. Introduction (special issue on telemetry). Bioscience 15(2):81–82.

Solomon, D. J., and Potter, E. C. E. 1988. First results with a new estuarine fish tracking system. Journal of Fish Biology 33:127–132.

Standora, E. A., Jr., T. C. Sciarrotta, D. W. Ferrel, H. C. Carter, and D. R. Nelson. 1972. Development of a Multichannel, Ultrasonic Telemetry System for the Study of Shark Behavior at Sea. Report to the Office of Naval Research, Department of Biology, California State University, Long Beach.

Stasko, A. B., and D. G. Pincock. 1977. Review of underwater biotelemetry with emphasis on ultrasonic techniques. Journal of Fisheries Research Board of Canada 34:1261–1285.

Stuehrenberg, L., A. Giorgi, and C. Bartlett. 1990. Pulse-coded radio tags for fish identification. Pages 370–374 *in* N. C. Parker, A. E. Giorgi, R. C. Heidinger, D. B. Jester, Jr., E. D. Prince, and G. A. Winans, editors. Fish-marking techniques. American Fisheries Society, Symposium 7, Bethesda, Maryland.

Trefethen, P. S. 1956. Sonic equipment for tracking individual fish. U.S. Fish and Wildlife Service Special Science Report. Fisheries 179:1–11.

Trefethen, P. S., J. W. Dudley, and M. R. Smith. 1957. Ultrasonic tracer follows tagged fish. Electronics 30(4):156–160.

Trefethen, P. S. and D. F. Sutherland. 1968. Passage of adult Chinook salmon through Brownlee Reservoir, 1960–62. U.S. Fish and Wildlife Service Fisheries Bulletin 67:35–45.

Voegeli, F. A., M. J. Smale, D. M. Webber, Y. Andrade, and R. K. O'Dor. 2001. Ultrasonic telemetry, tracking and automated monitoring technology for sharks. Environmental Biology of Fishes 60:267–281.

Winter, J. D. 1996. Underwater biotelemetry. Pages 555–590 *in* L. A. Nielsen and D. L. Johnson, editors. Fisheries techniques, 2nd edition. American Fisheries Society, Bethesda, Maryland.

Section 3

Designing Studies Based on Acoustic or Radio Telemetry

John D. Koehn

INTRODUCTION

Telemetry (both radio and acoustic) is a widely used methodology that can remotely transfer information from tagged fish, allowing physiology, behavior and ecology to be studied largely undisturbed for long uninterrupted periods (Cooke et al. 2004). Studies can be undertaken using a variety of equipment, and conducted across a wide range of environmental conditions and scales (both spatial and temporal). Consequently, there is not just one approach that can be applied to all studies; rather, telemetry is an extremely powerful tool that offers a range of options (especially equipment and techniques) when designing a study. It is important to explore how telemetry technology can best be applied to meet your research objectives.

As with any technique, telemetry has advantages and limitations. There are many techniques, adaptations, and lessons learned from previous studies (often unreported in the literature) that can be used to improve the effectiveness and efficiency of this research method. While this Section, along with others in this book, will provide some of these lessons, more importantly, it will try to encourage you to think more conceptually about what you are trying to achieve and how to best use telemetry as a method for your study. It also aims to assist the researcher when considering the key decisions, trade-offs and compromises that need to be made during study design and equipment selection, through the provision of information and a process to assist making these decisions. Additionally, this Section emphasizes the importance of proper planning to provide a practical approach for designing and implementing an efficient, effective, and successful field study based on telemetry. It will not provide all the answers, but will prompt you to ask the right questions. It provides information and directions that will help you make better decisions about study design, points to other key sources (including other Sections in this book) that can provide information on individual topics in more detail, and highlights limitations of methods and concerns that may arise. This Section focuses on the need to address the principal considerations associated with designing a study based on telemetry and assessing the technical requirements for its implementation. These include:

1. Development of clear study objectives (i.e., the questions being asked);
2. Determination of the methods and data needed to answer these objectives;

3. Selection of suitable telemetry equipment and techniques that can effectively collect the data; and
4. Assessment of the feasibility and practicality of this approach.

Of course, in reality, designing your study occurs as the result of an iterative and dynamic mix of these considerations, with revisions, compromises, and trade-offs being made between what is desired, what is needed, and what is possible and practical.

DEVELOPING STUDY OBJECTIVES

It is easy to get excited about telemetry and just want to go tag and track fish without sufficient planning. Such an approach, however, often leads to unsatisfactory results. As with any research, clear study objectives are essential and will determine the utility of telemetry as an effective method (Priede 1980). Clear objectives will help direct the study design, study location, gear type, field procedures and techniques, and determine the quality and quantity of data that needs to be collected. Poorly planned studies can waste time, money, and effort and result in erroneous conclusions (Winter 2000). Even basic studies require clear aims and a sound design. The idea of using SMART guidelines (Specific, Measurable, Attainable, Relevant, and Time-framed) (www.projectsmart.co.uk) to formulate objectives should apply to any project. The relevant questions that apply to many studies based on telemetry revolve around these guidelines and recognize the need to define the scope of the problem to be addressed and then set testable hypotheses. Studies may have multiple areas of interest (e.g., habitat and movements), and each should have clearly stated objectives. The use of telemetry can then be related to the objectives and assessed in terms of its suitability.

WHAT TYPE OF INFORMATION IS NEEDED?

In general, studies based on telemetry may be descriptive, correlative, or manipulative (Winter 2000) and typically are used to study animals in the field and to transmit information about their ecology (e.g., location), behavior, or physiology (Kenward 2001; Cooke et al. 2004). While the interpretive power of descriptive studies is often limited, they are nevertheless, useful and often necessary to determine unknown life history aspects such as movements or spawning times and locations. They can also often lead to other, more detailed studies. Correlative studies aim to determine the relationships between measured variables and can provide valuable data for management. Manipulative studies on the other hand, involve altering some factor or factor(s) (e.g., flow and temperature) then testing the impacts of these changes on the study animal can provide data to support or reject hypotheses. Each of these types of study has different data requirements and approaches (see Burnham and Anderson 2002). Consequently, determining individual studies need to be prompted by typical questions such as: is a control(s) necessary? How many treatments should/can we test without overcompromising statistical power? How many individuals (replicates) do we need to tag? Can we use a Before-After Control-Impact (BACI) design to analyze data?

The study objectives, locations, data requirements and funding will determine the type and amount of equipment that is needed. For example, behavioral studies that require fine-scale information may need more equipment or specific equipment characteristics (e.g., rapid transmitter pulse rates that allow for better resolution, but reduced operational life). Presence/absence studies may only require information at a coarser scale. Such studies tend to require less equipment and transmitters with longer pulse rates (e.g., 30 pulses per minute versus 60 pulses per minute), resulting in extended tag life. Survival study estimates may need high levels of precision and require large numbers of tags, with the impact of premature tag failure or loss being significant. Less efficient tracking systems, although undesirable, may be less of a factor for behavioral studies, but could seriously complicate data interpretation in survival studies. Determining the adequate sample sizes for each study is an important part of the design.

DETERMINING AN APPROACH THAT WILL PROVIDE DATA NEEDED

Telemetry is a widely applicable research tool that has been used to study a wide range of species (Eiler et al. 2000; Cooke et al. 2004; Jellyman 2009). Common types of ecological studies include: movements (large, small, diel) (e.g., O'Connor et al. 2005; David and Closs 2001; David et al. 2007), homing (e.g., Tyus 1985; Koehn et al. 2009), home ranges (e.g., Daugherty and Sutton 2005), activity patterns, spawning (e.g., Eiler et al. 1992; Weatherley et al. 1996), habitat use (macro, micro, reconstructed, protected areas) (e.g., Koehn 2009a, 2009b; Nicol et al. 2004; Cooke et al. 2005), tracking pest fish (e.g., Diggle et al. 2004), physiological and environmental monitoring (e.g., Miller et al. 1980; Johnsen 1980b; Lucas et al. 1993), behavior (such as schooling) (e.g., Johnsen 1980a), assessment of population sizes (e.g., Hasbrouck et al. 2000), stock assessment (e.g., Eiler et al. 2006) and management issues around river structures (including weirs and power stations) (e.g., O'Connor et al. 2006; Stuart et al. 2010; Sections 6.5 and 8.1). Such a diversity of studies necessitates a range of approaches, techniques, equipment and study requirements.

Before embarking on a study based on telemetry, the first question that must be asked is: is telemetry the best method to answer the project objectives, or are there other methods that are more effective and/or cost-efficient? This can only be answered following a close examination of the advantages and disadvantages of the techniques relevant to the study situation. For instance, traditional mark–recapture studies tend to provide more cost-effective and reliable population estimates compared to telemetry. However, telemetry may be advantageous estimating fish abundances in cases when you can't see them, they are difficult to capture, or you need information about their movements in relation to abundance. Additionally, telemetry can be used to test some of the basic mark–recapture assumptions. One advantage of telemetry is that a variety of detailed information can be collected over time (e.g., location, depth, activity, environmental parameters), without the need to recapture or directly observe animals. It can, however, be relatively complicated (implanting tags, tracking, logging) compared to other techniques (e.g., tag and recapture), may be gear and time intensive (and hence costly) and can adversely impact study animals. There are also practical limitations when using telemetry: can you afford to purchase enough suitable equipment? Are there sufficient resources to monitor the required number of fish as frequently as is needed (e.g., hourly, daily, weekly, monthly) across the spatial and temporal scales to obtain the needed information? Equipment limitations (e.g., range, tag size, tag longevity) and physical restrictions on its use need to be recognized (e.g., water conductivity; Section 5.1) and addressed.

STUDY SPECIES

There are a range of considerations that need to be made associated with the species of interest for the study. While the purpose of the study may be to discover new aspects of the biology of the study species, an assessment of existing knowledge will facilitate project design. Body type, behavior (demersal, midwater, schooling), migratory behavior, spawning season/gonad development, physiological differences between sexes, size at maturity, and habitat patch size of the study species, can all influence the techniques used. Such information can also be used to help assess and avoid any likely study biases.

Is telemetry suitable to study your species? Potential effects from capture, handling and tagging can all result in erroneous/poor data. These include: fish mortality, impaired performance and health (stress, infections, etc), or altered behaviors (tagged versus untagged fish) or tag loss over the short or longer terms (Section 4). Although it may not be possible to avoid all impacts, these factors need to be explicitly considered and assessed in the study design and interpretation of data. Transmitter implantation and attachment techniques must be optimal for the study species, and body size, type and shape can influence the tag type, shape and attachment technique (Daniel et al. 2009; Section 4). Familiarization with the known mobility and movement patterns of the species may help set the spatial limits of the study, but if the species has not been well studied, then researchers should be prepared for a range of possibilities. Tracking nocturnal species during daylight hours may not give a true indication of the daily movement patterns (e.g., Koster and Crook 2008). While night tracking poses significant logistic difficulties and safety issues, study objectives need to consider whether or not this is a requirement.

In many studies, the objectives are based on specific questions pertaining to the management or ecology of a particular species; hence the subject species is already selected for you. In other instances, there may be some latitude in your selection of species and study animals, making it possible to avoid particular problems (e.g., tagging larger individuals to avoid issues related to transmitter size and operational life). While this may bias the data collected, it must be weighed against the study objectives and logistics, and taken into account when interpreting results. Alternatively, a species may need to be selected to answer a broad ecological, rather than a species-based, objective. For example, are fish movements affected by changes in flows? While this type of study requires a slightly different assessment in the selection of a suitable study species to meet this objective, many of the same questions need to be addressed. The physiological and behavioral traits, conservation status (stable versus threatened or endangered) or abundance of the species may limit your study.

STUDY AREA CONSIDERATIONS

The environmental parameters of the study area can pose limitations on techniques and the collection of data, and need to be thoroughly assessed prior to the design of the study. Is this the most suitable site at which to answer the questions asked or does it pose unreasonable limitations that may compromise the study objectives (e.g., barriers that restrict fish movement)? Does the study area allow the ability to capture and tag adequate numbers of fish that are a representative sample of the population or that meet the data requirements of the study? Will you be able to effectively monitor these tagged fish after they are released? Does the study area provide a sufficient number or range of representative habitat types to make useful comparisons?

Aspects such as water type (marine, estuarine, freshwater), water dimensions (width, depths), area geomorphology (i.e., potential obstructions from formations), climatic conditions (maximum and minimum temperatures, snow, ice), water characteristics (conductivity, turbidity, turbulence), habitat structure (substrate type, wood or coral formations, vegetation), and noise or signal interference (e.g., from power stations, airports, boat traffic) all need consideration. Each study area is different and presents different challenges. For example, when working in rivers, river form (straight, meandering, multi channel), floodplain width, presence of floodplain or anabranch channels, and tree coverage must be taken into account. The likely spatial extent of the study also needs to be defined, although in some cases this may need to be expanded if fish unexpectedly move outside this range. Are the environmental conditions consistently uniform across the spatial extent of the study area and across the seasons? Are management actions likely to cause any unreasonable environmental changes (e.g., changes to river flows or lake levels)? What are the extremes of these changes (e.g., floodplain width, maximum depth)? Environmental variables have the potential to limit the technical operation of a tracking study. For example, water conductivity may increase in downstream reaches or at the bottom of deep, stratified pools, especially under low flow conditions. Both water conductivity and depth of the transmitter can decrease the reception range of radio tags significantly (Kenward 2001; Section 5.1). Extremely high temperatures can compromise the stability of some data loggers and laptop computers, while low temperatures can reduce the efficiency of batteries and hence the performance of data loggers or the longevity of tags.

Optimally, suitable (and efficient) access to the study site should be available so appropriate tracking can occur throughout the study period under all conditions, however, this may not be possible and such logistical considerations should be factored into the overall study design (see Section 5.3). A desk top investigation of a new study site is an important starting point, together with a review of previous studies that may have occurred in the area. Other information and data such as historical and current flow gauging, climate and temperature records, GIS layers of land use and aerial photos, may all be of use. There is no substitute, however, for a thorough reconnaissance, especially for long-term studies, as time spent refining logistics is usually repaid through increased time efficiencies later. The study area must be home to adequate numbers of fish that can be captured and tagged to provide adequate data to answer your study objectives. The tagged fish must be a representative sample of the population relevant to your study question, and you must have the ability to effectively and efficiently monitor your fish once they are released (see Sections 4 and 5.3).

DATA COLLECTION

Data collection requirements are driven by the study objectives as the study must aim to collect satisfactory data to answer the principal questions. Determining what you need to collect (what parameters, how many fish, are monitored how often and for how long) will determine the type of tagging and tracking (e.g., manual or automated station) to be used. Minimum data requirements will determine how many tagged fish are required and what your sampling strategy (daily, all months, seasonal, multiple years) should be. Consideration needs to be given to the logistics of fish capture and tracking: specifically, can sufficient numbers of fish be captured, tagged and tracked to provide enough data so that robust conclusions can be drawn to answer the project objectives? The study of rare or threatened species, for example, can often limit the number of study individuals (e.g., Koehn et al. 2008). The units of measurement and what constitutes a sample needs to be determined. There are two measures of data that need to be considered: the number of fish to be tracked; and the number of relocations of each fish. Individual animals or their locations can only be used as a sampling unit when they act independently and the factors associated with them can be measured independently (White and Garrott 1990). 'Groups' of fish (schools, age or size classes, sexes) may not exhibit independent behavior and this needs to be considered in design and analysis and interpretation of data.

Location accuracy requirements (see Sections 7.1 and 7.2), directionality and spatial scale of the study will determine the type of tracking data that needs to be collected. Maximum reception range may influence tag size, signal configuration (see Sections 5.1 and 7.1) and tracking methods (see Sections 5.3, 7.1, and 7.2). These assessments may influence study duration and frequency of sampling, and decisions on tag characteristics (type, size) which in turn can affect fish response. There is a need to establish a tracking timeframe and timetable with the provision for increased tracking at vital times (e.g., spawning season) if needed.

There is often an unavoidable premature loss of study animals or tags in studies based on telemetry (Winter 2000), and as tags/fish go 'missing' there may be the need for replacement fish to be tagged or for additional tracking to be undertaken (e.g., by air to cover a wider area) to maintain adequate sample sizes. Assessment of tag rejection needs to be made either from previous work or from data collected as part of your study. Tag retention/rejection rates vary between species and the use of mortality switches is a useful addition to alert the researcher of tag rejection or fish death (Eiler 1995; O'Connor et al. 2009). Winter (2000) suggested the following observations for potential tag loss: 10% to transmitter failure; 10% to animal disappearance; 10% to animal deaths; and >30% to animal harvest. This, of course will not apply to nontarget species but for popular angling species or commercially harvested fish the number of individuals lost to harvest may well be >30%, especially over long-term studies. Given these losses, it is possible that only about 40% of tagged fish may provide good data. This, however, may be a worst case scenario, as during large-scale tagging of Pacific salmon *Oncorhynchus* spp. on the Yukon river <3% of tagged fish did not move up river (Eiler et al. 2006). The amount of each type of data required to answer the study objectives should be determined by a priori testing, with these potential losses incorporated into the study design.

SELECTION OF SUITABLE EQUIPMENT AND TECHNIQUES

Equipment limitations can restrict the type and amount of data collected. Equipment needs to be suitable for the tasks set out in the study design and be able to cope with the environmental conditions to be encountered (e.g., reception range, depth) and tracking methods used (e.g., aerial versus boat tracking). Transmitter types and options need to be considered including physical (size, shape, weight, internal/external aerials) signal (frequency, strength, duty cycle, pulse rates), output patterns (identification code, activity, mortality, depth or other signal outputs), and performance attributes (reception range, longevity) (Sections 5.1 and 7.1). While selection of these attributes will depend on the objectives of your study, there is also the need to consider the range of trade-offs between these factors and equipment types and capabilities (e.g., larger tags give better range and longer life, but also an increased potential to adversely affect the fish). It is important to use equipment that addresses study objectives and not tailor your study objectives around equipment that is either already available or has already been purchased. Because of the set up costs of telemetry this can be a common potential pitfall in project design, e.g., 'Joe has some equipment we could use'—but in reality isn't well suited to the study question.

The range of gear and its capabilities can, initially, be quite bewildering, and as equipment can be costly, careful assessment is needed. Beware of the temptation to construct your own equipment to save costs, as it is likely that the time will be better spent collecting data for the study. As a result of technological advances, equipment capabilities and models are regularly updated and it is necessary to keep abreast of new developments via manufacturers and researchers. This information often lags behind what is described in the published literature. Most manufacturers have informative websites with fact sheets, frequently asked questions, equipment brochures and manuals, software updates, data sheets, references and technical and published papers. Contact should be made with several equipment suppliers to ensure an unbiased view, and guidance can be given by manufacturer's consultants. Researchers need to be aware that the purchase of a particular brand of equipment (e.g., receivers) can also potentially limit compatibility with equipment (e.g., tags or data loggers) from other suppliers. Contact with other researchers is a good way of seeking 'independent' advice from those who have used and tested the equipment. Be sure to ask site-specific questions regarding tags and tracking equipment. Manufacturer representatives may be able to assist with project assessments to solve particular problems related to equipment use and performance. Adequate time should also be allowed to order and receive gear and tags throughout the study as it is easy to underestimate the time delays in equipment selection (Kenward 2001). Consideration should be given to purchasing good quality equipment as equipment failures can potentially be very costly. Many manufacturers also supply service and repairs and may provide replacement equipment to cover such 'down time.' Be sure to order a spare of everything.

FREQUENCY SELECTION

The choice of telemetry frequency has always been the subject of controversy among researchers. Researchers are, however, often not fully aware of the impact that frequency choice can have on overall equipment performance, and ultimately the quality of the data that their study generates (Sisak and Lotimer 1998).

Acoustic Versus Radio

Both acoustic and radio telemetry are suitable for use in aquatic environments, but there are costs and benefits associated with both (Winter 1983; Priede 1992; Sisak and Lotimer 1998; Nielson et al. 2009). Acoustic tags for fish studies generate sound waves from 50 to 416 kHz (Section 7.1), whereas radio tags generally emit signals in the 30–300 MHz range (Section 5.1). For the majority of freshwater studies, radio telemetry is the method of choice (Sisak and Lotimer 1998) due to advantages in antenna size and reception characteristics (Section 5.1). Increasingly acoustic tags are also being used in freshwaters (e.g., Section 8.1) and are used exclusively in seawater due to the higher water conductivity (Winter 1983). Difficulties can arise with radio telemetry in areas such as estuaries and lowland rivers where salinity levels may increase, areas with intermittent high salinities, or areas that have saline pools or haloclines. Combined acoustic/radio tags may be a useful option in such situations. Radio tags can be tracked from greater distances, by a range of methods (e.g., land, boat, air) (Section 5.3) whereas submerged hydrophones are necessary for tracking of acoustic tags which limit reception range when tracking by boat or fixed data logger stations (Section 7.1). A comparison of the general performance of acoustic and radio tags under different study circumstances is given in Table 1.

Unfortunately, there is no one frequency that is optimal in all cases (Sisak and Lotimer 1998), and assessing the benefits and drawbacks of a particular frequency is complicated and involves not only the tags, but the other components of the telemetry system. For example, a comparison of the performance of 'high' (>50 MHz) and 'low' (<50 MHz) radio frequencies under different circumstances is given in Table 2, and illustrates how the different attributes of the system may impact field operations and research objectives. The selection of radio frequencies must be undertaken carefully, however, as this may lock future research into these frequencies as receivers and antennas are frequency specific. While the choice of frequency is an important one, this is only one part of the telemetry system, with each component being inter-related and the effectiveness of tracking will be as a result of your whole system (Sisak and Lotimer 1998).

TAGGING

The potential impacts of tagging and the resultant tag retention rates on study animals will depend on the study species, the type of transmitter used, and tagging method, and tagging trials may need to be conducted to assess these factors. Tag rejection can be problematic for some species and may limit the use of telemetry as a study tool (e.g., O'Connor et al. 2009) and tag retention rates appear to be species-specific (Butler et al. 2009) and can be negatively influenced by advanced reproductive state. Options for transmitter attachment and potential impacts are described in numerous articles and subsequent Sections (e.g., Adams et al. 1998a, 1998b; Martinelli et al. 1998; Moser et al. 1990; Schrader and Jones 2000; Crook 2004; Sections 4 and 7.1) and it is important to recognize that atypical behavior may be observed after release and that the time before resumption of "normal" behavior will vary with species and location (Guy et al. 1994; Adams et al. 1998a, 1998b).

Simple considerations such as the time of day (cooler early mornings and late afternoons) and provision of overhead shade are considerations to reduce stress from surgical procedures in hotter climates. To minimize the risk of infection and fish mortality or tag rejection, surgical skills need to be learned and practiced. This may require some training in veterinary techniques. Each tag should be tested to ensure its functionality before attachment or implantation. As tag life is largely dependant on battery life, manufacturers should give both a guaranteed and an expected tag life for each transmitter type. There is the need to allocate a sufficient amount of time to capture enough fish to provide sufficient data and to avoid sample bias, especially if studying rare or threatened species. In such cases it may be that tagging techniques have not been applied to the species before, and while other previous studies may provide proven techniques, the transferability of techniques such as tagging from other similar species need to be treated with caution as there are often species specific reactions (Ebner et al. 2009; O'Connor et al. 2009).

Table 1. Comparison of the general performance of radio and acoustic transmitters under different study conditions.

Characteristic	Acoustic	Radio
Site conditions		
Saline/high conductivity water (> 600 EC)	Excellent	(not usable)
Low conductivity water	Excellent	Excellent
Deep water (> 20 m)	Excellent	Poor
Turbulent water	Poor	Excellent
Dense aquatic vegetation	Poor	Very good
In water obstructions	Poor	Poor to marginal
Turbid water	Good	Very good
Algae	Poor	Excellent
Thermocline/Temperature gradient	Poor to marginal	Good
Ice	Poor	Good
Study species		
Fast/mobile species	Poor (mobile tracking) Good (stationary arrays)	Excellent
Long migrations	Poor (mobile tracking) Good (stationary arrays)	Excellent
Number of animals	Good	Good
Equipment		
Tracking options	Hydrophone in water Boat, stationary array	Antenna usually in air on foot, vehicle, boat, aircraft
Automated stations	Good	Excellent
3 D positioning	Established methods	Difficult
Archival tags	Established methods	Possible
Directionality	Poor to fair	Good to excellent (dependent on frequency and antenna type)
Power usage	Good	Good

TABLE 2. Relative performances of low (<50 MHZ) and high (>50 MHz) radio frequencies for a range of characteristics (Winter 1983; Eiler 1990; Sisak and Lotimer 1998).

Characteristic	< 50 MHz	> 50 MHz
Attenuation	Less (better signal)	Greater
Depth reception	Greater	Less
High conductivity (> 600 EC)	Greater	Less
Signal obstruction by topography and vegetation	Less	More
Signal efficiency	Less	More
Transmitter efficiency	Less	More
Antenna	Bulky[1]	Smaller
Complex habitats	Better	Worse
Directionality	Worse[1]	Better

[1]Directional antennas often unsuitable for field use due to large size, particularly in remote settings or when mobile tracking.

Tagging options can often be constrained by the characteristics of your species and its surroundings but your choice will affect the data you can collect. For example: if the species of interest require the tag to be relatively small to maintain the appropriate tag to body weight ratio, then archive tags are unlikely to be suitable. If the main objective is to measure the affect of predation on the species of interest, then the use of transmitters with external antennas might not be preferable because it could provide a predation cue and bias results. External antennas can also pose problems with entanglement for species that reside among dense vegetation and instream debris, and can result in increased infection rates (Broadhurst et al. 2009). Conversely, transmitters with internal (coiled) antennas have reduced range and add to tag size and weight. If the species of interest is benthic and the water is deep, then lower radio frequencies or acoustic tags might be advantageous (Sections 5.3 and 7.1). Some of these considerations are given in Table 3.

TRACKING

Tracking needs to be undertaken effectively and time efficiently, with accuracy and minimal disturbance to the study animal (Kenward 2001). Tracking procedures will be influenced by the study objectives, equipment and site conditions and are discussed in detail in Sections 5.3 and 7.1. Detection of fish can occur manually with tracking by land, boat, air (fixed-wing aircraft or helicopter) (or in combination) or via automated logger devices. Each method is capable of providing data that can vary greatly in terms of spatial and temporal resolution. Fish presence or absence, or fish passing a particular location is the basic information gained remotely from a fixed tracking station. While there is often a need for manual tracking, this should be carefully assessed since it usually entails a greater allocation of time, and can come to dominate workloads. Accurate manual tracking of individuals can be an expensive, time consuming and at times frustrating process (Amlaner and MacDonald 1980), with the effort needed to locate individuals often being underestimated (Winter 2000). Automated logging towers (Section 5.3) can reduce these costs, although manual tracking will be necessary for some studies (particularly where finer spatial scale information is necessary e.g., detailed microhabitat use). Contingency plans (e.g., the addition of tracking by aircraft or additional tracking trips) may be needed to locate 'lost' fish, remembering that studies often provide new and unexpected (but exciting) results. Many movement studies report 'errant' fish that do not conform to the overall movement patterns. Data on those fish (which are often harder to detect), may be ecologically more important than the 'easier to track' majority. Conversely, data on such 'exciting' movements are often highlighted and over emphasized compared to the majority trend and caution needs to be exercised when interpreting results and relating them to the general population.

Determining the detection range needed for your study is important to avoid collecting biased data. Too little range will decrease the chance of detection of study animals while too much range can provide unnecessary data. This can be particularly problematic with fixed position data loggers if fish locate within range of the logger and are repeatedly recorded, hence filling the logger memory (Section 5.2). Detection range is a combination of signal strength (largely determined by the power output of the transmitter) and reception range (determined by how this is detected by the receiving antenna). Different antenna options have different patterns of reception (Kenward 2001; Sections 5.1 and 7.1) but ultimately reception can be affected by many components of your tracking system (Sisak and Lotimer 1998; Section 5.2)). Reception range is greatest from the air (e.g., tracking by aircraft) and reduced at water level (e.g., tracking by boat) and signals can also be blocked or altered by physical obstructions. Detection ability will also vary depending on tracking method, receiver scan times, number of fish in the area, operator tracking efficiency/skill, noise and hearing thresholds of the operator (Section 5.3). Manual tracking can also be compromised by factors such as site access, instream structures and water velocity. General tracking guidelines and considerations for radio and acoustic tracking are discussed in Sections 5.1, 5.2, 5.3, 7.1, and 7.2.

ACCURACY AND PRECISION

Some location error inevitably occurs whenever fish positions are determined by radio or acoustic tracking. Location error can be simply described as the linear distance between the actual and estimated location and can include data variability (e.g., 95% confidence intervals) to report overall error (Zimmerman and Powell 1995). Unfortunately, errors associated with tracking techniques are rarely reported (Samuel and Kenow 1992; Simpkins and Hubert 1998), even though they can cause limitations in the accuracy of data collected (White and Garrott 1990). Ignoring measurement errors (or other nonsampling errors such as animal movements during location determination) may result in a serious underestimation of variability (Samuel and Kenow 1992). Saltz (1994) reported that 77% of studies using triangulation either did not mention error (29%) or provided incomplete information (48%). As fish are typically less visible than terrestrially tracked species, under-estimating precision and accuracy can easily occur. Nevertheless, the risk of ignoring errors associated with location accuracy is that there is potentially undetected and unknown bias in the results (Parker et al. 1996), either in locating the fish, mapping their positions, or both (Brenden et al. 2004). Total location error dictates the spatial scale at which reliable inferences can be made from the data (Brenden et al. 2004) and a failure to account for these errors can lead to inaccurate data interpretation (Simpkins and Hubert 1998).

The importance of accuracy can be dictated by the scale, needs of the study and the parameters measured. Typically, errors are of greater concern when data are being used to determine fine-scale movements or microhabitat use (Niemela et al. 1993). When habitat patches are relatively large, location error may have little influence, but when habitat patches are small with a high degree of interspersion, and the error large, misclassification of habitat type is more likely (Samuel and Kenow 1992). Similar considerations also need to be given to the scales needed for movement studies. While there are many methods by which error can be incorporated into data analysis (e.g., Garrott et al. 1986), an ideal approach is to attempt to minimize this error and to quantify it so that any biases in data can be assessed. Errors should be estimated before the study so they can be accommodated in the study design and in the analysis and interpretation of results. Unacceptable errors, not determined at the start of the study, have potentially costly impacts (Parker et al. 1996) in terms of lost time, effort, and money expended, and misinterpretation or an inability to address study objectives with data obtained.

Errors related to tracking fish can be attributed to equipment performance, tracking technique, and site specific factors. Brenden et al. (2004) attributed position accuracy to two essential factors: location error (which includes both tracking error and movement error) and mapping error. Tracking error relates to error in determining location direction ($\alpha°$) and increases with distance between the transmitter and receiver (E_1 compared to E_2 in Figure 1a). Hence, reduced reception range decreases the distance between the receiver and the transmitter and reduces this error (Simpkins and Hubert 1998). Lee et al. (1985) and Garrott et al. (1986) both attributed radio-tracking error of greater than 10° in directionality to signal bounce and removed them from the data. Error can also be caused by animal movements that occur during tracking (E_3 in Figure 1a), especially between the different fixes needed for triangulation (Section 5.3). Note that movement error may have a greater proportional effect when the fish is closer (Figure 1a). This may apply to particular, fast-moving species or species undertaking migrations, and transmitters programmed with pulse rates of less than 50 ppm can be more difficult to detect and therefore less accurate. Other potential influences contributing to location error include: electromagnetic interference of power lines (Parker et al. 1996) or electric fences causing distortion on fixes (Ward et al. 1986). Additionally, all signals can become essentially omni-directional depending on the distance from the source, water depth, signal bounce and site specific factors (e.g., geomorphology) (Winter 1983). Radio waves can be reflected off cliffs, hillsides, wood, buildings, rocks (Kenward 2001) potentially causing location error or false readings. Errors due to mapping (E_4 in Figure 1a) may be less common than in the past due to the use electronic location (GPS) and mapping devices compared to manual methods. This, however, still needs to be accounted and GPS locations accuracy can vary with location, time (number of available satellites), terrain (e.g., steep valleys, overhanging vegetation) and weather conditions. Total location error = (E_1 or E_2) + E_3 + E_4 (Figure 1a).

Triangulation is a common method used to locate radio tagged fish (White and Garrott 1990), however, this approach can be time consuming and introduce substantial location errors (Stasko and Pincock 1977; White and Garrott 1990; Simpkins and Hubert 1998). Triangulation error can occur due to: inaccuracy of the direction receiving antenna; operator error; the distance between the transmitter and receiver; the distance between the two receiving points and/or; the angle between the transmitter and receiver points (90° being optimal). After determining these errors an error polygon can then be formed to illustrate the uncertainty of measurements (Figure 1b).

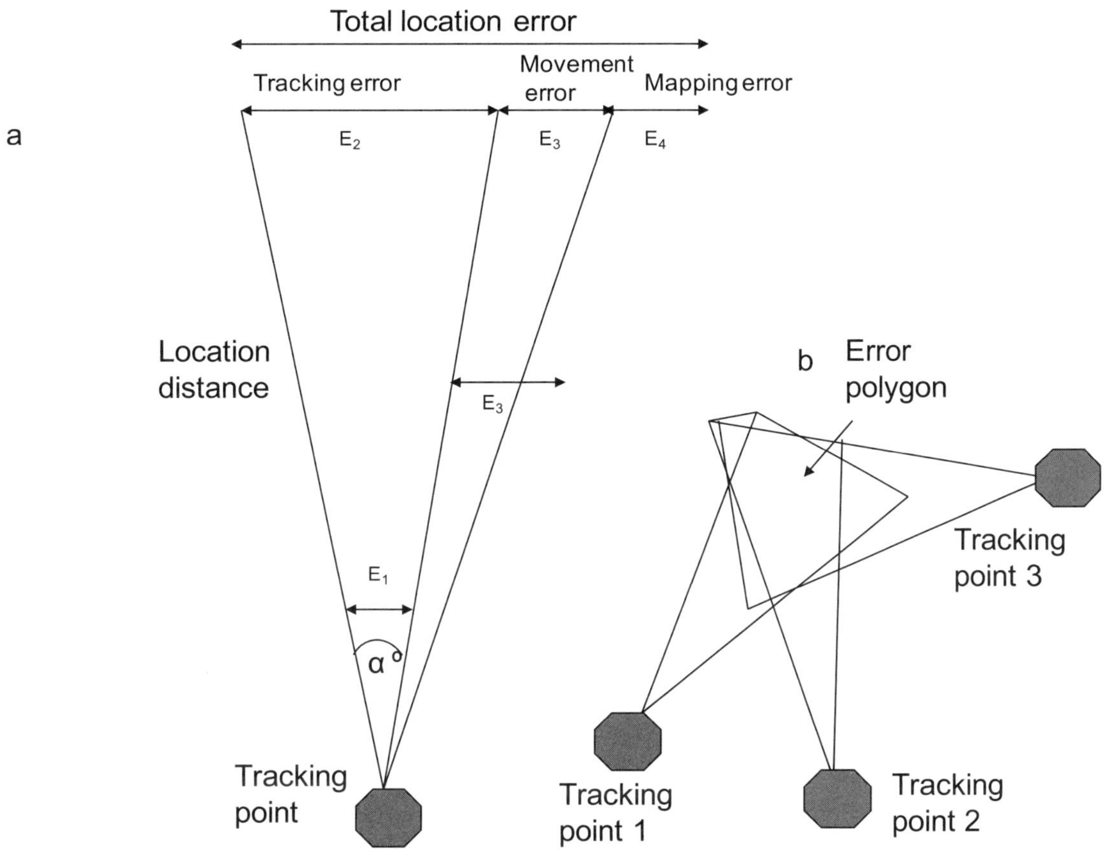

Figure 1. Illustration of the potential components of (a) total location error and (b) an error polygon for radio and acoustic tracking.

Negating the need for triangulation and any associated mapping avoids the two greatest causes of location error (Stasko and Pincock 1977; White and Garrott 1990; Simpkins and Hubert 1998). The 'antenna reduction method' described by Koehn et al. (2012) was found to be faster and more accurate at locating transmitters compared to triangulation. This type of method is used by many researchers and it is particularly applicable when tracking more sedentary or benthic species by boat. Similar results have been reported by Niemela et al. (1993), Bruno et al. (1990) and by Broadhurst and Ebner (2007) in small streams. The following options can be used to minimize location error (see Figure 1):

- Reducing the distance between the receiver and the transmitter
- Avoiding movement error by minimize time between fixes
- Minimizing the use of mapping
- Minimizing the use of triangulation

- Using the same trained and experienced personnel
- Using pulse rates >50 ppm

Two other types of tracking error can also occur: false negatives which underestimate the number of animals in a particular habitat type, and false positives, which overestimate them. Difficulty in detecting the tagged animals is likely to result in a bias toward false negative results. Inadequate reception range and short scan times can possibly result in non fish detection, especially when many fish occur in the same area. Conversely, some false positives have occurred through electrical interference with loggers (Section 5.3). Both errors need to be quantified (Section 9.5).

DATA ANALYSIS

There is often a need to match tracking data with associated environmental information (e.g., climate, temperature, flow) and there may also be the need for different types of tracking data to be collected concurrently such as water and fish depth and habitat type (Koehn 2009a). Early consideration should be given to likely options for data management and analysis (Sections 9.4 and 9.5). For example, there are many approaches to determining a species' home range and analysis of some representative data collected in a pilot study is useful to ensure that it can be effectively analyzed (Kenward 1992; Section 9.4). Analysis of telemetry data is a rapidly developing field that now often utilizes nonstandardized statistical approaches, including analytical models (Burnham and Anderson 2002; Patterson et al. 2007). In some instances, analytical techniques for telemetry data are more advanced in terrestrial ecology and options such as the use of state-based models that can link the process of movement to its purpose, using observations over time and space (Patterson et al. 2007) can to be explored. Some care should be taken with the use of software that is designed for the analysis of terrestrial data (e.g., for home ranges) as not all packages are adept at the analysis of data in linear systems such as rivers.

As a large portion of telemetry is undertaken on terrestrial wildlife rather than fish, it is worth mentioning some notable differences that may apply. In rivers, fish may be seen to inhabit essentially, long, linear landscapes (depending on the study scale and purpose). While this may be an advantage when tracking (one dimension rather than two), sometimes this data are less adaptable to interpretation using software packages designed for terrestrial wildlife data (e.g., home ranges). For example, when used to describe riverine movements many home range programs have difficulty accounting for inaccessible areas such as islands or gravel bars or three dimensional aspects associated with depth. Linear riverine habitats are also often subjected to barriers such as weirs which may limit movements. Unlike terrestrial habitats, many components of aquatic habitats are hidden under the water, making tracking and measurement difficult. Aquatic habitats also have the potential to change rapidly over time according to the amount of water (e.g., changes in depth or velocity or area of floodplain inundation). Such issues should be kept in mind when reading general telemetry literature.

MAKING COMPROMISES AND TRADE-OFFS

Trade-offs are inevitable in the design of any telemetry field study. A common compromise is tag size and longevity versus signal strength and range (Table 3). For example, when

Table 3. Trade-offs to be considered when selecting radio and acoustic transmitters.

Transmitter options	Advantages	Disadvantages	Alternative options/potential solutions
Increased operating life	Longer-term studies	Increases size and weight	Limit studies to larger species or individuals (e.g., older life stages) Reduce transmitter range Use varied duty cycle (e.g., 12 h on, 12 h off), but this will limit tracking options e.g., stationary loggers, diel Reduce signal pulse, width, and rate, but this will impact tracking efficiency and location resolution
Increased reception range (power output)	Increases signal range and study scale Increases chance of detection	Increases size and weight	Limit studies to larger species or individuals (e.g., older life stages) Reduce transmitter longevity (operating life) Use varied duty cycle (e.g., 12 h on, 12 h off), but this will limit tracking options e.g., stationary loggers, diel Reduce signal pulse, width, and rate, but this will impact tracking efficiency and location resolution
Smaller transmitter size	Studies of smaller fish and early life stages	Limits transmitter longevity and range	Limit research objectives and field methods to account for reduced longevity and range • Shorter study duration • Smaller study scale Use varied duty cycle (e.g., 12 h on, 12 h off), but this will limit tracking options e.g., stationary loggers, diel Reduce signal pulse width and rate, but this will impact tracking efficiency and location resolution
Internal antenna	Reduces impacts of tangling Reduces infection rates Simplified tagging procedures	Reduces range (radio) Increases size and weight	Use larger, more powerful transmitters, but this will impact size Limit studies to species with restricted home ranges or in small, accessible waters

Table 3. Continued.

Transmitter options	Advantages	Disadvantages	Alternative options/potential solutions
Additional sensors (e.g., depth, temperature)	Additional information, both positioning and environmental data	Increases size and weight	Reduce transmitter longevity and range Limit studies to larger species or individuals (e.g., older life stages) Use varied duty cycle (e.g., 12 h on, 12 h off), but this will limit tracking options e.g., stationary loggers, diel
Activity/mortality circuit[1]	Efficient method for monitoring live fish Reduces errant data	Increases size and weight	Reduce transmitter longevity and range Limit studies to larger species or individuals (e.g., older life stages) Use varied duty cycle (e.g., 12 h on, 12 h off), but this will limit tracking options e.g., stationary loggers, diel

[1]Used to instantaneously signal a period of inactivity or mortality.

considering options to conserve battery life, pulse rates of 35, 40 and 55 ppm can give expected transmitter life of 792, 540 and 400 d respectively (www.atstrack.com) and micro controlled transmitters (delayed tag initiation, time on and off) can offer additional power saving options. Such options should be considered carefully to ensure that they will not compromise tracking efficiency and location resolution. Figure 2 provides a systematic approach to decision making and assessing trade-offs. This may be used in combination with other Sections in this book and other information sources, to help guide decisions when determining the most suitable options for key aspects of your study. These tracking and equipment options all need to be balanced with cost.

TRAINING PERSONNEL AND TESTING EQUIPMENT

There is often considerable pressure on researchers to just get the radio tracking gear and get started in the field, assuming that all equipment and techniques will be suitable and work within the given conditions. While manufacturer's specifications provide aspects of the technical capabilities of the tags and tracking equipment, they cannot quantify the effects of the combination of all field and study conditions. Despite manufacturer's specifications, and the

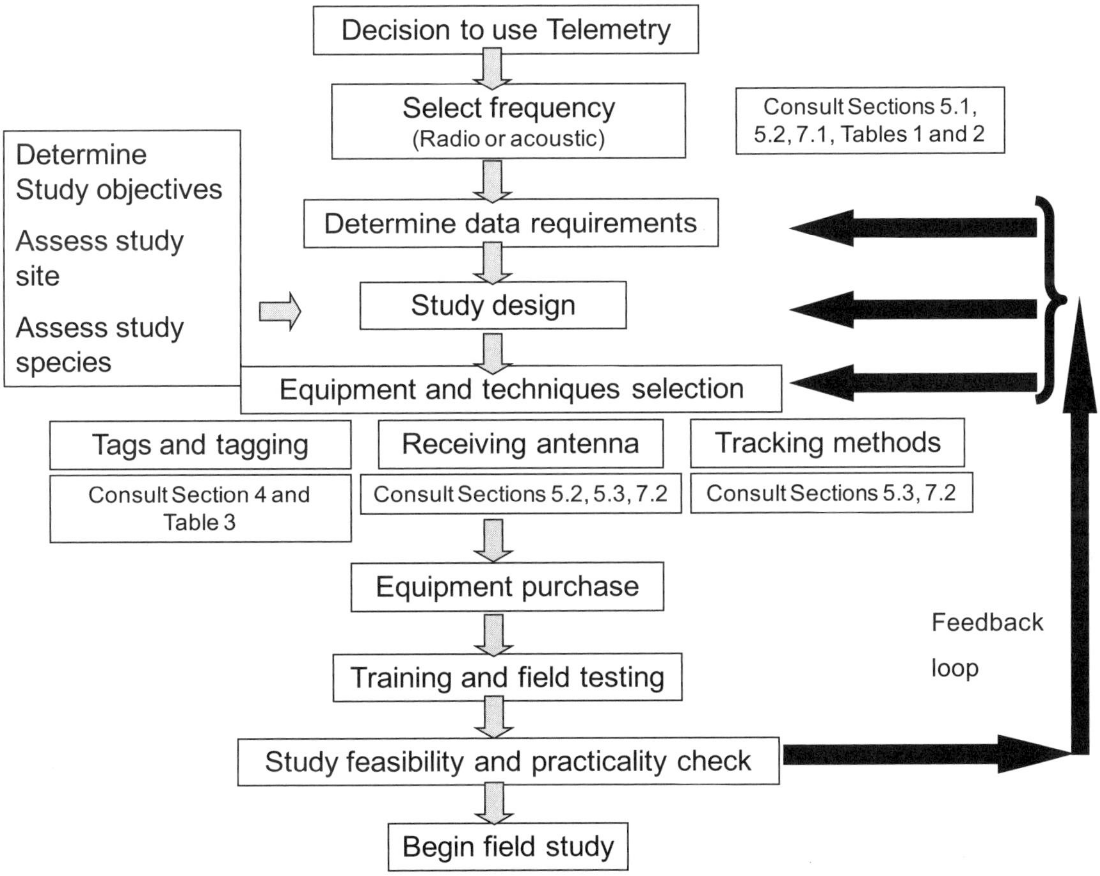

FIGURE 2. A systematic approach to decision making and assessing trade-offs in designing a telemetry field study.

extensive consideration given to equipment purchase, compromises and techniques, things often change under field conditions requiring equipment to be tested on site to determine actual performance.

While the development of efficient field techniques can be time consuming, this is especially important for researchers less experienced in telemetry techniques. It is important to emphasize the need to allow enough time to get all the equipment in hand well in advance of the start of the study so operators can familiarize themselves with it, set it up, and test it. A properly designed study will have this time built into the study timelines. Good advice on equipment choice is available from manufacturers (Kenward 2001) and from other practitioners. Training courses are periodically available (although at limited locations), and while information from this book will hopefully assist with theoretical training, there is no substitute for practical training and use of the gear in field trials to hone techniques. Many U.S. Federal agencies offer training to other State and Federal agencies. Workshops and symposiums are also periodically offered by specific training and equipment vendors to educate the user. There is the need for field readiness in practical, technical and specialized skills (e.g., surgery) (Section 4) and the importance and value of such training should not be underestimated. It is recommended that trials are undertaken at the beginning of the study to determine transmitter ranges and quantify location errors (Figure 1).

Field testing (including collection and analysis of test data) can be an important component of the establishment phase of a telemetry study. Testing reduces risks of equipment failure, improper use of equipment, hones techniques and increases efficiencies, reducing potential liabilities, especially in long-term studies. Testing needs to occur at the appropriate spatial and temporal scales. A pilot study, however, is not the only time to optimize data collection as there is the scope for refinement throughout the study (Kenward 2001). Nevertheless, adequate planning allows the study to be comprehensively developed so it can be undertaken properly at the first 'real' attempt.

DETERMINING IF THE PROPOSED APPROACH IS FEASIBLE AND PRACTICAL

The feasibility and practicality of the project can have major impacts on the success of study and quality of data obtained. This is the reality check of your wonderful ideas and plans—is the study plan physically and logistically possible? Does it provide the type of data needed (i.e., meet study objectives)? This is of particular importance if the information obtained is to be used to support controversial management decisions (e.g., harvest allocations, threatened species). Often there may be a need to verify interpretation of the telemetry data collected to show that the results reflect normal behavior rather than being an artifact of the techniques used. This may require some additional aspects to be included in the study design (e.g., behavioral trials) and further illustrates the benefit of feasibility studies. Feasibility trials allow for feedback (what's proposed versus what the field tested experiences were) and reassessment of the study objectives and equipment selection before the main study is undertaken. The need for flexibility and revision should be built into project design.

In any area of research that is new, there can be surprises. A good rule may be to 'expect the unexpected.' Telemetry studies often open up new areas of study for species and, together with the normal operational difficulties of field studies and the fragility of technical and electronic equipment, things can, will, and do go wrong. To keep a project on track, there is the need to consider and plan for contingencies. Adequate resources are needed to ensure the purchase of appropriate gear including back up equipment (Sections 5.2 and 5.3), an adequate number of tags (including replacements) and enough resources for tracking time and other costs (e.g., unexpected aircraft hire). Unfortunately, funding is all too often a major limitation to study design. It is important to recognize this early in the study and if the funding is insufficient to adequately obtain the data needed, then study should be postponed or the study objective scaled back to a reasonable level. Frank discussions and realistic expectations are essential in the planning phase. A likely example may be using borrowed radio telemetry equipment with an inadequate number of tags to attempt to answer questions that really need many more tags, or perhaps acoustic telemetry equipment, to provide adequate data. Some cost savings may, however, be possible through the loan of suitable equipment, as many researchers with 'seasonal' projects have equipment sitting on the shelf for months at a time until they need it for their study. Most researchers are happy to loan equipment to a trustworthy colleague and this can help overcome the project-start up costs or allow initial trials to be undertaken. This still may not solve the problem of an under-funded project however, and the consequence of getting marginal information with the wrong technology or an inadequate sample size may be an incorrect interpretation or management decision that could be to the detriment of the species.

Field conditions can change, with droughts, floods, fires, storms, high and low temperatures, snow, ice and water level fluctuations posing limitations on field gear, study animals, field researchers and tracking ability. Intentional vandalism and damage by birds, animals, or insects (e.g., ants in loggers) can cause the loss of equipment and data (Section 5.3). Tag 'losses' will always occur through rejections, deaths, capture by anglers, tag failure, or dispersal outside the study detection zone or into inaccessible areas. Data storage and capacity for automated downloads of data need to be planned for. Study of threatened species may require special permits and animal ethics permits are likely to be needed for radio tagging techniques in some jurisdictions. Occupational health and safety considerations must be included in field studies based on telemetry. The limitations of boat and aircraft tracking outlined by Eiler (Section 5.3) (noise from outboard motors or birds, motor interference, motion sickness) need to be considered. If the species under study is also sought by anglers or commercial fishers, the return of transmitters can provide additional information on the fate or location of fish. This may require a publicity campaign and liaison with particular groups to facilitate such returns.

POTENTIAL SOURCES OF BIAS

While telemetry can avoid many observational or capture biases, as with any technique, there are other potential biases that need to be considered so that steps can be taken to minimize them. Avoiding biases is overlooked in many studies, and often cannot be assessed by traditional statistical methods. The aim is to have a nonbiased study for the parameters that have been identified in your study objectives. Bias may occur with:

- the capture of the fish to be tagged—are they representative of the population that you wish to study? For example, have the tagged fish been more susceptible to capture by a particular method (such as bank-oriented fish being more susceptible to capture);

- fish sex; are the sexes evenly represented (especially given difficulties of sexing some species easily in the field); behavior; will individually tagged schooling fish act independently?

- fish size; as larger fish can accommodate larger transmitters, this may pose a size (and possibly maturity) bias or limitations (e.g., adult and sub adult fish may occupy different habitats);

- the impact of the tags; performance or behavior bias on the animal are difficult to determine but should be considered (Section 4).

Issues such as limitations in the detection range of tags can pose a bias against large-scale movements (as they may move outside the detection zone), fish located in deep water or on the distant floodplain. A systematic sampling schedule is important to cover all sampling times evenly, including the potential for nocturnal sampling. Some biases, however, may be unavoidable (e.g., tagging a larger size component of the population due to transmitter size) and while this doesn't mean that the study should not be conducted, it is important to recognize and address such issues, and ensure that the interpretation of data reflects these limitations.

Careful consideration should be given to the analysis of pooled data as pseudo replication can occur as a result of biased representation of data from individuals, age classes, sexes, temporal scales (diel cycles, seasons), habitat patches or inadequate spatial separation. Data from multiple relocations of the same individual may not be considered independent and autocorrelated data needs to be considered in analyses.

CONCLUSION

By necessity, this Section has posed many questions and potential obstacles for the researcher to consider. These should not be viewed in a pessimistic way however, as in reality telemetry provides a wonderful and very powerful tool for the study of both marine and freshwater fishes. Methods (both radio and acoustic) can be applied to a wide range of questions and studies based on telemetry have contributed to many major discoveries in fish biology. However, the power and attraction of such equipment and techniques, should not seduce the researcher into thinking that it can answer all questions. Perhaps because of the attractiveness of its high-tech nature, it, more than other methods, needs to be carefully assessed and incorporated into the design to best meet the study objectives. The vast array of equipment and techniques makes such assessment a difficult one, but it is hoped that this Section, along with the remainder of the book provides a pathway to maximizing study outcomes.

ACKNOWLEDGMENTS

The authors thank all those who I have worked with and assisted with fish tracking and movement and migration and habitat projects. Valuable comments on manuscript drafts were kindly provided by John Eiler, Justin O'Connor, Noah Adams, John Beeman, and the two anonymous reviewers.

REFERENCES

Adams, N. S., D. W. Rondorf, S. D. Evans, J. E. Kelly, and R. W. Perry. 1998a. Effects of surgically and gastrically implanted radio transmitters on growth and feeding behavior of juvenile Chinook salmon. Transactions of the American Fisheries Society 127:128–136.

Adams, N. S., D. W. Rondorf, S. D. Evans, J. E. Kelly, and R. W. Perry. 1998b. Effects of surgically and gastrically implanted radio transmitters on swimming performance and predator avoidance of juvenile Chinook salmon (*Oncorhynchus tshawytscha*). Canadian Journal of Fisheries and Aquatic Sciences 55:781–787.

Amlaner, C. J., and D. W. MacDonald, editors. 1980. A handbook on biotelemetry and radio tracking. Proceedings of an international conference on telemetry and radio tracking in biology and medicine, Oxford, 20–22 March 1979. Permagon Press, Oxford, U.K.

Brenden, T. O., B. R. Murphy, and E. M. Hallerman. 2004. Estimating total error in locating radiotelemetry transmitters by homing in a riverine environment. Journal of Freshwater Ecology 19:295–304.

Burnham, K. P., and D. R. Anderson 2002. Model selection and multimodel inference: a practical information-theoretic approach. Springer-Verlag, New York.

Broadhurst, B., and B. Ebner, 2007. An improved technique for small-scale radio-tracking of crayfish and benthic fishes in upland streams. Transactions of the American Fisheries Society 136:423–427.

Broadhurst, B. T., B. C. Ebner, and R. C. Clear. 2009. Radio-tagging flexible-bodied fish: temporary confinement enhances radio-tag retention. Marine and Freshwater Research 60:356–360.

Bruno, N. A., R. W. Gregory, and H. L. Schramm. 1990. Nest sites used by radio-tagged largemouth bass in Orange Lake, Florida. North American Journal of Fisheries Management 10:80–84.

Butler, G. L., B. Mackay, S. J. Rowland, and B. C. Pease. 2009. Retention of intra-peritoneal transmitters and post-operative recovery of four Australian native fish species. Marine and Freshwater Research 60:361–370.

Cooke, S. J., S. G. Hinch, M. Wikelski, R. D. Andrews, L. J. Kuchel, T. G. Wolcott, and P. J. Butler. 2004. Biotelemetry: a mechanistic approach to ecology. Trends in Ecology and Evolution 19:334–343.

Cooke, S. J., G. H. Niezgoda, K. C. Hanson, C. D. Suski, F. J. S. Phelan, R. Tinline, and D. P. Philipp. 2005. Use of CDMA acoustic telemetry to document 3-D positions of fish: relevance to the design and monitoring of aquatic protected areas. Marine Technology Society Journal 39:31–41.

Crook, D. A. 2004. Is the home range concept compatible with the movements of two species of lowland river fish? Journal of Animal Ecology 73:353–366.

Daniel, A. J., B. J. Hicks, N. Ling, and B. O. David. 2009. Acoustic and radio transmitter retention in common carp (*Cyprinus carpio*) in New Zealand. Marine and Freshwater Research 60:328–333.

Daugherty, D. J., and T. M. Sutton. 2005. Seasonal movement patterns, habitat use, and home range of flathead catfish in the lower St. Joseph River, Michigan. North American Journal of Fisheries Management 25:256–269.

David, B. O., and G. P. Closs. 2001. Continuous remote monitoring of fish activity with restricted home ranges using radiotelemetry. Journal of Fish Biology 59:705–715.

David, B. O., G. P. Closs, S. K. Crow, and E. A. Hansen. 2007. Is diel activity determined by social rank in a drift-feeding stream fish dominance hierarchy? Animal Behaviour 74:259–263.

Diggle, J., J. Day, and N. Bax. 2004. Eradicating European carp for Tasmania and implications for national European carp eradication. Inland Fisheries Service, Hobart.

Ebner, B. C., M. Lintermans, M. Jekabsons, M., Dunford, and W. Andrews. 2009. A cautionary tale: surrogates for radio-tagging practice do not always simulate the responses of closely related species. Marine and Freshwater Research 60:371–378.

Eiler, J. H. 1990. Radio transmitters used to study salmon in the glacial rivers. Pages 364–369 *in* N. C. Parker, A. E. Giorgi, R. C. Heidinger, D. B. Jester, Jr., E. D. Prince, and G. A. Winans, editors. Fish-marking techniques. American Fisheries Society, Symposium 7, Bethesda, Maryland.

Eiler, J. H. 1995. A remote satellite-linked tracking system for studying Pacific salmon with radio telemetry. Transactions of the Amrican Fisheries Society 1245:184–193.

Eiler, J. H., D. J. Alcorn, and M. R. Neuman, editors. 2000. Biotelemetry 15. Proceedings of the 15th international symposium on biotelemetry. Juneau, Alaska, 10–15 May 1999. International Society on Biotelemetry, Wageningen, The Netherlands.

Eiler, J. H., B. D. Nelson, and R. F. Bradshaw. 1992. Riverine spawning by sockeye salmon in the Taku River, Alaska and British Columbia. Transactions of the American Fisheries Society 121:701–708.

Eiler, J. H., T. R. Spencer, J. J. Pella, and M. M. Masuda. 2006. Stock composition, run timing, and movement patterns of Chinook salmon returning to the Yukon River basin in 2004. US Department of Commerce, NOAA Technical Memorandum NMFS-AFSC-165.

Garrott, R. A., G. C. White, R. M. Bartmann, and D. L. Weybright. 1986. Reflected signal bias in biotelemetry triangulation systems. Journal of Wildlife Management 50:747–752.

Guy, C. S., D. W. Willis, and J. J. Jackson. 1994. Biotelemetry of white crappies in a South Dakota glacial lake. Transactions of the American Fisheries Society 123:63–70.

Hasbrouck, J. J., S. L., Hammarstrom, J. A., Carlon, and R. A. Clark, 2000. Two methods to estimate abundance of salmonids using biotelemetry. Pages 129–138 *in* J. H., Eiler, D. J. Alcorn, and M. R. Neuman, editors. Biotelemetry 15. Proceedings of the 15th International Symposium on Biotelemetry. Juneau, Alaska, 10–15 May 1999. International Society on Biotelemetry, Wageningen, The Netherlands.

Jellyman, D. 2009. A review of radio and acoustic telemetry studies of freshwater fish in New Zealand. Marine and Freshwater Research 60:321–327.

Johnsen, P. B. 1980a. The use of longterm ultrasonic implants for the location and harvest of schooling fish. Pages 777–779 *in* C. J. Amlaner, and D. W. MacDonald, editors. A handbook on biotelemetry and radio tracking. Proceedings of an international conference on telemetry and radio tracking in biology and medicine, Oxford, 20–22 March 1979. Permagon Press, Oxford, U.K.

Johnsen, P. B. 1980b. The movements of migrating salmonids in the vicinity of a heated effluent determined by a temperature and pressure sensing radio telemetry system. Pages 781–784 *in* C. J. Amlaner, and D. W. MacDonald, editors. A handbook on biotelemetry and radio tracking. Proceedings of an international conference on telemetry and radio tracking in biology and medicine, Oxford, 20–22 March 1979. Permagon Press, Oxford, U.K.

Kenward, R. E. 1992. Quantity versus quality: programmed collection and analysis of radio-tracking data. Pages 421–422 *in* I. G. Priede and S. M. Swift, editors. Wildlife telemetry: remote monitoring and tracking of animals. Ellis Horwood, Chichester, U.K.

Kenward, R. E. 2001. A manual for wildlife radio tagging. Academic Press, London.

Koehn, J. D. 2009a. Using radio telemetry to evaluate depths inhabited by Murray cod *Maccullochella peelii peelii*. Marine and Freshwater Research 60:317–320.

Koehn, J. D. 2009b. Muliti-scale habitat selection by Murray cod (*Maccullochella peelii peelii*) in two lowland rivers. Journal of Fish Biology 75:113–129.

Koehn J. D., S. J., Nicol, J. A., Leischke, J. P., Lyon, and K. Pomorin, 2008. Spatial ecology of an endangered native Australian Percichthyid fish, the trout cod *Maccullochella macquariensis*. Endangered Species Research 4:219–225.

Koehn, J. D., J. H. Eiler, J. A. McKenzie, and W. G. O'Connor. 2012. An improved method for obtaining fine-scale location of radio tags. Pages 379–384 *in* J. McKenzie, B. Parsons, A. C. Seitz, R. K. Kopf, M. Mesa, and Q. Phelps, editors. Advances in fish tagging and marking technology. American Fisheries Society, Symposium 76, Bethesda, Maryland.

Koehn, J. D., J. A., McKenzie, D. J., O'Mahony, S. J., Nicol, J. P. O'Connor, and W. G. O'Connor. 2009. Movements of Murray cod (*Maccullochella peelii peelii*) in a large Australian lowland river. Ecology of Freshwater Fish 18: 594–602.

Koster, W. M., and D. A. Crook. 2008. Diurnal and nocturnal movements of river blackfish (*Gadopsis marmoratus*) in a south-eastern Australian upland stream. Ecology of Freshwater Fish 17:146–154.

Lee, J. E., G. C., White, R. A., Garrott, R. M. Bartman, and A. W. Alldredge. 1985. Assessing accuracy of a radiotelemetry system for estimating animal locations. Journal of Wildlife Management 49:658–663.

Lucas, M. C., A. D. Johnstone, and I. G. Priede. 1993. Use of physiological telemetry as a method of estimating metabolism of fish in the natural environment. Transactions of the American Fisheries Society 122:822–833.

Martinelli, T. L., H. C. Hansel, and R. S. Shively. 1998. Growth and physiological responses to surgical and gastric radio transmitter implantation techniques in subyearling Chinook salmon (*Oncorhynchus tshawytscha*). Hydrobiologia 371/ 372:79–87.

Miller, W. F., Sandwell, F. W. and Gorton, P. A. 1980. A monitoring chamber for trout behaviour and physiology. Pages 307–311 *in* C. J. Amlaner, and D. W. MacDonald, editors. A handbook on biotelemetry and radio tracking. Proceedings of an international conference on telemetry and radio tracking in biology and medicine, Oxford, 20–22 March 1979. Permagon Press, Oxford.

Moser, M. L., A. F. Olson, and T. P. Quinn. 1990. Effects of dummy ultrasonic transmitters on juvenile coho salmon. Pages 353–356 *in* N. C. Parker, A. E. Giorgi, R. C. Heidinger, D. B. Jester, Jr., E. D. Prince, and G. A. Winans, editors. Fish-marking techniques American Fisheries Society, Symposium 7, Bethesda, Maryland.

Nicol, S. J., J. Lieschke, J. Lyon, and J. D. Koehn. 2004. Observations on the distribution and abundance of carp and native fish, and their responses to a habitat restoration trial in the Murray River, Australia. New Zealand Journal of Marine and Freshwater Research 38:541–551.

Niemela, S. L., J. B. Layzer, and J. A. Gore. 1993. An improved radiotelemetry method for determining use of microhabitats by fishes. Rivers 4:30–35.

Nielson, J. L., H. Arrizabalaga, N. Fragoso, A. Hobday, M. Lutcavage, and J. Sibert editors. 2009. Tagging and tracking marine mammals with electronic devices. Springer, New York.

O'Connor, J. P., D. J. O'Mahony, and J. M. O'Mahony. 2005. Movement of *Macquaria ambigua*, in the Murray River, south-eastern Australia. Journal of Fish Biology 66:392–403.

O'Connor J. P., D. J. O'Mahony, J. M. O'Mahony, and T. J. Glenane. 2006. Some impacts of low and medium head weirs on downstream fish movement in the Murray-Darling Basin in southeastern Australia. Ecology of Freshwater Fish 15:419–427.

O'Connor, J. P., J. D. Koehn, S. J. Nicol, D. J. O'Mahony, and J. A. McKenzie. 2009. Radio-tag retention trials for golden perch *Macquaria ambigua*, silver perch *Bidyanus bidyanus* and carp *Cyprinus carpio*. Marine and Freshwater Research 60:334–340.

Parker, N., H., Pascoe, H. Moller, and R. Maloney. 1996. Inaccuracy of a radio-tracking system for small mammals: the effect of electromagnetic interference. Journal of Zoology, London 239:401–406.

Patterson, T. A., L. Thomas, C. Wilcox, O. Ovaskainen, and J. Matthiopoulos. 2007. State–space models of individual animal movement. Trends in Ecology and Evolution 23:87–94.

Priede, I. G. (1980). An analysis of objectives in telemetry studies of fish in the natural environment. Pages 105–118 *in* C. J. Amlaner, and D. W. MacDonald, editors. A handbook on biotelemetry and radio tracking. Proceedings of an international conference on telemetry and radio tracking in biology and medicine, Oxford, 20–22 March 1979. Permagon Press, Oxford, U.K.

Priede, I. G. 1992. Wildlife telemetry: an introduction. Pages 3–25 *in* I. G. Priede, and S. M. Swift, editors. Wildlife telemetry: remote monitoring and tracking of animals. Ellis Horwood, New York.

Saltz, D. 1994. Reporting error measures in radio location by triangulation: a review. Journal of Wildlife Management 58:181–183.

Samuel, M. D., and K. P. Kenow. 1992. Evaluating habitat selection with radio-telemetry triangulation error. Journal of Wildlife Management 56:725–734.

Schrader, W.C., and Jones, M.D. 2000. Use of radio telemetry to describe movement patterns of adult trout in the Teton River, Idaho. Pages 59–68 *in* J. H., Eiler, D. J. Alcorn, and M. R. Neuman, editors. Biotelemetry 15. Proceedings of the 15th International Symposium on Biotelemetry. Juneau, Alaska, 10–15 May 1999. International Society on Biotelemetry, Wageningen, The Netherlands.

Simpkins, D. G., and W. A. Hubert. 1998. A technique for estimating the accuracy of fish locations identified by radiotelemetry. Journal of Freshwater Ecology 13:263–268.

Sisak, M. M., and J. S. Lotimer. 1998. Frequency choice for radio telemetry: the HF vs. VHF conundrum. Hydrobiologia 371/ 372:53–59.

Stasko, A. B., and D. G. Pincock. 1977. Review of underwater biotelemetry, with emphasis on ultrasonic techniques. Journal of the Fisheries Research Board of Canada 34:1261–1285.

Stuart, I. G., J. D., Koehn, T. A., O'Brien, J. A., McKenzie, and G. P. Quinn. 2010. Too close for comfort: a fishway exit and a hydro power station inlet. Marine and Freshwater Research 61:23–33.

Tyus, H. M. 1985. Homing behaviour noted for Colorado squawfish. Copeia 1985:213–215.

Ward, G. D., A. A. C. Fairweather, and R. E. Brockie. 1986. Possum (*Trichosurus vulpecular*) biology on Hawkes Bay farmland. Ecology Division Report No. 5. DSIR, Lower Hutt, New Zealand.

Weatherley, A. H., P. A. Kaseloo, M. D. Gare, J. M. Gunn, and B. Lipicnik. 1996. Field activity of lake trout during the reproductive period monitored by electrocardiogram radiotelemetry. Journal of Fish Biology 48:675–685.

White, G. C., and R. A. Garrott. 1990. Analysis of wildlife radio-tracking data. Academic Press, San Diego, California.

Winter, J. D. 1983. Underwater biotelemetry. Pages 371–395 *in* L. A. Nielsen, and D. L. Johnson, editors. Fisheries Techniques. American Fisheries Society, Bethesda, Maryland.

Winter, J. D. 2000. Designing telemetry studies and other technical and analytical considerations. Pages 229–248 *in* J. H. Eiler, D. J. Alcorn, and M. R. Neuman, editors. Biotelemetry 15. Proceedings of the 15th International Symposium on Biotelemetry. Juneau, Alaska, 10–15 May 1999. International Society on Biotelemetry, Wageningen, The Netherlands.

Zimmerman, J. W., and R. A. Powell. 1995. Radio-telemetry error: location error method compared with error polygons and confidence ellipses. Canadian Journal of Zoology 73:1123–1133.

Section 4

Techniques for Telemetry Transmitter Attachment and Evaluation of Transmitter Effects on Fish Performance

Theresa L. Liedtke and A. Michelle Wargo Rub

INTRODUCTION

One assumption of nearly every biotelemetry study is that the tagged animals are representative of the untagged population. That is, that the processes by which study animals are captured, handled, and tagged, as well as the act of carrying a transmitter, will have minimal effect on their behavior and performance. This assumption, commonly stated as a lack of transmitter effects, must be valid if telemetry studies are to describe accurately the movements and behavior of an entire population of interest, rather than only of a subset of that population.

Considering the sequence of events necessary to implement telemetry studies (i.e., collection, handling, transmitter attachment), as well as the intrusive nature of many transmitter attachment techniques, it is likely that there will be some effect on the study animals. These potential impacts can range from mild to severe, from transitory to permanent, and may be manifest immediately or not for several days to weeks after tagging.

Direct physical impacts can include elevated stress levels, injury, or even death (Knights and Lasee 1996; Jepsen et al. 2001; Lacroix et al. 2004). Interrupted integrity of the scales, mucus, or skin can place fish at increased risk of infection (Mellas and Haynes 1985; Swanberg et al. 1999; Bauer et al. 2005; Harms 2005). Effects on fish behavior include altered buoyancy compensation ability (Gallepp and Magnuson 1972; Fried et al. 1976), reduced swimming performance (McCleave and Stred 1975; Counihan and Frost 1999; Makiguchi and Ueda 2009), reduced feeding, or changes in dominance status (Greenstreet and Morgan 1989; Armstrong and Rawlings 1993; Welch et al. 2007). Changes in behavior such as these can affect a fish's growth, rate of maturation or migration, or increase its vulnerability to predation. At some level, these transmitter effects will be present in nearly every study using telemetry and if substantial, they can violate the critical assumption that tagged fish are representative of the untagged population.

Careful consideration of fish collection, handling, and tagging techniques will help minimize transmitter effects and ensure that valid inferences can be made about the population of interest. This Section describes fish handling, tagging, and performance evaluation procedures to help researchers anticipate and evaluate potential effects on fish performance, so that they can be minimized.

STUDY DESIGN CONSIDERATIONS

Some general considerations for study design should be applied to all telemetry studies in order to minimize transmitter and handling effects. The study design will drive decisions such as the choice of telemetry technology (e.g., radio or acoustic), the specifications of the transmitter (required battery life and sensors), and the tagging method. Common tagging methods include external attachment, insertion into the stomach (gastric implantation), or placement within the abdominal cavity (surgical implantation). Each study will have unique requirements and limitations, and researchers must carefully weigh the available options to optimize results.

A good first step during the study design phase is a review of the telemetry literature, matching as well as possible the target species, life stage, and size of the planned study. A wide range of potential negative consequences and pitfalls can be avoided if properly anticipated and managed. In addition, we encourage researchers to contact experienced professionals to gather insights into their current and past practices and a summary of their "lessons learned." In cases where the literature is limited or not directly applicable, researchers may need to experiment with different techniques to ensure the best outcome.

Transmitter Selection

It is important to use the smallest transmitter that will meet study objectives. In general, larger and heavier transmitters are needed when longer battery life and improved signal strength are required. Transmitter dimensions will be further increased if integrated sensors (e.g., pressure, temperature) are needed to meet study objectives. Starting with the smallest transmitter, researchers must ask themselves whether needs for increased detection range or frequency, battery life, or sensor packages will offset the increased risk posed to the fish by use of a larger transmitter. This tradeoff can reduce the ability to make inferences from the study results.

An estimate that can be used to evaluate the risk of transmitter effects is the transmitter-to-body weight ratio (or tag burden) for the fish. The most commonly accepted limit for tag burden is from Winter (1996) who recommended that transmitter weight in air not exceed 2% of fish body weight in air. Some studies have used tag burdens above 2% on captive animals (Moser et al. 1990; Martinelli et al. 1998; Anglea et al. 2004; Welch et al. 2007; Brown et al. 2010), however researchers should apply caution when extrapolating published results, especially from laboratory to field studies, where conditions under which the fish must perform may be more challenging. There is clearly an upper limit to the tag burden that a fish can reasonably accommodate, and the most defensible approach is to take all possible steps to minimize transmitter size.

Transmitter effects correlated with high tag burdens include elevated mortality and transmitter expulsion (Moore et al. 1990; Lacroix et al. 2004; Jepsen et al. 2008b; Hall et al. 2009), reduced growth (Jepsen et al. 2008b), and reduced swimming performance (Adams et al. 1998b; Zale et al. 2005). No "safe" tag burden has been established, so compliance with the

"2% rule" does not eliminate the potential for transmitter effects (Jepsen et al. 2005). Therefore, the responsibility lies with individual researchers to evaluate and minimize these risks, or to state that they are unknown.

In addition to tag burden, other considerations may influence transmitter selection including transmitter volume, length, diameter, shape, density, and coating, any of which may also influence fish behavior and performance (Sakaris et al. 2005; Penne et al. 2007). Consideration of the physical dimensions of a transmitter is especially important for small or anguilliform fishes or for fish with compressed body forms because they may be more limiting than tag burden (Moser et al. 2007).

The choice of attachment method and monitoring technology (radio versus acoustic telemetry) can also have large effects. For example, a heavier transmitter might have less impact on fish when implanted surgically (near the center of mass) than it would if externally attached off-midline to one side of the dorsal fin. Unlike acoustic transmitters, radio transmitters use an antenna to provide signal output. The radio antenna may be external (extending from the body of the transmitter), or internal (coiled and incorporated into the body of the transmitter). The external antenna requires a tagging method that accommodates the antenna, whereas use of internal antennas simplifies the tagging process. The choice of antenna configuration will influence the choice of attachment method as well as signal propagation (see Section 5.1).

Transmitter selection involves simultaneous consideration of several variables specific to each study. The goal is to work through all of the considerations in choosing the technologies and methods that best support study goals while limiting transmitter effects. If support for transmitter selection is lacking in the literature, or if the implications of study results are especially important, researchers should consider a formal evaluation of transmitter effects. The implications posed by transmitter effects are so powerful that some researchers recommend that each study conduct controlled evaluations of effects (Jepsen et al. 2005). Techniques for such evaluations will be addressed later in this Section.

Tagging Protocols

A standard operating procedure (SOP) should be developed to describe specifics of the handling and tagging methods. Having written guidance will enable efficient tagging operations (Smith et al. 2009) and ensure consistent application across individuals and through time. The principal attachment techniques have been described in the literature, so guidance is available on the specifics of most individual methods. Researchers should seek out colleagues who have performed these procedures and consult with veterinarians who can provide general input on equipment selection and methods (e.g., suture material and patterns) (Harms and Lewbart 2011). Protocols that have been developed by agencies and academic institutions, many of which can be found online or within the gray literature, should be reviewed and modified as needed.

The SOP should specifically define all aspects of fish handling, collection, transmitter attachment, and release. A good approach is to take each step of the field operation (from collection to release) and specifically define acceptable limits for the relevant conditions, for example holding density, water temperature, or fish recovery time. The SOP should be specific about such details, clearly stating when and where a given variable (e.g., water temperature) should be measured, and what actions should be taken if variables are not within an acceptable range. The SOP should be developed during the planning and design phase of a study and may be modified from the existing SOPs of various agencies and academic institutions. Once established, the SOP should be used for training and refresher sessions of all tagging personnel and to guide execution of the study.

Tagger Training

Personnel performing any method of transmitter attachment procedure (hereafter referred to as taggers) must exhibit a minimal level of proficiency in order to minimize stress to fish (Cooke et al. 2003). Some techniques require greater proficiency than others; for example, surgical implantation is generally more challenging than gastric implantation and requires a level of manual dexterity that may not be possessed by all taggers. Regardless of the attachment method, practice sessions are essential to good tagging outcomes (Smith et al. 2009). Ideally, taggers will practice on fish of the same species, size, and life stage that will be used for the study. The use of surrogates may be required when the study fish are protected or difficult to obtain. Surrogates should provide the best possible match to the target species and size. In cases where proficient taggers have been inactive for a period, or when a new species or size-class is being tagged, taggers should participate in a refresher session to reinforce proper technique.

A tagger-training program should include a knowledgeable mentor, the opportunity to practice on model systems, and continuous evaluation and feedback. Ideally, the mentor should have broad knowledge of fish handling and tagging procedures, and experience with the species and life stage of interest. Many veterinarians now have specialized training in fish medicine, making consultation and collaboration powerful approaches to developing or refining transmitter implantation techniques (Wagner and Cooke 2005; Harms and Lewbart 2011). The use of model systems, rather than live fish, can be useful during the initial tagging practice sessions. For example, when learning suturing techniques, taggers can practice on banana or orange peels until surgical knots are tied correctly and consistently. This approach is ideal because training objectives can be accomplished with fewer fish, thus reducing logistical and animal welfare concerns. Once proficient with the model system, taggers should advance to a more realistic surrogate such as dead fish where they can refine their skills without attention to fish handling. This allows the novice tagger to focus on the attachment procedure while they develop proficiency. More than one procedure can often be performed on a single dead fish. Once they have mastered basic tagging skills, trainees can advance to practice on live fish.

Mentors should evaluate proficiency and provide feedback to taggers throughout the training program. Tagging proficiency should be evaluated by holding practice fish and monitoring short-term survival, transmitter loss, and external and internal condition of the fish. For surgical and gastric tagging methods, necropsies should be conducted posttagging with both the tagger and the mentor present to visualize transmitter position within the body cavity, incision apposition (if applicable), and any potential impacts to internal organs. For external tagging methods, a similar external exam should be conducted.

Where suitable holding conditions can be arranged, practice fish can be held for days or weeks to allow more complete assessment of survival, wound healing, transmitter retention, and infection. Untagged fish (controls) can be included to evaluate potential holding effects not related to tagging. Photographs of practice fish can help document proficiency in delicate attachment procedures, and video recordings can be useful to document tagging methods and as a learning tool for future trainees.

One component of tagging proficiency that is easily quantifiable and linked to the stress response of fish is the length of time needed to complete an attachment procedure. Taggers should strive to become competent and efficient in a given technique in order to limit fish exposure to anesthesia and handling. However, procedures performed too quickly may compromise proper execution. The time required to execute a particular procedure should be monitored, and timing of the trainee can be compared to that of an experienced tagger as one indicator of proficiency.

Tagger Effects

When multiple taggers are used, the experimental design should anticipate "tagger effects," with potentially different outcomes for individual taggers. Even with experienced and proficient taggers, there are likely to be some differences, although these are often difficult to assess. The probability that fish tagged by different taggers will have different short-term mortality rates (the easiest response to measure) is low, but mortality is only a crude indicator of tagging success (Mulcahy 2003; Jepsen et al. 2008b). Differences in stress response or recovery (e.g., healing of incision or puncture wounds) among taggers are more likely to occur, but are more difficult to detect. For example, a tagger that needs 2 min to complete an attachment procedure that involves partially removing a fish from the water might stress a fish less than a tagger that requires 3 min to complete the same procedure.

Anticipating potential effects is the best approach to managing them. The most simplistic approach is to use a single, experienced tagger throughout the study. If a single tagger is not sufficient, minimize the number of taggers and train them all using the same training program and SOP. Another approach to avoiding tagger effects is to rotate operations so that each tagger contributes equally to all treatment groups. Be prepared to quantitatively assess tagger effects by documenting which fish were handled by each tagger and conducting a formal evaluation of response metrics by tagger.

FISH HANDLING

Researchers should use the utmost care when handling fish, starting with the method of capture and continuing through the release of tagged fish. Stress associated with the individual steps in handling and tagging can be cumulative and lethal (Maule et al. 1988; Wedemeyer et al. 1990; Portz et al. 2006), so the goal should be to provide the best possible care at every step in order to manage the overall impact. Uncontrolled stressors during handling may reduce survival and the capacity to handle additional stressors such as tagging (Kelsh and Shields 1996).

Although handling needs will be study specific (based on fish size, logistical concerns, etc.), some general principles apply. One basic guiding principle is that fish should be collected using the least destructive and stressful method that is effective. If fish are to be held prior to release, set up procedures to measure and minimize holding density prior to collection. Where holding containers are used, they should have dark interiors and covers to minimize disturbance (Portz et al. 2006) and should be monitored to maintain water quality. Dissolved oxygen should be maintained near saturation, and water temperature should be maintained within a few degrees of the water temperature at the collection and release sites (Kelsh and

Shields 1996). Differences in water temperature of more than a few degrees can be managed by mixing the source and destination water to produce a gradual change (Stickney and Kohler 1990; Kelsh and Shields 1996). Researchers should avoid handling and tagging fish near the upper end of their thermal tolerance (Cooke et al. 2011).

Stress response to handling is well documented (Wedemeyer 1972), so handling and netting activities should be performed as gently as possible (Kelsh and Shields 1996). Rough handling can damage the mucus layer of the fish, making it vulnerable to infection (Harms 2005). We recommend use of a sanctuary net, which is designed to transfer fish in water, rather than removing them from water in a standard net. Where possible, use a crowding device to aggregate fish gently in a container prior to removal rather than pursuing them aggressively with a net.

Tagging Station

A tagging station should be designed and constructed to hold fish in the proper orientation for tagging and to allow for easy access to equipment and supplies (e.g., transmitters and instruments). The tagging station can be based on published examples (Summerfelt and Smith 1990; LaVigne 2002; Harms 2005) with modifications, or can be designed to meet specific needs (Figure 1). The tagging station should be constructed with plastic, closed-cell foam, or other materials that do not absorb significant amounts of water, and can therefore be effectively disinfected. Any surfaces that will directly contact fish should be smooth and kept moist to prevent damage to the skin, scales, or mucus layer of the fish. Similarly, surfaces designed to hold equipment and supplies should be kept dry to prevent exposure to pathogens that may reside in the water source. Some tagging procedures require that fish be partially out of the water or held in confined spaces. In these cases the tagging station should accommodate either gravity feed irrigation or a pump system to allow consistent water flow over the gills during the implantation procedure. Irrigation is most easily accomplished when the fish is held in a reclined position, with the head at the lowest point. Good lighting is another consideration and is especially important for surgical procedures. Adjusting the height of the tagging surface so that the tagger is positioned comfortably will improve the efficiency of the tagging process.

Aseptic Techniques

Aseptic techniques should be practiced during all transmitter attachment procedures to mitigate the risk of infection (Mulcahy 2003, 2011; AFS 2004, Harms 2005). Fish will be rendered more vulnerable to infection by any handling procedure that interrupts the outer mucus layer, regardless of whether or not the skin is penetrated (Harms 2005). These include contact with nets, measuring, weighing, and transmitter attachment. In an ideal setting, such as a veterinary hospital or laboratory, the goal of aseptic technique would be to create a surgical environment that is completely sterile. In a field setting, there are obvious challenges to achieving this level of asepsis, and the aquatic environment poses additional challenges to keeping the equipment, tagging platform, site of attachment, and transmitter free of contamination. Even in controlled laboratory settings strictly sterile procedures cannot be accomplished on fish (AFS 2004; Harms 2005). Despite these challenges, there are several means available to help reduce the risk of infection, and some of these are discussed below. Mulcahy (2003) provides a thorough review of practical aseptic techniques for transmitter implantation procedures.

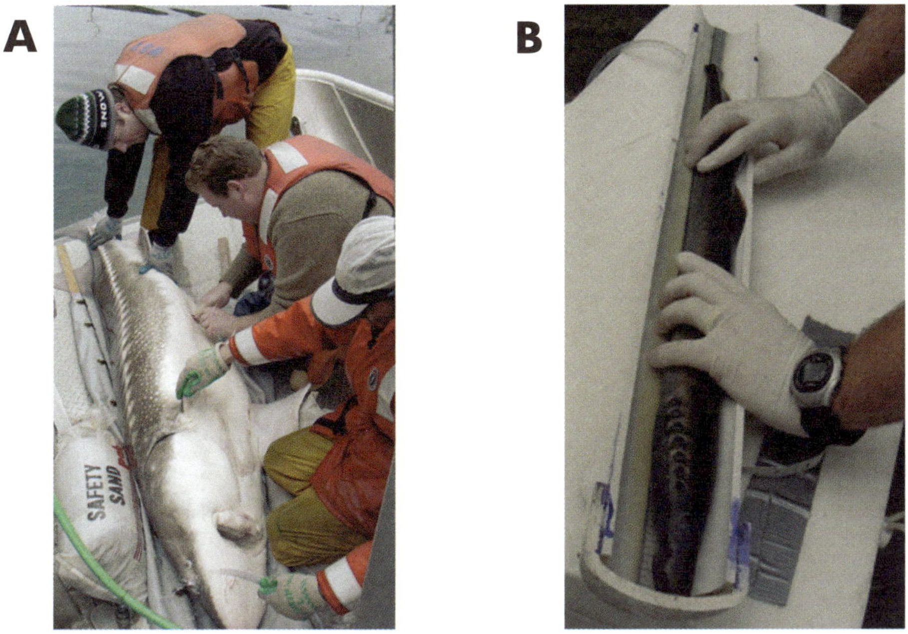

Figure 1. Photographs of example tagging operations for a) white sturgeon *Acipenser transmontanus*, and b) Pacific lamprey *Lampetra tridentata*.

Gloves.—Medical-grade exam gloves should be worn during invasive tagging procedures. For surgical implantation, these gloves should ideally be sterile and changed between fish. However, it is virtually impossible to avoid contamination from either the fish or the water source throughout a given procedure (AFS 2004; Harms 2005). The primary function of gloves is to reduce the transmission of potential pathogens from fish to fish, but they also protect the tagger from anesthetics, disinfectants, aquatic zoonoses and other potential waterborne pathogens.

Equipment and instruments.—Equipment such as tanks, buckets, nets, air stones, the tagging station, etc., should be disinfected between successive tagging sessions. Although often overlooked, a thorough drying period can be a useful addition to chemical disinfection for equipment (Harms 2005). During a given work session the environment should be kept as clean as possible. Very few disinfectants have documented efficacy against aquatic pathogens, so researchers should look for efficacy against the specific families of bacteria and/or fungi that might be encountered in a given setting. When selecting a chemical disinfectant, researchers should evaluate the potential toxicity of products, both to the fish and to personnel, and follow the label recommendations for dilution and exposure time to ensure effective disinfection. A final consideration is whether the disinfectant can be safely discharged into the environment. This is especially important in field settings where it is very difficult to ensure that chemicals will be completely contained.

Some standard veterinary techniques could be harmful to fish and therefore require modification. For example, precautions against infection that typically precede invasive procedures in warm-blooded animals include applying antiseptics to the skin and draping the skin with a sterile cloth. Antiseptics can be irritating to the skin of fish, as reported by Hart and

Summerfelt (1975), and their use is generally not recommended (Wagner et al. 1999; Wagner et al. 2011). Cloth draping, if applied to fish, should be sterile and moist to avoid removal of protective mucus that contributes to the fish's natural ability to fight off pathogens. Clear plastic sterile draping is recommended by Harms (2005) because it can help to keep the fish moist and provides a sterile working surface.

Irrigation of the gills during surgical implantation presents a risk of contamination of the peritoneal cavity. Since the water used for irrigation is likely pathogen-rich, researchers must take steps to avoid having it contaminate the incision area or enter the body cavity (Harms 2005; Chomyshyn et al. 2011). A good tagging station will meet this need by holding fish in a reclined position with the incision at an elevation greater than the head.

Ideally, all instruments should be sterilized prior to each procedure to reduce infection and tissue reaction (Marty and Summerfelt 1986; Moore et al. 1990; Mulcahy 2003; Harms 2005). Small autoclaves or pressure cookers can be used for this purpose (Mulcahy 2003). When tagging large numbers of fish in a field setting, there may not be enough instruments on hand to allow for prior sterilization and packaging of separate instruments for each procedure. However, it should still be possible to begin each tagging session with sterile instruments and to disinfect and rinse instruments between procedures (Lucas 1989; Lacroix et al. 2004). Sterilization or disinfection procedures must be strictly enforced for instruments that penetrate the body cavity because fish to fish transmission of pathogens has been documented (Elliott and Pascho 2001).

Careful consideration must be given to the required contact time for full efficacy of a disinfectant. For example, if a disinfectant requires 10 min of contact time, and the average procedure time is 2–3 min, sequential procedures must be separated by about 13 min or several sets of instruments must be available for rotation. Additionally, instruments disinfected with chemicals must be rinsed thoroughly with sterile saline or deionized water prior to use to avoid potentially toxic residue. When selecting chemical disinfectants for these applications, it is important to note that products used for disinfecting equipment such as tanks and nets may not be suitable or approved for disinfecting instruments that will be in direct contact with fish. Carefully review product labels for correct application, contact time, and disposal procedures.

Transmitters.—Transmitters should be disinfected prior to gastric implantation and sterilized prior to surgical implantation. However, similar to the instruments, complete sterilization of transmitters may be difficult or even impossible to achieve or maintain in a field setting. The goal in sterilization or disinfection is to reduce transmission of pathogenic organisms from the environment to the fish or from one fish to another. Such organisms will likely be found in highest densities on the skin and within the mucus layer of the fish itself, as well as within the source water. Thus, pathogens will be difficult to avoid entirely as transmitters are attached or implanted.

Disinfection of transmitters can be accomplished using the procedures established for disinfecting instruments. However, extra care should be taken to ensure thorough rinsing of disinfectant from the transmitter due to its prolonged close tissue contact. Researchers should also ensure that the selected disinfectant will not damage the transmitter coating. Sterilization of transmitters is challenging because the most common sterilization technique uses heat, which can damage transmitter components. Alternative approaches such as gas and chemical sterilization are available. However, these techniques must often be performed ahead of time in controlled laboratory settings due to their potentially hazardous nature. In these situations, presterilized transmitters may be packaged for transport into the field (Mulcahy 2003). Once transmitters have been disinfected or sterilized, they should be handled only with clean instruments or gloved hands, avoiding contact with any potentially contaminated surfaces or source water until they are attached or implanted.

Anesthesia

Fish are commonly anesthetized prior to transmitter attachment to allow ease of handling and reduced stress response. Anesthetics can be chemical or nonchemical agents used to induce calm, immobilize, eliminate pain sensations, and relax the musculature (Summerfelt and Smith 1990), and their proper use is necessary for the ethical treatment of animals (Mulcahy 2003). The different types of anesthetics and their effects are reviewed by Summerfelt and Smith (1990) and Ross and Ross (2008). The ideal anesthetic would have a rapid induction time (<3 min), a short recovery time (<5 min), no persistent effects on fish physiology or behavior, no harmful effects on humans, and would require no withdrawal time (Summerfelt and Smith 1990; Marking and Meyer 1985). Currently no anesthetics meet this ideal, but researchers should strive to meet these standards when selecting from available options.

The depth of anesthesia can be recognized by monitoring a series of physiological changes, beginning with a loss of reactivity to stimuli (stage 1), progressing to a loss of all reflex activity (stage 5), and eventual medullary collapse (stage 6; Summerfelt and Smith 1990). Invasive transmitter attachment procedures generally require stage 4–5 anesthesia and regular monitoring of the ventilation rate to prevent respiratory failure (stage 6). The concentration of the anesthetic should not be any higher than needed to achieve an induction time of about 3 min because the risk of mortality is inversely related to induction time (Summerfelt and Smith 1990). Total exposure time is also an important consideration, since prolonged exposure can lead to mortality (Mulford 1984; Summerfelt and Smith 1990).

Apart from efficacy, researchers must comply with drug regulations that dictate a withdrawal period for fish that may be consumed (Mulcahy 2003). After an anesthetic is selected for a given study, we recommend that researchers experiment with concentrations of the agent at study water temperatures: response to anesthesia can be highly variable at the level of the individual and species (Gilderhus and Marking 1987; Summerfelt and Smith 1990).

Chemical Anesthesia.—Tricaine methanesulfonate, also known as MS-222 and Finquel, is the only chemical anesthetic approved by the U.S. Food and Drug Administration (FDA) for use in fish. Label instructions for MS-222 indicate that it should only be used in the ictalurid, salmonid, esocid, and percid families and water temperatures during exposure should not exceed 10°C (Bowker and Trushenski 2011). In addition, a 21-d withdrawal period is required for fish that may be captured and used as food (Schnick et al. 1986; Summerfelt and Smith 1990; Schnick 2006). Exposure to MS-222 induces immediate physiological changes (Wedemeyer 1970; Houston et al. 1971; Strange and Schreck 1978), some of which can be long-lasting (Soivio et al. 1977), and the effects can be exacerbated by elevated water temperature (Schramm and Black 1984). A stock solution of 100 mg MS-222/mL water has a pH of about 2 in deionized water (Summerfelt and Smith 1990), but may be less acidic in other water sources, depending on their buffering capacity. To counteract the low pH and minimize the corresponding physiological effects on fish, the addition of sodium bicarbonate is recommended to buffer the anesthetic solution to near pH of 7 (Allen and Harman 1970; Wedemeyer 1970; Soivio et al. 1977; Harms 2005). Dosages between 50 and 100 mg/L have been effective in reaching stage 3 anesthesia in salmonids (Summerfelt and Smith 1990), but dosages should be determined for each individual study and set of conditions, as responses will vary. Gilderhus and Marking (1987) advise caution in determining concentrations of MS-222 because the difference between the effective and lethal doses can be small.

For many studies the requirement for a 21-d withdrawal period following use of MS-222 is too restrictive so alternative anesthetics must be considered. Carbon dioxide is listed as "generally recognized as safe" by the FDA for use with food, and is approved for aquatic use without a withdrawal period (Schnick et al. 1986; Summerfelt and Smith 1990), but induces a state of calm in fish by asphyxiation, and thus has several disadvantages. These include extended induction and recovery times and an overall shallow effect (Marking and Meyer 1985; Gilderhus and Marking 1987). Stage 4–5 anesthesia has been achieved in freshwater fish using carbon dioxide (Gelwicks et al. 1998; Peake 1998; Sanderson and Hubert 2007), but resulted in high mortality during lengthy procedures (Sanderson and Hubert 2007) or when high doses were applied at elevated water temperatures (Gelwicks et al. 1998). The use of carbon dioxide for marine species is especially difficult because the high concentration of ions in the environment reduce its solubility (Bowker and Trushenski 2011).

Other compounds that have been investigated as candidates for a zero withdrawal time anesthetic include clove oil (Anderson et al. 1997; Woolsey et al. 2004; Sink et al. 2007) and derivatives of clove oil called Aqui-S (containing isoeugenol) and Aqui-SE (containing eugenol)(Sanderson and Hubert 2007; Woods et al. 2008). Clove oil is typically composed of 85–95% eugenol, with the remaining fraction being isoeugenol, methyleugenol, or a combination of both (FDA 2007). Although clove oil and its components have "generally recognized as safe" status for use in some applications, they do not have that status for use as a fish anesthetic (FDA 2007). Benzocaine was identified as a potential candidate for an immediate release anesthetic in 1994, but did not undergo testing because no corporate sponsor could be identified to support the expensive approval process (Bowker and Trushenski 2011). Benzocaine is chemically similar to MS-222 and produces a physiological response in fish similar to buffered MS-222 (Ross and Ross 2008). There is a well-identified need for an immediate release anesthetic in fisheries and approval efforts are ongoing. Currently Aqui-SE and a commercial extract of benzocaine have corporate sponsorship and are being investigated as candidates (Bowker and Trushenski 2011).

Nonchemical anesthesia.—In addition to chemical forms of anesthesia, nonchemical approaches such as hypothermia and electroimmobilization have been investigated. Nonchemical approaches differ from true anesthetics in that they do not induce loss of sensation, but simply immobilize fish. The main advantage of nonchemical approaches is the lack of residual chemical effects and therefore the option for unrestricted release of exposed fish.

Hypothermia will immobilize fish and reduce general metabolism (Summerfelt and Smith 1990), but is not recommended where it is the only form of anesthesia applied because the level of anesthesia is incomplete (Ross and Ross 2008). The application of hypothermia may also be limited, depending on acclimation temperature of the fish, because cooling must be gradual in order to avoid thermal shock (Summerfelt and Smith 1990). Although it has been evaluated as a means of anesthetizing salmonids for basic handling in hatchery settings (Hovda and Linley 2000), it is unlikely to be suitable for transmitter attachment procedures.

In the absence of an approved chemical anesthetic with no withdrawal requirement, electroimmobilization (EI) is gaining popularity for fish handling applications. A basic EI system involves electrodes placed into tanks (Kynard and Lonsdale 1975) or nets (Gunstrom and Bethers 1985; Orsi and Short 1987) to produce a uniform electric field, and some commercial systems are available for purchase (Walker et al. 1994; Gaikowski et al. 2001; Zydlewski et al. 2008). Summerfelt and Smith (1990) provide a good summary of this technique. Although most reports of EI use are for short-term immobilization (<5 min), Jennings and Looney (1998) achieved sufficient immobilization to conduct surgical implant procedures. However, the efficacy of EI for telemetry tagging field procedures remains in question, as tagging requires deeper levels of anesthesia than simple handling procedures. At present, the literature does not provide many examples of its successful use. The technique, however, holds promise, and further experimentation is needed.

Holding Procedures

There may be compelling reasons to hold fish both before and after tagging, however, these must be balanced against logistical constraints and the risk of additional stress to fish due to holding, particularly if holding conditions are suboptimal. Each study must carefully consider which holding approach, if any, is most appropriate. Pretag holding occurs between fish collection and tagging, while posttag holding is the lapse between tagging and release. One rationale for pretag holding is to allow fish to recover from the stress of collection (Stickney and Kohler 1990; Kelsh and Shields 1996). A combination of stressors, for example collection and tagging, can be lethal even if fish can tolerate them independently (Portz et al. 2006), so it is sometimes necessary to separate them in time (Stickney and Kohler 1990). Another rationale for pretag holding is to restrict access to food so that fish enter a postabsorptive state prior to an invasive procedure such as surgery (Summerfelt and Smith 1990). This practice has the added benefit of aiding aseptic technique by minimizing regurgitation of food and defecation in the surgical area.

Holding fish after tagging allows for monitoring, but has risks as well. At a minimum, the posttag holding period should allow fish to fully recover from anesthesia. Once they have regained equilibrium and are swimming normally, fish can either be released or held for a more extended period. Hart and Summerfelt (1975) and Stasko and Pincock (1977) recommended immediate release of tagged fish rather than risk the stress of holding, although several researchers prefer a posttag holding period. Moore et al. (1990) suggested holding fish to ensure recovery, but they cautioned that extensive holding could outweigh this benefit.

Based on concerns that fish need time to regain neutral buoyancy after tagging, Gallepp and Magnuson (1972), Fried et al. (1976), and Moser et al. (1990) all recommend holding fish in shallow containers with easy access to air. A short posttag holding period may be useful for fish released into fast-moving water, or where predators are present (Jepsen et al. 2002). Fish that use complex habitat like holes or interstitial spaces have been shown to have improved transmitter retention if they are held in a simpler habitat system for a short period following tagging (Broadhurst et al. 2009). For gastric implantation, where transmitter loss can occur soon after tagging, a short holding period allows evaluation of transmitter loss and easy recovery of regurgitated transmitters. For both immediate and delayed release strategies, the tagging operation should be configured to allow release of tagged fish without additional handling. This might be accomplished by placing a fish into a portable container immediately after tagging so that no further netting or handling is required.

The research questions, species being studied, logistical considerations, and available holding facilities will largely determine which holding procedures, if any, are selected. Lack of adequate facilities or reliable water quality may make immediate release necessary, despite the potential benefits of a recovery and observation period. Likewise, extensive holding could be detrimental for migratory species at critical life stages (Adams et al. 1998a, 1998b; Martinelli et al. 1998). One suggested alternative to holding is to release fish immediately after recovery from tagging, but to defer data collection until some predetermined recovery period has passed (Moser et al. 1990; Martinelli and Shively 1997). Ultimately, each study must weigh the tradeoff between allowing fish to recover from acute stressors and chronic stress from being confined under artificial conditions. If both pretag and posttag holding periods are used, ensure that the cumulative holding period is considered.

TRANSMITTER ATTACHMENT TECHNIQUES

The choice of transmitter attachment method should be determined by the research objectives and the species or life stage being studied: there is no perfect tagging procedure. Considerations related to the research objectives include the period of monitoring, field conditions, and presence of any sensors on the transmitter. Studies where fish will be monitored for extended periods require tagging procedures that provide high transmitter retention and low transmitter effects through time. Field conditions such as high water temperatures or heavy vegetation can increase the risk of infection for external or surgical implants and the risk of entanglement for externally attached transmitters. Sensors associated with the transmitter (e.g., electromyogram, light) must be considered so that the tagging method accommodates any special requirements.

Considerations related to the species and life stage include body form and stage of gonad development. Fish with elongated body forms or laterally compressed fishes may not have enough space in the body cavity to permit surgical implantation. The presence of enlarged gonads can complicate surgical incision and suturing, may increase the risk of transmitter loss due to elevated pressure on incised tissues (Jepsen et al. 2002), or may cause elevated rates of mortality (Winter 1996). Below is an introduction to each of the three most common attachment techniques; advantages and disadvantages of each are summarized in Table 1. A more complete review of these three procedures can be found in Bridger and Booth (2003).

External Attachment

External attachment of transmitters can be a useful approach when the body cavity of the fish will not accommodate the transmitter, for short-term studies, when specialized sensors need to be outside of the body (Tanaka et al. 2001), or for monitoring gravid females where surgical implantation might lead to higher mortality (Winter 1996). External transmitters can be attached rapidly, sometimes without use of anesthesia, and this may be an important advantage when an anesthetic clearance period cannot be accommodated or when the potential physiological effects of the anesthetic cannot be tolerated (Thompson et al. 2008).

The most common method of external attachment is with surgical stainless steel wires or suture material passed through the dorsal musculature. A plastic or neoprene backing plate may be fixed to the transmitter and on the opposing side of the fish to aid attachment and

TABLE 1. Potential advantages and disadvantages of the three most commonly used transmitter attachment techniques for telemetry.

External attachment	Gastric implantation	Surgical implantation
Potential Advantages		
• Executed quickly • Light or no anesthesia • No attenuation of signal by fish body wall or internal organs • Can be retrieved without recapture (e.g. pop-up tags) • External sensors accommodated	• Executed quickly • Light or no anesthesia • Position at center of balance • Low risk of entanglement/drag • Low risk of infection • Some sensors accommodated	• Position at center of balance • No entanglement or drag • Internal sensors accommodated • Allows long-term monitoring
Potential Disadvantages		
• Risk of fouling, entanglement or transmitter loss • Source of drag/swimming affected • Irritation/infection at attachment site • Balance affected if asymmetric • Not suitable for growing fish!	• Interference with/cessation of feeding • Cough response • Rupture of stomach • Regurgitation or egestion of transmitter • Irritation at attachment/exit site	• Relatively slower procedure • Biologically invasive • Deep anesthesia • Requires a skilled surgeon/tagger • Risk of active/passive expulsion • Risk of infection • Irritation at radio antenna exit site
References		
Ross and McCormick 1981; Lewis and Muntz 1984; Mellas and Haynes 1985; Matthews et al. 1990; Thorstad et al. 2000; Collins et al. 2002	Adams 1998a; Jepsen et al. 2001; Hall et al. 2009; Smith et al. 2009; Dunning and Ross 2010	Kaseloo et al. 1992; Bridger and Booth 2003; Jepsen et al. 2002; Jepsen 2003; Okland et al. 2003; Lacroix et al. 2004; Hall et al. 2009

reduce irritation (Matthews et al. 1990; Cooke et al. 2003; Thompson et al. 2008). Some techniques have been proposed to reduce irritation or tissue damage from continuous contact between the transmitter and the surface of the skin. Herke and Moring (1999) developed a "soft" harness system that uses thick-walled tubing as a conduit for the wire or suture, and Crook (2004) reduced the amount of material in contact between transmitter attachment points. The entire external transmitter package (e.g., transmitter, wires, attachment plates, and pads) should be as small and streamlined as possible to reduce drag and to minimize potential effects on swimming performance. In addition, the use of anterior attachment points will reduce the risk of entanglement (Ross and McCormick 1981).

Potential risks of external attachment include entanglement, fouling, excessive drag, irritation at the attachment site, and balance issues if the transmitter package is mounted asymmetrically on the fish (Figure 2). Entanglement and fouling can be significant concerns, especially in habitats with vegetation (Mellas and Haynes 1985; Thorstad et al. 2001), and can contribute to excessive drag (Ross and McCormick 1981; Lewis and Muntz 1984; Mellas and Haynes 1985) and increased transmitter loss. The attachment mechanism may loosen through time, causing chafing, abrasion, or muscle damage and dorsal scale loss (Ross and McCormick 1981; Mellas and Haynes 1985; Matthews et al. 1990). Collins et al. (2002) noted increased active rubbing against the substrate in fish with externally attached transmitters, and Rikardsen and Thorstad (2006) report an increased risk of recapture. Flexible-bodied fish may be at higher risk for transmitter loss (Broadhurst et al. 2009). Several studies reported reduced swimming performance of fish equipped with external transmitters (McCleave and Stred 1975; Counihan and Frost 1999; Makiguchi and Ueda 2009), but in other cases no effect on swimming performance was observed (Thorstad et al. 2000).

The size, shape, and positioning of the transmitter are important factors. Sutton and Benson (2003) examined the effect of transmitter size and shape on retention in juvenile lake sturgeon *Acipenser fulvescens*. They compared several sizes of standard, cylindrical tags to compressed tags having a lower profile and found that small cylindrical tags were retained more frequently, despite a higher profile and increased drag. They attributed this difference to the uneven body surface of the sturgeon and the inability of the compressed transmitter to fit securely against the dorsal surface (Sutton and Benson 2003). Claireaux and Lefrancois (1998) attached the transmitter to the ventral surface of the fish, under the pectoral fins, and reported success in some preliminary trials.

Externally attached transmitters are typically mounted asymmetrically (e.g., on one side of the fish) rather than at the center of balance (Figure 3). One approach in alleviating problems resulting from asymmetrical attachment is use of a saddle-type transmitter, where the transmitter components are split into two packages of approximately equal weight in a sad-

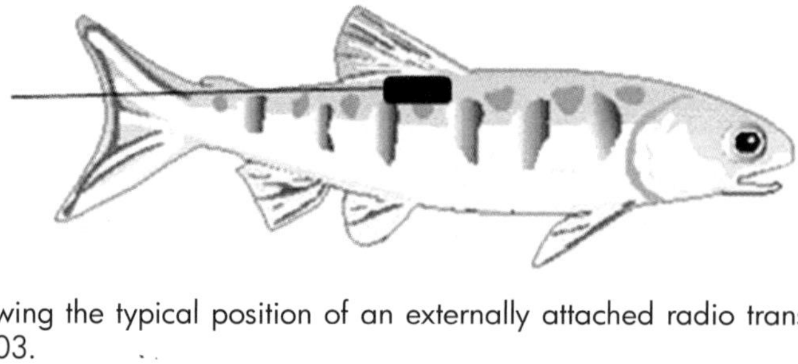

Figure 2. Sketch showing the typical position of an externally attached radio transmitter. Redrawn from Hockersmith et al. 2003.

Figure 3. Photograph of a radio transmitter externally attached to an Atlantic salmon *Salmo salar*. Photograph courtesy of N. Jepsen.

dle shape, which is then positioned across the center of balance. However, Brown and Eiler (2000) used a saddle-type transmitter on a migratory fish and reported delayed movements and reduced distance traveled. In addition, Weimer et al. (2006) compared asymmetrically attached external transmitters to saddle transmitters, and found that saddle transmitters took longer to attach, moved further from the initial attachment location, and had higher transmitter loss. The results of this study prompted the authors to concur with the recommendation of Ross and McCormick (1981) that streamlined, single-sided transmitters should be used rather than saddle-style transmitters.

As with other attachment procedures, there are advantages and disadvantages to external attachment. Due to the evidence of significant transmitter effects in a variety of species, external attachment may be most suitable in studies where neither gastric nor surgical implantation can be accommodated.

Gastric Implantation

Gastric implantation of transmitters, like external attachment, can be performed quickly, with limited or no use of anesthetics (McCleave et al. 1978; Pearcy 1992; Rivinoja et al. 2006; Smith et al. 2009; Dunning and Ross 2010). Internal placement of a transmitter eliminates many concerns related to external attachment because the transmitter is positioned at the center of balance, creates no additional drag, and the risk of entanglement is controlled. It also has advantages over surgical implantation in that there are no wounds or incisions and therefore a lower risk of infection. Nevertheless, options for the use of sensors with gastrically implanted transmitters may be limited, and gastric implants may have some effects on physiology (Jepsen et al. 2001) and behavior, depending on the species and life stage. Gastric implantation is applicable for life stages with minimal feeding activity, such as migrating anadromous adult salmonids (Keefer et al. 2004), and has been used successfully in studies that implant transmitters into large numbers of fish.

Gastric tagging procedures typically use a plastic tube, rod, or plunger system to insert the transmitter down the pharynx, past the cardiac sphincter, and into the stomach (Figure 4). To ease the insertion process, the transmitter and insertion tool may be coated in a lubricant such

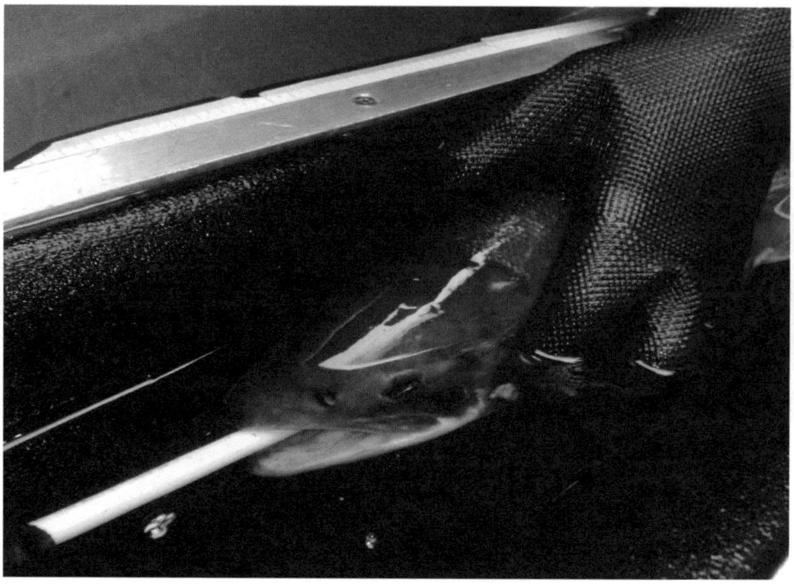

FIGURE 4. Photograph showing the gastric insertion of a radio transmitter into an adult coho salmon *Oncorhynchus kisutch*.

as glycerin (Bridger and Booth 2003; Ramstad and Woody 2003). Rivinoja et al. (2006) used mucus from the lateral surface of the fish to aid insertion. After the transmitter has been positioned the insertion tool is removed and the transmitter antenna, if present, exits the mouth of the fish (Figure 5). Antennas are typically bent so that they trail backwards along the body, and they can be anchored to the roof of the mouth with sutures (Bridger and Booth 2003). Alternatively, the antenna can be guided so that it exits through the operculum (McCleave et al. 1978; Bridger and Booth 2003). Some irritation may occur where the antenna exits the mouth (Adams et al. 1998a), and bending the transmitter antenna will likely cause some reduction in signal output (Beeman et al. 2007).

Negative outcomes of gastric implantation include reduced feeding, stomach rupture, and transmitter loss. The presence of the transmitter in the stomach may reduce feeding behavior or food intake (Armstrong and Rawlings 1993; Jepsen et al. 2001; Winger and Walsh 2001; Hall et al. 2009), and some researchers have reported "coughing" or flaring of the opercula (Mellas and Haynes 1985; Adams et al. 1998a). If the transmitter is too large or is inserted too far or too forcefully, stomach rupture can occur. Although a fish with a ruptured stomach may survive in the short-term (Armstrong and Rawlings 1993; Ramstad and Woody 2003; Smith et al. 2009), it will obviously have been compromised and can no longer be considered representative of the general population.

The most widely reported negative outcome of gastric implantation is transmitter loss. In field settings, when tagged fish are at large, transmitter loss can be difficult to document and is therefore likely underreported. A short-term holding period is useful after gastric implantation to monitor for transmitter loss (Winger and Walsh 2001). Transmitters inserted into the stomach can be lost through either regurgitation or egestion. Loss through egestion can be especially difficult to document (Hall et al. 2009), and has not been reported often (Moser et al. 1990). Armstrong and Rawlings (1993) used x-rays to confirm egestion by observing trans-

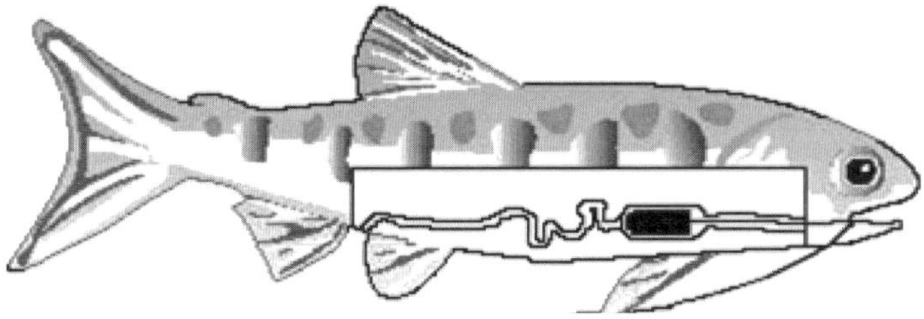

FIGURE 5. Sketch showing position of a gastric implantation of a radio transmitter with the external antenna exiting the mouth. Redrawn from Hockersmith et al. 2003.

mitters positioned in the intestine. Transmitter regurgitation rates can vary greatly, depending on the species (Mellas and Haynes 1985; Matthews et al. 1990; Smith et al. 1998; Winger and Walsh 2001; Keefer et al. 2004) and relative size of the transmitter (Koo and Wilson 1972; Moser et al. 1990).

The shape and texture of the transmitter may also influence rates of transmitter loss. Keefer et al. (2004) suggested the addition of rubber bands or surgical tubing to increase roughness and aid transmitter retention (Figure 6). Less commonly used approaches to increase transmitter retention include the addition of vulcanized tape (Rivinoja et al. 2006) and small hooks (Pearcy 1992; Parrish and Moffitt 1993). Although added roughness has been shown to increase retention (Pearcy 1992; Keefer et al. 2004), these techniques have not been shown to eliminate transmitter loss completely. Considering that transmitter loss is the most widely reported concern for gastric implantation, and because it can be difficult to assess in the field, we recommend the use of a reliable secondary mark or tag in studies where fish may be recaptured, detected by an alternate method, or seen after release (see Keefer et al. 2004). The secondary mark, combined with a recapture opportunity, will allow an assessment of the timing and rates of transmitter loss, which can be invaluable in interpreting study findings.

The potential for negative outcomes in fish with gastrically implanted transmitters emphasizes the need for researchers to refine their techniques to accommodate the species and life stage under study. For example, use of the smallest possible transmitter will reduce impacts to feeding behavior, but may increase transmitter loss in some fishes. Practice sessions to determine appropriate transmitter position are critical because shallow insertion increases the risk of regurgitation and deep insertion risks stomach rupture (Ramstad and Woody 2003; Smith et al. 2009). Researchers should experiment with added roughness to see if regurgitation rates can be minimized or controlled. Necropsies should be conducted on tagged fish to ensure transmitters are implanted at the appropriate depth; Armstrong and Rawlings (1993) suggest the use of x-rays to confirm ideal transmitter position (Figure 7). Once appropriate depth has been determined, create a mark on the insertion tool to help taggers to visually estimate this depth (Smith et al. 2009).

Surgical Implantation

Surgical implantation is recognized as the best attachment technique for long-term telemetry studies (Stasko and Pincock 1977; Winter 1996, Jepsen 2003; Cooke et al. 2011), primarily because many potential effects of the procedure diminish with time as the fish heals. Like gastric implantation, the transmitters can be positioned inside the body cavity at the center of

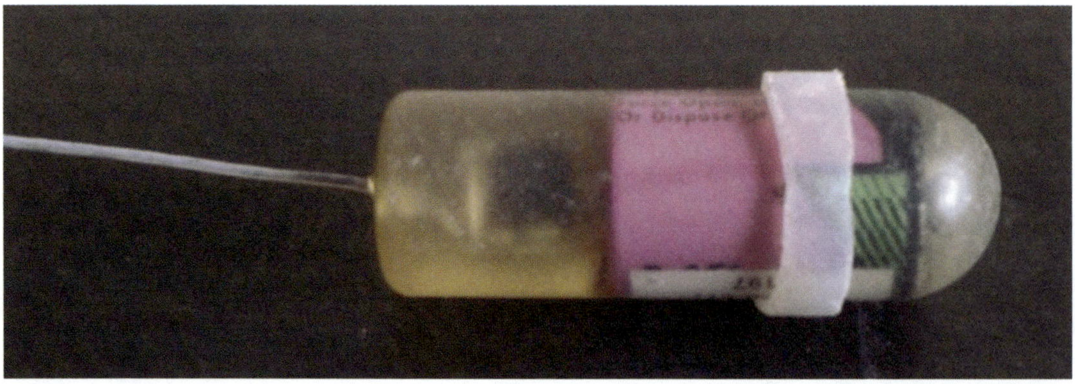

Figure 6. Photograph showing a radio transmitter with surgical tubing used to increase roughness and aid transmitter retention.

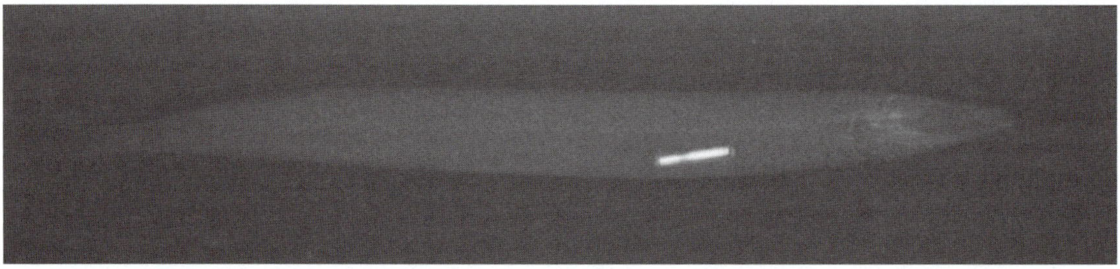

Figure 7. X-ray of a Pacific sand lance Ammodytes hexapterus showing the position of an implanted passive integrated transponder.

mass, eliminating drag or entanglement concerns. Transmitters with sensors such as the electromyogram can easily be accommodated (Kaseloo et al. 1992). However, like all transmitter attachment methods, there are also risks to surgical implantation.

Surgical procedures are invasive by nature, and will therefore generally require deeper anesthesia and a greater time investment to develop proficiency than other attachment methods. Potential adverse effects of surgical implantation include transmitter loss (both passive through an unhealed incision and through active expulsion), and changes in normal behavior and performance. Surgery is not a recommended attachment method in studies of gravid females due to observed elevated mortality, potential damage to gametes, and increased transmitter loss (Marty and Summerfelt 1986; Winter 1996).

There is strong evidence that the effects of surgical tagging differ depending on the species, size, and life stage of the study fish and the size of the implanted transmitter. Surgical procedures involve many small steps, each presenting the researcher with a variety of options. Clearly, there is no universal procedure that can be applied to all species and life stages with equal success. Therefore, it is the intent of this Section to provide a general review of the options available for each aspect of a surgical implantation procedure.

Incision.—In fusiform fishes transmitters are typically inserted through a skin incision made on or near the mid-ventral line of the fish. The ventral incision is preferred over a lateral location because it avoids interference with internal organs and reduces the risk of bleeding

(Schramm and Black 1984; Marty and Summerfelt 1986). A ventral incision also avoids any potential impacts to the swimming musculature. The decision to place the incision directly on or off the mid-line is species and life stage specific. A ventral mid-line incision is more symmetrical; however, an off-midline incision may be more advantageous in small fish or in fish with thin body walls because the transmitter will not be positioned directly above the incised tissue. Additionally, and particularly in large-bodied fish, incision site placement should avoid prominent ventral vasculature. Wagner and Stevens (2000) compared ventral incisions made on the mid-line and those made just off the mid-line. Although they hypothesized that the midline incision would heal more quickly with less irritation, they observed no significant differences in inflammation between the two incision sites. In nonfusiform fishes incision placement may be more variable. Researchers should identify the least traumatic entrance to an internal cavity large enough to accommodate the transmitter.

Once the incision is made, forceps are typically used to elevate the body wall as the incision is enlarged to reduce risk of damage to the underlying organs (Hart and Summerfelt 1975). In smaller fish, the use of specialized microsurgical scalpels is recommended to control penetration depth. Incisions should be only long enough to admit the transmitter (Hart and Summerfelt 1975) without tearing the skin. A good guide for incision length is slightly larger than the diameter of the transmitter, since the transmitter's length can be inserted through the incision once its head has entered the body. A short incision will minimize risk of organ damage, reduce the area vulnerable to infection, minimize tissue damage due to closure (e.g., sutures or staples), and reduce the time needed for closure.

Large scales may need to be removed in order to access the underlying skin to make the incision, close it, or create an exit site for an antenna (Mulford 1984; Petering and Johnson 1991; Knights and Lasee 1996). However, since scales are vital to the front-line defense of fish, and their removal can lead to bacterial or fungal infection, as well as muscle deterioration (Mellas and Haynes 1985; Harms 2005), this should be done with caution and only when procedures cannot be accomplished otherwise.

Transmitter positioning.—Once placed inside the body cavity, transmitters are generally positioned either directly beneath or posterior to the incision (Figure 8). Suturing can be performed with minimal risk to organs if the transmitter lies directly beneath the incision, and this location may minimize adhesion of viscera to the transmitter (Moore et al. 1990; Lacroix et al. 2004). Posterior positioning reduces pressure on the weakened tissues and may reduce the risk of transmitter loss (Ross and Kleiner 1982; Schramm and Black 1984). To control transmitter position and to reduce the risk of transmitter loss, some researchers have affixed transmitters to the pelvic girdle (Økland et al. 2003) or other internal structures (Schramm and Black 1984; Marty and Summerfelt 1986).

Radio transmitters typically have external antennas to enhance signal propagation, and these may complicate the surgical tagging procedure. One common approach is to have the antenna exit the body cavity posterior to the incision so that it trails back along the body of the fish (Figure 8; Adams et al. 1998a, 1998b; Bunnell et al. 1998; Martinelli et al. 1998; Walsh et al. 2000). The shielded needle technique (Ross and Kleiner 1982) or other similar procedures are commonly used to execute the antenna exit. There may be some long-term irritation at the exit site due to the motion of the antenna as the fish swims (Bauer et al. 2005; Jepsen et al. 2008b).

To avoid an additional puncture wound, some researchers use the posterior aspect of the incision as the antenna exit site (Knights and Lasee 1996; Bunnell et al. 1998). Walsh et al.

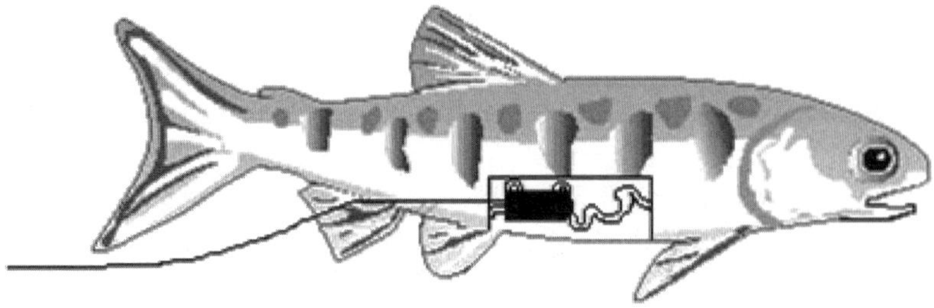

FIGURE 8. Sketch showing the position of a surgically implanted radio transmitter with the external antenna exiting the body wall. Redrawn from Hockersmith et al. 2003.

(2000) compared these two approaches and found no difference in mortality or transmitter loss. They recommend placing the exit site at the posterior edge of the incision since it was faster to perform. Bunnell and Isely (1999) used this exit site and reported that the antenna caused the posterior aspect of the incision to enlarge, allowing 12–27% of transmitters to be expelled. Another approach is to implant transmitters with the external antenna completely within the body cavity of the fish. Cooke and Bunt (2001) compared fish behavior and transmitter performance (i.e., signal reception range) when external antennas were either trailed outside the fish or coiled within the fish. They found no difference in fish behavior, but reduced transmitter performance for the coiled antenna group, as suggested by Winter (1996).

Incision closure.—The most variable and time-consuming aspect of surgical procedures is incision closure. The ideal closure technique would be quick, relatively easy to perform, and secure. Although a secure closure is the primary goal, the speed of the technique is an important consideration and has been implicated as a factor in postsurgical survival (Petering and Johnson 1991). Therefore, taggers should strive for speed in incision closure as well as overall surgical time, as long as speed does not compromise competency. The most common closure technique is suturing. Other techniques have been investigated, including leaving the incision open. This may be a viable approach when the incision is very small (Jepsen et al. 2002; Skov et al. 2005) but is less likely to be successful for the longer incisions required to implant most transmitters. As smaller transmitters become available, incisions will likely be small enough to be considered secure without closure.

Staples.—While not as common as suturing, surgical staples have been used to close incisions, and this method can reduce surgery time by about 50% (Mulford 1984; Mortensen 1990; Swanberg et al. 1999; Sanderson and Hubert 2007). In addition, the tagger can become proficient quickly, and the staples are not likely to loosen over time. However, staples can be difficult to position properly, and staple size must be carefully selected to prevent tearing and subsequent necrosis (Wagner and Cooke 2005). The literature reports mixed success when researchers have used this closure method for fish. Mulford (1984), Mortensen (1990), and Sanderson and Hubert (2007) reported good outcomes closing incisions with staples, whereas Starr et al. (2000) reported elevated mortality and transmitter loss. Swanberg et al. (1999) compared wounds closed with silk sutures versus staples and recommended the use of staples, with a reminder that they are permanent.

Adhesive.—Like surgical staples, the use of cyanoacrylate adhesive has the advantage of reducing handling and surgical time, and requires no specialized skills. Nemetz and MacMillan (1988) used cyanoacrylate to close a small lateral body wall incision, without implantation of a transmitter, and reported good wound healing. However, when transmitters were inserted, incisions closed with cyanoacrylate adhesives tended to reopen, leading to significant transmitter loss (Petering and Johnson 1991; Kaseloo et al. 1992). Kaseloo et al. (1992) also reported severe tissue necrosis at the incision site and did not recommend use of cyanoacrylates for wound closure. The use of cyanoacrylates, alone or in combination with sutures, is associated with a higher rate of incision re-opening and subsequent transmitter loss (Harms 2005). Some researchers advocate the use of cyanoacrylates to make the incision watertight. However, a well apposed and appropriately sutured incision will exclude water exchange, and the epithelial later will rapidly reestablish (Harms 2005).

Sutures.—A number of factors should be considered when suturing incisions, including the suture material and needle, the knot, and the suturing pattern. In fish the most commonly used suturing technique is a simple interrupted pattern and common suture materials include synthetic monofilament or polyfilament sutures (Hart and Summerfelt 1975; Mellas and Haynes 1985; Chisholm and Hubert 1985; Thoreau and Baras 1997; Martinelli et al. 1998). Synthetic sutures are generally recommended over organic materials, as several investigators have reported wound inflammation and delayed healing of incisions closed with silk and other organic suture materials (Gilliland 1994; Thoreau and Baras 1997; Swanberg et al. 1999; Wagner et al. 2000; Wagner and Stevens 2000; Hurty et al. 2002; Figure 10). Of the synthetic sutures, monofilament materials are often recommended over polyfilament (Thoreau and Baras 1997; Wagner et al. 2000) because they are noncapillary and therefore cannot conduct waterborne pathogens (Boothe 2003). Another advantage of monofilament is that the total surface area available for bacterial or fungal colonization is less than that of braided suture material. However, use of monofilament sutures does not guarantee that such colonization will not occur (Figure 9).

Manufacturers offer both absorbable and nonabsorbable suture materials. Fisheries researchers need to be aware that the design and testing of suture materials and their corresponding classification as absorbable or nonabsorbable were determined in mammalian systems. Although the same classification system developed for mammals is used in fish surgeries, there is no evidence that sutures perform comparably in fish. Under some circumstances, absorbable sutures have dissolved or weakened before wound healing was complete, leading to transmitter loss (Bunnell et al. 1998; Jepsen et al. 2002). Conversely, nonabsorbable (and potentially some absorbable) suture materials may persist well beyond wound healing, and this outcome may also have negative impacts on fish (Gilliland 1994). For example, we have observed drag on sutures, which over time caused them to loosen and tear from tissue near the suture entry and exit points. This tearing left secondary wounds perpendicular to the incision, presenting additional risk of infection (Figure 10). The recommendation of Harms (2005) to remove sutures several weeks after surgery in order to eliminate the risk of delayed inflammation is not realistic for most field studies, but highlights a potential challenge associated with extended suture retention.

The choice of a surgical needle will depend on the species, size, and life stage of the fish. The shape of the needle will vary depending on the thickness of the tissue, but curved needles

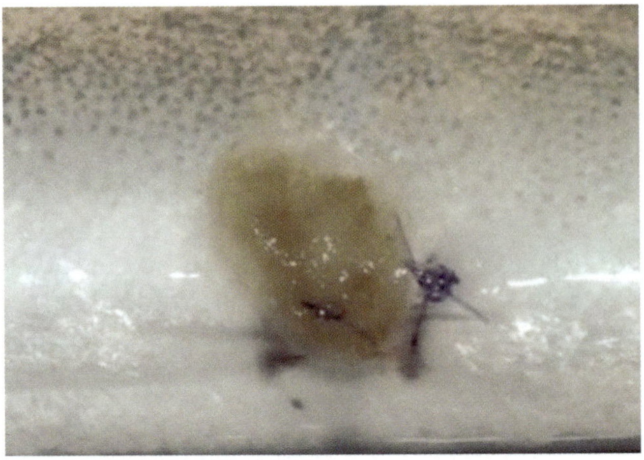

Figure 9. Photograph of flocculent organic material attached to monofilament sutures.

Figure 10. Photograph of secondary wounds from suture torn completely away from tissue (at left) and partially torn away from tissue (right).

such as a one-half circle are commonly used. The needle should be long enough to span the incision, and the diameter should be no larger than needed to accommodate the suture material. Needles with cutting tips have been recommended for use in fish (Harms 2005) to facilitate penetration of the skin, which can be thick in some species and life stages. However, for juvenile fish or species with delicate skin, needles with tapered tips may be preferable. Although a wide variety of needles are available, suture material generally comes prepackaged with a specific needle attached. This prepackaging need not limit the choice of needle since separate needles can be purchased and manually threaded with the preferred suture material in advance of the procedure.

Two knots are commonly used to secure sutures in implantation procedures: the square knot and the surgeon's (or friction) knot (Figure 11). Both knots are secure and straightforward to execute. The square knot is recommended for most applications because of its

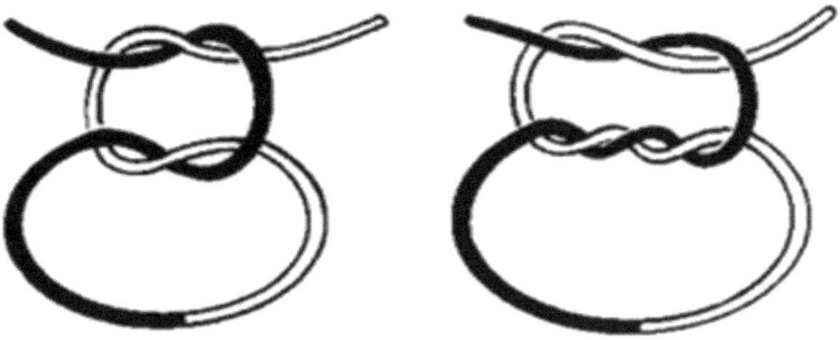

Figure 11. The square knot (left) and the surgeon's knot (right) are commonly used in transmitter implantation procedures. Reproduced from Toombs and Clarke 2003.

simplicity and corresponding strength. The surgeon's knot is preferred when the incision is under tension or when the suture material has a low coefficient of friction (Toombs and Clarke 2003; Trout 2003). A square knot consists of two single-wrap throws (i.e., a single wrap of the suture material around the surgical instrument). A surgeon's knot is simply a square knot with a double wrap within the first throw to prevent slippage. One or two additional throws are typically added to either knot to increase knot security and ensure that the ligature stays in place long enough to accommodate healing. The number of additional throws required will depend on the suture material and the predicted healing time.

It is important to note that a simple knot (e.g., the square knot) is stronger than a more complicated knot made with additional wraps and throws because every bend in the knot diminishes the strength of the suture material. In general, high-profile knots comprised of more than four throws, or complicated knots involving multiple wraps on a series of throws, are not recommended for use in fish. These knots can deter wound healing by placing undue drag on the ligature, which increases the likelihood that sutures will pull or tear out prematurely. Additionally, the extra suture material used for these complicated knots will increase the surface area available for bacterial and fungal colonization. Knot security decreases with increasing suture diameter (Toombs and Clarke 2003), so researchers should choose the smallest diameter suture that will effectively keep the incision apposed over time without tearing through the tissue.

The simple interrupted suture pattern is most commonly used for closing incisions (Figure 12). This pattern is preferred over a continuous pattern (e.g., simple continuous or vertical mattress) because each ligature is an independent closure effort. As such, the integrity of the entire incision will not be affected if one ligature becomes loose or untied (Harms and Lewbart 2011). Placement of short, single ligatures will also prevent a long strand of suture material from trailing behind the fish in case knots become untied. Wagner et al. (2000) tested a vertical mattress pattern and confirmed Hart and Summerfelt's (1975) finding that the simple interrupted suture pattern provided the best apposition and reduced inflammation.

The decision to close the body wall with one or more layers of suture should be species and life stage specific. Hart and Summerfelt (1975) compared closing an incision with a single layer of sutures (integument and peritoneum together) versus suturing these layers separately. They reported that single-layer suturing was faster than, and just as effective as, suturing layers separately and they recommended single-layer suturing to minimize stress and handling. Most surgical procedures reported in the literature use Hart and Summerfelt's (1975) recommended approach.

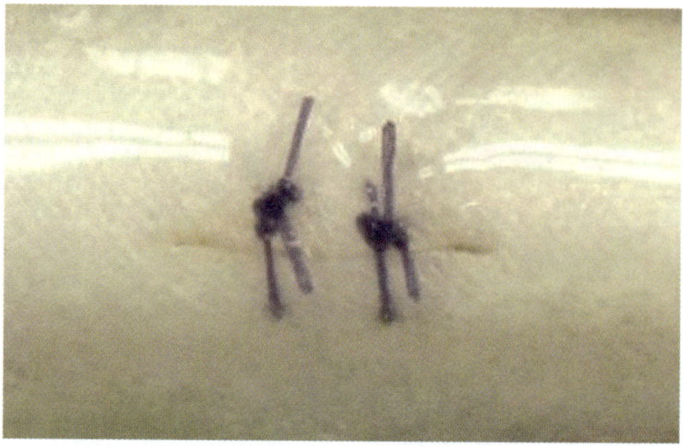

FIGURE 12. Photograph of a simple interrupted suture pattern shortly after surgery.

Sutures are the most commonly used incision closure method in fish surgeries, and there are diverse choices for suture and needle materials, knots, and patterns. We recommend that researchers consult an experienced veterinarian or medical professional for specific instruction in knot tying techniques. Additionally, surgery text books (see Slatter 2003 for example) should be reviewed for the principles underlying suture material selection and knot tying as these are complex topics.

Prophylactic treatments.—Due to the known risk of infection associated with surgical implantation procedures, a variety of prophylactic treatments have been tested. Some investigators have used a topical antiseptic at the incision site before and/or after making the incision (Moore et al. 1990; Knights and Lasee 1996; Lacroix et al. 2004). We recommend caution with this approach, since some disinfectants can irritate or damage the skin (Hart and Summerfelt 1975; Nemetz and MacMillan 1988; Wagner et al. 2011). Alcohol-based products are astringent and should never be used on fish (Mulcahy 2003). Wagner et al. (1999) compared wound healing of incisions with and without the use of an antiseptic and recommended against their use because there was no proven benefit.

Another approach to prevent bacterial infections is the use of antibiotics. Hart and Summerfelt (1975) recommended a postsurgical recovery bath containing oxytetracyline, and interperitoneal injection of this antibiotic in field studies. More recently, Isely et al. (2002) compared transmitter loss and mortality of fish tagged with and without the use of interperitoneal antibiotic. They reported that the antibiotic treatment was effective in delaying mortality, but had no effect on overall mortality. Navarrete et al. (2008) studied the effect of oxytetracyline on the intestinal flora of fish and reported that antibiotic use can enable opportunistic pathogens. Researchers considering the use of antibiotics in fish need to evaluate the true contribution of the treatment, considering that fish can likely receive only a single dose, and weigh this against concerns of developing antibiotic resistance and the possibility of opportunistic infection. Mulcahy (2011) suggests that antibiotic use during fish surgeries should be a rare event and that further efforts be made toward aseptic technique as the primary approach to preventing infection.

Complications.—The three primary concerns associated with surgical tagging are infection, potentially leading to mortality, reduced performance, and transmitter loss. Infections in surgically-tagged fish can be common and serious, even under ideal conditions, because the technique is invasive (Mellas and Haynes 1985). Some species are likely more susceptible to surgical complications than others, so methods should be tailored to species and life stage and tested for effectiveness. Use of aseptic technique is important for all transmitter implantation methods, but is paramount in surgical techniques, especially for the transmitter, which will remain in close contact with fish tissues (Lucas 1989; Moore et al. 1990; Wagner et al. 2000; Lacroix et al. 2004).

A second potential complication of surgical tagging is the effect of the transmitter on fish behavior or performance. Like the other implantation techniques, surgically tagged fish may show elevated mortality, reduced growth, changes in physiology, reduced swimming performance, or altered buoyancy compensation. These kinds of effects have been documented in a number of species (Schramm and Black 1984; Adams et al. 1998b; Walsh et al. 2000; Perry et al. 2001; Lacroix et al. 2004; Hall et al. 2009), and several reviews on the topic are available (Stasko and Pincock 1977; Jepsen et al. 2002; Bridger and Booth 2003; Mulcahy 2003).

A third potential complication of surgical implantation is transmitter loss. This is certainly a serious concern since there is ample evidence in the literature, across a variety of teleost taxa, to suggest that transmitter loss is not a rare phenomenon (Lucas 1989). There are three main mechanisms by which surgically implanted transmitters are expelled from the body cavity: 1) through the incision or antenna exit, 2) through the body wall, or 3) through transintestinal expulsion. A potential contributor to all three loss mechanisms is the development of a fibrous capsule surrounding the implanted transmitter (Figure 13). The fibrous capsule serves a protective function and is a typical foreign body response (Marty and Summerfelt 1986). Fish that have expelled transmitters had fibrous capsules that were more fully developed, with vasculature, compared to fibrous capsules in fish that retained transmitters (Marty and Summerfelt 1986). Although the fibrous capsule in and of itself does not lead to transmitter expulsion, its location and influence on adjacent tissues may dictate whether or not the transmitter will be lost (Marty and Summerfelt 1986; Jepsen et al. 2008b).

The path of least resistance for expelling transmitters, at least early in the healing process, is through the incision or the antenna exit site. Researchers report suture failure (Bunnell et al. 1998; Jepsen et al. 2002) or inflamed and necrotic tissue at the incision site as the primary causes of transmitter losses (Knights and Lasee 1996; Økland et al. 2003). Transmitters expelled through the incision are more common in gravid females, possibly due to the increased pressure from the developing egg mass (Marty and Summerfelt 1986). The weight of the transmitter may be a more important factor in potential transmitter loss than its physical dimensions. Heavier transmitters are lost more readily due to the effect of the transmitter mass on the poorly supported ventral abdominal wall (Lucas 1989; Figure 14).

Transmitter expulsion through the intact body wall is likely due to pressure necrosis, where the pressure from the transmitter causes necrosis in adjacent muscle, weakening the body wall (Lucas 1989). Marty and Summerfelt (1986) suggest that the fibrous capsule surrounding an implanted transmitter may be responsible for directly weakening and rupturing the body wall and forcing out the transmitter. Since both expulsion through the incision and the body wall are caused by pressure necrosis, reduction of transmitter weight or volume would likely reduce transmitter loss (Marty and Summerfelt 1986). Lacroix et al. (2004)

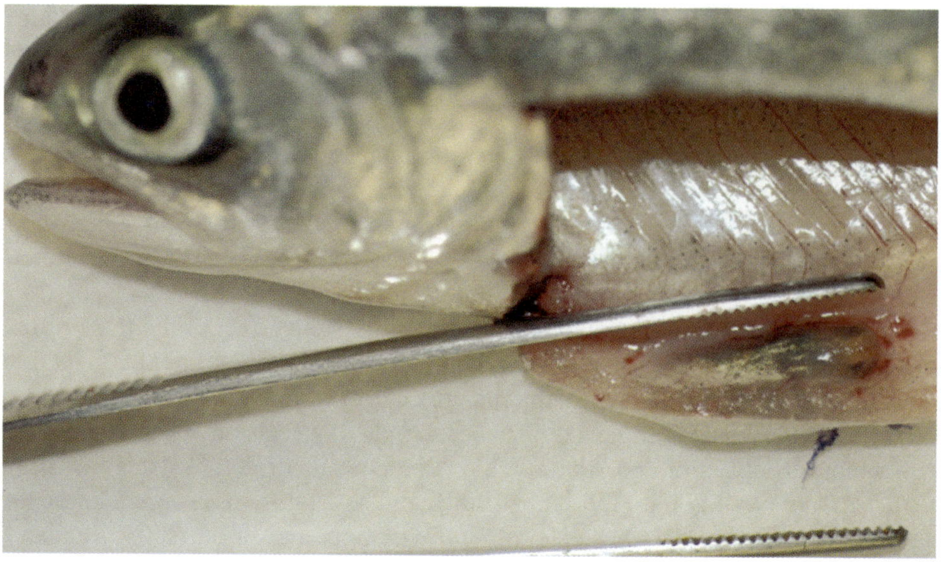

Figure 13. Photograph of a thin fibrous layer developed around an implanted acoustic transmitter.

suggested that transmitter loss through the body wall was caused by sharp edges on the implanted transmitter. Transintestinal expulsion of transmitters, where the transmitter enters the intestinal lumen and is then defecated, is not a common method of transmitter loss. It has been reported in only three species: rainbow trout *Oncorhynchus mykiss* (Chisholm and Hubert 1985), African catfish *Heterobranchus longifilis*, (Baras and Westerloppe 1999), and channel catfish *Ictalurus punctatus* (Marty and Summerfelt 1986).

Alternative Attachment Methods

Other approaches to transmitter attachment include oviduct insertion, tagging deepwater fishes at depth, and tagging by transmitter ingestion. Peake et al. (1997a) inserted transmitters into the urogenital opening of salmon (*Salmo salar* and *O. mykiss*), taking advantage of the fact that the oviduct in salmonids (and some other taxa) is not connected to the ovary. However, they noted 31% transmitter loss overall, with the highest transmitter loss occurring when transmitters were inserted into females during egg mass development (Peake et al. 1997a). Oviduct insertion is a good alternate technique to standard implantation methods if it is used prior to ovary development or after the eggs are shed. Pierce (2004) inserted transmitters into the oviduct of northern pike *Esox lucius*, hoping they would be shed when the fish spawned, allowing him to locate spawning sites. He reported that 40% of the transmitters were expelled during spawning, and noted that fish size and transmitter shape may have influenced the outcome.

Tagging deepwater fishes is challenging because of the extreme differences in water temperature and pressure between the surface and the fish's habitat. Bringing some deepwater species to the surface for tagging can result in severe barotrauma associated with rapid decompression (Parrish and Moffitt 1993). Often, the swim bladder must be punctured to allow the fish to return to depth. Since fish cannot safely be brought to the surface, telemetry

A

B

FIGURE 14. Photographs of transmitter loss through the body wall: a) an acoustic transmitter and b) a radio transmitter. (Photograph courtesy of N. Jepsen).

studies of these deepwater species should consider tagging fish at depth (Priede and Smith 1986). Grimes et al. (1983) designed a system with tags that break-away from a bottom long-line. One drawback of this approach, however, is that it is difficult for the investigator to confirm the species of fish that took the transmitter, so its application may be limited where multispecies fisheries occur. Parrish and Moffitt (1993) reduced barotrauma by using divers to perform gastric tagging of fish 30 m below the surface. Similarly, Starr et al. (2000) used divers at 20 m depth to surgically implant transmitters into fish captured at 100–200 m depth. Tagging fish at extreme depths (or even in shallow water) can be accomplished by enticing fish to ingest transmitters wrapped in bait. Successful tagging of deepwater fish species using transmitter ingestion has been done in both field (Priede and Smith 1986; Winger et al. 2002) and laboratory settings (Armstrong et al. 1992; Winger and Walsh 2001). These innovative tagging methods highlight the importance of reviewing transmitter attachment options during the design phase of a study and investigating techniques based on the needs of the fish and the goals of the study.

EVALUATION OF TRANSMITTER EFFECTS

There is substantial evidence that fish can suffer negative effects from both the attachment procedures and the presence of the transmitter itself (Mellas and Haynes 1985; Adams et al. 1998a, 1998b; Bridger and Booth 2003; Moser et al. 2007). Tagged fish may show elevated mortality, reduced growth, changes in physiology, reduced swimming performance, or altered buoyancy compensation. These kinds of transmitter effects have been documented in a number of species (Schramm and Black 1984; Adams et al. 1998a, 1998b; Walsh et al. 2000; Perry et al. 2001; Lacroix et al. 2004; Hall et al. 2009), and several reviews on the topic are available (Stasko and Pincock 1977; Jepsen et al. 2002; Bridger and Booth 2003; Mulcahy 2003).

Although a formal evaluation of transmitter effects may not be possible for every field study that uses telemetry, all researchers should review the literature for studies of the species and life stage of interest to see what transmitter effects have been documented. Steps can then be taken to avoid pitfalls such as exceeding specific recommendations for tag burden.

Even without a formal evaluation of transmitter effects, general findings from field telemetry studies can be used to help justify inferences made about the untagged population. For example, if tagged fish show signs of feeding activity such as capture on baited hooks, or sufficient numbers of recapture events allow confirmation of survival and growth of tagged fish, the researcher can have some confidence that transmitter effects are likely minimal. Tagged fish may show movement rates comparable to untagged fish, may school with untagged fish (Jepsen and Berg 2002), or may show strong homing behavior following tagging and release. Conversely, lack of significant movement of a tagged fish may suggest transmitter loss or mortality. Some studies have described "fallback" of tagged fish, suggesting that the downstream movement of an upstream migrating anadromous fish is evidence of adverse tagging or handling effects. Frank et al. (2009) evaluated "fallback" as an indicator of transmitter effects in the genus *Alosa* and reported that it is not reliable without significant understanding of typical movement patterns.

The value of recapturing tagged fish at the end of a monitoring period is not well recognized, but can be very helpful in assessing transmitter effects. Researchers should consider using the final period of active battery life of transmitters to locate and collect (e.g., via electroshocking) tagged fish to evaluate growth, wound healing, and risk of transmitter loss (Jepsen and Aarestrup 1999; Jepsen et al. 2000; Jepsen and Berg 2002).

In a formal evaluation of transmitter effects, a basic approach is to tag fish and compare their behavior and performance to untagged reference fish of similar size and condition. However, this approach is difficult to accomplish outside of the laboratory without identifying or marking reference fish in some way. Some researchers have used a nonelectronic tag or a small passive integrated transponder tag to identify reference fish (Hockersmith et al. 2003; Rivinoja et al. 2006). These secondary marks can also be applied to study fish. This will allow researchers to use less expensive "dummy" transmitters (i.e., transmitters without working components) that closely simulate the dimensions of active transmitters to carry out their comparisons. In addition, these secondary marks or tags can be used to identify study fish even if the transmitter is regurgitated or expelled.

Regardless of whether you are conducting a full-scale evaluation or monitoring only a few metrics in a small number of fish, the ideal animal for an evaluation of transmitter effects will be of similar size, origin, health status, physiological state, and stage of development as the fish in the planned study (Peake et al. 1997b). Laboratory studies are commonly used to evaluate transmitter effects, but there are always concerns about the relevance of such studies when their findings are applied to field studies. This is an important consideration and may warrant an approach where a field component is used to support a laboratory study (Martinelli et al. 1998; Moser et al. 2007; Jepsen et al. 2008b).

Transmitter effects can be monitored through time using the maximum anticipated transmitter life as the final planned sampling point, with additional sampling points reflecting significant events in the field, for example after a 24-h posttagging observation period. Other examples of potential sampling points include the beginning of migration, entry to seawater, or anticipated spawning. Monitoring over extended periods may be necessary to evaluate metrics such as complete wound healing or long-term transmitter retention.

Researchers have used a variety of indicators to evaluate transmitter effects, including survival, transmitter retention, feeding behavior and growth, physiology, swimming performance, buoyancy compensation, and predator avoidance ability. Direct transmitter effects include mortality and transmitter loss. Indirect transmitter effects impact behavior, physiology, and performance. These effects are more subtle than direct effects, and are more likely to occur (Mulcahy 2003; Jepsen et al. 2008b).

Survival

Survival is the most commonly reported metric associated with transmitter effects, probably because it is relatively easy to quantify in a controlled setting. Although poor survival of tagged fish is an obvious indicator of negative effects, survival is a simplistic indicator and may lead to the false conclusion of no transmitter effect, especially if it is the only performance metric evaluated (Mulcahy 2003). We recommend that survival be measured as one of several metrics used to evaluate transmitter effects. Researchers have reported variable survival rates following transmitter attachment, ranging from significant mortality (Walsh et al. 2000; Økland et al. 2003; Lacroix et al. 2004; Hall et al. 2009) to minimal or no mortality (Mellas and Haynes 1985; Martinelli et al. 1998; Close et al. 2003; Jepsen et al. 2000; Welch et al. 2007).

Transmitter Retention

Transmitter retention, like survival, is easy to assess in a controlled study. Loss of transmitters may not influence survival (Marty and Summerfelt 1986; Moore et al. 1990), so holding containers must be checked regularly for regurgitated or expelled transmitters. Both high transmitter loss (Chisholm and Hubert 1985; Økland et al. 2003) and high transmitter retention (Tyus 1988; Thoreau and Baras 1997; Walsh et al. 2000; Close et al. 2003) have been reported. Detailed studies of transmitter retention for surgically tagged fish include examination of wound healing and histology to predict outcomes and describe the mechanism and timing of transmitter loss (Marty and Summerfelt 1986; Lucas 1989; Knights and Lasee 1996; Thoreau and Baras 1997; Wagner et al. 2000).

Feeding Behavior and Growth

The presence of a transmitter, especially within the fish's stomach, likely influences feeding behavior or growth, so it is commonly reported as an indicator of transmitter effects. For external attachment or surgical implantation, feeding may be affected because the fish was stressed from anesthesia and handling, compromised by infection, or because the transmitter exerted pressure on the stomach. Findings vary widely with regard to transmitter effects on feeding behavior, ranging from no effect (Lucas 1989; Lucas and Johnstone 1990; Moore et al. 1990; Cote et al. 1999; Jepsen and Aarestrup 1999) to significant effects on feeding behavior and growth (Armstrong and Rawlings 1993; Adams et al. 1998a; Welch et al. 2007). Apart from directly measuring the growth of tagged fish through time, researchers have used other measures of feeding activity such as the presence of food in the gut or quantification of feeding events. Armstrong and Rawlings (1993) used x-rays to measure food intake of a commercial fish food labeled with iron.

Physiological Response

The physiological response of fish to handling, anesthesia, transmitter attachment techniques, and transmitter presence is the foundation for assessment of transmitter effects. The presence of a transmitter can be a chronic stressor, and handling and tagging procedures can produce acute stress, with the potential for long-term effects (e.g., systemic infection). A further concern is that the series of small, acute stressors may be cumulative (Maule et al. 1988; Sharpe et al. 1998). Physiological changes due to the presence of a transmitter have been documented (Jepsen et al. 2001; Close et al. 2003), and are likely present to some extent in all tagged fish.

Indicators of stress used to evaluate transmitter effects include plasma levels of cortisol, glucose, and lactate (Wedemeyer et al. 1990). More simplistic measures of fish health such as hematocrit and leucocrit can also be used (Adams et al. 1998a; Martinelli et al. 1998; Mesa et al. 2003). Evaluations of physiological response must be carefully planned and executed to limit the impact of handling and sampling on the response being measured. Lower et al. (2005) used a unique, noninvasive approach, where levels of cortisol released by individual fish were measured in the water. Physiological assessments typically use a time-series approach, where the condition of the fish is evaluated at increasing time intervals after the stressor event. Scaling the time series with finely separated intervals, particularly during the initial stages, may be useful in determining when to release tagged fish based on reduced stress response, or when reliable data collection might begin in the field.

Swimming Performance

Evaluating swimming performance is a reliable way to assess the overall physical condition and stress response of fish. This approach has been used to evaluate transmitter effects in a variety of species (Cote et al. 1999; Thorstad et al. 2000; Cooke and Bunt 2001; Mesa et al. 2003). Transmitter effects related to swimming performance vary widely with taxa and transmitter attachment method (Cote et al. 1999; Counihan and Frost 1999; Cooke and Bunt 2001; Koed and Thorstad 2001; Mesa et al. 2003; Makiguchi and Ueda 2009). When designing the evaluation of transmitter effects, consider measuring swimming performance shortly after transmitter attachment, to simulate the release of tagged fish during the study.

Several approaches have been used to evaluate swimming performance (see reviews by Beamish 1978 and Hammer 1995). The most commonly used measure is critical swimming speed (Moser et al. 1990; Cote et al. 1999; Counihan and Frost 1999; Lacroix et al. 2004; Makiguchi and Ueda 2009). Critical swimming speed is a measure of sustained swimming ability and takes several hours to measure because fish are put into a swim chamber and gradually stepped through increasing water velocities until they fatigue (Figure 15). Brett (1964) compared critical swimming performance to marathon level activity. On the other extreme, researchers have measured burst swimming performance, where fish are put into a "drag strip" and stimulated to encourage sprint swimming, as might be used to evade predators (Nelson et al. 2002; Nelson and Claireaux 2005).

Somewhere between these two extremes lies the prolonged swimming ability of fish. Measuring prolonged swimming generally takes less than an hour to accomplish because it measures fish fatigue at a single water velocity (Mellas and Haynes 1985; Maule et al. 1988; Close et al. 2003; Zale et al. 2005) and can be a useful approach in field settings. Using a combination of field and laboratory approaches, Koed and Thorstad (2001) measured the critical swimming performance of fish surgically implanted with transmitters and released to remain at large in the field for one year. They recaptured these fish through electroshocking, brought them into the laboratory, paired them with similarly sized controls, and evaluated swimming performance.

Buoyancy Compensation

Buoyancy compensation is an important consideration because attachment of the transmitter induces negative buoyancy. Fish that are negatively buoyant may demonstrate altered depth distribution (resting on the bottom) or increased metabolic requirements due to in-

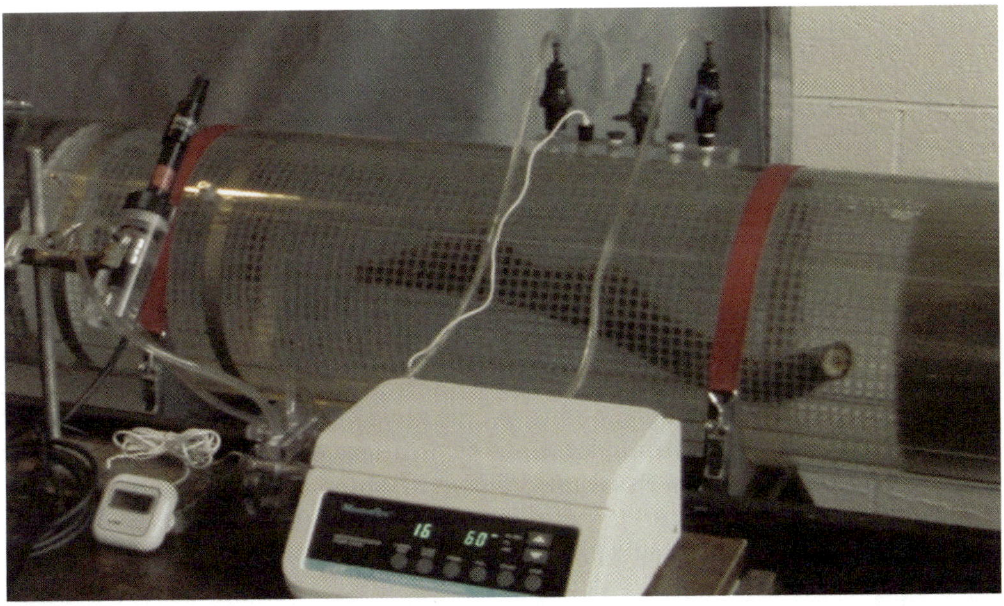

FIGURE 15. Photograph of a Pacific lamprey *Lampetra tridentata* in a swim tunnel to evaluate swimming performance.

creased fin use or faster swimming (Gallepp and Magnuson 1972). One way fish can counteract excess mass is by adjusting the volume of the swim bladder. Depending on the abilities of the fish, this process can be relatively rapid (gulping air by physostomes) or slow (gas diffusion by physoclists). Researchers should be aware that handling and other fish stressors can cause fish to expel air, changing their buoyancy (Harvey et al. 1968), so gentle handling in all aspects of fish collection and tagging should be emphasized. Fried et al. (1976) reported that tagged Atlantic salmon smolts (physostomes) denied access to air after tagging were unable to regain neutral buoyancy, even after 24 h. Alternately, tagged fish provided with access to air were able to adjust their buoyancy, compensating for the weight of the transmitter, within 6 h (Fried et al. 1976).

The goal of evaluating transmitter effects with respect to buoyancy should be to determine the time required to recover neutral buoyancy; this will vary with fish species and transmitter characteristics. Saunders (1965) used a method to assess buoyancy based on the Cartesian diver principle, where the pressure of neutral buoyancy of fish was determined. An anesthetized fish was placed into an air-tight chamber, and pressure in the chamber was reduced by use of a vacuum pump until the fish floated off the bottom (where it was neutrally buoyant) by expansion of the swim bladder.

Carrying a transmitter may have long-term effects on the buoyancy of fish. The swim bladder enlarges to adjust for the weight of the transmitter, and as long as the transmitter is present, the bladder of a tagged fish will be larger than that of a similar sized fish without a transmitter. This perpetually enlarged bladder may restrict the ability of tagged fish to execute rapid vertical movements relative to untagged fish (Gallepp and Magnuson 1972; Perry et al. 2001). To compensate for this effect, the use of neutrally buoyant transmitters has been suggested (Gallepp and Magnuson 1972; Perry et al. 2001).

Predator Avoidance

Predation challenges are an ecologically relevant method of measuring fish performance and the effects of sublethal stressors. Although not commonly applied, predator avoidance trials have been used to evaluate transmitter effects for small fish (Adams et al. 1998b; Anglea et al. 2004; Jepsen et al. 2008a). Mesa et al. (1994) noted several mechanisms that might increase prey vulnerability, including reduced ability to detect predators, decreased burst-speed capability, and ineffective schooling behavior. Tagged fish, especially those equipped with external transmitters or trailing antennas, may be more conspicuous to predators. Thus, predator avoidance capability is a useful metric for evaluating transmitter effects and planning field telemetry studies.

Predator avoidance ability is generally evaluated by exposing tagged fish and untagged controls to a group of acclimated predators to assess whether the predators prefer one group over another. Prey are typically given time in a protected area to acclimate to the tank and presence of the predators and then released to an area where they are vulnerable. The trial continues until approximately 50% of the prey have been consumed. Use of this threshold allows for detection of predator preferences, accounting for the changing availability of each prey type (Coutant 1973). Remaining prey are counted to determine if one prey type was more vulnerable to predation.

Several considerations will assist in the design of a robust study of predator avoidance ability. Firstly, we recommend that a relevant predator be selected (Mesa et al. 1994; Adams et al. 1998b; Marine and Cech 2004) rather than selecting a predator out of convenience. Select a predator from among those living in the same system as the study species (Figure 16). Such predators may initially require additional effort because they need to be acclimated to a laboratory setting and encouraged to consume the target prey, but their use will allow greater confidence in application of the results to natural settings. Secondly, treatment and control groups should be carefully selected so they are of comparable size because size selectivity can be an important driver in predation (Parker 1971). Finally, having refugia in the experimental tank may be an important consideration. With refugia present, predators may more readily adapt to the tank environment, and prey will have more options for evasive behavior (Jepsen et al. 2008a).

SUMMARY AND RECOMMENDATIONS

Telemetry is a useful tool to study the behavior of fish in their natural environment because individual fish can be intensively monitored over long periods. An objective of most telemetry studies is to gain insight from tagged animals that can then be extrapolated to the untagged population. This objective assumes that the attachment procedure or presence of the transmitter does not significantly alter normal behavior, but proactive effort is required to ensure the validity of this assumption. A review of the literature suggests that, although there are three common methods (external, gastric, and surgical), there is no single best transmitter attachment procedure; all of them have risks and benefits.

Reports of transmitter effects suggest that the suitability of attachment procedures must be evaluated for each species and life stage under study (Moore et al. 1990). In the absence of clear guidance for a particular species or life stage, the selection of a transmitter attachment technique should be evaluated. We recommend that researchers planning a study that will use

FIGURE 16. Photograph of smallmouth bass *Micropterus dolomieu* and juvenile Chinook salmon *Oncorhynchus tshawytscha* in an evaluation of predator avoidance ability.

telemetry first review the relevant literature (matching species, life stage, size, etc.), consult with knowledgeable colleagues such as veterinarians, conduct controlled studies of transmitter effects, and think critically about whether or not a planned study will meet the basic assumption of telemetry.

Sublethal transmitter effects are likely to occur and can be difficult to assess, so a prudent approach will include selection of the best possible techniques for capture, handling, anesthesia, and transmitter attachment (Bridger and Booth 2003; Mulcahy 2003). With so much emphasis on the procedures for transmitter attachment, best practices for capture, general handling, and anesthesia are often overlooked, or at least underemphasized. Good handling techniques will minimize stressors (Harvey et al. 1968) and potential bias in results. Limiting fish transfers and handling should be a priority, especially during the effort to release fish (Moore et al. 1990; AFS 2004). Anesthesia is commonly required to attach a transmitter, and some would argue that lack of anesthesia is inhumane (Mulcahy 2003), but its use can increase the risk of negative outcomes if not properly monitored (Mulford 1984).

Researchers should take advantage of any opportunity that arises to examine fish after tagging. The use of a secondary mark will aid this effort, and is essential to assess potential transmitter loss. Fish may be incidentally recaptured through recreational or commercial harvest or purposefully recaptured by first locating the tagged fish using the transmitter signal and then netting or electroshocking to recover it (Bunnell et al. 1998; Koed and Thorstad 2001; Jepsen 2003; Jepsen et al. 2000, 2002). In some cases the fish may not be physically recaptured, but may be observed visually or recorded using video surveillance.

Regardless of the means, recapture provides an opportunity to evaluate, in a real-life setting, the effectiveness of the transmitter attachment method. Conduct both external and internal exams on recaptured fish, looking for tissue or organ damage, infection, healing, and location of the transmitter relative to the original placement. Document findings with photography, and provide detailed descriptions during study reporting. Negative outcomes of transmitter attachment procedures are rarely reported, but they provide valuable insight. For example, based on recaptured fish, Bauer et al. (2005) found the location of the exit site for a trailing antenna had migrated 3 cm, and Thorstad et al. (2001) described an antenna with significant biofouling.

Researchers should report, in detail, the procedures used for transmitter attachment and the training and experience of the tagger. We encourage greater rigor in reporting of procedures such as pretag and posttag holding, fish handling, and efforts toward aseptic technique. The skill of the tagger can have a significant effect on the outcome of the tagging procedure (Wagner and Cooke 2005), so the number of taggers, their contribution to the study effort, and their experience and training should be reported in detail.

Since there is no single perfect transmitter attachment technique for all conditions, and the literature describes a wide range of negative consequences for fish that carry transmitters, telemetry researchers must carefully plan and execute their studies in order to obtain meaningful results.

ACKNOWLEDGMENTS

We thank USGS and NOAA Fisheries staffs that have conducted the large field studies where tagging and fish handling skills have been developed and refined. We appreciate the editorial review by JoAnne Butzerin with the NOAA Fisheries Northwest Fisheries Science Center. Reviews by Christopher Peery and Niels Jepsen improved the document. Reference to trade names does not imply endorsement by the federal government.

REFERENCES

Adams, N. S., D. W. Rondorf, S. D. Evans, J. E. Kelly, and R. W. Perry. 1998a. Effects of surgically and gastrically implanted radio transmitters on growth and feeding behavior of juvenile Chinook salmon. Transactions of the American Fisheries Society 127:128–136.

Adams, N. S., D. W. Rondorf, S. D. Evans, J. E. Kelly, and R. W. Perry. 1998b. Effects of surgically and gastrically implanted radio transmitters on swimming performance and predator avoidance of juvenile Chinook salmon (*Oncorhynchus tshawytscha*). Canadian Journal of Fisheries and Aquatic Sciences 55:781–787.

Allen, J. L., and P. D. Harman. 1970. Control of pH in MS-222 anesthetic solutions. The Progressive Fish-Culturist 32:100.

American Fisheries Society (AFS) 2004. Guidelines for the use of fishes in research. American Fisheries Society, Bethesda, Maryland.

Anderson, W. G., R. S. McKinley, and M. Colavecchia. 1997. The use of clove oil as an anesthetic for rainbow trout and its effects on swimming performance. North American Journal of Fisheries Management 17:301–307.

Anglea, S. M., D. R. Geist, R. S. Brown, K. A. Deters, and R. D. McDonald. 2004. Effects of acoustic transmitters on swimming performance and predator avoidance of juvenile Chinook salmon. North American Journal of Fisheries Management 24:162–170.

Armstrong, J. D., and C. E. Rawlings. 1993. The effect of intragastric transmitters on feeding behaviour of Atlantic salmon, *Salmo salar*, parr during autumn. Journal of Fish Biology 43:646–648.

Armstrong, J. D., A. D. F. Johnstone, and M. C. Lucas. 1992. Retention of intragastric transmitters after voluntary ingestion by captive cod, *Gadus morhua* L. Journal of Fish Biology 40:135–137.

Baras, E., and L. Westerloppe. 1999. Transintestinal expulsion of surgically implanted tags by African catfish *Heterobranchus longifilus* of variable size and age. Transactions of the American Fisheries Society 128:737–746.

Bauer, C., G. Unfer, and G. Loupal. 2005. Potential problems with external trailing antennae: antenna migration and ingrowth of epithelial tissue, a case study from a recaptured *Chondrostoma nasus*. Journal of Fish Biology 67:885–889.

Beamish, F. W. H. 1978. Swimming capacity. Pages 101–187 *in* W. S. Hoar and D. J. Randall, editors. Fish physiology. Volume VII Locomotion. Academic Press, New York.

Beeman, J. W., S. Juhnke, L. Dingmon, M. van den Tillaart, and T. Thomas. 2007. Effects of antenna length and material on output power and detection of miniature radio transmitters. Hydrobiologia 582:221–229.

Boothe, H. W. 2003. Suture materials, tissue adhesives, staplers, and ligating clips. Pages 235–244 *in* D. H. Slatter, editor. Textbook of small animal surgery, Volume 1, 3rd edition. Saunders, Philadelphia.

Bowker, J. and J. Trushenski. 2011. AFS policy statement regarding the need for an immediate- release anesthetic/sedative for use in the fisheries disciplines. Fisheries 36:132–135.

Brett, J. R. 1964. The respiratory metabolism and swimming performance of young sockeye salmon. Journal of the Fisheries Research Board of Canada 21:1183–1226.

Bridger, C. J., and R. K. Booth. 2003. The effects of biotelemetry transmitter presence and attachment procedures on fish physiology and behavior. Reviews in Fisheries Science 11:13–34.

Broadhurst, B. T., B. C. Ebner, and R. C. Clear. 2009. Radio-tagging flexible-bodied fish: temporary confinement enhances radio-tag retention. Marine and Freshwater Research 60: 365–360.

Brown, R. and J. H. Eiler. 2000. Performance of Yukon River inconnu tagged with external and internal transmitters. Pages 145–153 *in* J. H. Eiler, D. J. Alcorn, and M. R. Neuman, editors. Biotelemetry 15: Proceedings of the 15th International Symposium on Biotelemetry. Juneau, Alaska USA. International Society of Biotelemetry. Wageningen, The Netherlands.

Brown, R. S., R. A. Harnish, K. M. Carter, J. W. Boyd, K. A. Deters, and M. B. Eppard. 2010. An evaluation of the maximum tag burden for implantation of acoustic transmitters in juvenile Chinook salmon. North American Journal of Fisheries Management 30:499–505.

Bunnell, D. B., and J. J. Isely. 1999. Influence of temperature on mortality and retention of simulated transmitters in rainbow trout. North American Journal of Fisheries Management 19:152–154.

Bunnell, D. B., J. J. Isely, K. H. Burrell, and D. H. Van Lear. 1998. Diel movement of brown trout in a southern Appalachian River. Transactions of the American Fisheries Society 127:630–636.

Chisholm, I. M., and W. A. Hubert. 1985. Expulsion of dummy transmitters by rainbow trout. Transactions of the American Fisheries Society 114:766–767.

Chomyshyn, L., S. H. McConnachie, and S. J. Cooke. 2011. Evaluation of water entry into the coelom and different levels of aseptic technique during surgical implantation of electronic tags in freshwater fish. Reviews in Fish Biology and Fisheries 21:61–70.

Claireaux, G., and C. Lefrancois. 1998. A method for the external attachment of acoustic tags on roundfish. Hydrobiologia 371/372:113–116.

Close, D. A., M. S. Fitzpatrick, C. M. Lorion, H. W. Li, and C. B. Schreck. 2003. Effects of intraperitoneally implanted radio transmitters on the performance and physiology of Pacific lamprey. North American Journal of Fisheries Management 23:1184–1192.

Collins, M. R., D. W. Cooke, T. I. J. Smith, W. C. Post, D. C. Russ, and D. C. Walling. 2002. Evaluation of four methods of transmitter attachment on shortnose sturgeon *Acipenser brevirostrum*. Journal of Applied Ichthyology 18:491–494.

Cooke, S. J., and C. M. Bunt. 2001. Assessment of internal and external antenna configurations of radio transmitters implanted in smallmouth bass. North American Journal of Fisheries Management 21:236–241.

Cooke, S. J., B. D. S. Graeb, C. D. Suski, and K. G. Østrand. 2003. Effects of suture material on incision healing, growth and survival of juvenile largemouth bass implanted with miniature radio transmitters: case study of a novice and experienced fish surgeon. Journal of Fish Biology 62:1366–1380.

Cooke S. J., C. M. Woodley, M. B. Eppard, R. S. Brown, J. L. Nielsen. 2011. Advancing the surgical implantation of electronic tags in fish: a gap analysis and research agenda based on a review of trends in intracoelomic tagging effects studies. Reviews in Fish Biology and Fisheries. 21:127–151.

Cote, D., D. A. Scruton, L. Cole, R. S. McKinley. 1999. Swimming performance and growth rates of juvenile cod intraperitoneally implanted with dummy acoustic transmitters. North American Journal of Fisheries Management 19:1137–1141.

Counihan, T. D., and C. N. Frost. 1999. Influence of externally attached transmitters on the swimming performance of juvenile white sturgeon. Transactions of the American Fisheries Society 128:965–970.

Coutant, C. C. 1973. Effect of thermal shock on vulnerability of juvenile salmonids to predation. Journal of the Fisheries Research Board of Canada 30:965–973.

Crook, D. A. 2004. A method for externally attaching radio transmitters to minimize dermal irritation. Journal of Fish Biology 64:258–261.

Dunning, D. J., and Q. E. Ross. 2010. Effect of radio-tagging on escape reactions of adult blueback herring to ultrasound. North American Journal of Fisheries Management 30:26–32.

Elliott, D. G., and R. J. Pascho. 2001. Evidence that coded-wire tagging procedures can enhance transmission of *Renibacterium salmoninarum* in Chinook salmon. Journal of Aquatic Animal Health 13:181–193.

FDA (Food and Drug Administration). 2007. Guidance for industry 150: concerns related to the use of clove oil as an anesthetic for fish. U.S. Department of Health and Human Services, Center for Veterinary Medicine, Rockville, Maryland.

Frank, H. J., M. E. Mather, J. M. Smith, R. M. Muth, J. T. Finn, and S. D. McCormick. 2009. What is "fallback"?: metrics needed to assess telemetry tag effects on anadromous fish behavior. Hydrobiologia 653:237–249.

Fried, S. M, J. D. McCleave, and K. A. Stred. 1976. Buoyancy compensation by Atlantic salmon (*Salmo salar*) smolts tagged internally with dummy telemetry transmitters. Journal of the Fisheries Research Board of Canada 3:1377–1380.

Gaikowski, M. P., W. H. Gingerich, and S. Gutreuter. 2001. Short-duration electrical immobilization of lake trout. North American Journal of Fisheries Management 21:381–392.

Gallepp, G. W., and J. J. Magnuson. 1972. Effects of negative buoyancy on the behavior of the bluegill, *Lepomis macrochirus* Rafinesque. Transactions of the American Fisheries Society 101:507–512.

Gelwicks, K. R., D. J. Zafft, and J. P. Bobbitt. 1998. Efficacy of carbonic acid as an anesthetic for rainbow trout. North American Journal of Fisheries Management 18:432–438.

Gilderhus, P. A., and L. L. Marking. 1987. Comparative efficacy of 16 anesthetics chemicals on rainbow trout. North American Journal of Fisheries Management 7:288–292.

Gilliland, E. R. 1994. Comparison of absorbable sutures used in largemouth bass liver biopsy surgery. The Progressive Fish-Culturist 56(1): 60–61.

Greenstreet, S. P. R., and R. I. G. Morgan. 1989. The effect of ultrasonic tags on the growth rates of Atlantic salmon, *Salmo salar* L., parr of varying size just prior to smolting. Journal of Fish Biology 35:301–309.

Grimes, C. B., S. C. Turner, and K. W. Able. 1983. A technique for tagging deepwater fish. Fishery Bulletin 81:663–666.

Gunstrom, G. K., and M. Bethers. 1985. Electrical anesthesia for handling large salmonids. The Progressive Fish-Culturist. 47:67–69.

Hall, J. E., J. Chamberlin, A. N. Kagley, C. Greene, and K. L. Fresh. 2009. Effects of gastric and surgical insertions of dummy ultrasonic transmitters on juvenile Chinook salmon in seawater. Transactions of the American Fisheries Society 138:52–57.

Hammer, C. 1995. Fatigue and exercise tests with fish. Comparative Biochemistry and Physiology 112A:1–20.

Harms, C. A. 2005. Surgery in fish research: common procedures and postoperative care. Laboratory Animals 34:28–34.

Harms, C. A., and G. A. Lewbart. 2011. The veterinarian's role in surgical implantation of electronic tags in fish. Reviews in Fish Biology and Fisheries 21:25–33.

Hart, L. G., and R. C. Summerfelt. 1975. Surgical procedures for implanting ultrasonic transmitters into flathead catfish (*Pylodictis olivaris*). Transactions of the American Fisheries Society 104:56–59.

Harvey, H. H., W. S. Hoar, and C. R. Brothers. 1968. Sounding response of the kokanee and sockeye salmon. Journal of the Fisheries Research Board of Canada 25:1115–1131.

Herke, S. W., and J. R. Moring. 1999. "Soft" harness for external attachment of large radio transmitters to northern pike (*Esox lucius*). Fisheries Research 39:305–312.

Hockersmith, E. E., W. D. Muir, S. G. Smith, B. P. Sandford, R. W. Perry, N. S. Adams, and D. W. Rondorf. 2003. Comparison of migration rate and survival between radio-tagged and PIT-tagged migrant yearling Chinook salmon in the Snake and Columbia Rivers. North American Journal of Fisheries Management 23:404–413.

Houston, A. H., J. A. Madden, R. J. Woods, and H. M. Miles. 1971. Some physiological effects of handling and tricaine methane sulfonate anesthetization upon the brook trout, *Salvelinus fontinalis*. Journal of the Fisheries Research Board of Canada 28:625–633.

Hovda, J. and T. J. Linley. 2000. The potential application of hypothermia for anesthesia in adult Pacific salmon. North American Journal of Fisheries Management 62:67–72.

Hurty, C. A., D. C.Brazik, J. M. Law, K.Sakamoto, and G. A. Lewbart. 2002. Evaluation of the tissue reactions in the skin and body wall of koi (*Cyprinus carpio*) to five suture materials. Veterinary Record 151(11): 324–328.

Isely, J. J., S. P. Young, T. A. Jones, and J. J. Schaffler. 2002. Effects of antenna placement and antibiotic treatment on loss of simulated transmitters and mortality in hybrid striped bass. North American Journal of Fisheries Management 22:204–207.

Jennings, C. A., and G. L. Looney. 1998. Evaluation of two types of anesthesia for performing surgery on striped bass. North American Journal of Fisheries Management 18:187–190.

Jepsen, N. 2003. Long-term retention of surgically implanted radio transmitters in pikeperch. Journal of Fish Biology 63:260–262.

Jepsen, N., and K. Aarestrup. 1999. A comparison of the growth of radio-tagged and dye-marked pike. Journal of Fish Biology 55:880–883.

Jepsen, N., and S. Berg. 2002. The use of winter refuges by roach tagged with miniature radio transmitters. Hydrobiologia 483:167–173.

Jepsen, N., M. Christoffersen, and T. Munksgaard. 2008a. The level of predation used as an indicator of tagging/handling effects. Fisheries Management and Ecology 15:365–368.

Jepsen, N., L. E. Davis, C. B. Schreck, and B. Siddens. 2001. The physiological response of Chinook salmon smolts to two methods of radio-tagging. Transactions of the American Fisheries Society 130:495–500.

Jepsen, N., A. Koed, E. B. Thorstad, and E. Baras. 2002. Surgical implantation of telemetry transmitters in fish: how much have we learned? Hydrobiologia 483:239–248.

Jepsen, N., J. S. Mikkelsen, and A. Koed. 2008b. Effects of tag and suture type on survival and growth of brown trout with surgically implanted telemetry tags in the wild. Journal of Fish Biology 72:594–602.

Jepsen, N., C. Schreck, S. Clements, and E. B. Thorstad. 2005. A brief discussion on the 2% tag/body mass rule of thumb. Pages 255–259 in M. T. Spedicato, G. Marmulla, and G. Lembo, editors. Aquatic telemetry: advances and applications. Proceedings of the Fifth Conference on Fish Telemetry held in Europe. Ustica, Italy, June 9–13, 2003. FAO-COISPA, Rome.

Jepsen, N., S. Pedersen, and E. Thorstad. 2000. Behavioural interactions between prey (trout smolts) and predators (pike and pikeperch) in an impounded river. Regulated Rivers: Research and Management 16:189–198.

Kaseloo, P. A., A. H. Weatherley, J. Lotimer, and M. D. Farina. 1992. A biotelemetry system recording fish activity. Journal of Fish Biology 40:165–179.

Keefer, M. L., C. A. Perry, R. R. Ringe, and T. C. Bjorn. 2004. Regurgitation rates of intragastric radio transmitters by adult Chinook salmon and steelhead during upstream migration in the Columbia and Snake Rivers. North American Journal of Fisheries Management 24:47–54.

Kelsh, S. W., and B. Shields. 1996. Care and handling of sampled organisms. Pages 121–155 *in* B. R. Murphy and D. W. Willis, editors. Fisheries techniques, 2nd edition. American Fisheries Society, Bethesda, Maryland.

Knights, B. C., and B. A. Lasee. 1996. Effects of implanted transmitters on adult bluegills at two temperatures. Transactions of the American Fisheries Society 125:440–449.

Koed, A., and E. B. Thorstad. 2001. Long-term effect of radio-tagging on the swimming performance of pikeperch. Journal of Fish Biology 58:1753–1756.

Koo, T. S. Y., and J. S. Wilson. 1972. Sonic tracking striped bass in the Chesapeake and Delaware canal. Transactions of the American Fisheries Society 101:453–462.

Kynard, B., and E. Lonsdale. 1975. Experimental study of galvanonarcosis for rainbow trout (*Salmo gairdneri*) immobilization. Journal of the Fisheries Research Board of Canada 32:300–302.

Lacroix, G. L., D. Knox, and P. McCurdy. 2004. Effects of implanted dummy acoustic transmitters on juvenile Atlantic salmon. Transactions of the American Fisheries Society 133:211–220.

LaVigne, H. R. 2002. An improved portable surgical table for the field and laboratory. North American Journal of Fisheries Management 22:571–572.

Lewis, A. E., and W. R. A. Muntz. 1984. The effects of external ultrasonic tagging on the swimming performance of rainbow trout, *Salmo gairdneri* Richardson. Journal of Fish Biology 25:577–585.

Lower, N., A. Moore, A. P. Scott, T. Ellis, J. D. James, and I. C. Russell. 2005. A noninvasive method to assess the impact of electronic tag insertion on stress levels in fish. Journal of Fish Biology 67:1202–1212.

Lucas, M. C. 1989. Effects of implanted dummy transmitters on mortality, growth and tissue reaction in rainbow trout, *Salmo gairdneri* Richardson. Journal of Fish Biology 35:577–587.

Lucas, M. C., and A. D. F. Johnstone. 1990. Observations on the retention of intragastric transmitters, and their effects on food consumption, in cod, *Gadus morhua* L. Journal of Fish Biology 37:647–649.

Makiguchi, Y. and H. Ueda. 2009. Effects of external and surgically implanted dummy radio transmitters on mortality, swimming performance and physiological status of juvenile masu salmon *Oncorhynchus masou*. Journal of Fish Biology 74:304–311.

Marine, K. R., and J. J. Cech, Jr. 2004. Effects of high water temperature on growth, smoltification, and predator avoidance in juvenile Sacramento River Chinook salmon. North American Journal of Fisheries Management 24:198–210.

Marking, L. L. and F. P. Meyer. 1985. Are better anesthetics needed in fisheries? Fisheries 10(6):2–5.

Martinelli, T. L., H. C. Hansel, and R. S. Shively. 1998. Growth and physiological responses to surgical and gastric radio transmitter implantation techniques in subyearling Chinook salmon (*Oncorhynchus tshawytscha*). Hydrobiologia 371/372:79–87.

Martinelli, T. L., and R. S. Shively. 1997. Seasonal distribution, movements, and habitat associations of northern squawfish in two lower Columbia River reservoirs. Regulated Rivers: Research and Management 13:543–556.

Marty, G. D., and R. C. Summerfelt. 1986. Pathways and mechanisms for expulsion of surgically implanted dummy transmitters from channel catfish. Transactions of the American Fisheries Society 115:577–589.

Matthews, K. R., T. P. Quinn, and B. S. Miller. 1990. Use of ultrasonic transmitters to track demersal rockfish movements on shallow rocky reefs. Pages 375–379 *in* N. C. Parker, A. E. Giorgi, R. C. Heidinger, D. B. Jester, Jr., E. D. Prince, and G. A. Winans. American Fisheries Society Symposium 7, Bethesda, Maryland.

Maule, A. G., C. B. Schreck, C. S. Bradford, and B. A. Barton. 1988. Physiological effects of collecting and transporting emigrating juvenile Chinook salmon past dams on the Columbia River. Transactions of the American Fisheries Society 117:245–261.

McCleave, J. D., J. H. Power, and S. A. Rommel, Jr. 1978. Use of radio telemetry for studying upriver migration of adult Atlantic salmon (*Salmo salar*). Journal of Fish Biology 12:549–558.

McCleave, J. D., and K. A. Stred. 1975. Effects of dummy telemetry transmitters on stamina of Atlantic salmon (*Salmo salar*) smolts. Journal of the Fisheries Research Board of Canada 32:559–563.

Mellas, E. J., and J. M. Haynes. 1985. Swimming performance and behavior of rainbow trout (*Salmo gairdneri*) and white perch (*Morone americana*): effects of attaching telemetry transmitters. Canadian Journal of Fisheries and Aquatic Sciences 42:488–493.

Mesa, M. G., J. M. Bayer, and J. G. Seelye. 2003. Swimming performance and physiological responses to exhaustive exercise in radio-tagged and untagged Pacific lampreys. Transactions of the American Fisheries Society 132:483–492.

Mesa, M. G., T. P. Poe, D. M. Gadomski, and J. H. Petersen. 1994. Are all prey created equal? A review and synthesis of differential predation on prey in substandard condition. Journal of Fish Biology 45:81–96.

Moore, A., I. C. Russell, and E. C. E. Potter. 1990. The effects of intraperitoneally implanted dummy acoustic transmitters on the behavior and physiology of juvenile Atlantic salmon, *Salmo salar* L. Journal of Fish Biology 37:713–721.

Mortensen, D. G. 1990. Use of staples to close surgical incisions for transmitter implants. Pages 380–383 *in* N. C. Parker, A. E. Giorgi, R. C. Heidinger, D. B. Jester, Jr., E. D. Prince, and G. A. Winans. American Fisheries Society Symposium 7, Bethesda, Maryland.

Moser, M. L., D. A. Ogden, and B. P. Sandford. 2007. Effects of surgically implanted transmitters on anguilliform fishes: lessons from lamprey. Journal of Fish Biology 71:1847–1852.

Moser, M. L., A. F. Olson, and T. P. Quinn. 1990. Effects of dummy ultrasonic transmitters on juvenile coho salmon. Pages 353–356 *in* N. C. Parker, A. E. Giorgi, R. C. Heidinger, D. B. Jester, Jr., E. D. Prince, and G. A. Winans. American Fisheries Society, Symposium 7, Bethesda, Maryland.

Mulcahy, D. M. 2003. Surgical implantation of transmitters into fish. ILAR (Institute for Laboratory Animal Research) Journal 44:295–306.

Mulcahy, D. M. 2011. Antibiotic use during the intracoelomic implantation of electronic tags into fish. Reviews in Fish Biology and Fisheries 21:83–96.

Mulford, C. J. 1984. Use of a surgical skin stapler to quickly close incisions in striped bass. North American Journal of Fisheries Management 4:571–573.

Navarrete, P., P. Mardones, R. Opazo, R. Espejo, and J. Romero. 2008. Oxytetracycline treatment reduces bacterial diversity of intestinal microbiota of Atlantic salmon. Journal of Aquatic Animal Health 20:177–183.

Nelson, J. A., and G. Claireaux. 2005. Sprint swimming performance of juvenile European sea bass. Transactions of the American Fisheries Society 134:1274–1284.

Nelson, J. A., P. S. Gotwalt, S. P. Reidy, and D. M. Webber. 2002. Beyond U_{crit}: Matching swimming performance tests to the physiological ecology of the animal, including a new fish "drag strip." Comparative Biochemistry and Physiology 133A:289–302.

Nemetz, T. G., and J. R. MacMillan. 1988. Wound healing of incisions closed with a cyanoacrylate adhesive. Transactions of the American Fisheries Society 117:190–195.

Økland, F., C. J. Hay, T. F. Naesje, N. Nickandor, and E. B. Thorstad. 2003. Learning from unsuccessful radio tagging of common carp in a Namibian reservoir. Journal of Fish Biology 62:735–739.

Orsi, J. A., and J. W. Short. 1987. Modifications in electrical anesthesia for salmonids. The Progressive Fish-Culturist 49:144–146.

Parrish, F. A., and R. B. Moffitt. 1993. Subsurface fish handling to limit decompression effects on deepwater species. Marine Fisheries Review 54:29–32.

Parker, R. R. 1971. Size selective predation among juvenile salmonid fishes in a British Columbia inlet. Journal of the Fisheries Research Board of Canada 28:1503–1510.

Peake, S. 1998. Sodium bicarbonate and clove oil as potential anesthetics for nonsalmonid fishes. North American Journal of Fisheries Management 18:919–924.

Peake, S., R. S. McKinley, T. A. Beddow, and G. Marmulla. 1997a. New procedure for radio transmitter attachment: oviduct insertion. North American Journal of Fisheries Management 17:757–762.

Peake, S., R. S. McKinley, D. A. Scruton, and R. Moccia. 1997b. Influence of transmitter attachment procedures on swimming performance of wild and hatchery-reared Atlantic salmon smolts. Transactions of the American Fisheries Society 126:707–714.

Pearcy, W. G. 1992. Movements of acoustically-tagged yellowtail rockfish Sebastes flavidus on Heceta Bank, Oregon. Fishery Bulletin 90:726–735.

Penne, C. R., N. L. Ahrens, R. C. Summerfelt, and C. L. Pierce. 2007. Effect of relative volume on radio transmitter expulsion in subadult common carp. North American Journal of Fisheries Management 27:986–991.

Perry, R. W., N. S. Adams, and D. W. Rondorf. 2001. Buoyancy compensation of juvenile Chinook salmon implanted with two different size dummy transmitters. Transactions of the American Fisheries Society 130:46–52.

Petering, R. W., and D. L. Johnson. 1991. Suitability of a cyanoacrylate adhesive to close incisions in black crappies used in telemetry studies. Transactions of the American Fisheries Society 120:535–537.

Pierce, R. B. 2004. Oviduct insertion of radio transmitters as a means of locating northern pike spawning habitat. North American Journal of Fisheries Management 24:244–248.

Portz, D. E., C. M. Woodley, and J. J. Cech, Jr. 2006. Stress-associated impacts of short-term holding on fishes. Reviews in Fish Biology and Fisheries 16:125–170.

Priede, I. G., and K. L. Smith. 1986. Behavior of the abyssal grenadier, *Coryphaenoides yaquinae*, monitored using ingestible acoustic transmitters in the Pacific Ocean. Journal of Fish Biology 29 (supplement A):199–206.

Ramstad, K. M., and C. A. Woody. 2003. Radio tag retention and tag-related mortality among adult sockeye salmon. North American Journal of Fisheries Management 23:978–982.

Rikardsen, A. H., and E. B. Thorstad. 2006. External attachment of data storage tags increases probability of being recaptured in nets compared to internal tagging. Journal of Fish Biology 68:963–968.

Rivinoja, P., K. Leorardsson, and H. Lundqvist. 2006. Migration success and migration time of gastrically radio-tagged v. PIT-tagged adult Atlantic salmon. Journal of Fish Biology 69:304–311.

Ross, L. G., and B. Ross. 2008. Anesthetic and sedative techniques for aquatic animals, 3rd edition. Blackwell, Oxford, U.K.

Ross, M. J. and C. F. Kleiner. 1982. Shielded needle technique for surgically implanting radio-frequency transmitters in fish. The Progressive Fish-Culturist 44:41–43.

Ross, M. J., and J. H. McCormick. 1981. Effects of external radio transmitters on fish. The Progressive Fish-Culturist 43:67–72.

Sakaris, P. C., R. V. Jesien, and A. E. Pinkey. 2005. Retention of surgically implanted ultrasonic transmitters in the Brown Bullhead Catfish. North American Journal of Fisheries Management 25: 822–826.

Sanderson, T. B. and W. A. Hubert. 2007. Assessment of gaseous CO2 and AQUI-S as anesthetics when surgically implanting radio transmitters into cutthroat trout. North American Journal of Fisheries Management 27:1053–1057.

Saunders, R. L. 1965. Adjustment of buoyancy in young Atlantic salmon and brook trout by changes in swim bladder volume. Journal of the Fisheries Research Board of Canada 22:335–348.

Schnick, R. A. 2006. Zero withdrawal anesthetic for all finfish and shellfish: need and candidates. Fisheries 31:122–126.

Schnick, R. A., F. P. Meyer, and D. F. Walsh. 1986. Status of fishery chemicals in 1985. The Progressive Fish-Culturist 48:1–17.

Schramm, H. L., and D. J. Black. 1984. Anesthesia and surgical procedures for implanting radio transmitters into grass carp. The Progressive Fish-Culturist 46:185–190.

Sharpe, C. S., D. A. Thompson, H. L. Blankenship, and C. B. Schreck. 1998. Effects of routine handling and tagging procedures on physiological stress responses in juvenile Chinook salmon. The Progressive Fish-Culturist 60:81–87.

Sink, T. D., R. J. Strange, and R. E. Sawyers. 2007. Clove oil used at lower concentrations is less effective than MS-222 at reducing cortisol stress responses in anesthetized rainbow trout. North American Journal of Fisheries Management 27:156–161.

Skov, C. J. Brodersen, C. Bronmark, L. A. Hansson, P. Hertonsson, and P. A. Nilsson. 2005. Evaluation of PIT-tagging in cyprinids. Journal of Fish Biology 67:1195–1201.

Slatter, D. H., editor. 2003. Textbook of small animal surgery, 3rd edition. Saunders, Philadelphia.

Smith, G. W., R. N. B. Campbell, and J. S. MacLaine. 1998. Regurgitation rates of intragasgtric transmitters by adult Atlantic salmon (*Salmo salar* L.) during riverine migration. Hydrobiologia 371/372:117–121.

Smith, J. M, M. E. Mather, H. J. Frank, R. M. Muth, J. T. Finn, and SD. McCormick. 2009. Evaluation of a gastric radio tag insertion technique for anadromous river herring. North American Journal of Fisheries Management 29:367–377.

Soivio, A., K. Nyholm, and M. Huhti. 1977. Effects of anesthesia with MS-222, neutralized MS-222 and benzocaine on the blood constituents of rainbow trout, *Salmo gairdneri*. Journal of Fish Biology 10:97–101.

Starr, R. M., J. H. Heine, and K. Johnson. 2000. Techniques for tagging and tracking deepwater rockfishes. North American Journal of Fisheries Management 20:597–609.

Stasko, A. B., and D. G. Pincock. 1977. Review of underwater biotelemetry, with emphasis on ultrasonic techniques. Journal of the Fisheries Research Board of Canada 34:1261–1285.

Stickney, R. R., and C. C. Kohler. 1990. Maintaining fish for research and testing. Pages 633–663 *in* C. B. Schreck and P. B. Moyle, editors. Methods for fish biology. American Fisheries Society, Bethesda, Maryland.

Strange, R. J., and C. B. Schreck. 1978. Anesthetic and handling stress on survival and cortisol concentration in yearling Chinook salmon *Oncorhynchus tshawytscha*. Journal of the Fisheries Research Board of Canada 35:345–349.

Summerfelt, R. C., and L. S. Smith. 1990. Anesthesia, surgery and related techniques. Pages 213–272 *in* C. B. Schreck and P. B. Moyle, editors. Methods for fish biology. American Fisheries Society, Bethesda, Maryland.

Sutton, T. M., and A. C. Benson. 2003. Influence of external transmitter shape and size on tag retention and growth of juvenile lake sturgeon. Transactions of the American Fisheries Society 132:1257–1263.

Swanberg, T. R., D. A. Schmetterling, and D. H. McEvoy. 1999. Comparison of surgical staples and silk sutures for closing incisions in rainbow trout. North American Journal of Fisheries Management 19:215–218.

Tanaka, H., Y. Takagi, and Y. Naito. 2001. Swimming speeds and buoyancy compensation of migrating adult chum salmon *Oncorhynchus keta* revealed by speed/depth/acceleration data logger. The Journal of Experimental Biology 204:3895–3904.

Thompson, L. A., S. J. Cooke, M. R. Donaldson, K. C. Hanson, A. Gingerich, T. Klefoth, and R. Arlinghaus. 2008. Physiology, behavior, and survival of angled and air-exposed largemouth bass. North American Journal of Fisheries Management 28:1059–1068.

Thoreau, X. and E. Baras. 1997. Evaluation of surgery procedures for implanting telemetry transmitters into the body cavity of tilapia *Oreochromis aureus*. Aquatic Living Resources 10:207–211.

Thorstad, E. B., F. Økland, and B. Finstad. 2000. Effects of telemetry transmitters on swimming performance of adult Atlantic salmon. Journal of Fish Biology 57:531–535.

Thorstad, E. B., F. Økland, and T. G. Heggberget. 2001. Are long term negative effects from external tags underestimated? Fouling of an externally attached telemetry transmitter. Journal of Fish Biology 59:1092–1094.

Toombs, J. P., and K. M. Clarke. 2003. Basic operative techniques. Pages 199–207 *in* D. H. Slatter, editor. Textbook of small animal surgery, Volume 1, 3rd edition. Saunders, Philadelphia.

Trout, J. N. 2003. Principles of plastic and reconstructive surgery. Pages 274–292 *in* D. H. Slatter, editor. Textbook of small animal surgery, Volume 1, 3rd edition. Saunders, Philadelphia.

Tyus, H. M. 1988. Long-term retention of implanted transmitters in Coloradio squawfish and razorback sucker. North American Journal of Fisheries Management 8:264–267.

Wagner, G. N., and S. J. Cooke. 2005. Methodological approaches and opinions of researchers involved in the surgical implantation of telemetry transmitters in fish. Journal of Aquatic Animal Health 17:160–169.

Wagner, G. N., S. J. Cooke, R. S. Brown, and K. A. Deters. 2011. Surgical implantation techniques for electronic tags in fish. Reviews in Fish Biology and Fisheries 21:71–81.

Wagner, G. N., and E. D. Stevens. 2000. Effects of different surgical techniques: suture material and location of incision site on the behaviour of rainbow trout (*Oncorhynchus mykiss*). Marine and Freshwater Behavior and Physiology 33:103–114.

Wagner, G. N., E. D. Stevens, and P. Byrne. 2000. Effects of suture type and patterns on surgical wound healing in rainbow trout. Transactions of the American Fisheries Society 129:1196–1205.

Wagner, G. N., E. D. Stevens, and C. Harvey-Clark. 1999. Wound healing in rainbow trout following site preparation with a povidone-iodine antiseptic. Journal of Aquatic Animal Health 11:373–382.

Walker, M. K., E. A. Yanke, and W. H. Gingerich. 1994. Use of electronarcosis to immobilize juvenile and adult northern pike. The Progressive Fish-Culturist 56:237–243.

Walsh, M. G., K. A. Bjorgo, and J. J. Isely. 2000. Effects of implantation method and temperature on mortality and loss of simulated transmitters in hybrid striped bass. Transactions of the American Fisheries Society 129:539–544.

Wedemeyer, G. 1970. Stress of anesthesia with MS-222 and benzocaine in rainbow trout (*Salmo gairdneri*). Journal of the Fisheries Research Board of Canada 37:909–914.

Wedemeyer, G. A. 1972. Some physiological consequences of handling stress in the juvenile coho salmon (*Oncorynchus kisutch*) and steelhead (*Salmo gairdneri*). Journal of the Fisheries Research Board of Canada 29:1780–1783.

Wedemeyer, G. A., A. B. Barton, and D. J. McLeay. 1990. Stress and acclimation. Pages 451–489 *in* C. B. Schreck and P. B. Moyle, editors. Methods for fish biology. American Fisheries Society, Bethesda, Maryland.

Weimer, E. J., J. P Duehr, and M. L. Brown. 2006. Comparison of two external transmitter types on two sizes of bluegills and yellow perch. North American Journal of Fisheries Management 26:670–675.

Welch, D. W., S. D. Batten, and B. R. Ward. 2007. Growth, survival, and tag retention of steelhead trout (*O. mykiss*) surgically implanted with dummy acoustic tags. Hydrobiologia 582:289–299.

Winger, P. D., B. R. McCallum, S. J. Walsh, and J. A. Brown. 2002. Taking the bait: in situ voluntary ingestion of acoustic transmitters by Atlantic cod (*Gadus morphua*). Hydrobiologia 483:287–292.

Winger, P. D., and S. J. Walsh. 2001. Tagging of Atlantic cod (*Gadus morphua*) with intragastric transmitters: effects of forced insertion and voluntary ingestion on retention, food consumption and survival. Journal of Applied Ichthyology 17:234–239.

Winter, J. D. 1996. Advances in underwater biotelemetry. Pages 555–590 *in* B. R. Murphy and D. W. Willis, editors. Fisheries techniques, 2nd edition. American Fisheries Society, Bethesda, Maryland.

Woods, L. C., D. D. Theisen, and S. He. 2008. Efficacy of Aqui-S as an anesthetic for market-sized striped bass. North American Journal of Fisheries Management 70:219–222.

Woolsey, J., M. Holcomb, and R. L. Ingermann. 2004. Effect of temperature on clove oil anesthesia in steelhead fry. North American Journal of Fisheries Management 66:35–41.

Zale, A. V., C. Brooke, and W. C. Fraser. 2005. Effects of surgically implanted transmitter weights on growth and swimming stamina of small adult Westslope cutthroat trout. Transactions of the American Fisheries Society 134:653–660.

Zydlewski, G. B., W. Gale, J. Holmes, J. Johnson, T. Brigham, and W. Thorson. 2008. Use of electroshock for euthanatizing and immobilizing adult spring Chinook salmon in a hatchery. North American Journal of Aquaculture 70:415–424.

Radio Telemetry: Principles and Practice

Section 5.1

Radio Telemetry in Fresh Water: the Basics

VALERIAN B. KUECHLE AND PETER J. KUECHLE

INTRODUCTION

Radio telemetry has been used to monitor fish in freshwater since the late 1960s (Lonsdale and Baxter 1968). It followed the earlier use of sonic techniques (Trefethen 1956; Trefethen et al. 1957). Acoustic telemetry was initially used due to the widely held belief that radio waves would be totally attenuated by water. Much of the early research on radio wave propagation in water focused on communicating with submarines in the ocean, and the difficulties encountered undoubtedly contributed to this belief. Due to these disappointing results and the lack of published information on radio wave propagation in freshwater, it took some time for telemetry users to realize the differences between using radio telemetry in highly conductive salt water versus relatively low conductivity freshwater. Since that time, numerous papers on signal behavior in freshwater have been published by both fishery users and the electrical engineering community (Hafez et al. 1979; Lonsdale and Baxter 1968; Shen et al. 1976). Thus, we now have a much clearer understanding of the parameters which determine the success or failure of an aquatic radio telemetry application.

Radio telemetry is used primarily in freshwater, although it has also been applied to marine animals that break the surface of the water. We consider water with conductivity of less than 1000 microSiemens/centimeter (μS/cm) as freshwater. Most freshwater will be below 500 μS cm. In contrast, the conductivity of salt water can be as high as 50 S/cm. In order to develop a system that best addresses the research needs and to use it effectively, it is important to understand all elements that make up an aquatic telemetry system. In this Section we describe those elements and explain how they influence the outcome.

Radio telemetry systems are gear-intensive and dependent on the surrounding environment. The major components of the system include a:

1. *Transmitter* (also referred to as tag) that generates and emits a radio signal, and is attached to the study animal. The transmitter consists of a power source (battery), control circuits, components for generating the desired radio frequency and controlling signal modulation, assembly and encapsulation (potting) materials, and a transmitting antenna.

2. *Receiving antenna* used to capture (receive) the radio signal emanating from the transmitter and change it into a voltage that can be detected and recognized by a radio

receiver. Several different types of antennas are available, and each has different attributes depending on the frequency being used.

3. *Receiver* used to convert the radio signal captured by the receiving antenna into an audible or electronic form that can be used to identify the tagged animal.

4. *Data logger* used to operate the receiver when the operator is not present, and store the data collected.

5. *Cables and connectors* used to connect the receiving antenna to the radio receiver.

The operating environment will influence the suitability of the telemetry equipment used, and is a critical consideration during the gear selection and study design phase of the project. The radio frequency used will have a major impact on the performance and suitability of the equipment under field conditions. Signal propagation and signal path are also important consideration in relation to field operations and will affect signal loss from 1) signal spreading, 2) within the water column, 3) at the surface (air/water interface), and 4) signal polarization. These issues will be covered in more detail.

DEFINITIONS

Radio telemetry is a relatively technical discipline, and it is easy for the novice to be confused by the terminology. Several terms are basic to a discussion and understanding of this technique and its application, including:

Bandwidth: The difference between the lower and upper frequencies of a continuous set of frequencies. It has several refining definitions depending on its application. In radio waves it is the range of frequencies occupied by the wave; while in receivers or signal filters it is the range of frequencies passed (not screened out) by the device.

Conductivity: Measure of a material's ability to pass electric current. Conductivity for freshwater is normally given in centimeters, grams and seconds (cgs) units as µS/cm or in meters, kilograms, seconds (mks) units as mS/m. Care is needed to ensure consistency in unit usage. Conductivity is important in fish telemetry because radio waves are absorbed (attenuated) relative to the conductivity of the material.

Decibel (dB): The decibel is a logarithmic unit indicating the ratio of a physical quantity (most often power, voltage or sound intensity) to a reference level. In radio telemetry these are most often power ratios given by $10*\log_{10}P_1/P_2$ or voltage ratios given by $20*\log_{10}V_1/V_2$. If the ratio P_1/P_2 is greater than one, the level will be positive; if it is less than one it will be negative. Thus, if P_1 is the output power and P_2 is the input power, a positive dB level will indicate an increase in power from input to output and a negative dB number will indicate a decrease or loss. Decibel levels are also given with specified reference values. Two examples of power levels are dBm and dBW; where the power levels are referenced to one milliwatt and one watt respectively.

Dielectric constant: A measure of relative electrical density. In physical terms it is a measure of a material's ability to concentrate electrostatic lines of flux compared to a vacuum. The velocity of radio waves is inversely proportional to the square root of the dielectric constant. Thus, for freshwater with a dielectric constant of 80, the velocity of a radio wave is 9 times less than it would be in air. A radio wave will travel at 300×10^6 m/s in air, but only 33.3×10^6 m/s in water.

Dipole: In antennas, a dipole consists of two conducting elements separated by a feed point. It is the most basic real antenna.

Electromagnetic wave: Energy in the form of orthogonal electrical and magnetic waves (90° from each other and to the direction of travel, Figure 1). Radio waves are at the lower end of the electromagnetic spectrum which ranges from zero hertz to 10^{23} hertz. In a radio transmitter, electromagnetic waves (radio waves) are produced by alternating currents in the antenna.

Gain: Relative measure of signal level at one point when compared to another point. It may be an increase (positive gain) or decrease (negative gain or signal loss).

Impedance: Resistance to flow of alternating current in an electric circuit. Impedance is defined as the voltage divided by the current plus the phase angle between the voltage and current.

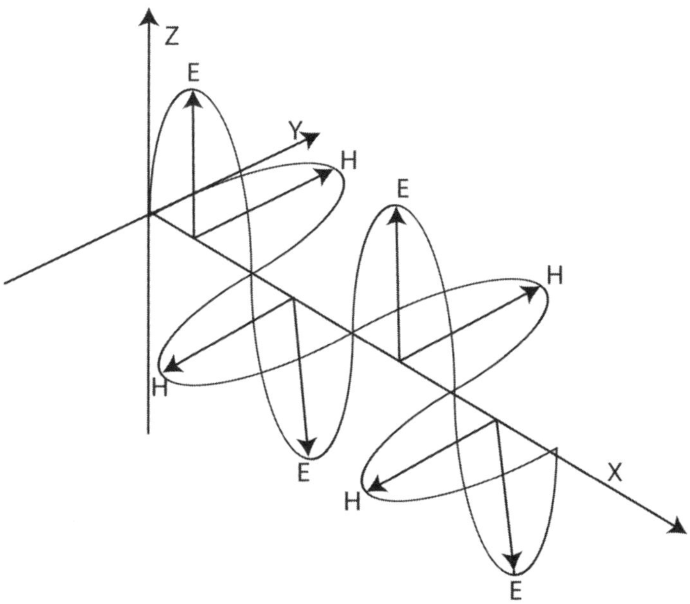

FIGURE 1. Components of an electromagnetic wave showing the relative directions of the electric field (E) and magnetic field (H). The two fields are orthogonal and travel direction is also orthogonal to the E and H field directions. Since the plane of the electric field is vertical in this case, the wave is considered vertically polarized. Although this figure shows the wave traveling in the X direction, it radiates in all directions from the source. The distance the wave travels in the X direction during one cycle is the wavelength.

Polarization: Radio waves are composed of an orthogonal electrical field and magnetic field (Figure 1). By convention the direction of polarization is the direction of the electrical field in relation to the earth. An antenna composed of wires with the wires parallel to ground is horizontally polarized, and an antenna with the wires vertical is vertically polarized.

Reflection: Redirection of radio waves that encounter a surface or boundary between two different materials. The angle of incidence and angle of reflection are equal.

Refraction: Is the bending of the radio wave when it enters a material where the wave speed is different. Wave speed is related to the dielectric constant; the higher the value the slower the speed.

Wavelength: Distance the electromagnetic wave emitted by the transmitter travels in one cycle (units are meters). It is given by the speed of the wave in the media divided by the frequency. For radio waves in air the wave speed is: 3×10^8 m/s (the speed of light).

TRANSMITTERS

The primary function of radio frequency transmitters in freshwater telemetry studies is to provide a mechanism for remotely locating the fish. The transmitter accomplishes this by converting energy stored in its internal battery to a radio wave for transmission to a distant receiving antenna. In addition to providing location information, the signal from the transmitter may also be modulated to telemeter data on the motion, temperature, depth (indicated by pressure), and identity of the transmitter so that an individual animal can be distinguished. Because the size and weight of the transmitter determines the burden it will have on the animal, the transmitter is often the most limiting element in the radio telemetry system. However, miniaturization of electronic components has enabled the development of transmitters weighing as little as 0.2 g. Although still influenced by other factors, such as operating life and reception range requirements, the size of the transmitter is less of a limitation now than it was in the past.

Almost all transmitters used today are purchased from commercial sources. Although most of these suppliers consider their designs to be trade secrets, they all basically operate in the same way. Understanding the basic functions can help researchers select the type of transmitter best suited for their study and understand the tradeoffs involved. For this discussion we have broken the transmitter into four components: power source, transmitter control circuitry, radio frequency generation, and antenna. Several examples of radio transmitters are shown in Figure 2. Tag attachment designs and methods are covered in a separate Section of this book (Section 4).

Power Source

Power has a major impact on the capabilities of the transmitter, ranging from tag life to transmitting range. The type of battery used can affect performance and transmitter design. Lithium batteries are the most commonly used power source in transmitters weighing more than three grams, while silver oxide cells are the prevalent power source in lower weight transmitters. Lithium batteries are the power source of choice for a number of reasons, includ-

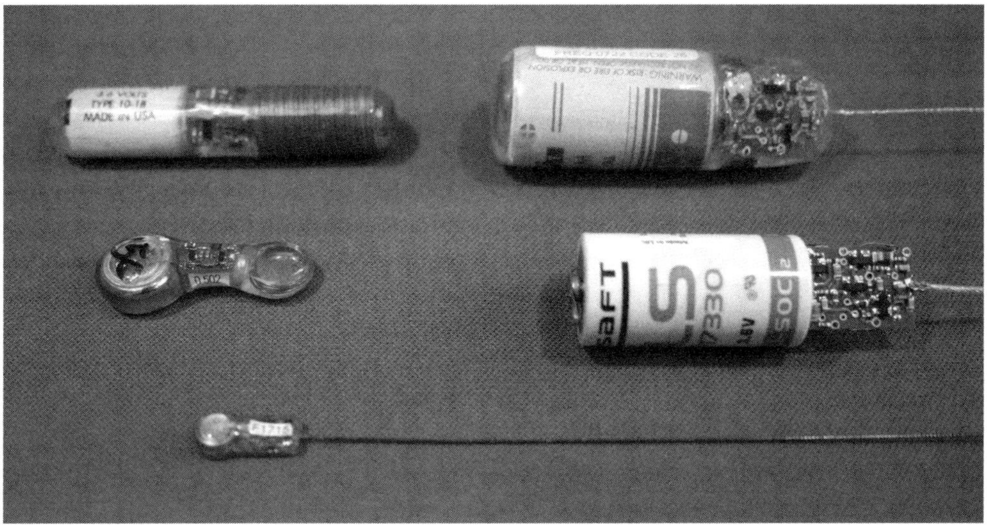

FIGURE 2. Sample of radio transmitters for aquatic telemetry. Upper left is a transmitter with a lithium battery and an internal coiled wire antenna. Mid left is a tuned loop transmitter with a single silver oxide cell. Lower left is a small microprocessor controlled transmitter with two silver oxide cells. On the right side is an encapsulated and un-encapsulated transmitter with a lithium battery designed for gastro insertion.

ing: higher voltage (range 2.8–3.6 V), high energy density, excellent performance over wide temperature ranges, less variable voltage discharge characteristics, and long shelf life. Several different chemistries and internal constructions are available in commercially produced lithium cells. These variations affect the output voltage, maximum current draw and battery capacity. Most batteries are cylindrical (although rectangular shapes are also available), which will influence the design of the transmitter as well as the attachment method.

The drain on the power source is at two levels, the first part is a quiescent drain which is continuous and powers the control circuitry. It is typically in the 2–20 microampere range. The second part of the battery drain is the power supplied to the radio frequency pulse. It may range as high as 20 milliamperes, however it is only on for a short time, typically 20 ms. The total drain is the sum of these two parts. Battery capacity is generally given as ampere hours or milliamphere hours. To determine transmitter operating time for a given battery the battery capacity is divided by the total drain yielding hours of operation.

Shelf life of a lithium battery is excellent, with an estimated 95% of their capacity remaining after five years. This long shelf life is the result of a passivation layer which builds up between the cathode and anode terminals during periods of no drain. The passivation layer reduces internal discharge in the cell. This may be a problem if a high startup current is required after the device has been off for a long period because the circuit may not start properly at reduced voltage. High startup currents are often required for transmitters equipped with a GPS receiver or satellite transmitters, and should not be an issue with most fish transmitters. Some transmitter manufacturers recommend starting transmitters at least once a month while they are on the shelf to prevent this problem, but we know of no data that indicates this is needed or effective, and it potentially reduces the shelf life of the transmitter.

While some localities regulate disposal of lithium batteries, there are currently no national restrictions on disposing of lithium cells in household or commercial waste streams. There are, however, shipping restrictions on cells and combinations of multiple cells where elemental lithium exceeds one gram. These cells are considered hazardous material due to the flammability of lithium. Since only C and D sized cells contain one gram or more of lithium this is probably not a significant concern with most fish transmitters.

Since low weight lithium cells are currently unavailable, transmitters weighing less than three grams are generally constricted to using silver oxide cells. The smallest commercially available silver oxide cell is 4.8 mm in diameter, 1.6 mm in height, weighs 0.13 g, and has a capacity rated at 8 milliampere (mA) hours. Silver oxide cells are available with either a potassium hydroxide or sodium hydroxide electrolyte. Of the two, potassium hydroxide batteries perform better with higher peak current loads and have better performance at low temperatures. The disadvantage is that potassium hydroxide batteries are less available and have a shorter shelf life due to higher self-discharge. Since silver oxide cells are small and have less reactive materials than lithium cells, their ability to supply high current is limited. Thus, the power output of transmitters with small silver oxide batteries for extended operating life or reception range may also be limited. Both potassium hydroxide or sodium hydroxide batteries have a voltage output of 1.5 V, which is a problem in applications using microprocessors with flash memory, as they require a voltage of 2.5 V or greater to function properly. Microprocessors are used in most transmitters to control the on and off times of the transmitter (see further discussion later in this Section). Thus, two cells in series must be used in transmitters using microprocessors with flash memory. When size requirements dictate a single silver oxide battery, transmitters controlled by Application Specific Integrated circuits (ASIC) or astable multivibrator control circuits are an option, as they can operate satisfactorily from a single cell. Read only memory (ROM) based microprocessors will also operate satisfactorily at these voltages, although the development costs associated with producing small quantities have made their use in fish telemetry too expensive. Peak current output capacity is further limited by decreases in temperature. Although the shelf life of silver oxide cells is not as good as that of lithium cells, data seems to indicate they should retain 90% of their capacity for up to three years (Linden and Reddy 2002).

Transmitter Control Circuits

All transmitters use a control circuit to turn the transmitter on and off in order to conserve power. The length of time the transmitter is on (referred to as pulse width) and how often the signal is transmitted (referred to as the pulse repetition rate or period), can be varied to achieve the desired goals of the study. If the objective is only to determine the location of the transmitter, a repetition rate of about one pulse per second works well. Pulse rates exceeding 60 pulses per minute (ppm) do not seem to increase the accuracy of bearing data; while pulse rates below 60 ppm make determining an accurate location more difficult. The accuracy of the bearing data are sometimes sacrificed in exchange for a slower pulse rate to conserve power and extend the life of the transmitter. Rates as low as 12 ppm are used in studies where transmitter life is the highest priority. The repetition rate can also be varied as a function of other parameters of interest, such as temperature, depth, or activity patterns. In this case both the location of the transmitter and the supplementary information are provided.

The pulse width may be optimized for either increased transmitter life or ease of signal detection. Shorter or narrower pulse widths save battery power, but are more difficult to detect, especially in noisy environments. A pulse width of 20 milliseconds (ms) is generally accepted as an optimum width for detection by the human ear (Kolz and Johnson 1981;

Plomp and Bouman 1959). Automated tracking systems also generally perform better with wider pulse widths which results in greater detection ranges. All pulsing signals distribute their power over a range of frequencies centered on their measured frequency. This frequency range is known as bandwidth and is dependent on the signal modulation. The bandwidth occupied by the transmitted signal is a function of the pulse width. Figure 3 compares the occupied bandwidth for 5 ms and 20 ms pulses. The null points (where dB level approaches –80 dB) occur on both sides of the center frequency (1000 Hz) at frequency multiples of 1/pulse width. Thus, the primary signal lobe for the 20 ms pulse (100 Hz) is four times narrower than the primary lobe of the 5 ms pulse (400 Hz). This occupied bandwidth is solely dependent on the pulse width and is not dependent on the operating frequency. Approximately 90% of the available power from the transmitter signal is contained within the primary lobe bandwidth. This becomes an important consideration in relation to the performance of the receiver, which must have bandwidth filters at least as wide as the bandwidth of the primary lobe to capture most of the transmitter's power. Receiver bandwidth is defined as the range of frequencies a receiver will pass, while rejecting all signals and noise outside this range. Thus, there is an inverse relationship between required receiver bandwidth and transmitter pulse width (wider pulses allow the receiver to use narrower bandwidths), which in turn is directly related to receiver bandwidth and noise rejection; narrower receiver bandwidths will reject more noise.

The type of control circuit used can impact transmitter performance under field conditions. The simplest control circuit used in fish transmitters is an astable multivibrator where the pulse rates and widths are determined by a combination of resistors and capacitors. This simple design results in a very low quiescent drain of about 1–2 microamperes (μA). Since resistors and capacitors are temperature dependent, the timing of the pulse rates and pulse width can vary with changes in temperature. This is an important consideration if the study animal is traveling through an area where temperatures change dramatically. Wide variation in temperature can alter the pulse width and pulse rate enough so that a receiver programmed to identify only certain pulse rates and widths will not be able to recognize the signal. Dramatic changes in temperature can also affect the life of the transmitter. Elevated temperatures can cause the pulse rate to increase, thereby using more power over a shorter period. Micro-

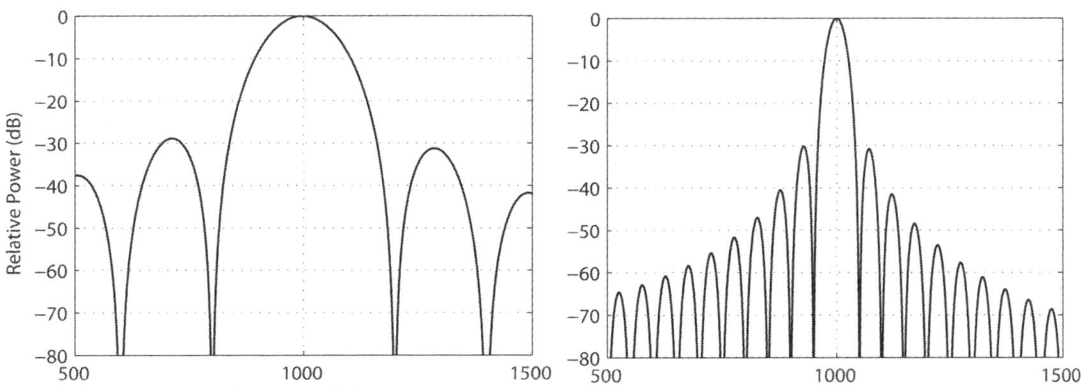

FIGURE 3. Power spectrum of a transmitter pulse, the left figure is for a 5 ms pulse and the right figure is a 20 ms pulse. The peak lobe centered on 1000 Hz contains 90% of the power and is 400 Hz wide for the 5 ms pulse and 100 Hz wide for the 20 ms pulse.

processor controlled transmitters (discussed below) can eliminate temperature variations. The effect of temperature on the transmitter should be discussed with the manufacturer to determine how the temperature in the intended study area might affect transmitter performance. Regulating transmitter voltage can also affect the consistency and dependability of the pulse width and rate. With existing methods, variations in pulse width and rate can be mitigated by careful transmitter design and can be kept within 5% of design values over a normal operating range of 0–30°C. While this level may not be accurate enough to transmit information such as temperature and depth, it has virtually no effect on location accuracy for most aquatic based studies.

Although the simple design of the astable multivibrator has the advantage of smallest size and lowest current drain, the more common control circuit in transmitters is a microprocessor used in combination with either a clock crystal or clock oscillator. The function of the clock crystal or oscillator is to provide very accurate timing. Accuracies of a few seconds per year are routine. While this design has a larger constant drain of 5–20 μA, the use of a microprocessor allows the transmitter to emit virtually any pulse rate or pattern, limited only by the imagination of the programmer. Perhaps the greatest advantage of programmable transmitters is the ability to optimize the length of time the transmitter is on (commonly referred to as the duty cycle) and conserve power. The frequency and duration of the duty cycle can be managed to maximize the life of the transmitter. For example, if location data are not needed continuously, the transmitter can be programmed to transmit for one hour and then turn off for two hours. Similarly, the transmitter can be programmed to transmit only on week days from 0800 to 1800 hours when researchers will be looking for the transmitters. The rest of the time it could be off, thereby saving a significant amount of battery power. These on and off times may be programmed to true time of day or may be designed to follow a fixed repeating cycle after startup. The former requires more circuitry and also requires some synchronization with real time and local time. When transmitters are programmed to repeat a fixed cycle after the initial startup, proper synchronization of all transmitters is obviously important. Although automated tracking stations typically require around-the-clock coverage, on-off cycling has been applied by having the transmitter on for several seconds, then off for several seconds. Of course, the transmitter must be within reception range of the tracking station for at least one complete on-off cycle in order for it to be detected.

In addition to flexibility in programming the duty cycle, the microprocessors may be used to accumulate input data and alter the output signal based on programmed decision criteria. One common application is to monitor fish activity. By incorporating a motion switch into the transmitter, fish movement can be indicated by inserting extra pulses in the normal pulse stream (i.e., repetitive signal pattern). An extra pulse can be added each time the switch closes, or it may be programmed to only add one pulse if it has been active during a predetermined time period; for example the previous second. The microprocessor can also be programmed to change the repetition rate of the signal to indicate that the transmitter has not moved for a specified period of time (e.g., hours, days, weeks). If the preprogrammed period is large, an "inactive" signal can be used to infer that the animal is no longer alive, although this assumption should be verified since some fish species have been known to hold for extended periods before resuming active movements. Typically the "inactive" condition is indicated by a doubling of the normal pulse repetition rate. The transmitter reverts to an "active" signal pattern if the fish starts moving again. Longer reset period can also be useful when the fish are not monitored on a regular basis by providing information that reflects an extended period of time. An option to consider when using transmitters to infer mortality, is to program the microprocessor to lock the transmitter in the "inactive" signal mode (or permanently pause the transmitter signal) after a predetermined period of no movement. If this is not done, something may eat the dead fish and the movement of the scavenger will reactivate the transmitter and introduce false positives into the data. One detriment to using microprocessors as the control circuit is the requirement of a minimum of 2.7 V to maintain the microprocessor memory. Thus, they cannot operate on a single silver oxide cell.

A third option for a transmitter's control circuit is an Application Specific Integrated Circuit (ASIC). Due to lower voltage requirements, this circuit has the advantage of being able to operate using a single silver oxide cell which means the overall size of the transmitter can be relatively small. The major deterrent to using ASICs is their high development cost. ASICs are custom integrated circuits and are produced in silicon. As such, they are costly to develop with costs typically in the $500,000 USD range. In addition, design and operational options such as different code sets and duty cycle are limited due to their operating program being fixed at the time of their manufacture. This makes ASICs practical only in long term, large scale studies where single silver oxide cell transmitters are required due to the size of the fish.

Radio Frequency Generation

The radio frequency (RF) generating portion of the transmitter consists of an oscillator, a frequency multiplier, if required, and one or more amplifiers. The oscillation frequency is normally controlled by a quartz crystal which is a very thin slice of quartz that oscillates at a very precise frequency. These crystals have very high frequency stability over time and over a wide temperature range. For frequencies below about 75 MHz, transmitters are operated with the quartz crystal at its primary frequency. Above this level, frequency multipliers are used to increase oscillations to the desired level. Reliable operation of a crystal oscillator above 75 MHz becomes more difficult, so these oscillators are operated at a sub-harmonic of the desired output frequency and then multiplied up to the desired frequency. For example, if the desired output is 150.150 MHz, the crystal oscillator will operate at 50.05 MHz and the frequency output is multiplied by three times to get the desired output. A crystal at 75.075 MHz could also be used with a multiplication factor of two to yield the same output frequency.

While the oscillator operates more reliably at its base frequency, the frequency multiplication necessary to generate a higher frequency has several undesirable consequences. The frequency multiplication process is inefficient, and the higher the multiplication the lower the conversion efficiency. Conversion efficiency is the ratio of desired output power to battery power. Secondly, the multiplication process generates harmonics at multiples of the base frequency. For example, a transmitter broadcasting a signal at 150.150 MHz (50.05 multiplied by three) will also produce harmonic signals at 100.100, 200.200 MHz, and higher multiples. These harmonics require power to generate, thereby reducing the life of the battery, and could cause potentially harmful interference to other users of the frequency spectrum. These undesired harmonics can be reduced by using filters in the transmitter; however, filters increase tag size, which can be a problem where tag size is an issue. The crystal is normally the second largest electronic component on the transmitter with the battery being the largest.

Temperature can alter the frequency of the transmitter in a predictable way. The variation in frequency as a function of temperature for the most common type of crystal used in fish telemetry (AT cut) is shown in Figure 4. This plot indicates transmitter frequency will increase

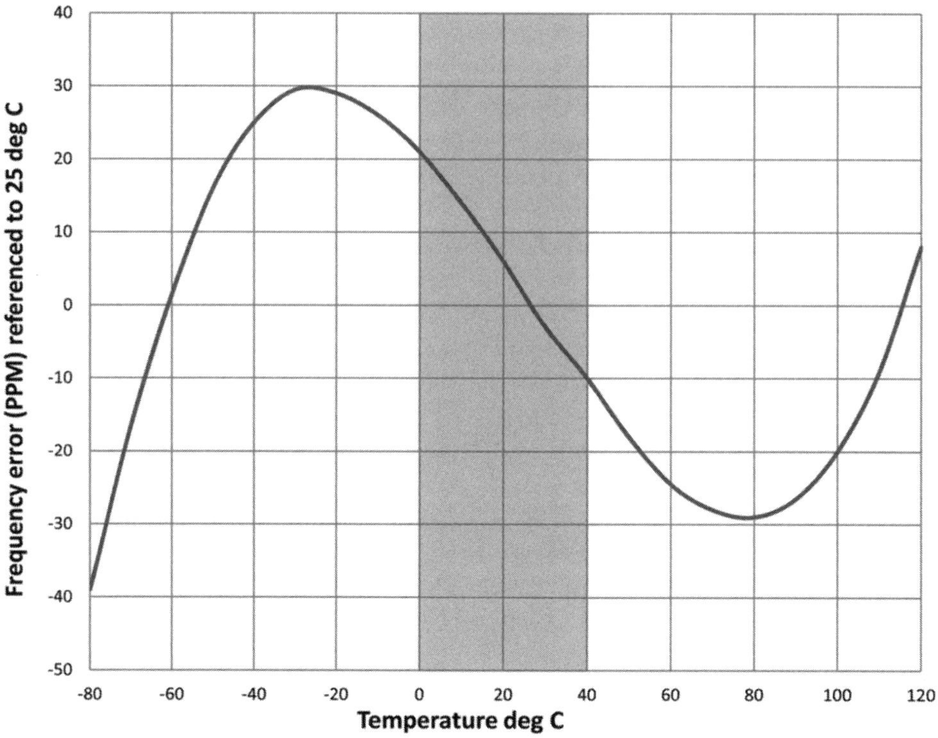

FIGURE 4. Frequency shift of an AT cut frequency control crystal as a function of temperature. Temperature range typically encountered in freshwater studies is indicated (shaded area). Shift is normalized to 25° C. The PPM on the chart is parts per million, thus a 150 MHz transmitter frequency will increase by 20 x 150 or 3 kHz when moved from 25°C to 0°C.

as temperature decreases and conversely will decrease as temperature increases when operating within the temperature range typically encountered in freshwater studies. Other types of crystal are available; however, the AT-cut shown here is the most stable over this range. At an extreme temperature variation range of 0–40°C, frequency would change approximately 30 ppm. At a frequency of 150 MHz, the frequency change would be 4.5 kHz (150 × 30); while at 50 MHz it would be 1.5 kHz (50 × 30). Normally, frequency variation with temperature is not an issue in fish radio telemetry because the fish temperature environment is tempered by the surrounding water and temperature changes are small.

Transmitter Assembly

The transmitter components (battery, control circuitry, frequency generator) must be assembled and packaged into a structure that can be used in fish. Most transmitters are manufactured by integrating the electronic components onto a printed circuit board (PCB). The PCB facilitates efficient integration of the parts in the least amount of space. Because efficient assembly of PCBs requires specialized "pick-and-place" machinery, most of the vendors supplying transmitters have this part of the process done by a third party that specializes in this type of manufacturing. These companies are better able to control issues such as solder temperature and circuit board cleaning.

Transmitter Encapsulation

Once all the parts are assembled, the final package must be protected from the water environment. This is usually done by applying an epoxy or urethane encapsulant. The specific gravity of these materials generally ranges from 1.1 to 1.2 and is slightly greater than the density of water, thus adding weight. The weight of a transmitter in water will range from 35% to 65% of its weight when in air. Larger transmitters usually have a greater weight reduction in water than smaller ones. While these materials typically provide adequate protection, they are still hydroscopic, meaning they transmit moisture at a finite rate. Thus, moisture will eventually reach the electronic components of the transmitter. Manufacturers of these materials generally give a moisture absorption rate; however it is difficult to translate these specifications to applications in fish telemetry. When moisture reaches the electronic parts on the PCB, it can combine with ionic compounds (which may have remained after the PCB was cleaned) and cause the transmitter to malfunction. Fortunately, in most fish telemetry studies the water is cool, thereby retarding the chemical reactions that could lead to malfunctions. Because the water is cool and moisture penetration takes time, most fisheries studies will have finished before problems occur. One method used to increase the reliability of transmitters is applying a conformal coating to the PCB before encapsulation. Parylene, a material commonly used on medical and military devices to provide moisture protection, is sometimes used on fish transmitters.

The research community's need for smaller transmitters continues to pressure manufacturers to produce even smaller products. To accomplish this, the circuit traces on the PCB are placed closer together, which increases the potential for ionic conductivity problems. Furthermore, the move to lead-free components and lead-free PCBs adds a new set of issues with the potential formation of "tin whiskers." These whiskers increase the potential of short circuiting on PCBs. These issues are not major quality issues in aquatic telemetry applications; however, users need to be aware of potential problems in long term studies.

Another issue which has attracted less attention is whether the encapsulation materials are bio-compatible when the transmitter is implanted in the fish. We know of no definitive study that addresses this issue. Most studies either do not mention the potential problem or simply report no effects. One way to mitigate any potential effect is to coat the encapsulated transmitter with Parylene or some other benign material. For example, the U.S. Food and Drug Administration Class IV materials have been tested for compatibility with animal tissue. While this may not relate directly to fishes, these materials are much less likely to contain contaminants than those materials in the more industrial class of encapsulants. These coatings are normally very thin and add little to the weight or bulk of tags.

Transmitter Antenna

There are several types of transmitter antennas, but all of them serve the same function and have certain characteristics in common. The antenna is the interface between the transmitter and the propagation media, in this case the water. It induces an electromagnetic field (radio wave) which radiates outward into the surrounding media. All antennas have two terminals and the electromagnetic field is established by developing a voltage difference between the terminals. Figure 5 illustrates this situation. The principle antenna properties are impedance (resistance to current flow), gain (relative signal level compared to a refer-

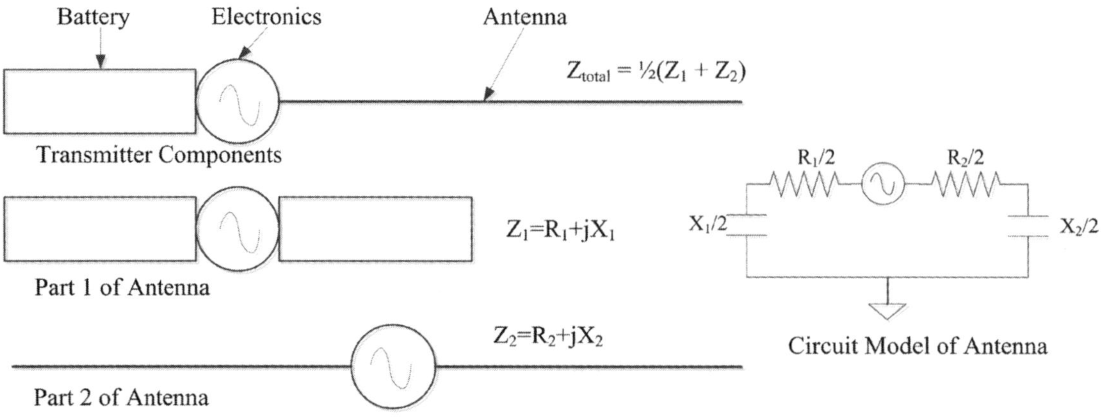

FIGURE 5. Schematic model of a transmitter and antenna. The upper drawing is a complete model which can be analyzed by separating it into dipole models as shown in Part 1 and Part 2. R_n is the magnitude of the resistive part of the impedance and X_n is the magnitude of the reactive (imaginary) part. The impedance of the two parts is then added back together to get the result Z_{total}. The circuit on the right side indicates the electrical components of the model.

ence), and directivity. Directivity of an antenna is defined as the ratio of power radiated in its maximum direction to the average power radiated into three dimensional space. Due to physical size limitations, the gain for antennas on aquatic transmitters is directly related to antenna impedance. The impedance of an antenna is complex (as in complex numbers), having a resistive part which is real and a reactive part which is imaginary. Ideally the reactive part should be zero (a resonant antenna) and the resistive part should be matched to the output impedance of the electronics. This would normally occur for antenna lengths at multiples of one half wavelength, resulting in maximum power transfer and increased detection ranges. Power radiates through the resistive part, while power is only temporarily stored in the reactive part. The behavior of the reactive part is explained by noting that the voltage and current are 90° out of phase (similar to the electromagnetic wave in Figure 1 with voltage indicated by E and current by H) which means the power is positive during one half of a cycle and negative during the other half. The result is no net power out from the reactive part. Thus the power flow to the resistive part, which accounts for the signal radiation is restricted by the reactive part. The ratio between the magnitudes of the resistive and reactive parts determines the power output and antenna efficiency.

The optimum minimum size of any antenna in any application is for the electrical length of the antenna to occupy a sphere of one-half wave length diameter in the media. The electrical length is the length of the antenna expressed as a fraction of wavelengths in the media. Since the wavelength in water is less than in air, the antenna electrical length will appear longer in water than in air. Efficiency, the measure of power delivered to the antenna compared to the power radiated, decreases as the size of the antenna decreases. Therefore, longer antennas would provide better reception range if no other factors were a consideration. It is often assumed that the antenna length is the only important part of the transmitter antenna specification and the role of the battery and electronics is typically and incorrectly ignored. This focus on antenna length is due in part to concern that a long antenna wire could physically affect the fish, but also because of a misunderstanding of the complete functionality of the antenna. In fact, the size of the battery and electronics has a large impact on transmitter performance.

This is because normally the battery and electronics are much shorter than the antenna, and thus make a greater contribution to the reactive part of the impedance.

Most fish transmitters are equipped with a dipole antenna (also known as a whip antenna), consisting of a straight length of wire. The wire can be bare, coated with a substance to reduce abrasions to the fish's skin, or insulated. As previously mentioned, the minimum optimal length of a wire antenna is one half wavelength; which is one meter at 150MHz. Transmitters with antennas this long would be unsuitable for most freshwater fish. The wavelength of bare wire antennas in a media other than air is reduced by the square root of the relative dielectric constant of the media. Conductivity of the surrounding media has a relatively minor influence on electrical length. In water with a relative dielectric constant of 80, the bare wire antenna acts as if it were approximately nine times longer than in air. Thus at 150 MHz, a half wave bare wire antenna will be approximately 0.11 m long in water, which is obviously more appropriate when tagging average-sized fish. If the antenna is insulated, as most are, the electrical length in the media is determined by the ratio of the antenna conductor diameter to antenna insulation diameter (see Appendix A). Depending on these ratios and the dielectric constant of the insulating material, the length of an insulated half wavelength antenna will be more than that of a bare wire in water, but less than it would be in air. An experimental determination of optimum antenna length that also considers matching of the electronics components can be used to achieve the most efficient power transfer.

Another transmitter antenna option is a loop antenna. This antenna is also referred to as a magnetic dipole or an internal antenna as the antenna is internal to the transmitter package or inside the fish. For maximum efficiency, loop antennas are tuned to operate as a resonant circuit (the sum of the reactive parts is zero at the output frequency). They have two advantageous characteristics; first, transmitters with internal antennas may be contained entirely inside the fish, which simplifies tagging procedures. Secondly, they are not affected electrically by submergence in water or when surrounded by fish tissue as are the more common whip antennas. However, because the volume occupied by the antenna is low, efficiency is also low. Thus, transmitter range will be substantially less (typically about 25% of the range) than that of a dipole.

Other forms of internal transmitter antennas are also used occasionally. One form is the helix; either as a distinctly formed helix or as a compressed coil. All of these antennas have very low efficiency and should only be used where very short range is acceptable and very small size is required. It is difficult to optimize these small antennas because a minor change in the other transmitter components has a significant impact on performance. The electrical characteristics of these antennas are also affected by the media surrounding them.

In addition to selecting transmitters that are appropriate in size and have adequate power and signal characteristics, choosing the best transmitter antenna is a difficult decision in specifying radio transmitters for aquatic use. The "bigger is better" rule is constrained by a "do no harm" to the animal requirement. In addition, it is difficult to make an accurate correlation between the antenna physical size and effect on detection range because battery and electronic circuit size is also a factor. Based on the preceding discussion we can make the following conclusions for transmitter antennas:

1. Except for the tuned loop antenna the entire package is part of the antenna.

2. The physical size of the entire package is important. Larger packages give significantly better reception range.

3. The physical size of the battery and electronics is a critical factor in determining the transmitter antenna optimum length and efficiency.

4. The optimum efficiency and power output of a transmitter antenna will be at its resonant length (the sum of the reactive parts is zero).

5. All transmitter antennas will have an optimum length, even if they are not resonant.

6. Antenna insulation thickness influences optimum antenna length. Thicker insulation requires a longer antenna.

7. Transmitters with tuned loop antennas or internal antennas have dramatically reduced range when compared to dipoles.

8. The material of the antenna wire has marginal influence on optimum antenna length or efficiency.

9. In many cases, requirements for compatibility with the study animal do not allow an antenna of optimum length.

10. It is difficult to make quantitative correlation between increased antenna size and increased range.

Transmitter Signal Modulation

The pulse sequences (signal modulation) of transmitters used in aquatic telemetry can be divided into either standard or coded ID types. In addition, temperature, motion, pressure (depth) and other parameters can be added to the pulse sequences.

Standard tags.—The term standard tag refers to tags having a single pulse that repeats at a constant rate. The pulse repetition rate, or pulses per minute (ppm), typically varies from 12 ppm to 60 ppm. Slower rates are chosen to save power while faster rates make directional location quicker and more accurate. Additional data can be transmitted from standard tags by modulating the pulse rate as a function of some defined parameter or by inserting extra pulses in the sequence. An example of the former is varying the pulse rate as a function of temperature. Activity is often indicated by adding pulses to the sequence based on the status of a tilt switch or accelerometer in the transmitter. Individual transmitters are identified by having each one on a separate frequency, or by varying the pulse rate substantially (e.g., 30 ppm versus 120 ppm). Standard tags have the advantage of simplicity and lowest power drain. In addition, they do not require special equipment to decipher their identity and they are easier to locate as only one tag will be detected on each receiver channel. In most cases, they are detected by listening to the receiver's audio.

Coded ID tags.—In studies where large numbers of tags are deployed, it may be advantageous to have more than one radio tag per frequency channel. This reduces the time needed to scan all the possible frequencies. For example, if 600 fish are tagged with transmitters broadcasting on individual frequencies and each frequency was scanned for 2 s, the resulting scan cycle would be 1200 s to listen for all 600 transmitters. In that amount of time it might be possible for fish to move past a detection site before the scan cycle is complete. One way to reduce the scan cycle is to have multiple transmitters broadcasting on the same frequency. If 200 transmitters could broadcast on a single frequency, the scan cycle could be reduced to 6 s (see Sections 5.2 and 5.3).

To accomplish having 200 or more transmitters broadcasting on the same frequency, a pattern of pulses unique to each transmitter is used. Although there is an effort underway to develop a standard way to code transmitters that would be in the public domain, nothing like this currently exists. The details of transmitter coding schemes are considered proprietary information by most of the transmitter manufacturers. As a result, each vendor has their own way of creating unique patterns of pulses to facilitate multiple transmitters on a single frequency. While these methods are unique, all operate similarly and involve a number of trade-offs. The goal in any code set is to make each individual code as different from any other code as possible. Greater differences mean less likelihood of one code being mistaken for another in the presence of noise or other misreads. This requirement may be expressed as code distance or cross-correlation index. Most codes use three or more pulses with a wide pulse at the beginning to indicate the start of a sequence followed by a series of narrower pulses. A typical pulse code sequence is shown in Figure 6. The time between pulses (t1, t2,

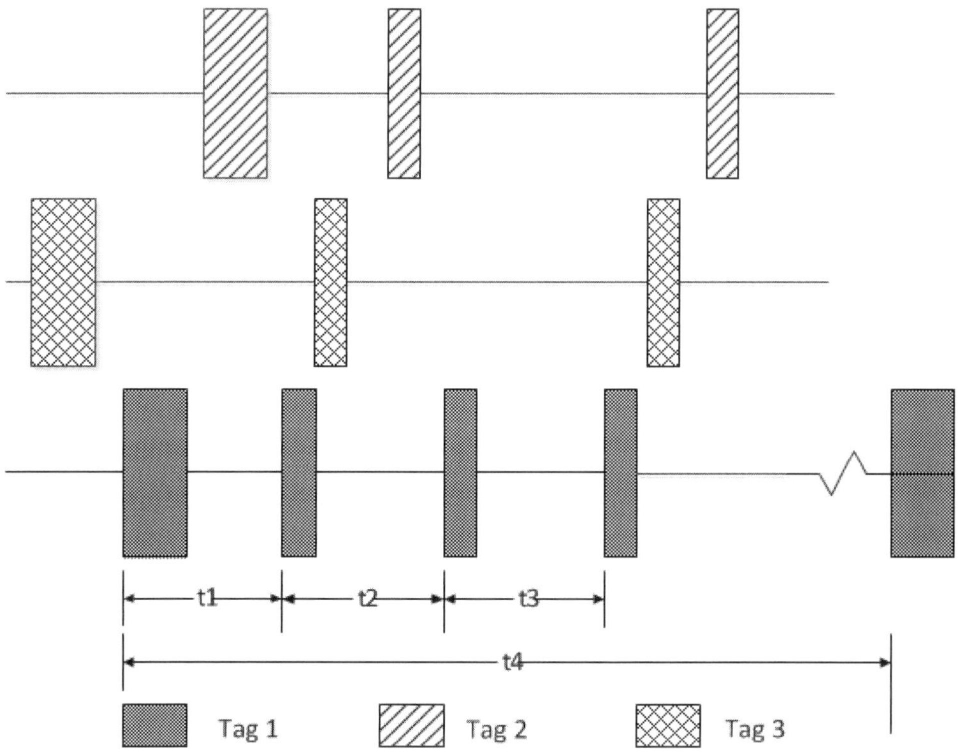

FIGURE 6. Pulse sequence of a coded ID transmitter. The code is identified by the wide initial pulse, the intervals between the pulses of the code burst (t1, t2, and t3), and the interval between code bursts (t4). Depending on the code set, identification may be based solely on the sequence of pulses t1, t2 and t3 or the value of t4 may be added to improve data reliability.

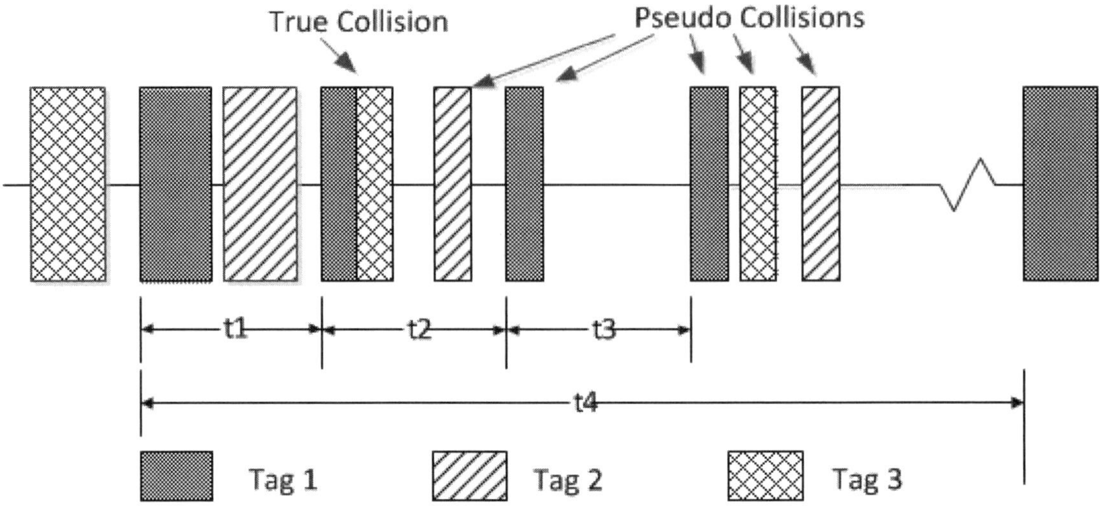

FIGURE 7. Illustration of a code collision preventing the ability to identify the code. The pulses from Tag #1 and Tag #3 overlap making the second pulse appear as a wide pulse. Tag #2 should not cause a problem as it does not overlap.

and t3) and the time between pulse groups (t4) is used to distinguish the individual code. The sequence of t1, t2, and t3 is called the ID code or code burst, and the time between the pulse sequences (t4) is the code period. The distance between pulses is varied to produce different codes. If more than one transmitter is in detection range of the receiver, collision between the signals may occur and degrade the ability to identify either transmitter. Figure 7 illustrates three transmitters being received simultaneously and two forms of signal collision. In the first example, two of the narrow pulses from transmitter #1 and transmitter #3 overlap, rendering the two codes undetectable. In this case the time reading for the start or end of the pulse will not fit a known pattern. In the second example, pulses from transmitter #3 are interspersed with those from transmitters #1 and #3, but do not overlap. Most detection algorithms can separate these interspersions by correlating known codes to the data stream, and methods can be used to reduce the possibility of collision occurring for the same transmitters in successive samples. Two common methods of mitigating these collisions are by varying the code periods (t4) slightly or by using a randomized code period where each successive period is different. The latter choice does, however, preclude using code period for validating code ID.

The operation of coded identification transmitters is similar to Radio Frequency Identification (RFID), (PIT tags are an example of RFID) and computer network ALOHA packet transmitters which can be modeled as a Poisson process (Finkenzeller 2003; Tanenbaum 1996). The probability of no code collisions during the code transmission is given by:

$$P = e^{-G}$$

$$G = \sum_{1}^{k} \frac{\tau}{T} k * n$$

where:
τ = *codetime*
T = *observation Period*
k = *number of repetitions of each tag during T*
n = *number of tags*

Using this estimate, the number of retries necessary to achieve a specified probability of detection is given by (Tanenbaum 1996):

$$P = 1 - (e^{-G}(1 - e^{-G})^{k-1}$$

The results for two probability levels are shown in Figure 8. This Figure is based on transmitters with code bursts of 3 pulses of that are 20, 5, and 5 ms in length. The number of tags that can be detected to a specified probability level per unit time can only be increased by reducing the on time of the transmitter; in this case 30ms (20 + 5 + 5 = 30). For example, if 15 tags were present and the code period (i.e., the interval between signal bursts) was one second, it would require 12 attempts, or 12 s, to reach a 99.9% probability level. If code period is increased to five seconds, five times as many tags can be deployed on the same frequency; however the scan time must be increased by five times as well. Thus, selection of code period depends on how long the fish are within reception range and the behavior of the fish (whether the tagged individuals are clustered together). Tag life is also a parameter to consider in tag

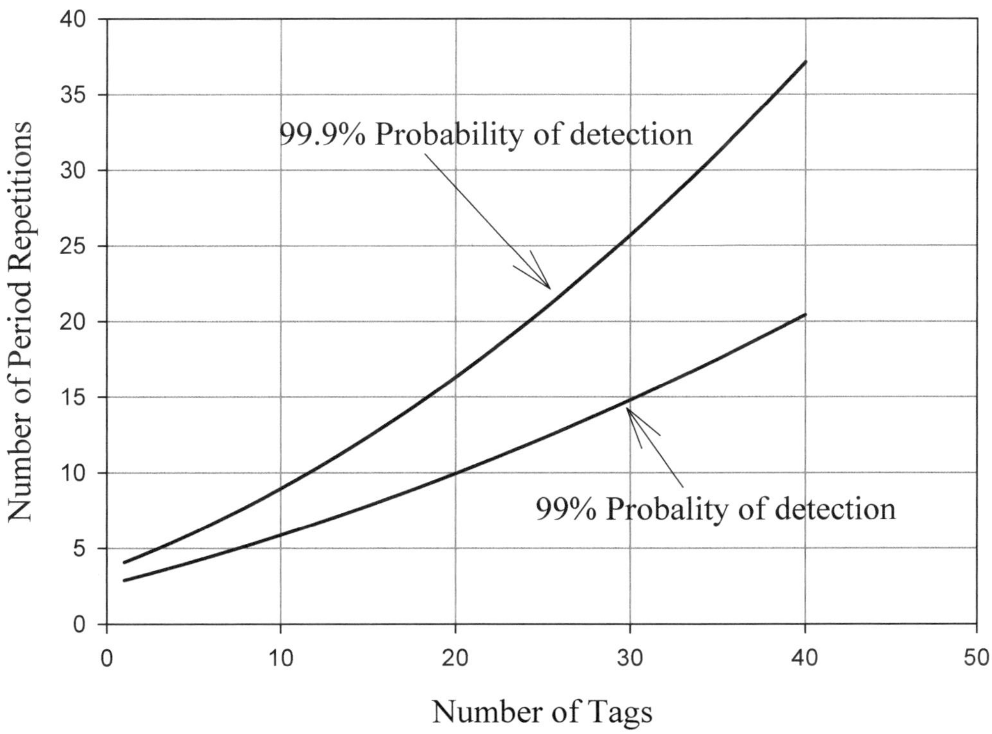

FIGURE 8. Plot of number of code repetitions necessary to reach a given detection probability level versus the number of tags present on the same frequency.

code period selection. Longer periods between code bursts can extend tag life, but the longer periods require longer scan times and can make it more difficult to locate tags in the field.

Digital signal processing (DSP) adds an additional dimension to the detection of coded ID tags. DSP allows the implementation of finer resolution in the frequency channel so that tags may be separated by time and frequency. During manufacture the transmitters are dispersed around their nominal frequency. This frequency distribution is a normal function of quartz crystal and manufacturing tolerances. The distribution can be used to separate tags in the frequency domain. They may also be intentionally distributed throughout the frequency channel. An illustration of DSP processing is shown in the Receiver portion of this Section (Figure 17).

Using coded ID transmitters allows greater numbers of transmitters to be scanned in less time, thus, reducing the possibility of missing transmitters. They also provide greater noise rejection as the additional pulses are used to verify the presence of a valid signal. A single pulse can be more easily interpreted as noise. The penalties in using coded ID transmitters is reduced life, because the transmitter is on longer to emit the additional pulses, and greater difficulty homing in on transmitters that are located near other transmitters in the same frequency channel (see Section 5.3). In the latter case, multiple transmitters will be heard simultaneously unless the receiver is designed to screen out selected ID codes.

Questions are frequently asked about the transmitter's output power under field conditions, since this can have a major impact on field logistics and data collection methods. Unfortunately there is no easy and accurate method to specify the power output of an aquatic radio transmitter. The transmitter must be submerged at a known depth and direction in a large body of water, preferably a freshwater lake, to accurately include the effect of the surrounding water on the antenna. The power can then be measured at some known distance away using an antenna and a power meter or spectrum analyzer. The power can then be normalized to a standard distance, typically one or ten meters. If a power meter or spectrum analyzer is not available an equally accurate method is to place a variable attenuator in the cable from the antenna to a standard tracking receiver. Attenuation is then added until the signal is too weak to detect. This results in a relative measurement, but is equally useful in calculating potential range using the equations given later in this Section. An advantage of measuring power with the transmitter in the water is that losses at the air water boundary are already included (see signal propagation in this Section).

In summary, proper transmitter selection is arguably the most critical element in the successful outcome of a freshwater radio telemetry project. The transmitter weight, configuration and attachment method must be chosen to minimize impact to the fish. At the same time electronic characteristics must be optimized to ensure adequate signal level at the receiver. In particular, antenna size and type has a major impact on transmitter power and thus, on reception range. Additionally, required transmitter operating life is also a significant factor in product selection. Knowledge of how these requirements interact is essential to proper transmitter selection. Transmitter vendors can be a good resource while examining the various trade-offs in the selection process.

SIGNAL PATH

The goal in any radio telemetry application is to be able to reliably detect the transmitter signals at a desirable distance. This is usually dictated by the research objectives and logistical considerations. To achieve this goal, the transmitter's signal must be strong enough to overcome the signal losses in the signal path and still be strong enough when reaching the receiving equipment to exceed the ambient noise within the area. In radio language, the signal to noise ratio (SNR) must be sufficiently positive to detect the transmitter. An understanding of the influence of the components in the system will help ensure a successful outcome. Signal transmission in aquatic telemetry is much more complex compared to terrestrial applications because the transmitter is immersed in water (a much denser media) and the signal must travel through this substance before reaching a receiving antenna. Signal level in aquatic telemetry is reduced through three main avenues; they are dispersion, absorption in water, and reflection and refraction at the air/water boundary (Figure 9). The unique aspects of signal propagation are discussed in the following paragraphs.

Frequency Selection

Almost all aquatic telemetry is done in the frequency range of 30 MHz to 225 MHz with most usage occurring in the 148 MHz to 172 MHz range. The actual frequency used depends on the country, user affiliation, and government regulations. Within the United States, federal government agencies are regulated by the Interdepartment Radio Advisory Committee

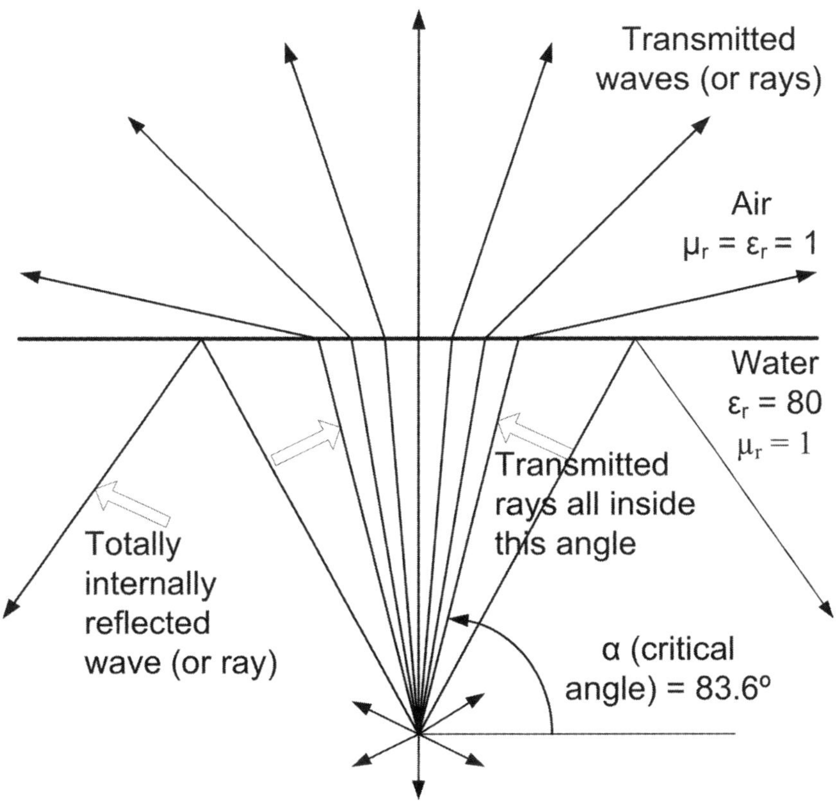

Figure 9. Radio waves from a submerged antenna at angle a less than 83.6° are totally reflected. For angle α greater than 83.6°, rays are transmitted through the surface into the air above and spread over almost 180°.

(IRAC), while nonfederal users are regulated by the Federal Communications Commission (FCC). In Canada the regulating agency is Industry Canada. There is no measurable performance difference within the 148 MHz to 172 MHz range. Tags in the 30 MHz to 50 MHz range are generally assumed to have less attenuation due to water conductivity. While this is true, as will be discussed later, there are also several mitigating factors.

Signal Loss from Signal Spreading

The signal travels out in all directions from the transmitter and occupies an ever increasing volume. Signal dispersion by spreading in aquatic telemetry occurs in two separate regions: between the transmitter and the water surface and from the water surface (directly above the transmitter) to the receiving antenna. For the distances typically found in fish telemetry, the transmitter may be considered a point signal source and thus, the radiation may be visualized as sphere with the transmitter in the center. At the receiving end, the signal received by the antenna (power) is dependent on the antenna's effective area of reception (aperture). Since the surface area of a sphere is proportional to the radius squared (R^2), we would expect the power level to also decease as a function of $1/R^2$. The complete equation for this spreading loss is given by the following Friis transmission equation (Friis 1946):

$$P_r = \frac{P_{tx} * G_{tx} * G_r * \lambda^2}{16 * \pi^2 * d^2}$$

where:

P_r = *received power*

P_{tx} = *transmitted power*

G_r = *receiver antenna gain*

G_{tx} = *transmitter antenna gain*

λ = *wavelength (in meters)*

d = *disance between transmitter and receiver antennas*

The above equation shows the interrelationship between parameters influencing reception range in an aquatic telemetry application. It is clear that this equation does not hold as d goes to zero. Also since P_{tx} cannot be measured at $d = 0$, it is normally measured at some close-in distance to establish a $P_{tx}(d_0)$ reference point. This equation can be reformatted to calculate path loss from spreading as a function of distance, ignoring the other parameters as they are constant in a given application. In logarithmic form this is:

$$Path\ Loss = 20 * \log_{10}\left[\frac{4 * \pi * d}{\lambda}\right] \ in\ dB$$

This equation is the basis for the rule of thumb where a 6 dB signal reduction reduces the detection range by half. Using the equation and setting $d = 2$ to indicate a doubling of distance, we have the log of 2 equal to 0.3, and 20 times 0.3 yields the 6 dB result. This can be used to give a quick estimate of potential range if the power is known. From a practical matter, the spreading loss in the water column can be ignored because the distance between the transmitter and the surface is short compared to other distances, and the loss due to water conductivity is much more of a factor. For example, increasing the transmitter depth from 5 m to 10 m would increase the spreading loss by 6 dB; whereas, the additional attenuation in the water column would typically be 6 dB per meter, or in this case an additional 30 dB. Thus, the losses from signal spreading in the water column are small compared to other losses in the system.

Signal Loss in Water Column

Signal attenuation by absorption in the water column is generally the most significant loss component in the signal path. This signal attenuation causes a dramatic reduction in reception range as a transmitter is required to transmit farther through water. As a comparison, moving a transmitter 10 m from the receiving antenna in a terrestrial environment will have no noticeable effect on the reception range; whereas, moving a transmitter in water from 1 m to 10 m from the receiving antenna may result in a total loss of reception. A number of authors have discussed signal reduction or range as a function of depth (transmission distance through water). This loss is a function of signal frequency, water conductivity and distance. This loss component is generally described using the Debye model which gives the attenuation factor (Biggs and Swarm 1968; Hafez et al. 1979; Saran and Held 1960; Shen et al. 1976) as:

$$Attenuation\,Rate = 8.686 * \alpha \;(dB/m)$$

$$\alpha = \frac{2\pi f}{c}\sqrt{\frac{\varepsilon'}{\cos\delta}} * \sin\left(\frac{\delta}{2}\right)$$

where:

$$\delta = \tan^{-1}\left(\frac{\varepsilon''}{\varepsilon'}\right)$$

$$\varepsilon'' = \frac{2\pi f(\varepsilon_s - \varepsilon_\infty)\tau}{1+(2\pi f\tau)^2}$$

$$\varepsilon' = \frac{\varepsilon_s + \varepsilon_s * (2\pi f * \tau)^2}{1+(2\pi f * \tau)^2} + \frac{\sigma}{2\pi f \varepsilon_\eta}$$

where:

$$\varepsilon_s = relative\,static\,permitivity\,of\;water\,(80.4)$$

$$\varepsilon_\infty = relative\,permitivity\,at\,end\,of\;dipole\,dispersion\,region\,(4.9)$$

$$\varepsilon_\eta = permitivity\,of\;free\,space\,(8.854\times10^{-12}\,F/m)$$

$$\sigma = ionic\,conductivity\,(measured)$$

$$\tau = relaxation\,time\,(10.1\times10^{-12}\,s)$$

It should be noted that the ionic conductivity typically measured or given in the literature is normalized to 25°C. Since conductivity has a positive temperature dependency of about 2% per °C, this factor should be taken into consideration in determining the attenuation. Using the previous equations, the relationship between attenuation and frequency for a number of different water conductivities is shown in Figure 10, with less attenuation in water with lower conductivity. Figure 10 also demonstrates how attenuation increases as frequency increases and it indicates that a lower operating frequency would be a better choice. Whether the difference is significant or not depends on the depth of the transmitter; a difference of 2–3 dB/m is minor if the depth is 1–2 m, but would be 20–30 dB if the depth was 10 m. This improvement is, however, mitigated by the requirement of much larger transmitting and receiving antennas to achieve comparable efficiencies at lower frequencies.

Signal Loss at the Surface (Air/Water Interface)

There is also a significant signal loss as the radio waves travel from the water to the air above. Some waves are either partially or entirely reflected at the surface and do not appear above the water (Figure 9). Even part of the wave penetrating the surface is reflected back due to the impedance difference between water and air. The signal that penetrates the surface is refracted (bent) when it travels from water to air. This behavior is dependent on the physical properties of water and air. Radio wave behavior at the air water interface is explained by Snell's Law(Kraus 1988). The physical parameters determining wave behavior in air and water are defined by their electrical permittivity and magnetic permeability. Permittivity defines the electrical properties of a material while permeability defines the magnetic properties of the material. The magnetic permeability in both media is the same and therefore has no effect on the outcome. The relative electrical permittivity (dielectric

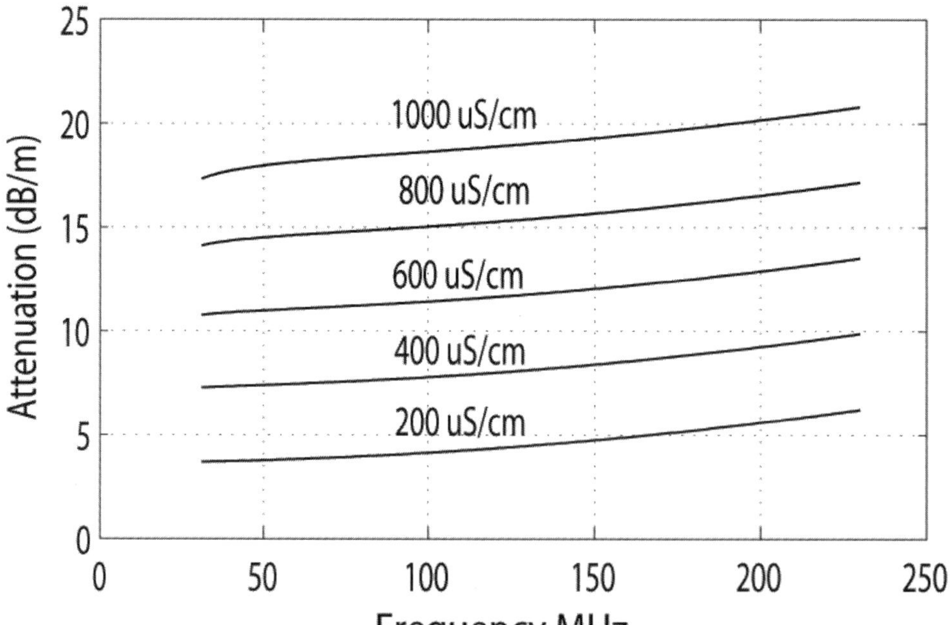

FIGURE 10. Attenuation rate in water as a function of frequency and water conductivity.

constant) for air is 1 and for water is approximately 80, with a slight dependency on temperature. The variation of permittivity with temperature can be ignored for the temperature ranges inhabited by fish. The radio transmitter can be considered a point source radiating in all directions. All waves leaving the transmitter at an angle greater than 6.4° from vertical are totally reflected (Figure 9, 12) (Kraus 1988). The angle of reflection as given by Snell's Law is:

$$\theta_{air} = asin\left(sin\theta_{water}\right)\sqrt{\frac{\varepsilon_{water}}{\varepsilon_{air}}}$$

$$where:$$

$$\varepsilon_{water} = relative\ dielelectric\ constant\ of\ water\ (80)$$

$$\varepsilon_{air} = relative\ dielectric\ constant\ of\ air\ (1)$$

In addition to the refraction at the surface, there is also a reflection of part of the wave based on the impedance mismatch between water and air. Thus, not all the energy is transmitted into the air above, some is reflected back. This behavior is polarization dependent and is given by Cheng (1993):

$$for\ perpendicular\ polarization: \frac{E_{reflected}}{E_{incident}} = \frac{\eta_a \cos\theta_w - \eta_w \cos\theta_a}{\eta_a \cos\theta_w + \eta_w \cos\theta_a}$$

$$for\ parallel\ polarization: \frac{E_{reflected}}{E_{incident}} = \frac{\eta_a \cos\theta_a - \eta_w \cos\theta_w}{\eta_a \cos\theta_a + \eta_w \cos\theta_w}$$

$$where: \eta = \sqrt{\frac{\mu}{\varepsilon}}$$

$$\eta = intrinsic\ impedance\ of\ the\ respective\ media$$

$$\mu = magnetic\ permeability\ of\ the\ respective\ media$$

$$e = dielectric\ constant\ of\ the\ respective\ media$$

As shown in Figure 9, only the signal intersecting the air water interface at less than 6.4° from vertical (total of only 12.8° of the signal radiating out from the transmitter) is transmitted into the air above the water. This signal, however, spreads out in all directions above the water. This spreading is given by (Velle 1979):

$$Spreading\ loss = \frac{\sqrt{\frac{\varepsilon_a}{\varepsilon_w}} * \cos\theta_a}{\sqrt{\frac{\varepsilon_w}{\varepsilon_a} - \sin^2\theta_a}}$$

$$where: \theta_a\ is\ the\ angle\ of\ transmission\ in\ air$$

The combined loss from signal reflection due to impedance mismatch and spreading at the surface (the signal occupies a small cone below the surface and a hemisphere above the surface) is shown in Figure 11. This combined loss will be at least 19 dB. Perpendicular polarization is the case where the transmitter antenna is perpendicular to the plane of the paper (you are looking at the end of the antenna), while in the parallel polarization case the transmitter antenna is parallel to the plane of the paper (you are looking at the length of the antenna). Only the reflected component of the signal loss is polarization dependent, while the spreading loss component is independent of polarization. A consequence of refraction at the surface is that the strongest signal will appear off the end of a transmitter antenna when the receiving antenna is also near the surface. Due to how most fish are tagged (with the transmitter parallel to the body and antenna horizontal), the reception range will be greater when the fish is moving toward or away from the receiving antenna. For aerial tracking where the elevation angle approaches 90 degrees, the differences related to orientation are minor (Figure 11).

Polarization also influences the optimum orientation of the receiving antenna. In almost all cases, the antenna of a transmitter used in fish telemetry is horizontal; thus, the electric field is horizontally polarized. However, better reception is normally obtained when the receiver antenna is oriented vertically. The phenomenon explaining this is illustrated in Figure 12. The signal starts as horizontally polarized in the water, but with refraction becomes predominately vertically polarized, at least near the water surface. This explains the conflict to the general rule stating that the polarization of transmitting and receiving antennas should be the same.

A related issue is the case where the water may have a layer of ice and snow. In most practical cases, the signal loss in ice or snow can be ignored, especially where it is less than one meter thick. The conductivity and dielectric constants of the ice and snow are not the same as those of water. The ionic conductivity of ice is much lower than the conductivity of the water it is made from and continues to decrease as temperature decreases. The relative dielectric constant of ice or snow at aquatic telemetry frequencies is much less than that of liquid water (King and Smith 1981). Since most of the signal loss is from ionic conductivity, the signal loss in ice will be much less than in water. The conductivity of ice will typically be ten times less than the water it is made from.

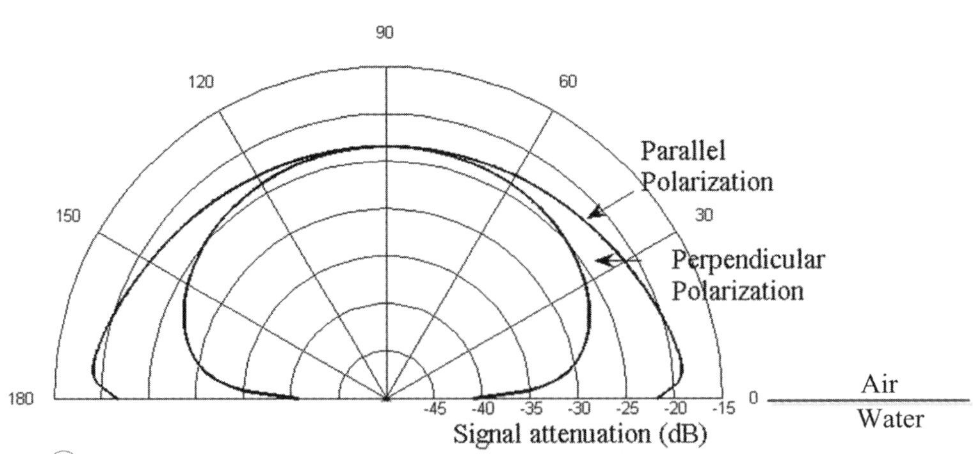

FIGURE 11. Signal power level in the air at the water surface as a function of elevation angle for parallel and perpendicular polarization. The pictorials below the plot show the corresponding direction of the transmitter.

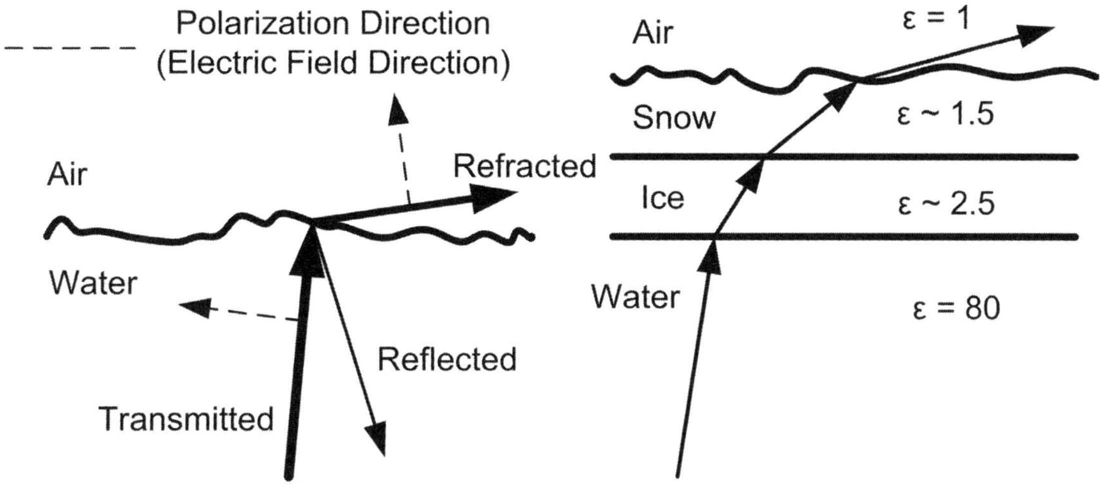

FIGURE 12. Left diagram shows how refraction changes polarization (direction of electric field) from horizontal in the water to near vertical in the air. Right diagram shows the direction of the signal as it passes from water through ice and snow to the air above. The amount of refraction is dependent on the dielectric constant ε.

A second factor to consider is the signal refraction as it passes from the water through the ice and snow above. Again the ice and snow can be ignored since the index of refraction depends only on the characteristics of the material where the signal originates; in this case water, and the characteristics of the final material, air (Karmel et al. 1998). The direction of the wave is shown in Figure 12; and is the same as that emerging directly from the water.

The last component of signal propagation is the signal loss as it travels from the point above the water to the receiver antenna, and again the Friis propagation equation for spreading loss applies. The simple Friis free space or line-of-site equation applies in aircraft tracking situations, and for surface tracking at distances of less than 0.25–0.5 km. At greater distances, reflection from the surface and the receiver antenna height become important and influence signal attenuation. The Friis equation then becomes:

$$\frac{P_r}{P_t} = \left(\frac{4\pi h_t h_r}{\lambda d}\right)^2 \cdot \left(\frac{\lambda}{4\pi d}\right)^2 \cdot g_t g_r$$

In log form and converting to standard dB format :

$$P_r = P_t G_t + 20\log(h_t) + 20\log(h_r) - 40\log(d) + G_r$$

where: P_r = *received power in dBm*
$P_t G_t$ = *effective transmitted power at the water surface*
G_r = *gain of the receiving antenna relative to an isotropic radiator*
h_t = *effective height of transmitting antenna (one in this case)*
h_r = *effective height of receiving antenna (meters)*
d = *distance in meters*

In the distance term, 40*log*d, the factor 40 could vary in the range from 20 as it does for free space (short ranges with no obstructions), to as much as 60 or more depending on whether there are other objects such as vegetation in the path (Goldman and Swenson 1999). These equations have more value in fish telemetry applications if we specify a frequency and estimate maximum distance. In that case we can measure a power $P_r(d_0)$ at distance (d_0). Then the received power will be proportional to:

$$P_r(d_1) = P_t(d_0)\left(\frac{d_1}{d_0}\right)^{-N}$$

This can be arranged in standard logarithmic form as :

$$P_r(d_1)(in\, dBm) = P_t(d_0)(in\, dBm) - N \bullet 10 \bullet \log\left(\frac{d_1}{d_0}\right)$$

This can be arranged in logarithmic form where N will vary from two for free space to six or more depending on terrain and surface conditions (Goldman and Swenson 1999). A plot of signal attenuation versus distance for N of two, four and six is shown in Figure 13. Choosing the correct N number will depend on conditions in the field and whether the user wants

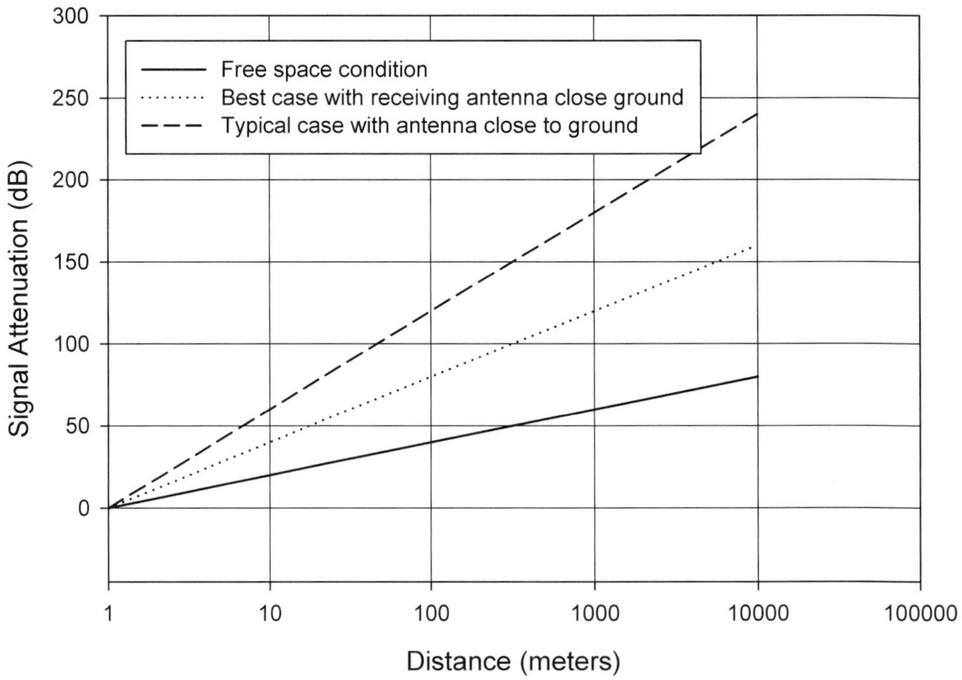

FIGURE 13. Radio wave attenuation versus distance (in meters) for free space condition (close range with no obstructions) (N = 20), best case with receiving antenna close to the surface (N = 40) and a typical case with antenna close to the surface (N = 60) (Goldman and Swenson 1999).Best case considers the earth as perfectly conducting and flat, while typical considers the earth as than perfectly conducting and the terrain as irregular.

to evaluate a worst case scenario. The additional gain achieved by increasing the receiving antenna height is also given by the Friis equation and is shown in Figure 14.

Figure 15 shows the possible signal paths for a radio signal to travel from the transmitter to a receiver antenna. These possible paths are:

1. Direct: From the transmitter to a receiver antenna in water (Path TE).

2. Reflected: From the transmitter to the water surface reflected back to a receiver antenna in the water (Path TBE).

3. Direct from the transmitter, through the surface of the water, to an antenna in air (Path TAC).

4. Via a lateral wave to an antenna near the surface or in the water (Path TDE).

The lateral wave is not explained by the previous discussion based on geometrical optics. Yet we know we can receive a signal near the water surface and also in the water beyond distances where the direct signal would be attenuated to a level where it would not be detectable. There are a number of papers in the technical literature describing the lateral wave (King 1985; Shavit and Rosen 1995; Wait 1998). An explicit solution to the integral equations describing lateral wave behavior is complex and beyond the scope of this Section. In summary, the wave can be described as an up, over, and down path. The wave is launched

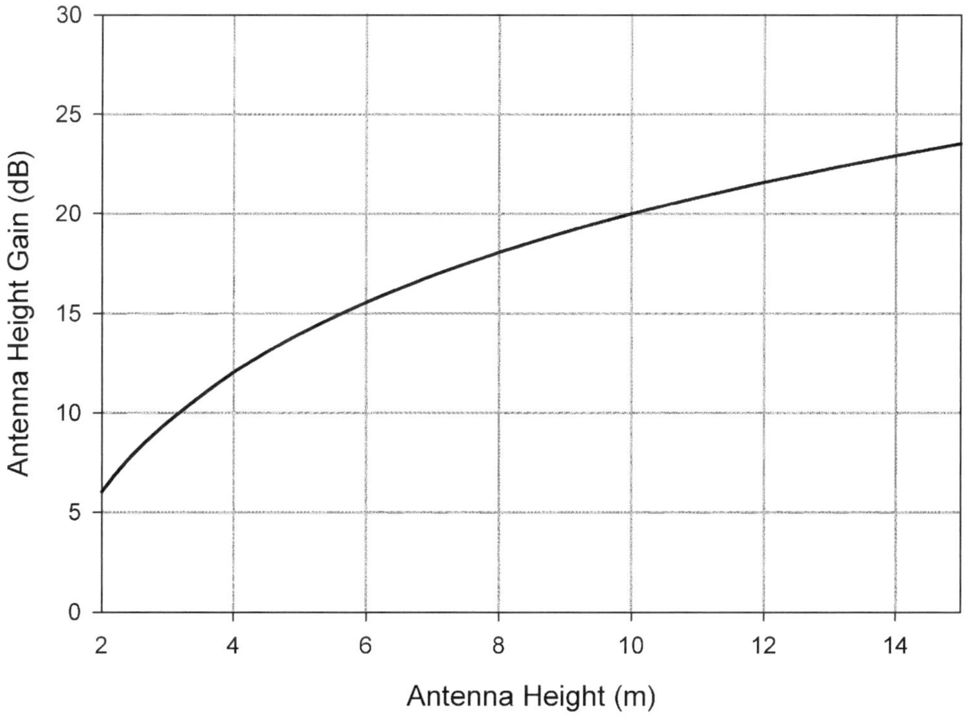

FIGURE 14. Additional system gain achieved by elevating the receiving antenna above the surface.

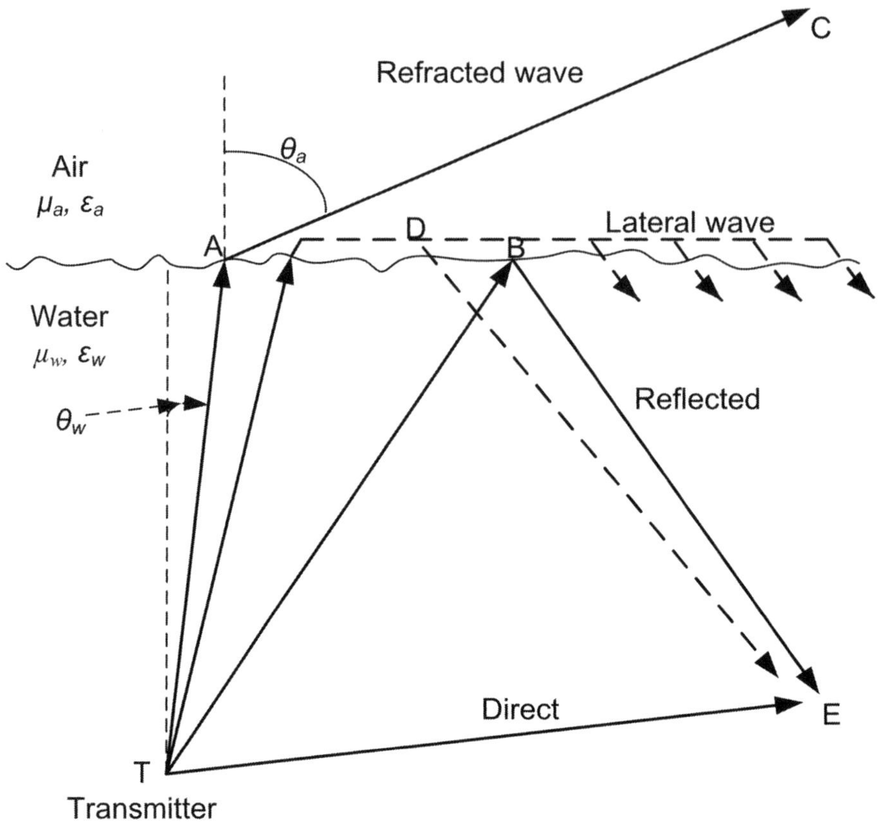

FIGURE 15. Possible signal paths from a submerged transmitter (T) to a receiving antenna positioned at C or E. Paths are described in the text.

only at the critical angle, travels along the air water interface and leaks back into the water at a critical angle. It is only present if both the transmitting and receiving antennas are in the water. In some cases, this wave may sometimes complicate the goals of some fish telemetry users. An example is where users may want to detect only the fish near an underwater antenna. The signal has three possible paths; the direct wave which is the wave of primary interest, the reflected wave which is generally weaker than the direct wave and the lateral wave which travels along the surface. If the fish is far from the underwater antenna, the direct wave TE will have faded away, but the lateral wave may still be detectable. Thus, the detection range cannot be accurately predicted because the signal attenuation will vary with the pathway and will also depend on the fish's depth.

Signal Polarization

The direction of signal polarization is indicated by the direction of the electrical field. For our purposes, the electric field can be assumed to be in the same direction as the direction of the antenna elements. Thus, if the antenna elements are vertical, it is vertically polarized. As shown in Figure 12 the direction of polarization does change as the signal is refracted. Other forms of polarization such as circular and elliptical exist, however, they are not significant in freshwater telemetry. Most often a signal is some combination of vertical and horizontal polarization due to signal refraction and reflection. In most cases where the receiving antenna is near the earth's surface, orienting the receiver antenna with the antenna elements vertical to match the signal's vertical polarization optimizes signal reception. As Figure 11 indicates, antenna orientation for aerial tracking is less critical as the polarization shifts from vertical to horizontal and as the elevation angle from the transmitter to receiver increases.

RECEIVER

The receiver performs the functions of translating the signal produced by the transmitter from the megahertz frequency range down to the audible frequency range. In addition, it separates signals on one frequency from signals on adjacent frequencies. It also reduces the effect of noise by allowing the desired signals to pass while rejecting (screening out) adjacent noise.

Frequency translation is accomplished by multiplying the incoming frequency with an oscillator offset from the desired frequency. The result is given by the trigonometric identity:

$$\sin(\alpha) \bullet \cos(\beta) = \tfrac{1}{2}[\sin(\alpha + \beta) + \sin(\alpha - \beta)]$$

Yielding the sum and difference of the two frequencies α and β, the receiver separates the desired signal from adjacent signals. It also reduces the noise by means of a band pass filter. The band pass filter passes a select range of frequencies while (ideally) rejecting all those outside of that range (Figure 16). Signal detection is only possible if the signal power exceeds the noise power. The minimum receiver bandwidth is determined in part by the pulse width of the signal (see explanation in transmitter section). Typically, these receiver bandwidths are 4–5 kHz even though a 20 ms wide pulse has 90% of its energy within 100 Hz. The receiver bandwidth is made wider than the minimum required in order to accommodate transmitter frequency drift and possible receiver frequency drift. The receiver bandwidth is less critical

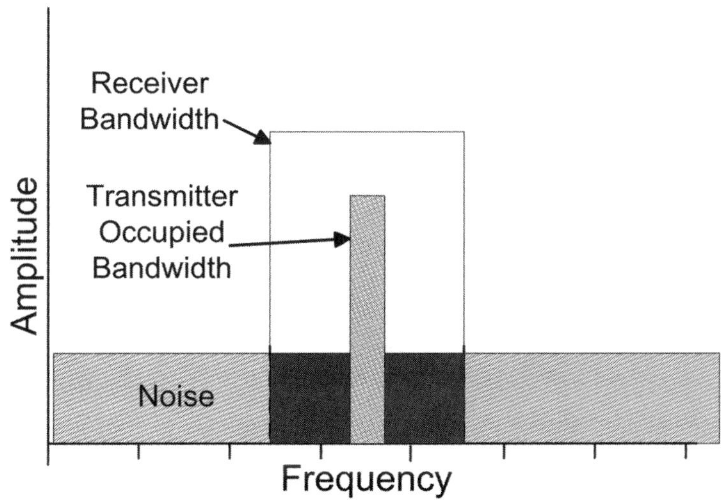

FIGURE 16. Comparison of receiver filter and transmitter bandwidths. Total power of a signal or of noise is given by the area under their respective envelopes. The filters defining receiver bandwidth limit the noise power while passing all of the desired signal power which increases the signal to noise ratio.

when listening to the signal because the ear also acts as a band pass filter with a band pass of about 50 Hz. This significantly reduces the influence of noise. Receiver bandwidth is more critical with automated tracking (Section 5.3) where detecting the transmitters depends exclusively on the receiver settings.

Receiver sensitivity is a measure of how well a receiver can detect a signal in the presence of noise. The two most common measures of receiver sensitivity for radio telemetry are minimum discernible signal (MDS) and signal to noise ratio (SNR). For listening by ear, MDS is the more meaningful measure since it is less dependent on receiver bandwidth. The MDS level is the lowest signal level that can be detected by ear. The minimum level is limited by the noise figure, a measure of the internal noise of the receiver. Fortunately, most receivers approach the theoretical noise figure limit, so receivers from most vendors will be equally sensitive. Signal to noise ratio, or more accurately, the signal to signal plus noise ratio is an electronic measure and is most often used in combination with a signal level and/or an error rate. For example, a signal of –130 dBm may be specified as giving a 6 dB SNR meaning the signal level is 6 dB above the noise level. Another example would be where a 10 dB SNR is required to achieve an error rate of less than 1 wrong or missed detection in 10,000 samples.

In addition to the internal noise of the receiver, external noise sources also influence detection range. This noise may be from the sun and stars which produce a background noise level. More prevalent, however, are man-made noise sources such as computers, automobile and boat ignition systems, electrical power lines and electric switch gear. Electrical noise is particularly problematic near electric power generating stations due to the presence of electrical arcing from high voltage lines, and switch gear. Electrical storms, even those thousands of miles away, also add noise, particularly at lower frequencies. Large cities are also noise sources and may reduce range; especially in aerial tracking.

Most modern receivers used in data logging use Digital Signal Processing (DSP) to improve data quality. The DSP process divides the receiver channel into a series of narrow filters or bins. The width and number of these bins is determined by the processing power available and is also limited by the occupied bandwidth of the signals shown in Figure 3. There is a trade-off between the number of bins, the width of each bin, and processing time. For example, if processing power is available for 256 bins and each bin is 100 Hz wide we could cover 25 kHz. If we wanted to cover 1 MHz range, each of the 256 bins would have to be 3900 Hz wide. The trade-off is the ability to cover a greater number of tags, but this allows more noise with a decrease in sensitivity. In addition, with narrow bins the signal may not fall exactly in the center of the bin meaning its available power will be shared by the adjacent bins (Figure 17). The software in the receiver must be able to accommodate the spread over several bins. A greater challenge exists when a strong signal may spread over to a bin also occupied by a weak signal. In those cases it becomes difficult to separate the two signals.

DATA LOGGERS

Data loggers are used in aquatic telemetry to monitor the status of transmitters without the need for operator presence. They are most often used to monitor the presence or absence of a signal or estimate signal direction when multiple antennas are used. They can, however, monitor any data the transmitter can send. Data loggers consist of an internal or external controller that is linked to internal memory to store the data. Data loggers may be integrated into the receiver

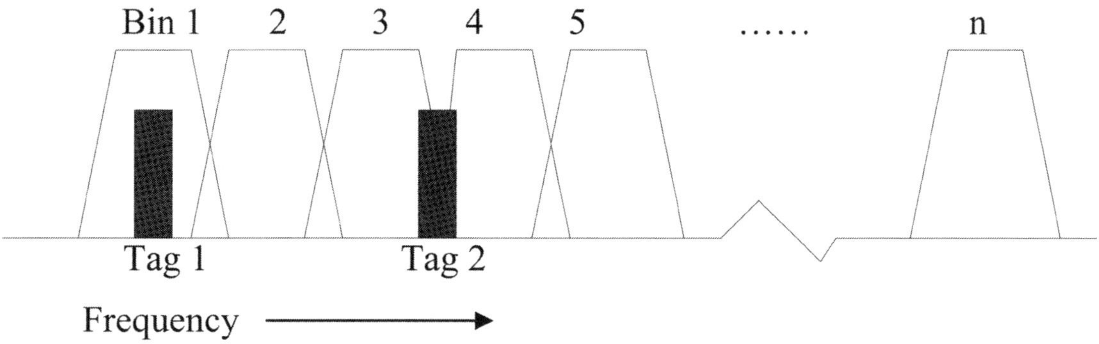

Figure 17. Illustration of a DSP dividing the receiver bandwidth into sub-bandpass filters. The width of each channel is determined in software and is a compromise between computation time and number of channels.

or may be separate units. The controller is used to set the sample rate, control the frequencies monitored, verify the data, and store the data. Although they can be designed to record signal strength, signal strength is not a good indicator of transmitter range since many factors such as transmitter depth will affect range and thus affect signal strength. Relative signal strength can be used successfully to indicate which of several antennas have the highest signal level. This technique is commonly used on rivers and at hydroelectric dams to indicate passage (Section 5.3, Section 6.5). The signal level required for accurate detection with a data logger is significantly higher than is required for detection by listening with the human ear. Typically signal levels of –130 dBm or greater are required for data loggers; whereas a signal level of –145 dBm is adequate for listening with the human ear. Data loggers operate by detecting signal levels above a set threshold. If the threshold is set too low, noise will introduce invalid data. If it is set too high, valid data may be missed. Coded ID tags greatly enhance the quality of data that the data logger collects by using additional signal parameters (e.g., signal pulse patterns) to validate data.

Data loggers usually require power for periods longer than the 12–24 h that can be provided by the receiver's internal battery. For longer operating times a supplementary power source must be provided (Section 5.3). Near hydroelectric dams, supplementary power is often available and provides an economical and reliable solution. In remote locations, larger capacity storage batteries can be used to increase operation for extended periods, but they are typically heavy and difficult to move in and out of remote sites. Solar panels can be used to charge the batteries and are one option for extending the operating time nearly indefinitely.

Several different methods are in use for data retrieval from data loggers. Data can be directly downloaded from the internal memory or compact flash card and this method is the most secure and lowest cost. The disadvantage is waiting to collect the data until you visit the site where the receiver is located. Quicker data retrieval is available using a cell phone or satellite modem . Alternatively, the receiver can be programmed to transmit the data to one of several satellites such as GOES, Iridium, or Globalstar.

RECEIVING ANTENNAS

The function of the receiving antenna is to capture the propagating radio wave and change it into a voltage for detection by the receiver. Three types of receiver antennas are in general use for fish telemetry. These are the loop antenna, dipole antenna and the Yagi antenna. In addition, inefficient antennas as sometimes needed when the transmitters are close so the signal does not overload the receiver. Essentially any metallic object can be used as an inefficient antenna for locating transmitters at very short ranges. There is also the isotropic radiator, a hypothetical antenna with equal sensitivity in all directions. Although an isotropic antenna cannot be built, it is used to compare the performance of real antennas to an antenna radiating equally in all directions. In addition to the mechanical characteristics of the receiving antenna, three electronic characteristics must be considered. These are frequency bandwidth, directivity, and gain. Frequency bandwidth is the frequency range over which the antenna will perform within specification. Directivity is the ratio of radiation or reception sensitivity in a given direction from the antenna to the radiation intensity or reception sensitivity averaged over all directions. Directivity is an antenna term; directionality may be a better description for radio telemetry. Higher directivity antennas increase precision when used for direction finding by increasing the signal from the forward direction while reducing it from peripheral directions. Gain is the ratio of radiation intensity or reception sensitivity in a given direction compared to a reference, usually an isotropic radiator or dipole. The gain of a dipole is 2.15 dB; thus when the gain of an antenna is given compared to an isotropic radiator, it will be 2.15 dB greater than when compared to a dipole antenna. An antenna can have high directivity, but low gain. An example is the loop antenna which has good directivity, but low gain. Since antennas are passive devices they can only achieve gain in one direction by reducing gain in other directions.

Dipole

The dipole is an elementary antenna consisting of two conductors with a feed or receiver port between the two conductors. It has a donut shaped reception pattern with a maximum perpendicular to the axis of the antenna. It is also omnidirectional around its vertical axis, meaning its sensitivity is equal over an azimuth of 360° (i.e., equal horizontal reception). It does not, however, have equal sensitivity over its elevation angle, with reduced reception with increasing height and no reception (null) immediately above the donut (Figure 18). The gain of a dipole with a length of one half-wave is 2.15 dB greater than that of an isotropic radiator. It achieves this gain by reducing the gain in the axial direction. Dipoles with lengths greater than one half-wavelength are also available. Their gain exceeds the 2.15 dB of the half wavelength dipole; however, the added gain is at the expense of additional nulls in the vertical plane antenna pattern.

Another modification of the half wave dipole is the magnetic mount quarter wave dipole. In this case, the antenna is only one quarter wavelength long, but depends on creating a mirror image of the antenna when it is mounted on a large metal surface such as an automobile roof or the bottom of a boat. The metal surface acts as a mirror and thereby creating an image of the antenna below the metal surface. The antenna does not need to make electrical contact with the metal surface; the metal surface merely needs to act as a conducting reflective surface. To operate correctly, this metal surface should have a radius of at least a quarter wavelength (0.5 m for 150 MHz). Dipoles are used in applications where directionality is not required, (Section 5.3) for example on a boat in a river (Figure 19).

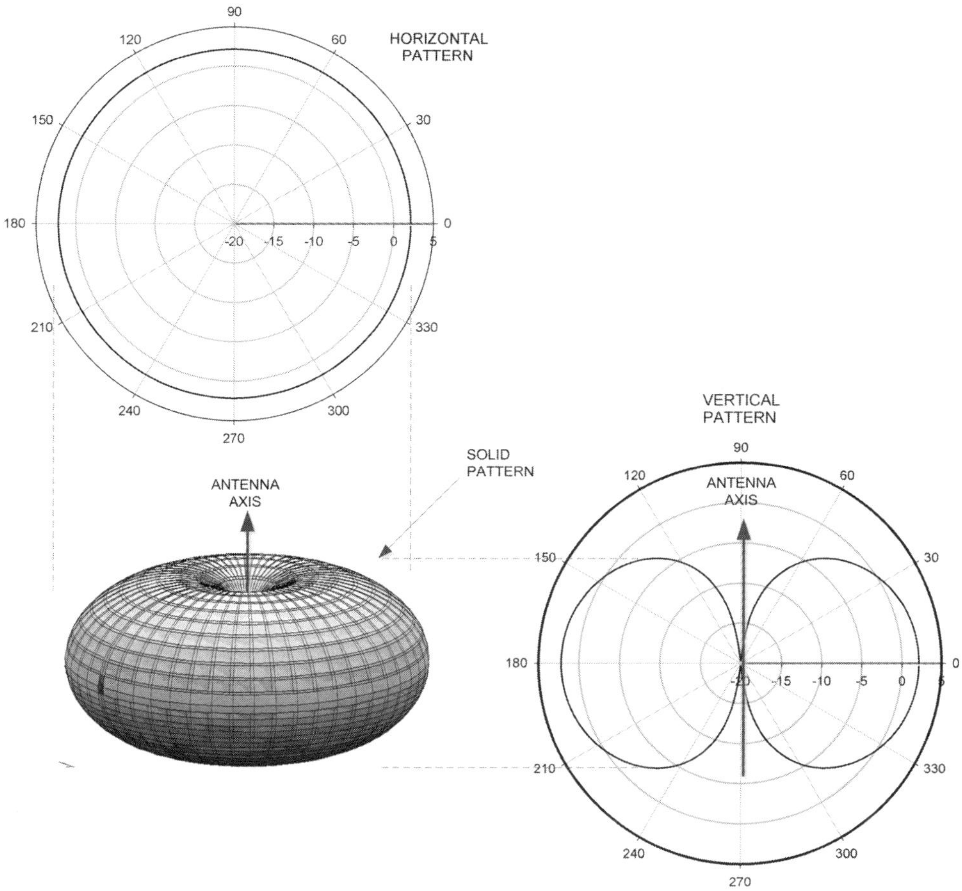

Figure 18. Directivity (gain) patterns of a vertically and a horizontally polarized half wave dipole antenna. Gain is 2.15 dB when compared to an isotropic radiator.

Loop Antenna

Directional loops are another receiver antenna option, particularly at lower frequencies where Yagi antennas may become too large for practical use. The circumference of a loop antenna is generally limited to one half wave length or less to ensure reliable direction finding. These loops may be either broad band or tuned, with tuned loops being the most prevalent. They are always used to receive vertically polarized waves.

Loop antennas have a number of limitations which impact their use in radio telemetry. First, they have low radiation resistance which makes it difficult to effectively match them to a receiver input. This mismatch results in increased noise going into the receiver, which can adversely affect signal detection and degrade tracking accuracy. This effect can be partially offset by using a tuned loop. However, this results in an antenna with good gain over only a narrow frequency range. Therefore, a tuned loop must be adjusted during field use if it is to cover more than about a 300 kHz frequency range. Directionally does not change even through the loop may not be optimally tuned. Loop antennas also are affected by their prox-

FIGURE 19. A vertically polarized dipole mounted on a boat for tracking on a river.

imity to ground. This will be noticed as a greater gain to one side of the loop biasing the null toward the other side. This problem is mitigated if a balanced loop design is used.

Since loops have a bimodal reception pattern the direction can only be determined as either plus or minus 180° from the null angles (Figure 20). Thus, some other method must be used to resolve the front or back ambiguity (Section 5.3). The sensitivity of a one-half wave circumference loop antenna is about 6 dB less than a half wave dipole.

Yagi Antenna

Yagi antennas are the most commonly used antenna in fish telemetry. They are relatively low cost, robust and have good directivity and gain. The optimum size of all antennas, including Yagi antennas, is frequency dependent. Optimal elements are approximately one half wave length long, and the spacing between elements is approximately one quarter wavelength. Thus, a four element Yagi at 150 MHz would be approximately 0.75 m long and have elements of 1.0 m length. A four element Yagi at 30 MHz would be approximately 3.75 m long and have elements 5 m long, illustrating the advantage of the higher frequency range. Although both antennas would have the same gain and directivity, the substantially larger size of the 30 MHz Yagi would be much more cumbersome to use under field conditions. Element spacing need not be uniform as a number of parameters may be varied to achieve an optimum outcome.

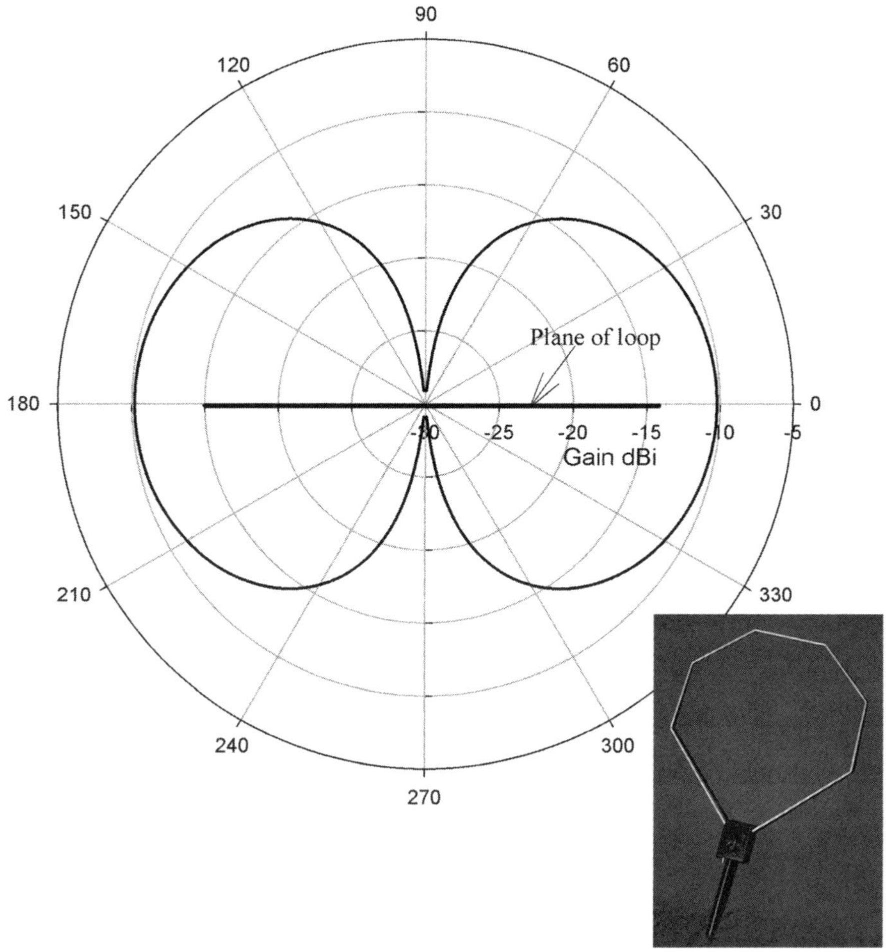

FIGURE 20. Directivity pattern of a loop antenna. Gain of a loop antenna is about 10 dB less than the gain of an isotropic antenna.

At frequencies above 120 MHz, Yagis with two to four elements may be hand held. Yagis with greater than four elements are generally considered too unwieldy for hand held use and are used only on permanent towers or temporary masts. Yagis used in fish telemetry are considered single frequency antennas even though they cover a range of 4 MHz or more; for example 150MHz to 154MHz. It should be noted that the pattern of a Yagi in a horizontally polarized position has a narrower beam than the same antenna vertically polarized (Figures 21 and 22). This is the result of the multiplication of the pattern of the dipole with the phasing pattern of the Yagi. Figures 21 and 22 and Table 1 summarize the characteristics of some typical Yagi antennas. While the H-antenna (also known as HB9CV) is not a Yagi in a strict sense, but a phased array, it has similar reception characteristics and is grouped with Yagi antennas in this discussion. The advantage of using a Yagi antenna with a greater number of elements is twofold. First, the gain is greater for longer antennas, and secondly, the major lobe of the antenna narrows (Figures 21 and 22). The narrower main lobe increases direction finding accuracy (Cochran and Pater 2001).

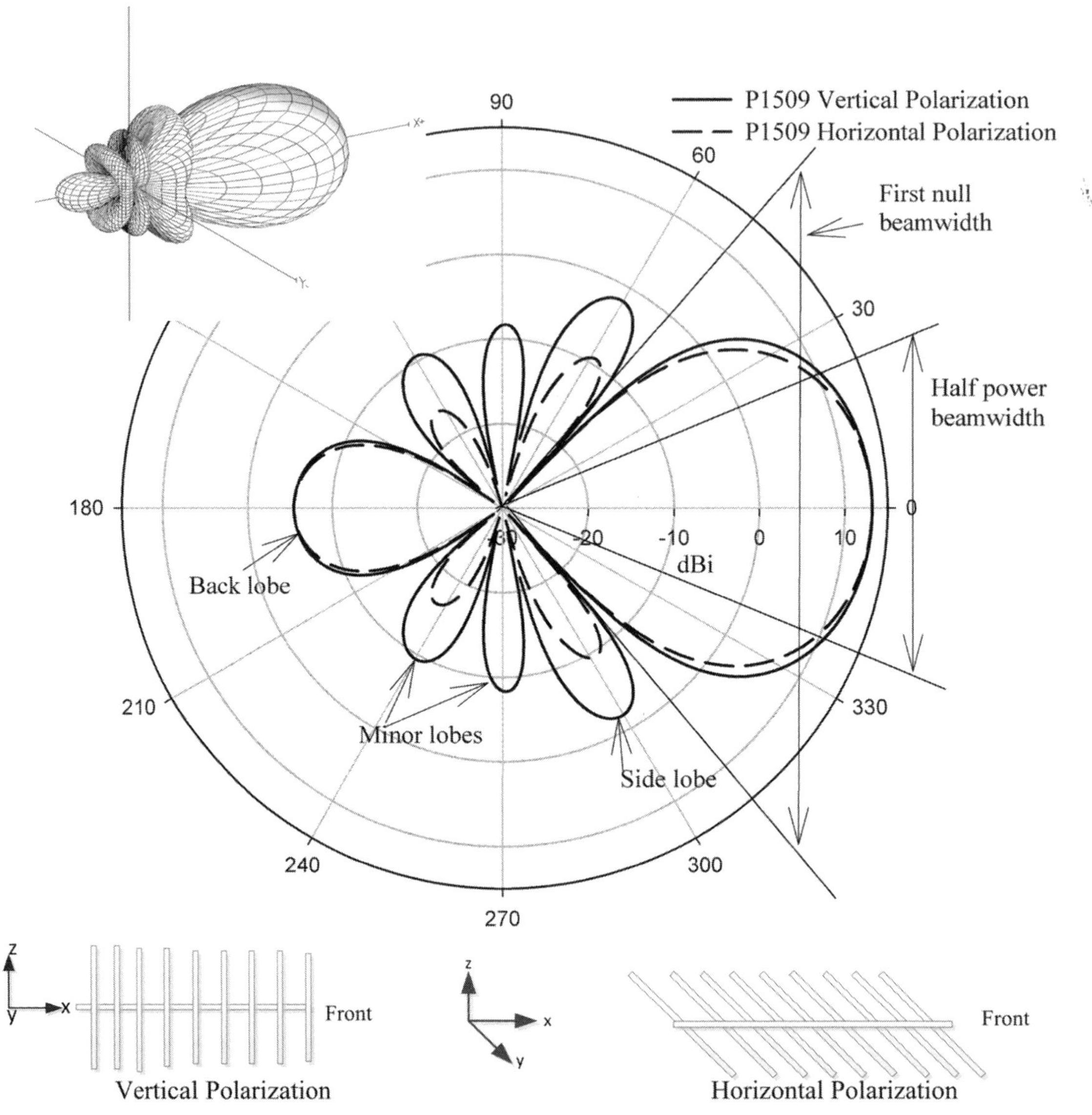

FIGURE 21. Vertical and horizontal directivity of a nine element Yagi antenna. Commonly used antenna descriptive terms are shown on the diagram.

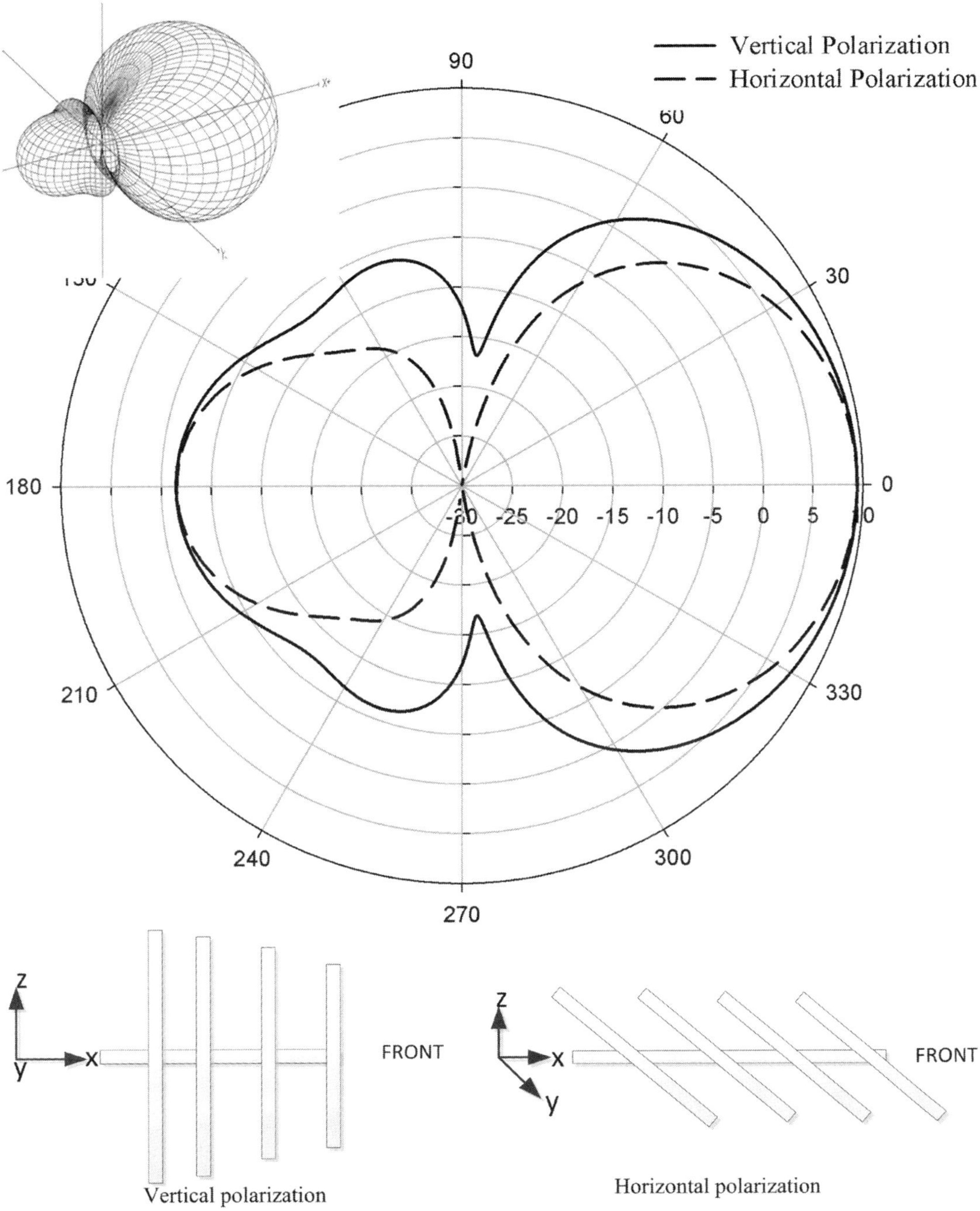

FIGURE 22. Vertical and horizontal directivity (gain) of a commonly used four element Yagi antenna. The horizontal beamwidth is less than the vertical beamwidth.

TABLE 1. Yagi antenna characteristics in relation to number of elements.

Antenna	Gain (dBi)	3 dB Beam Width Vertical (degrees)	3 dB Beam Width Horizontal (degrees)
H	7.0	108	66
3 Element	8.5	90	54
4 Element	9.4	81	58
6 Element	10.2	73	56
9 Element	13.0	49	43

Antenna Arrays

Antennas may also be combined into arrays to enhance gain or pattern characteristics. The most common of these is the null or peak arrangement shown in Figure 23. This arrangement is used to enhance accuracy when determining the direction of the transmitters from the tracking site. The resulting gain pattern is shown in Figure 24. The null (180° out-of-phase) arrangement is chosen for its deep null in the direction of the signal. The peak (in-phase connection) results in 3 dB higher gain. The in-phase arrangement also has deep nulls; however, they are offset from the direction of the signal. The antennas should be connected together using a hybrid combiner which may also be called a splitter or combiner. The function of this device is to combine the signals from the two or more antennas, and to isolate the antennas from each other. An ordinary tee connector should NOT be used, as half the power would go to the receiver and half to the other antenna. Phasing the antenna properly is also critical. The antennas must be exactly the same, be properly spaced and the cables must be of exactly equal lengths. Proper phasing is relatively easy to accomplish in stacked arrays. Many other array configurations make it possible to achieve other desired antenna patterns. In nonstandard arrangements such as corner reflector, proper phasing and pattern can only be determined by modeling or measurements. Keep in mind that increased gain in one direction can only be achieved by reducing gain in other directions.

CABLES AND CONNECTORS

Cables

There are many types of coaxial cable available, however aquatic telemetry systems are designed to use 50 ohm cable for the best match between components. Another factor that must be considered is the signal loss in the cable connecting the antenna to the receiver. In most cases, such as a hand held antenna, the cable is short and the cable loss is negligible when compared to other factors. Where the antenna is remotely located from the receiver, losses in the connecting cable will have an influence on system performance and should be mitigated. Figure 25 shows the signal loss versus distance for two commonly used cables. As shown, their signal loss is also a function of frequency. Cable loss is equivalent to increasing the noise factor of the system by an amount that is equal to the cable loss. A consequence of this signal loss is a reduction in reception range. In general, we do not want to increase the

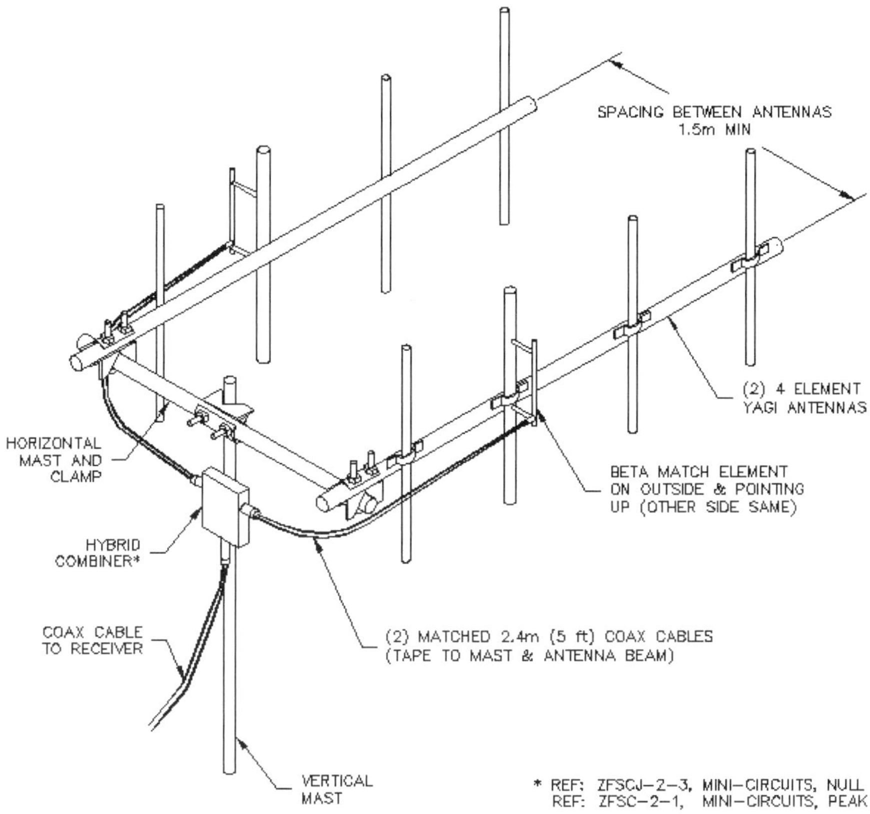

FIGURE 23. Stacked four element Yagi array used in null-peak antenna systems. The two halves of the system must be precisely matched for best directional accuracy.

noise factor by more than 5 dB. Choosing RG-8 cable over RG-58 is one way to reduce signal loss, but at the expense of increased weight and cable cost. Another way to mitigate signal loss in the cable is to place a low noise amplifier at the antenna to compensate for losses in the cable (see Section 5.2). Power for the amplifier can be fed through the coaxial cable. The impedance of coaxial cables is dependent on the space between the inner conductor and outer shield. Thus, kinks and sharps bends should be avoided to prevent pinching off the signal.

Connectors

The Bayonet Neill-Concelman (BNC) is the most commonly used coaxial cable connector in aquatic telemetry. It has a bayonet locking mechanism and is easy to connect and disconnect. Most receivers use a BNC connector as the antenna connection. Although they do not have a weather resistant construction, BNC connectors are commonly used outdoors without problems. Because of its size it can only be used on RG-58 or smaller diameter cables. For larger cables such as RG-8, UHF screw-on connectors are commonly used. There are many adaptors available to interconnect between similar types and between different series. There are many more connector types available, however, BNC and UHF are the most prevalent.

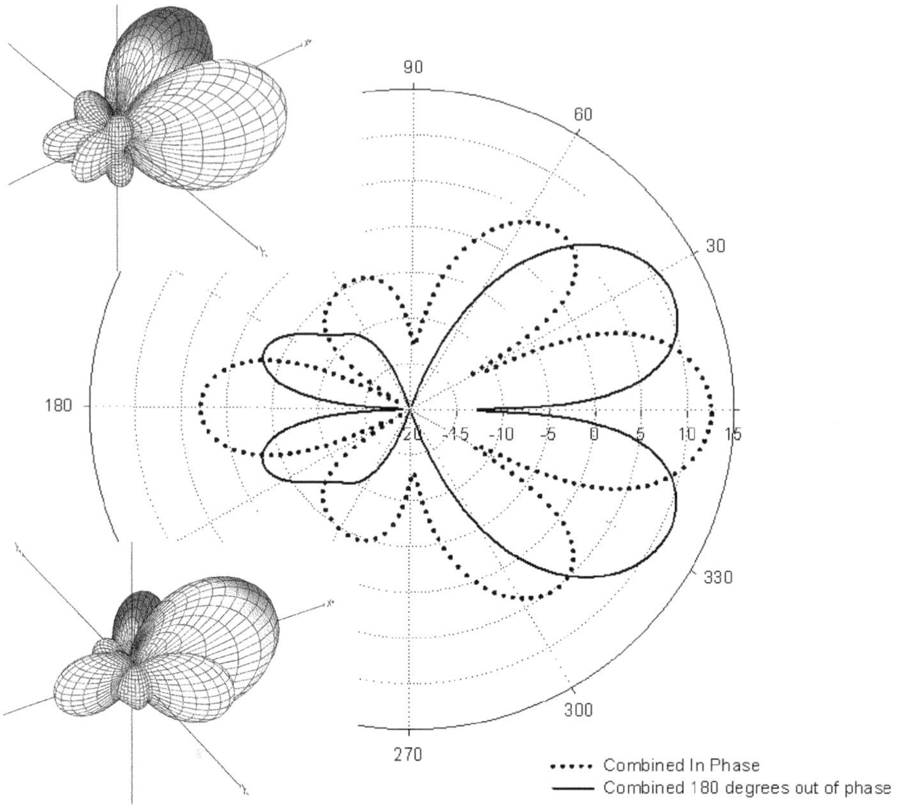

Figure 24. Directivity (gain) pattern of a peak (combined in phase) and null (combined 180 degrees out of phase) staked four element Yagi antennas spaced one wavelength. The upper 3D image is the null (180 degree) combination and the lower 3D image is the peak (0 degree) combination.

Both the BNC and UHF connectors are commonly available at consumer electronic stores in either crimp with a tool style or assemble without tools style.

Practical application

The information provided in this Section makes it possible to anticipate how different telemetry gear will perform in the field and help plan field logistics to take these considerations into account. The following example indicates how this information can be used to estimate reception range. Assume you are conducting a study under the following conditions:

1. Frequency of 150 MHz

2. Water conductivity of 400 µS/cm

3. Fish depth is 5 m

4. Receiving antenna is a handheld four element Yagi

5. Open lake view with no obstructions

6. Transmitter power is 0 dBm

7. Minimum detectable signal for the receiver is –145 dBm

With those assumptions, we can make the following calculations:

1. From Figure 10, loss due to attenuation at 150 MHz and 400 μS/cm is approximately 8 dB/m; this represents a loss of 40 dB for fish at 5 m depths.

2. From Figure 11, signal loss at the interface is 19 dB (i.e., signal attenuation ~ –19 dB).

3. From Table 1, the gain of a four element Yagi is 9.4 dB.

4. With this information the signal level can be calculated as:

0 dBm (tag power) + –40 dB (depth attenuation) + –19 dB (surface attenuation) + 9.4 dB (antenna gain) = –49.6 dBm.

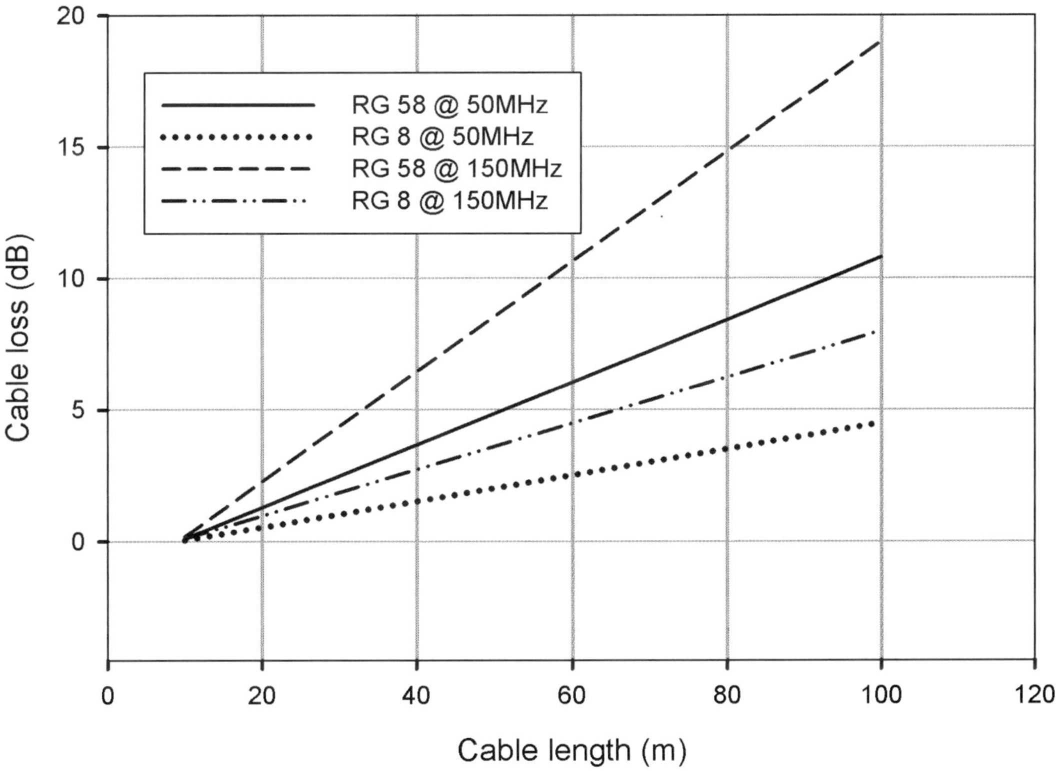

FIGURE 25. Signal loss in coaxial cable for two frequencies and two cable types in relation to cable length.

5. Thus, we can sustain approximately 95 dB loss (–145 dB – –49.6) before the signal is too weak to hear.

6. We can estimate that a 100 dB loss will occur at 500 m from Figure 13 and best case estimate. Best case for an antenna close to the ground ($N = 40$) would be conservative estimate as it is in the transition range between line-of-sight conditions and ground wave conditions.

7. Based on these calculations, reception range under these conditions will be slightly less than 500 m. We could increase the range by increasing the transmitter power or by increasing the antenna height above ground (Figure 14).

SUMMARY

Radio telemetry is a powerful tool, but it is critical to understand how the different system components, the various types of equipment available, and the surrounding environment can impact field methods and the data collected. These elements can have a major effect on the outcome of an aquatic radio telemetry project. Radio telemetry follows well defined physical principles which have been detailed in published scientific literature and emphasized in this Section. Several principal considerations include:

1. There is no free lunch. Equipment designed to address one consideration (e.g., reception range, transmitter size, signal characteristics, antenna type) may have a negative impact on another. Equipment selection and use needs to take this factor into account.

2. Telemetry systems are made up of several integrated components, including the transmitter, receiving antenna, receiver, and associated peripheral gear. All of these components need to perform well for the system to be effective. The factors affecting signal path are also critical considerations.

3. Water conductivity and transmitter depth in the water are the two most critical parameters in determining the signal reception range.

4. Transmitter antenna size matters; larger antennas mean greater reception range.

5. In terms of performance, there is an optimum transmitter antenna length. The optimum length is dependent on the size of the battery and electronics, length of the antenna wire, and the antenna insulation thickness. Achieving optimum performance must be balanced with other factors (see Item 7).

6. Transmitters with small internal antennas have much shorter range than longer external trailing wire antennas.

7. In most applications the appropriate transmitter and antenna size is dictated by the study organism rather than by electronic performance.

8. Yagi antennas are the receiving antennas of choice for their ruggedness and better gain characteristics. The number of antenna elements used will depend on whether the antenna will be hand held or permanently placed, and on the type of data being collected (e.g., 4 element Yagi for broader coverage, 9 element Yagi for fine-tuned locations).

9. Receiver choice for standard tags is based on cost and desired features. Most will have approximately equal reception range. Receivers for coded ID tags must be matched to the tag codes and can be either narrow band or wide band. Narrow band receivers detect fewer tags per unit time than wide band models, however, range on narrow band models is greater.

10. Stationary receivers equipped with data loggers have reduced signal reception range compared to listening by ear, but provide continuous coverage at a site.

With an understanding of the basic principles, a researcher can select the most appropriate radio telemetry equipment and use it to its best advantage. An understanding of these considerations will also enable the researcher to better estimate the time and budget required to complete the study objectives

REFERENCES

Beeman, J., and five coauthors. 2007. Effects of antenna length and material on output power and detection of miniature radio transmitters. Hydrobiologia 582(1):221–229.

Biggs, A., and H. Swarm. 1968. Radiation fields from an electric dipole antenna in homogeneous Antarctic terrain. IEEE Transactions on Antennas and Propagation 16(2):201–208.

Cheng, D. K. 1993. Fundamentals of engineering electromagnetics. Addison-Wesley, Reading, Massachusetts.

Cochran, W. W., and L. L. Pater. 2001. Direction finding at ultra high frequencies (UHF): improved accuracy. Wildlife Society Bulletin 29(2):594–599.

Finkenzeller, K. 2003. RFID handbook : fundamentals and applications in contactless smart cards and identification, 2nd edition. Wiley, Chichester, England; Hoboken, New Jersey.

Friis, H. T. 1946. A Note on a Simple Transmission Formula. Proceedings of the IRE 34(5):254–256.

Goldman, J., and G. W. Swenson, Jr. 1999. Radio wave propagation through woods. IEEE Antennas and Propagation Magazine 41(5):34–36.

Hafez, H. M., W. J. Chudobiak, and J. S. Wight. 1979. The Attenuation Rate in Fresh Water at VHF Frequencies. IEEE Transactions on Instrumentation and Measurement 28(1):71–74.

Hertel, T. W., and G. S. Smith. 2000. The insulated linear antenna-revisited. IEEE Transactions on Antennas and Propagation 48(6):914–920.

Johnson, R. C., and H. Jasik, editors. 1984. Antenna engineering handbook, 2nd edition. McGraw-Hill, New York.

Karmel, P. R., G. D. Colef, and R. L. Camisa. 1998. Introduction to electromagnetic and microwave engineering. Wiley, New York.

King, R. 1985. Electromagnetic surface waves: New formulas and applications. IEEE Transactions on Antennas and Propagation 33(11):1204–1212.

King, R. W. P., and G. S. Smith. 1981. Antenna in matter: Fundamentals, theory and applicatins. MIT Press, Cambridge, Massachusetts.

Kolz, A. L., and R. E. Johnson. 1981. The human hearing response to pulsed audio tones: implications for wildlife telemetry design. Pages 27–34 *in* F. M. Long, editor. Third International Conference on Wildlife Biotelemetry. University of Wyoming, Laramie.

Kraus, J. D. 1988. Antennas, 2nd edition. McGraw-Hill, New York.

Linden, D., and T. B. Reddy. 2002. Handbook of batteries. McGraw-Hill, New York.

Lonsdale, E. M., and E. T. Baxter. 1968. Design and field test of radio wave transmitters for fish tagging. The Progressive Fish-Culturist 30:47–52.

Plomp, R., and M. A. Bouman. 1959. Relation between hearing threshold and duration of tone pulses. Journal of the Acoustical Society of America 31(6):749–758.

Saran, G. S., and G. Held. 1960. Field strength mesurements in fresh water. Journal of Research of the National Bureau of Standards 52D(5):435–437.

Shavit, R., and E. Rosen. 1995. Lateral wave contribution to the radiation from a dielectric half medium. IEEE Transactions on Antennas and Propagation 43(7):751–755.

Shen, L., R. W. King, and R. Sorbello. 1976. Measured field of a directional antenna submerged in a lake. IEEE Transactions on Antennas and Propagation 24(6):891–894.

Tanenbaum, A. S. 1996. Computer networks, 3rd edition. Prentice Hall PTR, Upper Saddle River, New Jersey.

Trefethen, P. S. 1956. Sonic Equipment for Tracking Individual Fish. U.S. Fish and Wildlife Service Special Scientific Report 179.

Trefethen, P. S., J. W. Dudley, and M. R. Smith. 1957. Ultrasonic Tracer Follows Tagged Fish. Electronics 40(4):156–160.

Velle, J. I. 1979. An investigation of the loss mechanisms encountered in propagation from a biotelemetry transmitter submerged in fresh water. Doctoral Dissertation. University of Wyoming, Laramie, Wyoming.

Wait, J. R. 1998. The ancient and modern history of EM ground-wave propagation. IEEE Antennas and Propagation Magazine 40(5):7–24.

Zhongxiang, S., and R. H. MacPhie. 1996. Input impedance of a multilayer insulated monopole antenna. Pages 318–321 *in* Antennas and Propagation Society International Symposium, 1996. AP-S. Digest, Volume 1.

APPENDIX: TRANSMITTER ANTENNA DETAILED DISCUSSION

As discussed in the section on transmitter antennas and shown in figure 5, the impedance of all antennas is complex with a real and imaginary part. Power radiates through the resistive part, while power is only stored in the reactive part. The behavior of the reactive part is explained by noting that the voltage and current are 90° out of phase (similar to the electromagnetic wave in Figure 1 with voltage indicated by E and current by H) which means the power is positive during one half of a cycle and negative during the other half. The result is no net power out. The ratio between the magnitudes of the real and imaginary parts determines the power output and antenna efficiency.

The most commonly used antenna in aquatic radio telemetry is a wire antenna with a polymer coating. It is considered an insulated antenna. Antennas similar to these have been analyzed experimentally (Beeman et al. 2007) and analytically (Hertel and Smith 2000). From an electronic point of view, the typical fish transmitter may be analyzed as an asymmetrical dipole antenna with a long side and a short side as shown in the upper diagram of Figure 5. This can be modeled by considering two dipoles, a short one consisting of the battery and electronics, and another consisting of the wire antenna itself. The resulting impedance (Z) is determined by adding the impedances of both and dividing by two (Johnson and Jasik 1984). Each of the dipole elements may be modeled as resistive and reactive components connected in series. Another way of explaining this is to consider the "main" antenna as one antenna, and the battery and electronics combination as another equally important antenna. In addition, the resistive component may be separated into an ohmic loss resistance and a radiation resistance. The radiation resistance accounts for the generation of the radio wave. Except for very small antennas the ohmic resistance is much smaller than the radiation resistance and can be ignored. Thus, the ohmic resistance of the antenna material is not important; stainless steel will perform as well as lower resistance copper. Power delivered to the antenna is divided between the resistive and reactive components in proportion to their size. Power in the reactive component is only stored temporally and does not contribute to useful signal radiation. The value of the resistive and reactive components is dependent on antenna element size (length and radius), the antenna coating thickness and on the media that surrounds them. If the electrical length, which is the length of the antenna compared to a wavelength; is less than one quarter wave length long in the media, the reactance X, is negative (capacitive). If it is longer than one quarter wavelength, the reactance is positive (inductive). The antenna will be optimum when the reactance is zero.

The insulated antenna may be modeled as a transmission line to predict its behavior (Hertel and Smith 2000; Zhongxiang and MacPhie 1996). Modeling the battery and electronics is difficult as the ratio of insulation diameter to conductor diameter is not uniform. The optimum length may also be determined experimentally by cutting the antenna length by small increments and measuring the reception range (Beeman et al. 2007). This method does not, however, consider potential loss of efficiency resulting from a mismatch of the transmitter components to the changing antenna impedance.

(Beeman et al. 2007) correlated output power to antenna length and material, and demonstrated differences probably related to differences in antenna type and coating material. We modeled their Type I antenna using a transmission line analysis. The results shown in Appendix Figure 1, indicate the antenna did not reach a zero reactance resonant length; it is always negative. As a consequence, only a limited amount of the available power is transferred to the radiation load. This is shown by the antenna efficiency in Figure 6. The very small volume occupied by the battery and electronics (14.5 mm long by 6.3 mm by 4.5 mm) results in high reactive component which cannot be compensated without an impractically long antenna. This is contrasted with the larger ATS Model F1850 (55 mm long by 19 mm diameter) shown in Appendix Figures 1 and 2. In this case the larger battery and electronics volume results in a lower reactance which can be mitigated by a longer antenna. The antenna is resonant, as indicated by zero reactance, at about 17 cm and at 27 cm. Maximum efficiency is also achieved at these resonant lengths because more power is going to the load rather than being stored in the reactive part. The efficiency indicated is only the efficiency of the antenna, not the entire transmitter, and relates the power in the reactive load versus power in the radiated load. These comparisons do not indicate a poor design in the Beeman model, but rather an indication of the penalty of a small battery and electronics volume on transmitter output.

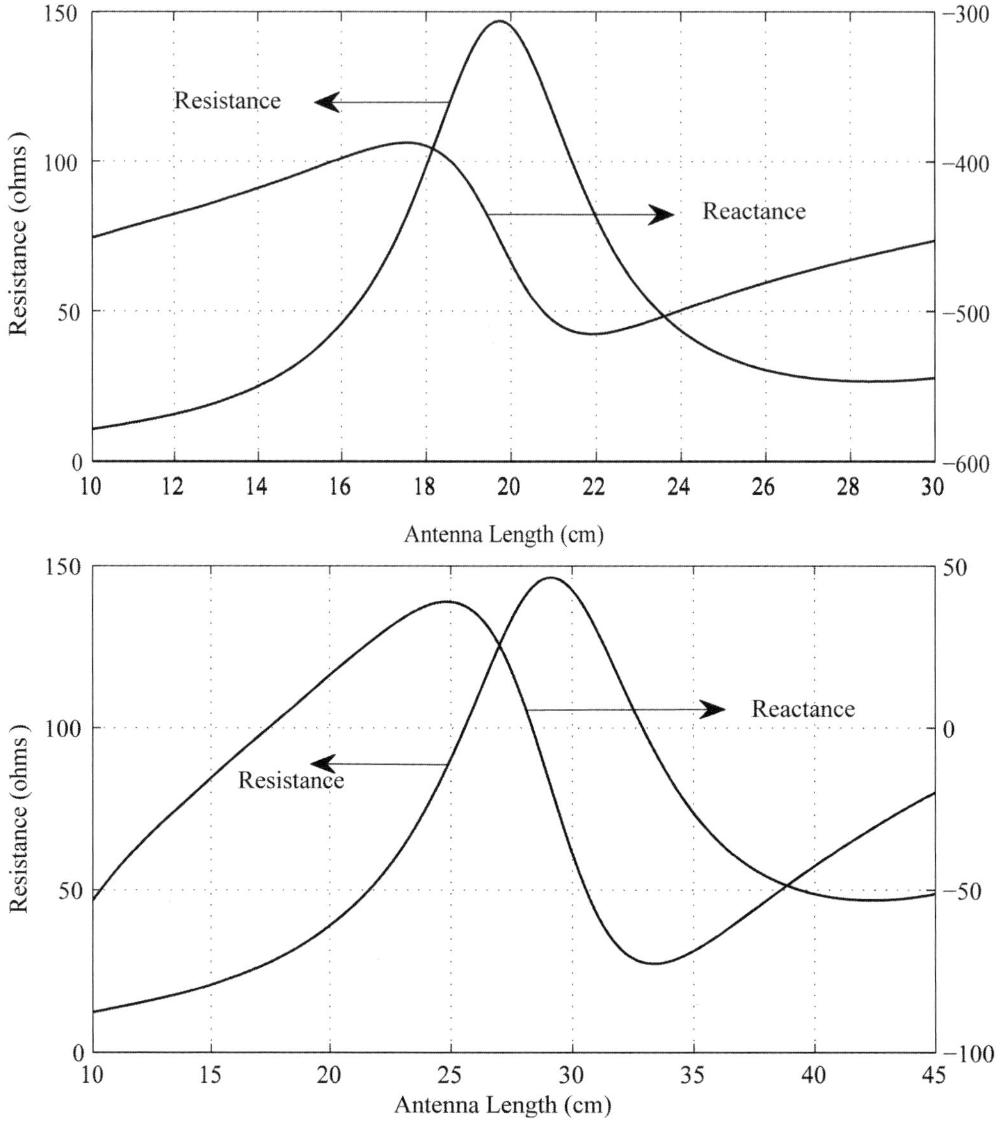

APPENDIX FIGURE 1. Upper figure is antenna impedance of the transmitter described by Beeman and the lower is the antenna impedance of the larger ATS model. Both are based on a transmission line model. Optimum performance will occur at resonance which is defined as zero reactance. This occurs at two points for the larger ATS model, but remains negative for the smaller Beeman transmitter.

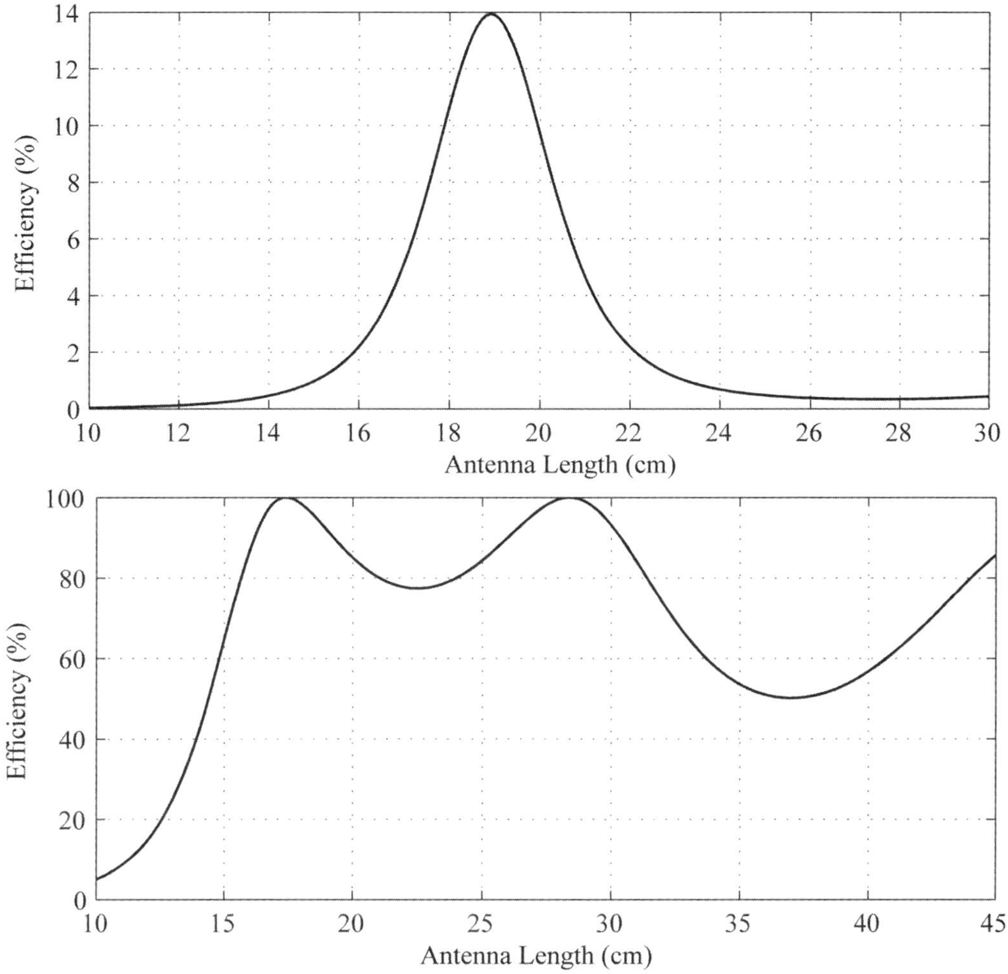

Appendix Figure 2. Transmitter antenna efficiency of the two transmitter models shown in Appendix Figure 1. The upper Beeman transmitter reaches its best efficiency where the resistive part of the impedance is maximum. The larger ATS model is most efficient at the zero reactance points.

Section 5.2

Optimization of Radio Telemetry Receiving Systems

Scott D. Evans and John R. Stevenson

INTRODUCTION

The performance of radio telemetry receiving systems can be affected by numerous factors, thus it is important for researchers to understand these factors when designing a radio telemetry study. One approach that has been used to describe these factors is the radio system equation which defines six variables that affect radio receiving systems (Sisak and Lotimer 1998; Section 5.1). This equation is:

$$P_r = P_t - L_w - L_{awi} - L_{air} + G_r - L_{rtl}$$

where P_r is the received power level, P_t is the transmitted power level, L_w is the signal loss in water, L_{awi} is the loss at the air-water interface, L_{air} is the loss in air, G_r is the gain of the receive antenna, and L_{rtl} is the loss of the receive transmission line (Figure 1). Three of these variables (L_w, L_{air}, and L_{awi}) are affected by factors that researchers have little control over (i.e., distance traveled by a given radio signal, water conductivity and temperature, angle of refraction of the signal as it exits the water). However, P_t, G_r, and L_{rtl} can be influenced by the user and planning for the effects of each is an essential first step towards designing a radio receiving system that operates at the highest level possible. The first of these factors P_t, is addressed while choosing which transmitter to purchase, a topic which was covered in a previous Section. Much of the information that we present will focus on maximizing the strength of the signal that is received by the radio receiver which is affected by G_r and L_{rtl}. Additionally, we will cover techniques to minimize the effects of ambient (or background) noise which is also an important aspect of radio receiving systems. Ambient noise is important because low signal-to-noise ratios will result in signals that cannot be detected by the receiving system. In other words, tagged fish within range of the receiving system are not detected because ambient noise levels exceed the received signal at the receiver.

This Section will provide the information and techniques necessary to optimize a radio telemetry receiving system. Previous Sections have provided a general understanding of radio theory and have presented the various types of equipment that are available, their technical

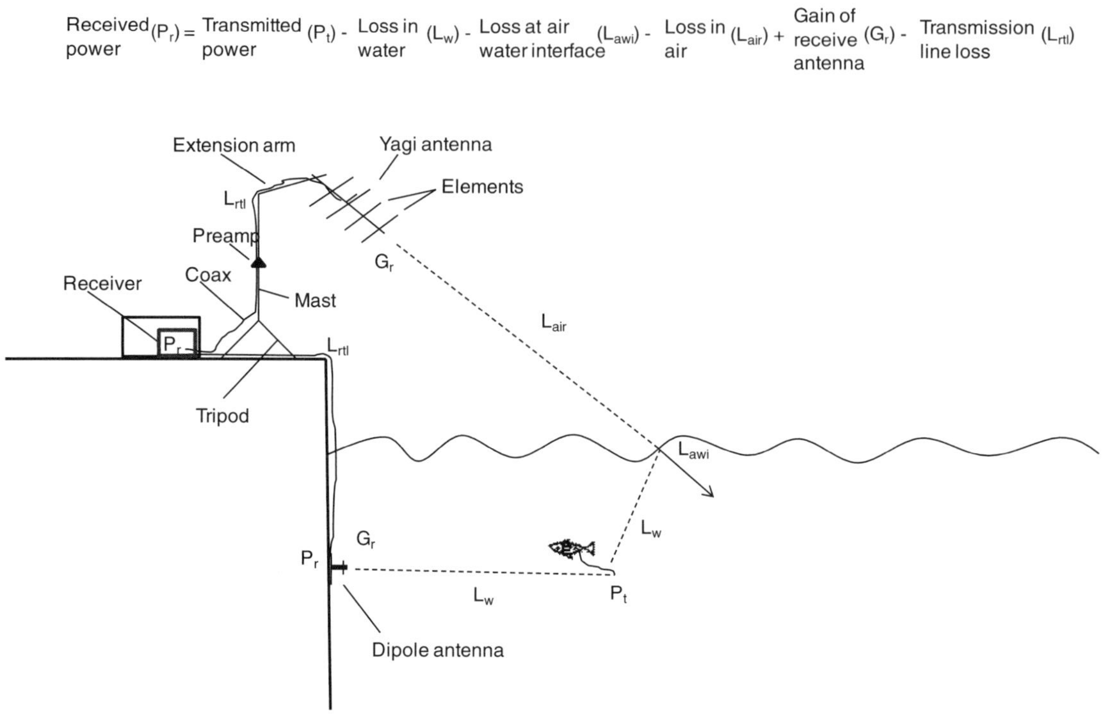

Figure 1. Schematic showing how the radio system equation affects a radio telemetry receiving station.

characteristics and capabilities, and how to design and deploy receiving systems. To facilitate clarity, a minimal amount of the same information will be presented, but we will focus on discussing how equipment selection and installation methods can be used to maximize system performance, how to optimize individual components within the system, how to optimize the system as a whole, and finally, how to verify and maintain system performance. Although the information presented in this Section is applicable to both large and small receiving systems, and to studies using both large and small transmitters, some techniques will be most beneficial to large studies using multiple arrays and receivers (such as at dams or fish passage structures), and to studies using low output miniature transmitters. It is up to you as a researcher or manager to evaluate the advantages and disadvantages of available choices and, ultimately, determine the level of effort and expense dedicated to a particular study.

PROPER PLANNING

Optimizing a telemetry system begins with proper planning before the equipment is deployed. It is difficult, if not impossible, to optimize a system that was poorly installed. To properly install a telemetry system, it is beneficial to create a schematic of each system and this can be facilitated by using existing blueprints of the structure that is being monitored. Diagrams of the telemetry system can then be overlaid onto the blueprints to create a visual template that can be used during the design phase of the study. The preparation of schematics accomplishes a number of tasks such as identifying the extent of telemetry coverage, the type and quantity of various telemetry components, and power sources available for operation of the telemetry system. This initial phase assists the researcher in organizing and planning the most efficient deployments strategy and fully appreciating the logistics of the study location. Often, specific challenges that were not apparent at the onset of the study are identified. A good example is in determining the best route to run cables to the receiver. This is particularly true when installing a telemetry system at a hydroelectric dam. Often, telemetry equipment at a dam is subjected to heavy equipment and vehicle traffic on a regular basis, and it is wise to route cables where they will be protected from damage. It is useful to consider other activities that may take place during the study. For example, if a construction project is to be conducted concurrently at the same location as the telemetry evaluation, the power supply could be disrupted, access to the system may be limited, or telemetry equipment could be damaged. After the schematics and design of the telemetry systems have been developed, it is a good idea to physically walk through each telemetry system and mark the location of all telemetry components. An easy way of accomplishing this is to use colored electrical tape to mark the location of all components. Different colors of tape can be used to color code different components. Another option is to label each component, identifying what it is and where it goes. During this layout phase, it is useful to identify any concerns that have been overlooked and implement any necessary changes. With the layout and any potential alterations completed, the procurement of the various telemetry components can begin.

Clearly, the physical surroundings must be considered when designing a telemetry system, but it is not the only factor that must be addressed to ensure the system can be optimized for maximum performance. Poorly installed systems often result from not considering how environmental conditions affect the performance of the system. For example, the location and extent of ambient electrical noise at the study site can have a significant effect on the performance of the system. Hydroelectric dams are one place where ambient noise can significantly affect system performance. Hydroelectric projects produce a large volume of electricity, which is transmitted via local and intertie transmission cables. Electricity transmitted through the cables, insulators, and transformers often generates ambient background noise that may occur within the range of frequencies being used in radio telemetry research. Should the ambient background noise occur on the same frequency as used in a telemetry study, it may reduce the detection range of the system, increase potential "false-positive" detections, and fill the receiver memory faster. A "false-positive" is a false reading and associated data record resulting from environmental noise; not from a valid transmitter. In a worst-case scenario, the noise may be so prolific as to mask valid transmitter reception entirely. In this case, a tagged fish within detection range of a given receiving system would not be "heard," and its signal not recorded by the receiver. While it is not possible to eliminate ambient electrical noise at a hydroelectric project, the impact of this interference can be minimized. The most obvious option is to refrain from using frequencies that are most problematic. While this strategy will reduce the potential problem of ambient electrical noise, it will likely not eliminate it. As such, a common strategy is to shield telemetry equipment from the source of the interference. Whenever possible, it is a good idea to place equipment susceptible to interference (e.g., receivers, preamplifiers, antennas, etc.) behind physical structures such as concrete or metal, which are quite effective at shielding the effects of ambient noise. While hydroelectric dams are an extreme example of a place where ambient noise can impact system performance, all researchers should consider the potential effect of noise regardless of where the system is deployed.

MOUNTING OPTIONS AND PROTECTING EQUIPMENT

To maximize the performance of a telemetry system, equipment must be properly mounted and protected for the duration of the study period. In this Section we will use examples from systems that were installed at hydroelectric dams to demonstrate ways to mount and protect equipment, however, many of the concepts are applicable to all telemetry systems regardless of where they are installed.

Powerhouse

Monitoring the movements of fish through the powerhouse of a hydroelectric dam usually requires mounting equipment in each turbine intake. To ensure the system is only detecting fish that are passing through the turbines, antennas need to be mounted well within the intakes so that only the fish that are committed to passing that route are detected. The physical structure of a turbine intake and the flows that are typically encountered can present a challenging environment for installing and maintaining antennas. One strategy is to utilize existing structures as mounting platforms. For example, underwater antennas can be mounted to the downstream side of trash rack structures (to avoid contact with debris), assuming the antennas are situated far enough downstream of the turbine intake entrance to ensure that a tagged fish swimming by the entrance is not detected. Another option is to mount underwater antennas to a fabricated metal frame that can be lowered, like a trolley, into the intake or any other area of the dam that needs to be monitored (Stevenson et al. 2000 and Skalski et al. 2001; Figure 2). A more permanent option is to construct antenna mounts that can be bolted to the walls within the turbine intake or affixed to the fish-guidance screens that are lowered into the turbine intake (Perry et al. 2007; Figure 3). Attachment of antennas to the walls of the turbine intake or to trash racks are attractive methods since the antennas do not have to be removed when various structures are installed or removed from the turbine intakes for dam maintenance. However, antennas mounted to turbine intake walls must be installed, repaired, and removed when the unit is dewatered, or by deploying divers into the turbine intake. The difficulty in scheduling times to dewater the turbine unit, or deploy a dive team, may prevent using this method of mounting the monitoring equipment.

Spillway

Monitoring fish movements through a spillway at a dam presents unique challenges. Water velocities through individual spill bays commonly reach 9–11 m/s. As a result, tagged fish move through the area more quickly, reducing the opportunity to log radio transmissions. The higher water velocities and passing debris can also damage coaxial cable. Despite these challenges, a well designed and balanced telemetry system at a spillway can result in relatively high detection probabilities (Skalski et al. 2001). A common strategy is to mount multiple underwater antennas on the inside of each pier nose associated with a specific spill bay (Figure 4). Multiple antennas within a spillway are then combined to function as a single antenna. The individual antennas within each bay should be deployed at a depth close enough to the spill bay opening to detect tagged fish as they pass the structure, yet far enough away from the opening to protect against shear forces associated with the water passing through the bay. The advantage of deploying multiple antennas in each bay is that should one fail, other antennas remain to detect fish passing through the bay.

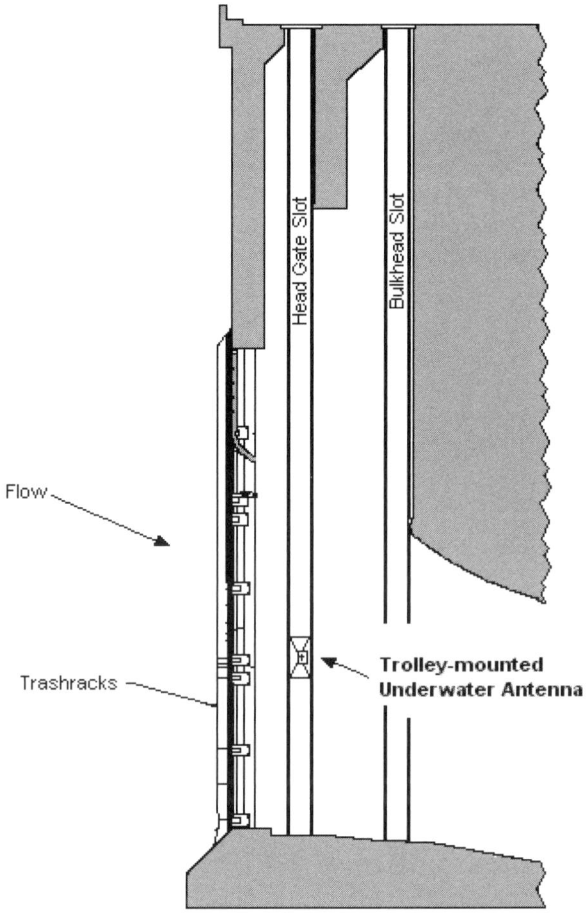

FIGURE 2. Schematic showing cross-section of the turbine intake of a hydroelectric dam and the location of a trolley-mounted dipole antenna within the head gate slot.

Alternative Passage Routes

Other routes of passage may need to be monitored at hydroelectric dams. These include, but are not limited to, juvenile bypass structures, ice and trash sluiceways, and adult fish ladders. Similar to the powerhouse and spillway, one of the most important considerations is to place the monitoring equipment at a location where it only detects fish that are committed to passing the route being monitored. The monitoring equipment should be mounted in a place where fish cannot swim upstream (for downstream migrating salmonids) and exit the passage route. Equally important is to select a location where water velocities will not be so high as to reduce the likelihood of detection. Finally, the mounting location should allow regular inspection, and if necessary, repair of the equipment. To increase the probability of detecting fish moving through a passage route with high water velocity, multiple antennas can be installed along the length of the passage route and spaced approximately twice the detection radius of the antennas. These antennas can then be combined together to function as a single antenna. For example, if the detection radius of the antenna is 5 m, then three antennas would each

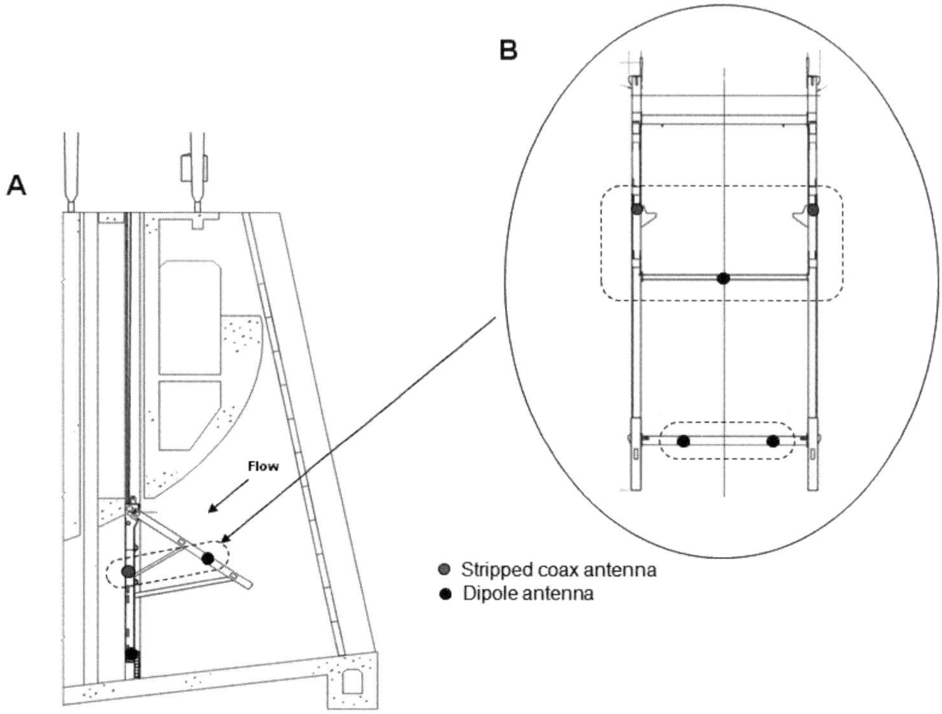

Figure 3. Schematic showing location of underwater radio telemetry antennas mounted to fish guidance screens. Cross-section view (A) shows guidance screen deployed within the turbine intake and inset (B) shows front view of guidance screen.

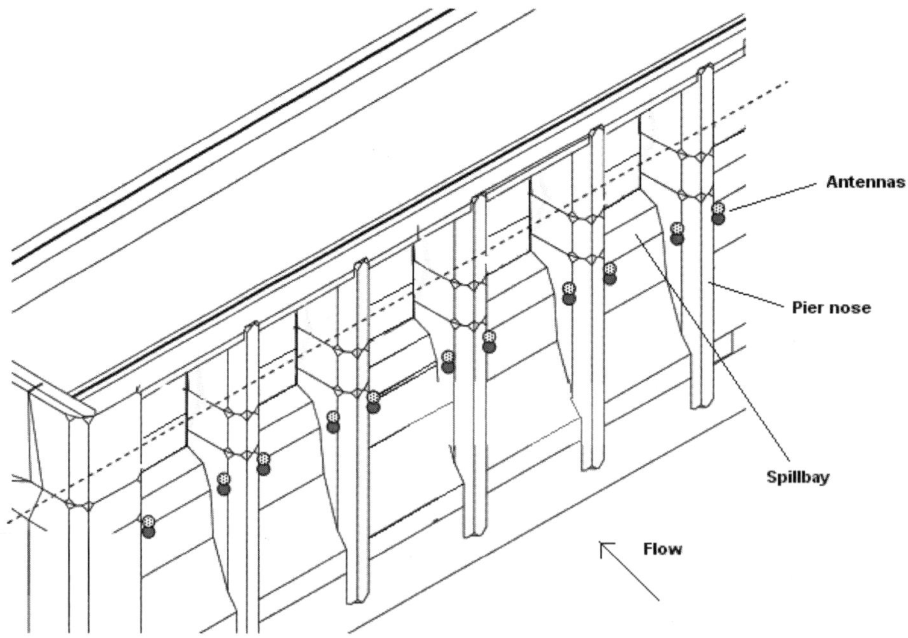

Figure 4. Schematic showing the upstream face of the spillway of a hydroelectric dam and the location of underwater antennas mounted on each pier nose near the crest of each spill bay.

be spaced about 10 m apart (some overlap is desirable). When combined and monitored as a single antenna, the detection field of this array would be nearly 30 m in length, rather than 10 m as with a single antenna (Figure 5). In this case, using three antennas nearly triples the detection area and provides redundancy in the event of antenna failure.

EQUIPMENT SELECTION

Clearly, maximum performance of a monitoring system depends on proper installation of equipment and careful consideration of how the physical surroundings and environmental conditions affect system performance. Equally important is selecting the appropriate type of equipment for each part of the monitoring system. In this Section we discuss how the performance of the telemetry system can be affected by decisions regarding which transmitter, antenna, and receiver to use.

Transmitter Power

Transmitter selection is based on numerous factors, such as the size and species of test fish, type of transmitter (digitally encoded or beeper type), transmitter frequencies and codes, tag longevity, burst rate, and transmitter strength. Procedures to select the proper transmitter are addressed in detail elsewhere in this book, and as such won't be discussed here (Section 5.1). With regard to system performance, the transmitted power level of the transmitter is one of the most important factors in selecting a transmitter. As described by the radio system equation ($P_r = P_t - L_w - L_{awi} - L_{air} + G_r - L_{rtl}$), the received power level is increased when the transmit power is increased. Unfortunately, the user's choice of transmitters is somewhat limited because transmitter size and weight must be limited for implantation into fish, especially when used in smaller juveniles (Brown et al. 1999; Jepson et al. 2005, Perry et al. 2001; Winter 1983). Since size and weight reductions are achieved through the use of smaller batteries and less circuitry, smaller transmitters have reduced power and shorter transmission ranges

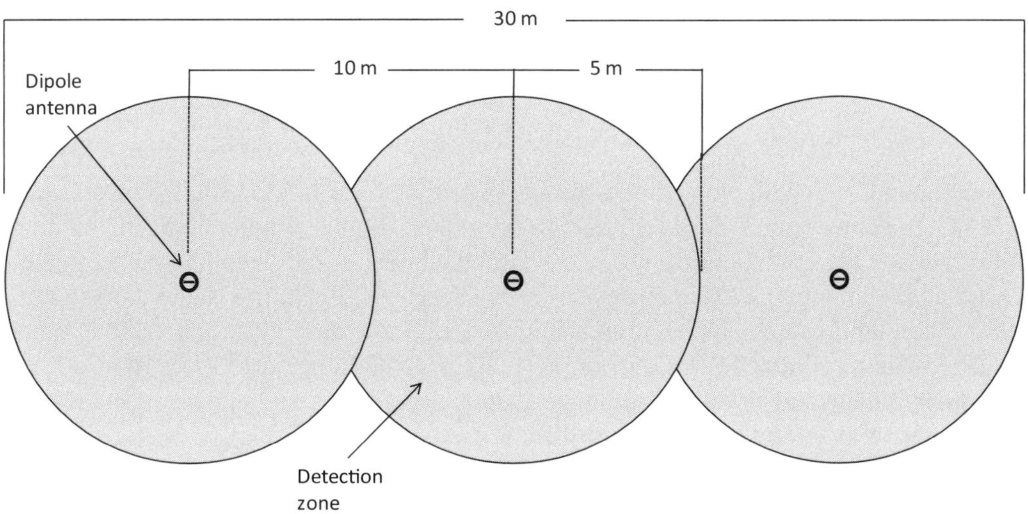

FIGURE 5. Schematic showing the appropriate spacing and resulting detection area of three underwater dipole antennas.

compared to larger transmitters. It is therefore crucial, especially when limited to the use of smaller transmitters, to ensure that the transmitter you select has an adequate transmit power. Transmitter suppliers can provide the transmit power level for available models and users should select the model with the highest transmit power that also meets other study criteria.

System performance is also heavily dependent on the burst rate of the transmitter. Burst rates vary, but typically range between 1 and 10 s. The burst rate of the transmitter, in conjunction with the desired number of detections over a specified amount of time, dictates the location and number of antennas that should be deployed. For example, consider a scenario where research objectives require that a minimum of three detections from each tagged fish are received in a flume where water velocities average 1 m/s. If transmitters for this study burst every 5 s, than at least two antennas would be required and they would need to be spatially separated by at least 5 m. This is because each tagged fish would travel approximately 15 m during the time interval in which the transmitter burst 3 times (one burst per 5 s with the fish traveling at 1 m per second). These calculations are based on the assumption that each antenna has a detection radius of 5 m, resulting in an overall detection range between the upstream and downstream boundaries of 10 m. Therefore, two antennas spaced 5 m apart would be necessary to monitor a 15 m section of the flume. This scenario illustrates the importance of understanding many physical factors associated with the environment being monitored (i.e., burst rate of the transmitter, behavior of the fish, velocity of the water, detection range of the antenna).

Receive Antennas

The receive antenna is one of the most critical components of the receiving system. Antenna performance determines the quality and continuity of the received signal. Unfortunately, price and tradition often drive antenna selection rather than quality and performance. Many factors must be considered when selecting an antenna, and these will be addressed later in this Section, but perhaps the most important is the gain of the antenna. As the radio system equation ($P_r = P_t - L_w - L_{awi} - L_{air} + G_r - L_{rtl}$) shows, increasing the gain of the receive antenna will have the resultant effect of increasing the received power level. Functionally, this means that the transmitter will be heard from a farther distance as antenna gain is increased. Antenna manufacturers increase gain by changing the design of the antenna to concentrate, or focus, the received energy over a narrower area. Antenna gain is then a measure of the amount of focus that an antenna can apply to the incoming signal relative to one of two omni-directional references: a half-wave dipole and an isotropic radiator (Section 5.1). A half-wave dipole is a simple antenna constructed of two conductors having an overall length of half the wavelength of the frequency being used with a radiation pattern similar to a doughnut. An isotropic radiator is a theoretical antenna that radiates (or receives) equally well in all directions and has a spherical radiation pattern. When referenced to a half-wave dipole, antenna gain is specified in dBd and when referenced to an isotropic radiator, antenna gain is specified in dBi. It is important to understand that an antenna gain specified in dBi will appear to have 2.15 dB higher gain than an antenna specified in dBd, (i.e., 0 dBd = 2.15 dBi). Therefore, beware of an antenna supplier that specifies antenna gain only in "dB." Using the wrong reference could cost you 2.0 dB in performance.

Although gain is a primary consideration in the selection of an antenna, there are other factors that contribute to the performance of an antenna and it's suitability to a particular study. The second most important factor to consider during antenna selection is the radiation pattern (a.k.a. directivity, beamwidth, or angle of reception) of the antenna. As mentioned previously, to increase gain the antenna is built to focus the received energy over a narrower area. This is accomplished through making the antenna longer, and in the case of Yagi style antennas (the most commonly used antenna in aerial aquatic telemetry applications), adding more elements (Figure 1). So, as gain increases, size, directivity (the angle of reception becomes narrower), and range increase. For example, a 4-element Yagi antenna measures about 1 m long (at 150 MHz), has a gain of about 7 dBd, and a relatively wide angle of reception of about 80° in the horizontal plane and 65° in the vertical plane. Conversely, a 9-element Yagi measures about 4 m long (at 150 MHz), has a gain of about 11 dBd, but has a narrower reception angle of about 45° in the horizontal plane and 35° in the vertical plane. Usually, too narrow a reception angle is not desirable because it restricts the detection zone too much, decreasing the likelihood of detecting rapidly moving fish. Three- or 4-element Yagi antennas usually provide the best value (gain, directivity, size, and cost). High-gain antennas that are relatively large may be difficult to install and maintain, especially if they experience high winds or ice. Additionally, the narrower angle of reception of high-gain antennas may not be suitable for some study areas or may require the purchase of additional antennas to cover the entire area. That being said, the additional range that a high-gain antenna provides can be advantageous in the appropriate situation, such as monitoring fish from the shoreline of a wide river. One should carefully consider the tradeoffs and select the antenna that offers the highest gain within the study requirements for size, directivity, range, and cost (Table 1).

Receive antennas used underwater are designed after the half-wave dipole, and therefore, have zero gain and, consequently, no directivity. Although these antennas have no gain, they are still effective because they can be placed at water depths greater than surface mounted antennas can detect and which are nearer the radio-tagged fish. This is advantageous because the signal reaching an underwater antenna isn't subject to loss at the air-water interface (L_{awi}) or in air (L_{air}), as for an aerial antenna, and only experiences loss in water (L_w). An additional advantage of underwater antennas, because of their reduced gain and range compared to aerial

Table 1. Types of receive antennas used in radio telemetry applications and their benefits and suggested uses.

Type, elements	Gain (dBd)	Length[1] (m)	Benefits/uses
Yagi, 3 or 4	6.5	1.1	Smallest antenna with largest beamwidth. Mobile or fixed use.
Yagi, 6	7.1	1.7	Higher gain but less beamwidth and larger. Mobile or fixed use.
Yagi, 9	11	3.6	Highest gain but least beamwidth and very large size. Fixed use.
Stripped	0	0.11	Inexpensive and easy to make. Used underwater in low flow.
Dipole	0	0.11	More durable and consistent radiation pattern. Used underwater.

[1]Based on model built for 150 MHz.

antennas, is their ability to provide increased position resolution of the transmitter. This benefit is often necessary in underwater environments such as fish passageways or other structures where fine-scale fish behavior is desired. Although commercially unavailable, two styles of underwater antennas are typically used in aquatic telemetry and are described by Beeman et al. (2004): 1) the stripped coaxial cable (coax) antenna and 2) the rigid dipole antenna. The stripped coax antenna is made by stripping all the material around the center conductor of the coaxial cable a distance of half the wavelength of the frequency of use underwater, (e.g., 11.4 cm at 150 MHz). This antenna is the easiest to construct but its flexibility results in variable and inefficient radiation patterns and eventual breakage under high flow conditions. The rigid dipole antenna is constructed of epoxy-filled Schedule 40 polyvinyl chloride (PVC) pipe and stainless steel bolts and fasteners, providing a resilient antenna that can be mounted to a fixed surface. The overall dimensions of the antenna (length of elements and distance from mounting surface) are equal to half the wavelength of the frequency of use underwater (11.4 cm at 150 MHz) and thereby minimize any effects of nearby reflective surfaces on the radiation pattern of the antenna. By mounting the antenna to a fixed surface, the user can maximize the range of the antenna by orienting the antenna elements parallel to the expected position of the transmitter antenna.

Transmission Line

Radio signals are carried from the receive antenna to the receiver along the transmission line. The transmission line consists primarily of coaxial cable but also includes other components such as connectors, splitter/combiners, antenna switches, preamplifiers, and attenuators. These additional components will be discussed in greater detail later, but it must be understood that these are also part of the transmission line because signal loss (or gain in the case of the preamplifier) occurs as it passes through these devices just like it does through the coaxial cable. Increases in the received power level (P_r) can be accomplished by decreasing the signal loss through the receive transmission line (L_{rtl}). Therefore, it is important to select a quality coaxial cable. Coaxial cable comes in several impedances (a measure of the load matching characteristics) determined by the size and spacing of the conductors and the type of dielectric (insulation) used between them. The two most common impedances used in coaxial cable are 50 and 75 ohms. The impedance of a particular coaxial cable is indicated by the RG (Radio Guide) number. For example, RG-59 and RG-6 are 75-ohm cables, typically used for television and other video applications. Radio applications use RG-58, RG-8, and RG-213 cables which have an impedance of 50 ohms. Since the impedance of all components of a radio system must match, it is important that all components of the radio system be 50 ohms. RG designations may also include a/U or A/U following the number (e.g., RG-58/U or RG-58A/U). Coax labeled RG-58 or RG-58/U (U stands for universal) indicates that the coax has a solid center conductor. Coax labeled RG-58A/U (A stands for aerial and U stands for universal) indicates that the coax has a center conductor made up of many (usually 17) thin wire strands. Either coax is suitable for radio applications, with the solid center conductor having slightly less loss and preferable for fixed applications (e.g., a shore or land-based receiver) where little to no movement of the coax will occur. Coax with a stranded center conductor is well suited for mobile applications (Section 5.3) due to its greater flexibility. Although all coax is shielded by definition, some have additional levels, or a greater percentage, of shielding which can be helpful in reducing interference caused by external sources of radio waves.

Perhaps the most important specification to consider when selecting coax is the nominal attenuation, usually expressed by the manufacturer as signal loss in dB per 100 m of coax. A good quality RG-58 coax has a nominal attenuation of about 15 dB per 100 m at 150 MHz. In comparison, RG-8, a larger diameter 50 ohm coax provides a nominal attenuation of about 6 dB per 100 m at 150 MHz; a savings of 9 dB per 100 m of coax which is an 8x increase in power for every 100 m of coax. By using a larger diameter coax, transmission line loss can be substantially reduced, although at the expense of additional cost and less flexibility of the coax. Different cable types used in radio applications and the benefits and suggested uses for each are summarized in Table 2.

INSTALLING AND OPTIMIZING COMPONENTS

Specific installation methods can vary greatly depending on the length, size, and objectives of the study and the conditions of the study area. However, there are many general guidelines or recommendations that should be followed during installation of equipment to ensure that the equipment operates as designed and is optimized to the fullest extent possible.

Receive Antennas

Commercially available antennas are designed for use at different frequencies based primarily on the lengths of the elements. However, some antennas may cover a wide range of frequencies and may be field tuned. It is important to ensure that antennas are ordered from the supplier tuned to the specific frequency, or the center of the range of frequencies, to be used. Alternatively, antenna designs that are field tunable can be tuned by the end user with the proper equipment, or by a hired radio technician or engineer.

Proper deployment of the antenna is critical to optimizing performance. Due to properties of signal propagation from underwater transmitters, the detection range of antennas can be increased by increasing the height of the antenna above the water. More specifically, since loss of signal occurs at the air-water interface (L_{awi}) due to refraction, increasing the height of the antenna above the water decreases the refraction angle, thereby resulting in less signal loss. The antenna height can be increased through the use of masts, towers, using available structure or topography, or a combination of these methods. Since the majority of radio sig-

TABLE 2. Types of 50 ohm coaxial cable used in radio telemetry applications and their benefits and suggested uses.

Type	Overall diameter (in.)	Center conductor	Benefits/uses
RG-58/U	0.195	Solid	Light, flexible, lower cost. Fixed use.
RG-58A/U	0.195	Stranded	Light, ultra-flexible, lower cost. Fixed or mobile use.
RG-8/U	0.405	Stranded	Lower loss than RG-58. Fixed use and long runs.
RG213/U	0.405	Stranded	Lower loss than RG-58. Fixed use and long runs.
LMR-400	0.405	Solid	Lowest loss but less flexible. Fixed use and long runs.

nals emitted from underwater transmitters are oriented vertically in air, it is important that the elements of the antenna be oriented vertically. This places the antenna in a vertically polarized position and optimizes signal reception. All elements of the antenna should be kept straight since any malformation of the antenna can distort the antenna's radiation pattern and will likely change the antenna's impedance, consequently degrading the reception of the signal. Similarly, the antenna's radiation pattern can be affected by any nearby structure, especially if it is made of metal and is parallel with the antenna's elements. Consequently, a good guideline is to maintain a distance of at least one wavelength from any structure, especially those that are conductive. Using the formula: wavelength (m) = velocity (m/s)/frequency (Hz), it can be determined that in air (since radio waves travel at the speed of light in air) at 150 MHz, one wavelength is 2 m, (i.e., 3×10^8 m/s/150,000,000 Hz). Radio waves travel much slower in water (33×10^6 m/s) so one wavelength underwater is only 21.8 cm at 150 MHz. An effective way to maintain the desired separation between the antenna and other structures (such as the mast) is to use an extension arm attached to the top of, and at a 90° angle to, the mast (Figure 1). Since noise interferes with signal reception, antennas should be pointed away from possible noise sources. A final consideration during deployment of antennas is to maintain tight and waterproofed connections at the antenna feedpoint. A quality coax sealing tape, wrapped around the coax connection at the antenna will prevent water intrusion into the connection. Waterproofing all outdoor connections is good practice because any moisture that gets into the connection will cause corrosion and increase the attenuation of the connection, resulting in signal loss and a reduction in system gain.

Transmission Line

Installation of coaxial cable usually involves simply unspooling cable from the antenna to the receiver and making sure that a sufficient length is unspooled before cutting the coax to avoid any unnecessary splices. If rodents or bears are known to inhabit the study area, it is highly advisable to protect the coax from being chewed on by placing it in conduit. Otherwise, securing the coax to handrails or other existing structure with cable ties or electrical tape will keep the coax off the ground and reduce the possibility of being damaged by foot or vehicular traffic. In areas where structure and access by people and animals is limited, simply laying the coax on the ground can usually be done with low risk. When handling, routing, and securing coax it is critical that the coax not be pulled too hard, bent beyond its bending radius (as specified by the manufacturer), pinched, or smashed, (e.g., by attaching cable ties too tightly). This is important because the electrical properties of the coax can change if the center conductor of the cable does not remain in the physical center of the cable, (i.e., if the insulation separating the center conductor and outer conductor is compressed or stretched).

When possible, installing new coax is best, but coax can usually be reused reliably for several years. If coax is reused, it should be inspected thoroughly before the start of the study for any damage such as cuts, abrasions, or malformations. In addition to the physical inspection, used coax should be tested for continuity (i.e., no resistance or a closed circuit) and to make sure there are no shorts between the two conductors. These tests can be made quickly and easily with an inexpensive multi-meter (also called a volt-ohm meter) or continuity tester and will be discussed in more detail near the end of this Section. Other tests, such as insulation resistance and Distance-To-Fault (DTF), can also be performed on the coax to identify and locate damage or areas of degradation. Specialized, relatively expensive equipment is necessary to conduct these tests but the advantages gained are often worth the extra cost for larger scale studies that use substantial amounts of coax.

Connectors

To facilitate attachment, coax needs to be terminated at the antenna and receiver ends with the proper type of connector. Connectors are available in several different types, attachment methods, and qualities. The BNC (Bayonet Neill-Concelman) connector is probably the most popular for above-water applications using RG-58 coax. The BNC connector has a twist-lock mechanism that is simple to use and provides a secure and positive connection. For above-water applications using RG-8 coax, the N connector (named after its inventor, Paul Neill) is typically used. Type N connectors are larger than BNC connectors to accommodate the larger diameter RG-8 coax, have a threaded fitting, and also provide a very secure connection. Most connectors are available in three different methods of attachment: crimp, clamp, and twist-on. The crimp method is the most common and most secure method of attachment. It involves the use of a crimping tool to crimp the cable shield to the connector with a crimp sleeve. The center conductor of the connector is either crimped or soldered to the coax center conductor. Clamp method connectors involve soldering the center conductor and clamping the cable shield between the connector and a threaded back nut. Clamp method connectors are useful when crimp tools are not available and, due to their design, are useful in weather exposure applications. Specialized clamp method connectors are even available for underwater use. The twist-on method involves simply twisting the entire connector body onto the stripped cable end. It can be popular in field applications due to its simplicity and no need for special tools but is probably the most unreliable method of attachment. Connector quality, as with most things, increases with price and can be identified by highly conductive plating such as silver and gold and either a Delrin or Teflon dielectric. As mentioned previously, the connector impedance must match the impedance of the coax, antennas, and all other components which in the case of biotelemetry applications is 50 ohms. As with antenna connections, all exposed connections should be waterproofed to prevent water intrusion and the resulting signal degradation due to corrosion. The preferred method of waterproofing connections is using a self adhering waterproofing tape. Alternatively, connections can be placed in a small enclosure. At a minimum, several wraps of a quality electrical tape should be used to protect the connection. Although connectors rated for underwater use are available, it is extremely difficult to prevent water intrusion in underwater connections so they should be avoided if possible. If underwater connectors are necessary, they should conform to the IP68 Standard for Watertight Connectors and must be installed to the manufacturer's specifications (IEC 60529:1989)

Splitter/combiners, antenna switches, preamplifiers, and attenuators

Other often-used components that are positioned between the antenna and the receiver and, therefore, can be considered part of the "transmission line," include splitter/combiners, antenna switches, preamplifiers, and attenuators (Figure 6). Splitter/combiners are used to split the signal among many receivers or combine the signal from multiple antennas. Antenna switches allow capable receivers to switch among multiple antennas, as well as listen to all connected antennas simultaneously. Preamplifiers are used to mitigate signal attenuation

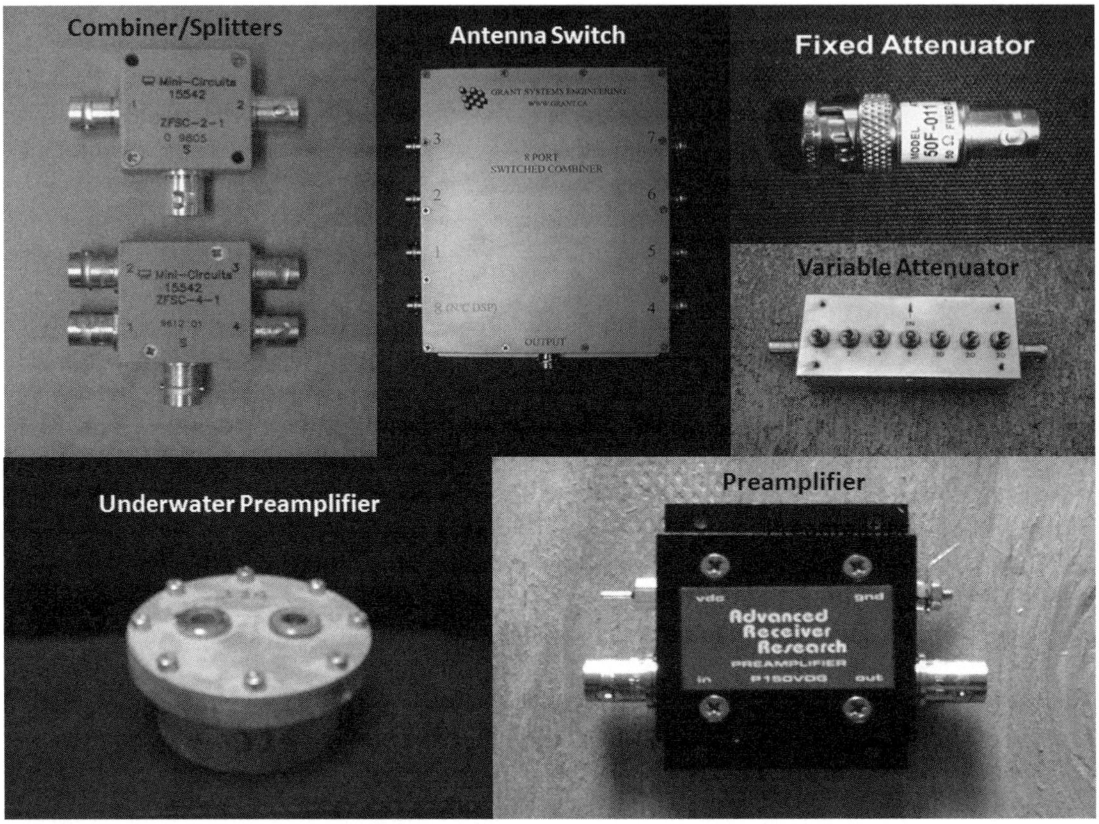

Figure 6. Photographs of splitter/combiners, an antenna switch, fixed attenuator, variable attenuator, and preamplifiers.

caused by long cable runs and the use of splitter/combiners and antenna switches by increasing the signal level. Fixed attenuators, commonly available in values ranging from 1 to 30 dB, are used at the end of the transmission line to decrease the amplified signal so there is no net gain or loss at the receiver input. Before or during installation, all components should be tested to confirm they are working to specification (or at least to an acceptable level) and that excessive loss is not occurring through any portion or signal pathway of the component. The simplest way to conduct these tests is by using a relatively inexpensive signal generator (often available from transmitter manufacturers) and a receiver. If desired, more accurate test equipment such as spectrum analyzers and bench-top signal and waveform generators are available and reduce the time and variability in signal strength that can be experienced (usually due to temperature changes) with the less stable signal generators and receivers. If a signal generator made by a tag manufacturer and a receiver are used, both should be allowed to warm-up for at least 15 min prior to taking measurements to limit variability in signal strength readings. To conduct the test, the signal generator is first connected directly to the RF input of the receiver or spectrum analyzer with two short (1 m) equal lengths of coax, coupled together with a barrel connector, and the signal strength is recorded. The signal strength should be adjusted to a medium level either by adjusting the gain on the receiver or adjusting the output of the signal generator. If the signal generator's output is not adjustable, fixed or manually variable attenuators (Figure 6) can be used to adjust the output to the appropriate level. This is important because, if the signal is too strong, the receiver will become saturated (reach its maximum) and will display a signal strength reading at a maximum limit. Alternatively, if the signal is too weak, any further signal attenuation caused by a faulty component could result in no signal being received when the component is tested. After recording the reference signal strength, the signal generator is connected to the RF input of the test device with one of the two coax leads and the test device is connected to the RF input of the receiver or spectrum analyzer with the other coax lead. If the device being tested is a splitter/combiner or antenna switch, the signal strength received while testing the device will be less than the reference signal strength. These components have an inherent attenuation (also called insertion loss) that increases with the number of inputs and slightly increases with the frequency (typically only 0.1–0.5 dB over the 30–200 MHz range). For example, a 2-way splitter/combiner which has one input port and two output ports (or two input ports and one output port if used as a combiner), or a total of 3 ports, typically has an insertion loss of a little over 3 dB. An 8-way splitter/combiner has a typical insertion loss of a little over 9 dB. A good rule of thumb to determine the insertion loss of splitter/combiners is that the insertion loss in dB will be about equal to the total number of ports (both input and output). Specific insertion loss values can and should be obtained from the manufacturer. These expected values should then be measured against the actual loss values obtained from the test. If a spectrum analyzer is used for testing, the loss in signal strength will be displayed in dB and the attenuation through the component can easily be calculated. If, however, a receiver is used that displays signal strength in some relative range of numbers, then the most effective way to determine insertion loss through the tested component is by using a manually variable attenuator. This is accomplished by replacing the device under test with the variable attenuator and then adjusting the attenuator until the signal strength equals that received while testing the device. The number of decibels required on the variable attenuator to achieve the signal strength received while testing the device is the insertion loss. Preamplifiers are tested in a similar fashion but, if working correctly, will produce a stronger signal at the receiver compared to the reference, rather than a weaker signal. Therefore, the variable attenuator is placed between the preamplifier and receiver (while the preamplifier is operating) and used to determine the number of decibels required to attenuate the signal to the reference signal strength. The value recorded is the gain of the preamplifier. Preamplifiers are available in different gains, typically 15 dB or 24 dB, with the higher gain model costing more than the lower gain model but at about the same cost per dB. It is advantageous to test the output of each amplifier prior to use since small differences (usually ± 2 dB) in gain often exist from that specified by the manufacturer. Knowing the expected increase in signal (measured in dB) for a given preamplifier facilitates correct placement in a multiple antenna array and serves as a benchmark for evaluating preamplifier condition.

Receivers

The first step to optimizing receivers is to ensure they are maintained and in good working condition. Receivers should be tuned and internal batteries should be replaced by the manufacturer at least every few years, or at the interval recommended by the manufacturer. It is often a good idea to have the receiver manufacturer, or a qualified radio technician, bench check the receivers before the start of a study to ensure they are performing to specification. To prevent inaccurate date and time stamps of transmitter locations in your data, the correct date and time should be synchronized on the internal clock of all receivers. This is especially important for receivers in close proximity to each other or those that are part of a multi-receiver array. The clocks of many brands of receivers will drift over time (usually up to 1 min per week) and must be checked and updated during the study as necessary. Another step that should be taken is to ensure that all receivers to be used in the study provide the same signal strength at a given gain, preferably at or near the gain to be used during the study. Depending on the brand of receiver, there may be a user changeable setting (the name of the setting can be different among manufacturers) that can be adjusted or the receiver may need to be adjusted by the manufacturer. To check that each receiver is hearing a standard signal identically at a given gain, the user should set the gain of each receiver to the same gain setting (preferably the gain to be used during the study) and then verify that the signal from a signal generator connected to the RF input of each receiver provides the same signal strength across all receivers. Calibrating the receivers in this way does not change the sensitivity of the receiver, but allows the optimum gain value to apply universally to any receiver. This is most important for receivers that will be deployed adjacent to one another and from which relative signal strengths will be compared to determine transmitter location. This step will allow receivers to be set at the same gain and will prevent the possibility of incorrectly positioning a transmitter due to one receiver incorrectly hearing a signal more strongly than another receiver.

The performance of a receiving system can also be optimized by the type of receivers used and the way in which they are deployed. The risk of not detecting transmitters increases when a high number of frequencies must be scanned. The risk becomes even higher when multiple antennas must be scanned or when transmitters are expected to have little capture time due to high water velocities, for example. Probably the most effective way to manage these circumstances is by using a broadband receiver. Broadband receivers can simultaneously listen to a broader band of frequencies (typically 25), thereby eliminating the need to scan through multiple frequencies like a narrowband receiver. One drawback to the broadband receiver is its susceptibility to noise. Because it listens to a wider frequency range, environmental ambient noise can flood the receiver, potentially resulting in the loss of data. Due to this limitation, broadband receivers are used with aerial antennas only in environmentally quiet areas. The most popular and effective use of broadband receivers is to monitor underwater antennas where detection ranges are smaller and environmental noise is minimal.

The time required to scan through multiple frequencies can also be reduced by using multiple narrowband receivers. By using a splitter/combiner to split the signal coming from each antenna to two receivers, and scanning one half of the frequencies in one receiver and the other half of frequencies in the second receiver, the scan time can be reduced by half. Although more exhaustive on receiver resources, this can be an effective means to reduce scan time if a sufficient number of receivers are at your disposal.

A final consideration toward optimization during receiver deployment is in how antennas are assigned and connected to receivers. Because of the difference in the amount of environmental noise present in the air versus underwater, it is important that both aerial antennas and underwater antennas not be connected to the same receiver. Since more noise will likely be heard by aerial antennas, compared to underwater antennas, the optimum gain setting on the receiver will be lower for aerial antennas than for underwater antennas. By using separate, dedicated receivers for aerial and underwater antennas, the gain can be optimized for each type of antenna, rather than compromised to work for both types.

OPTIMIZING THE SYSTEM

As mentioned previously, preamplifiers are used to compensate for signal attenuation that occurs as the signal travels through the transmission line and any inline devices installed between the antenna and receiver. Preamplifiers should be installed at the appropriate locations and at the appropriate intervals in order to ensure an optimized receiving system. After preamplifiers are installed, all antenna inputs to each receiver must be balanced, or made equal to each other, by adding the appropriate amount of attenuation. This is necessary because, in practice, the signal is usually over amplified due to the gain of the preamplifier not exactly matching the loss in the cable and inline devices. Additionally, preamplifiers introduce a small amount of additional noise to the system, which must be counteracted. This can be accomplished by establishing and maintaining a small amount (10 dB) of over amplification between preamplifiers, and then ultimately attenuating the signal back down to the reference signal strength at the end of the transmission line. As a general rule, amplifying and balancing antenna inputs with less than about 2 dB of loss may not be required if comparisons of signal strength among antennas will not be made to determine transmitter location. All antenna inputs, of systems designed to identify transmitter locations by comparing relative signal strength among inputs, should be amplified and balanced. It is important to ensure all receiver inputs are balanced for the same reason it is important that all receivers hear a standard signal identically at a given gain: to prevent the possibility of incorrectly positioning a transmitter due to having antennas with unequal detection zones. If, for example, comparisons of transmitter signal strength are made between two adjacent antennas with very different detection zones, an erroneous location might be assigned. This is possible because the antenna with the larger detection zone could have recorded a greater signal strength for the transmitter than the antenna with the smaller detection zone, even when the transmitter was closer to the weaker antenna. This scenario could result in bias in data as a result of skewed detection probability and could possibly indicate certain passage routes were more heavily used when they were not. Another critical reason for attenuating the amplified signal to the reference signal strength and ensuring equal detection zones among antennas is to avoid saturating the receiver with noise. If even one antenna is not appropriately attenuated after being amplified, excessive noise can degrade the performance of the entire receiving system. To get a better understanding of how a radio signal is effected as it travels from the antenna to the receiver and why using preamplifiers and attenuators is important, refer to Figure 7, which plots the received signal strength over distance from the antenna to the receiver for the same receiving system but under three different scenarios: 1) no amplification and no attenuation, 2) amplification but no attenuation, and 3) amplification and attenuation.

The process of amplification and balancing begins by determining preamplifier location. As a starting point, one preamplifier should be placed at each antenna, as close as possible to the antenna to avoid signal loss through the coax jumper required to connect the preamplifier and antenna. Typically, for aerial antennas, preamplifiers are connected to the antenna with a 2 m length of coax. This places the preamplifier at an accessible elevation when mounted to a 3 m antenna mast. For underwater antenna arrays, placing the amplifier underwater in a watertight housing is most efficient but, as mentioned previously, underwater housings and connections are very difficult to keep dry and often fail. For this reason, most underwater antennas used in the 150 MHz frequency range are built with up to 33.3 m of coax and the

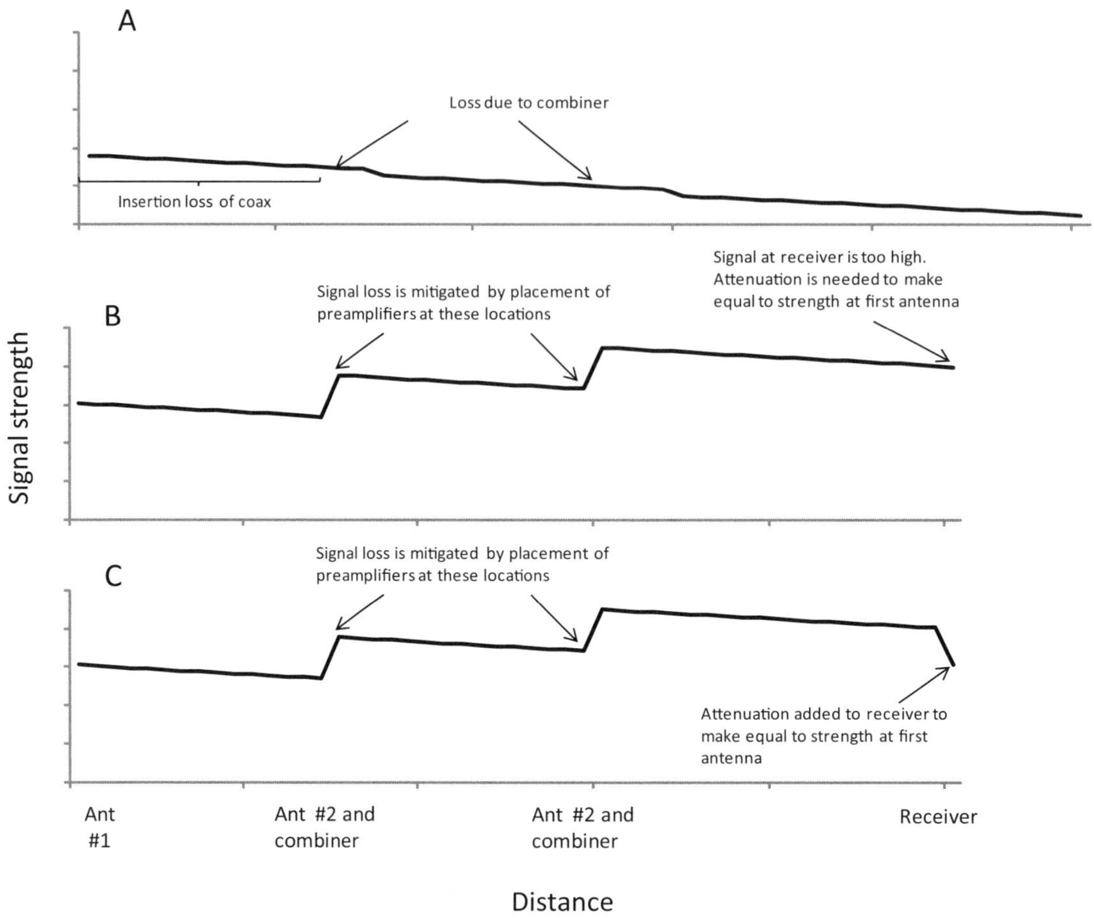

FIGURE 7. Graph showing signal strength over distance for a radio telemetry array that has not been optimized using preamplifiers and attenuators (A), an array that has had preamplifiers installed but no attenuation at the receiver (B), and an array that has been optimized by the proper placement of preamplifiers and attenuation at the receiver.

first preamplifier (the one closest to the antenna) is placed above water. This configuration will typically result in about 9 m of detection range for the underwater antenna, depending on water conductivity and the output of the selected transmitter. To determine where to place the second preamplifier (or if one is even needed), the gain of the first preamplifier and the cable loss must be known. For example, if a preamplifier with a gain of 25 db is used and the coax has an attenuation specification of 5 dB per 33.3 m, then the signal will be amplified for a distance of 166.5 m (25 dB/5 dB × 33.3 m). However, since we want to establish an over amplification of 10 dB, as discussed above, we would need to place the second preamplifier 100 m after the first preamplifier ((25 dB – 10 dB)/5 dB × 33.3 m) where there is still 10 dB of amplification. Thereafter, if coax length warrants, preamplifiers would be placed at 166 m intervals. The length of the remaining coax, as well as any splitters, combiners, or antenna switches, between the last preamplifier and the receiver, will determine the amount of attenuation required at the antenna input. For example, if 33.3 m of coax having 5 dB loss remains between the last preamplifier (having a gain of 25 dB) and the receiver, then 20 dB of attenuation would be required at the receiver's RF input. If, however, an antenna switch with an insertion loss of 11 dB was being used, then the attenuation needed in this instance would be only 9 dB (25 dB –5 dB –11 dB = 9 dB). If the insertion loss of any combiners or antenna switches at the receiver exceeds the amplification provided by the last preamplifier, then an additional preamplifier would need to be added at the receiver location. The additional preamplifier should be placed before the combiner or switch, and then any remaining over amplification attenuated at the receiver RF input. To ensure that theoretical and actual amplification and attenuation are correct, all antenna inputs should be tested. The method to test the balancing of the inputs is similar to that described earlier for testing splitter/combiners, antenna switches, and preamplifiers and includes: 1) introducing a signal from a signal generator to the receiver or a spectrum analyzer to determine the reference signal strength, 2) moving the signal generator to the input of the first amplifier (nearest the antenna) and introducing the same signal used in step 1, 3) determining the resulting signal strength at the receiver or spectrum analyzer, and 4) adjusting the amount of attenuation at the receiver's RF input until the received signal strength equals the reference signal determined in step 1. If the system was installed correctly and all components are operational, step 4 usually involves minor adjustments in attenuation (to the attenuation added based on theoretical calculations) due to slight differences between advertised and true output from preamplifiers (this is why it is good practice to test preamplifier output before installation) and errors in measuring the length of the coax and its associated loss. Systems should be balanced to within 1 dB for optimum performance.

Multiple receiving systems, (i.e., several receivers each with inputs from multiple antennas), used to collect data together can be balanced in a similar manner to individual systems, providing the benefit of direct comparison of results among systems. In this case, all receivers must first be calibrated (as described previously) so they record and display the same signal strength when given a reference signal at a given gain. The systems are then individually balanced as described above. Finally, the gains of all receivers are set equal to each other by setting the gain equal to the highest gain achievable on the system with the most ambient noise. This is necessary to avoid frequent recording of erroneous data from ambient noise but means that some receivers may be set to a lower than optimum gain if one receiving system is in a particularly noisy area. Although this practice produces balanced receiving systems with antennas having relatively equal detection zones, it is important to note that setting the receiver gains equal to each other can result in some systems having less gain than ambient noise allows.

MAINTAINING THE SYSTEM

Verification of system performance

Following installation and optimization of a telemetry system, it is critical that the entire system be tested to verify it is operating in a consistent and reliable manner and that detection ranges are adequate and documented. Since the amplification and balancing process did not involve the antennas or the length of coax connecting the antenna to the preamplifier, any problems with these components could go unnoticed if not tested. A system test also verifies that the receiver is configured and working correctly and that transmitters are detected and recorded within the desired area. To conduct the test, transmitters with the same output as those to be used during the study should be used. Researchers often order a few extra transmitters to be used for testing during setup and occasionally throughout the study. A quick way to verify operation of the system is to simply hold a test transmitter in the air near each aerial antenna. However, this is not the best method since the transmitter is more powerful in air than in water and could mask some problems with equipment. Underwater antennas will not hear a transmitter in air due to the signal attenuation caused by the water. To test underwater antennas, a transmitter needs to be placed underwater within range of each underwater antenna. The best method for verifying system performance for both aerial and underwater systems is to deploy one or more test transmitters underwater within the study area. Ideally, the tags can be towed through the study area with a boat or, if conditions permit, suspended from a buoy and allowed to drift. The latter is usually the preferred method since motor noise from boat use can cause system interference and may reduce detection ranges from what they are in the absence of motor noise. Additionally, the buoy can be outfitted with a GPS, if desired, to provide an accurate track of the transmitter as it drifts through the study area. The GPS, or alternatively, a rangefinder, can also be useful in determining the maximum range for each antenna. Several drifts should be conducted at various water depths to determine maximum depth of detection. When combined, these tests will not only verify how the system is performing, but will provide documentation of the coverage area of the receiving system. Any gaps or weak areas in coverage should be identified by the testing, allowing the opportunity to add or move antennas as necessary to improve system coverage.

In addition to the system performance testing conducted at the start of the study, routine checks of the system should be made periodically throughout the study period. This routine, in-season maintenance ensures that the receiving system continues to perform at the desired level and facilitates identification of any failed components or damage to the system. The level of effort necessary for in-season maintenance can vary depending on the size, complexity, and location of the system. Large, complex receiving systems that use a large number of electronic components will require more time and effort than simpler, smaller-scale systems. Similarly, receiving systems in remote areas usually demand more resources to adequately verify system performance than systems that are close and easy to visit. The frequency of site visits should be often enough to prevent loss of data if a malfunction or damage occurs. To substantially reduce the number of site visits required for maintenance, as well as data retrieval, most receivers have the ability to be interfaced with radio, cellular, or satellite modems that will allow remote access and control of the receiver (Section 5.3). Remote access of receivers, combined with the proper use of beacon transmitters, enables the user to assess system performance without visiting the site. Beacon transmitters are transmitters with large, often replaceable, batteries that can last the duration of the study. If placed within range of all antennas, beacon transmitters provide a consistent radio signal that can be used to ensure that all antennas and other components of the receiving system continue to operate. Steps, however, should be taken to avoid the collection of overabundant and unnecessary data from the beacon transmitters. Probably the most effective way to manage the extra data created by beacon transmitters is by having the manufacturer program the beacon to transmit only periodically, (e.g., for the first minute of every hour), or to transmit at a reduced pulse rate, (e.g., every 5 min instead of every 5 s). Some beacon transmitters available from various tag manufacturers are user programmable, providing an easy and adaptable means to managing beacon transmitters.

Troubleshooting and Maintenance

When problems do arise, they can usually be diagnosed by concentrating efforts on specific areas. Most issues are due to failed or drained power supplies, bad connections, or physical damage to the transmission line. Efforts should start at the receiver and proceed by working out toward the antenna as testing is successful. This enables you to verify operation of the least number of components at a time, making it easier to locate the problem. For total system failures, power supplies for the receiver and preamplifiers should be checked first. If the problem exists in one or more antennas (but not all), troubleshooting should start at the common component closest to the receiver, (e.g., the antenna switch or splitter/combiner). The first step to troubleshooting a lack of signal is to conduct a continuity test. This procedure checks for a continuous or closed circuit and verifies that there are no breaks or shorts in the cable or signal pathway of the device being tested. The continuity test is easily performed with an inexpensive continuity tester or multi-meter switched to the Ohm setting. One probe of the meter is placed on the center conductor at one end of the coax, and the other probe on the center conductor at the other end of the coax. If the center conductor of the coax is not broken, the meter will give a reading of zero, or very near zero, indicating zero resistance and good continuity. If the center conductor is broken, the meter will give a reading of O.L. (open loop) or 1 on a digital meter, or the needle will not move on an analog meter, indicating no continuity. The same test can be made on the outer conductor of the coax (or the body of one connector to the body of the other). If the continuity test needs to be performed on a very long length of coax that exceeds the length of the meter's test leads, a handy trick is to short the center pin to the body of the connector on one end of the cable and then measure continuity across the center pin and connector body on the opposite end of the cable. If the meter measures no resistance then the cable has continuity. To check for a short, (i.e., a connection between the inner and outer conductors), one of the meter's probes is placed on the center conductor and the other probe is placed on the body of the connector. The probes can be placed on the same end of the coax to check for a short but it is important that the other end of the coax not be connected to anything, (i.e., open). Otherwise, a reading of something other than zero will be obtained leading to the incorrect conclusion that there is a short in the line. If a short is found, it is likely due to a bad connector. To avoid wasting one or two connectors if the short is in the cable, the connectors should be cut off and the cable retested for a short without connectors installed. If the short, or any other damage, is in the cable, either the cable can be replaced or a testing device with Distance-To-Fault (DTF) measurement capabilities can be used to determine the location of the short. Distance-To-Fault is a performance verification and failure analysis tool used for antenna and transmission line service and maintenance. One of two measurement techniques, either Time Domain Reflectometry (TDR) or Frequency Domain Reflectometry (FDR), is used to conduct DTF measurements. Frequency Domain Reflectometry employs a user selectable RF signal to conduct the test, rather than pulsed DC signals as with TDR. For this reason, FDR is more sensitive than TDR and cannot only locate faults but can pinpoint small degradations in performance caused by such things as corroded conductors, loose connections, and moisture, thereby allowing a level of predictive maintenance. The greater sensitivity of FDR compared to TDR, however, comes at a higher cost; instruments using FDR start around U.S.$2500 and instruments using TDR start around U.S.$1500. The extra cost for cable and antenna analyzers with DTF capabilities may not be justifiable for most small-scale studies. Studies utilizing large amounts of coax and antennas, however, can realize substantial savings in troubleshooting and maintenance time by the use of TDR instruments.

Regardless of the size and budget of the study, the knowledge and experience of field crews, or the availability of test equipment, most researchers conducting biotelemetery studies have one thing in common: an adequate supply of spare equipment. Having enough spare items of every piece of equipment used is the best way to ensure that the system can quickly be back in operation, with little loss of data, if damage or failure occurs. Most electronic components are not field-serviceable, or researchers lack the expertise to make the repairs, and should be returned to the manufacturer for repair. If problems are realized soon after they occur, are located in a timely manner, and the faulty equipment is repaired or replaced without delay, the receiving system will maintain a high level of performance with little loss of data.

SUMMARY

The success of a radio telemetry study depends largely on receiving system performance. An understanding of how the system functions as a whole and how the system variables interact (as described by the system equation: $P_r = P_t - L_w - L_{awi} - L_{air} + G_r - L_{rtl}$) will enable the user to maximize receiving system performance by making informed choices during selection of equipment and utilizing installation techniques aimed at optimizing performance. By maximizing transmitter power and receive antenna gain and by minimizing loss through the receive transmission line; the received power can be increased. Each individual component in the receiving system should be calibrated and tested before use to ensure optimum performance. Other techniques, such as using preamplifiers to compensate for transmission line loss, balancing antennas in multi-antenna arrays, and setting receiver gains equal to each other in multi-receiver installations, can be used to optimize the receiving system. Finally, preseason system performance verification, in-season performance monitoring, and knowledge of how to troubleshoot and maintain the system, are essential to keeping an optimized receiving system functioning at maximum performance.

ACKNOWLEDGMENTS

We thank Noah Adams, John Eiler, Toby Kock, and Bob Magie for their critical review of this Section. Our appreciation is also extended to those individuals who assisted in developing the techniques discussed which were improved upon over many years of research. We would also like to express our gratitude to Cam Grant for passing along his knowledge of radio telemetry system design and optimization over the years.

REFERENCES

Beeman, J. W., C. Grant, and P. V. Haner. 2004. Comparison of three underwater antennas for use in radiotelemetry. North American Journal of Fisheries Management 24:275–281.

Brown, R. S., S. J. Cooke, W. G. Anderson, and R. S. McKinley. 1999. Evidence to challenge the "2% rule" for biotelemetry. North American Journal of Fisheries Management 19:867–871.

IEC 60529:1989+A1:1999(E). 1999. Degrees of protection provided by enclosures (IP Code). International Electrotechnical Commission, Geneva, Switzerland.

Jepson, N., C. Schreck, S. Clements, E. B. Thorstad. 2005. A brief discussion on the 2% tag/body mass rule of thumb. Pages 255–259 *in* M. T. Spedicato, G. Lembo, and G. Marmulla, editors. Aquatic telemetry: advances and applications. Proceedings of the Fifth Conference on Fish Telemetry held in Europe. Ustica, Italy. COISPA Technology and Research and Food and Agriculture Organization of the United Nations. FAO, Rome.

Perry, R. W., N. S. Adams, and D. W. Rondorf. 2001. Buoyancy compensation of juvenile Chinook salmon implanted with two different size dummy transmitters. Transactions of the American Fisheries Society 130:46–52.

Perry, R. W., A. C. Braatz, M. S. Novick, J. L. Lucchesi, G. L. Rutz, R. C. Koch, J. S. Schei, N. S. Adams, and D. W. Rondorf. 2007. Survival and migration behavior of juvenile salmonids at McNary Dam, 2005. Final Report of Research by the U.S. Geological Survey to the U. S. Army Corps of Engineers, Walla Walla District, Contract W68SBV50407617, Walla Walla, Washington.

Sisak, M. M., and J. S. Lotimer. 1998. Frequency choice for radio telemetry: the HF vs. VHF conundrum. Hydrobiologia 371/ 372:53–59.

Skalski, J. R., J. R. Stevenson, J. Lady, R. Townsend, A. E. Giorgi, M. Miller, and K. English. 2001. An assessment of project, pool, and dam survival for Chinook and steelhead smolts at Rocky Reach and Rock Island dams using radiotelemetry and PIT-tag techniques, 2000. Report prepared for the Public Utility District No. 1 of Chelan County, Wenatchee, Washington.

Stevenson, J. R., J. R. Skalski, J. Lady, R. Townsend, A. E. Giorgi, and R. McDonald. 2000. A pilot study assessing the feasibility of using radiotelemetry and PIT-tag techniques to estimate project, pool, and dam survival of steelhead smolts at Rocky Reach and Rock Island dams, 1999. Report prepared for the Public Utility District No. 1 of Chelan County, Wenatchee, Washington.

Winter, J. D. 1983. Underwater biotelemetry. Pages 371–395 *in* L. A. Nielsen and J. D. Johnsen, editors. Fisheries techniques. American Fisheries Society, Bethesda, Maryland.

Section 5.3

Tracking Aquatic Animals with Radio Telemetry

John H. Eiler

INTRODUCTION

Previous Sections have discussed aspects related to conducting a telemetry study, including the technical attributes of the transmitters and receiving equipment, overall study design, and tagging methods that minimize adverse impacts on the animals being studied. In addition to these considerations, an essential requirement of any successful project is the ability to effectively relocate the individuals that have been tagged.

A variety of aquatic fauna (hereafter referred to as fish, although our discussion also applies to other animals) have been tracked using radio telemetry since the early 1960s, although the methods used have evolved substantially. Initially studies were limited to small numbers of individuals (usually <40) in confined areas. Technological advances have made it possible to track large numbers of fish, wide-ranging species, and smaller individuals. Telemetry is increasingly being used to obtain quantitative as well as qualitative information ranging from fish movements, distribution, and habitat use (Eiler et al. 1992; Hilderbrand and Kershner 2000; Zigler et al. 2003; McCleave et al. 2006) to physiology and bioenergetics (Hinch and Rand 1998; Cooke et al. 2006; Weatherley et al. 2006). Tracking methods can be relatively straightforward when dealing with a few localized individuals, but monitoring large numbers of fish over vast expanses can be extremely challenging. Although the methods used can have a major impact on study success, the technical details used to address these challenges are often not included in published accounts.

This Section discusses the technical aspects and practical considerations associated with tracking fish. Understanding the technical capabilities and limitations is essential for selecting methods that are effective and provide usable information. Methods suitable for determining habitat use in small, urban streams will differ substantially from those used to study migratory patterns over substantial distances. Although this Section focuses primarily on tracking methods used in rivers, these techniques also applied to lacustrine environments. Tracking fish in lakes occasionally requires special considerations, which are discussed separately (Section 6.4). Many of the examples used in this Section are from telemetry studies on adult salmon (*Oncorhynchus* spp.) in Alaska. Although the research objectives, scale, life stage, and field conditions encountered in other studies may differ, the underlying principles are applicable to the problems faced by most researchers.

Tracking methods can be grouped into two basic categories: stationary tracking, where the fish are recorded when they move within reception range of a fixed location; and mobile tracking, where the researcher actively searches within the study area to determine the location of the fish. Both approaches have advantages, and the suitability of the method relates directly to the study objectives and logistical constraints.

STATIONARY TRACKING

Basic Approach

Stationary tracking typically uses one or more tracking stations placed at fixed locations to detect, identify, and record radio-tagged fish that move within reception range. Stations have been used to document residency patterns in areas of high use, and to detect fish passage at prominent man-made or natural structures, such as dams, fish passageways, or sections of river with high velocity flows (Gerlier and Roche 1998; Hinch and Bratty 2000; Boggs et al. 2004). Multiple stations, placed on primary travel routes and major tributaries, have been used to determine the distribution and migratory patterns of fish in large river drainages (Eiler 1995; Evenson and Wuttig 2000; Keefer et al. 2004; Ebner and Thiem 2009).

Advantages of Stationary Tracking

Stationary tracking has a number of advantages (Table 1). Stations provide continuous coverage, making it possible to concentrate tracking efforts at locations of particular interest. It is also possible to monitor isolated sites where regular access would be logistically difficult or prohibitively expensive. Stationary tracking can be extremely effective. Tracking success was substantially greater when using stations to track migrating salmon in large river drainages compared to repeated aerial surveys (Eiler 1995). Stations can operate under a wide range of environmental conditions, including inclement weather (e.g., heavy rain, low cloud cover, and high winds), extreme temperatures, and variable day length—conditions that can adversely affect or disrupt mobile tracking. Although initial costs for equipment and installation are relatively high, once in place the stations can be used for years, substantially reducing operational costs. Since the stations are automated, researchers are able to spend more time on other tasks.

The type of information provided by stationary tracking has several distinct advantages. By collecting information at standardized sites, it is possible to accurately determine timing and calculate movement rates that are directly comparable for different reaches of the study area or different groups of fish. With mobile tracking, fish are not recorded at standardized locations, making direct comparisons difficult. This problem can be minimized by conducting more frequent surveys, but this requires additional time, effort, and money, and the comparisons are still not truly standardized. Stationary tracking also provides a more comprehensive assessment of site residency than would be possible with periodic mobile tracks.

Station functions are software-based. The resulting flexibility makes it possible to fine tune the type of data collected to better address study objectives. Determining in advance what information is needed and working with vendors to develop the necessary software functions can greatly enhance the end result. The ability to access the data in a computerized form,

Table 1. Attributes of stationary and mobile tracking methods.

Attribute	Stationary tracking	Mobile tracking
Coverage	Continuous	Intermittent (resolution dependent on time interval between surveys)
	Limited to specific site	Able to survey extended area
	Able to operate under wide range of environmental conditions	Impacted by adverse environmental conditions
	Able to effectively monitor remote sites	Limited access with boat/foot surveys, good access with aerial surveys
Data types	Presence/absence at site Site passage Movement rates between sites	Location estimates[a,b] Movement rates between sites[c] Movement patterns between sites Use of transmitter sensors[b]
Equipment	Gear intensive	Moderately gear intensive
	Gear must be continually maintained over extended period	Minimal maintenance, gear checked and tested prior to survey
Costs	High initial costs (gear, site selection/preparation, and station installation)	Moderate initial costs (gear)
	Reduced operating costs over extended period (long field season, mult-year studies)	Low operating costs for boat/foot tracking, moderate to high operating costs for aerial tracking (dependent on number of surveys)

[a] Accuracy and precision affected by tracking techniques used.
[b] Can integrate with other information, including habitat type, environmental conditions, and activity patterns.
[c] Increased error when calculating movement rates due to variable tracking locations and time intervals.

either from regular visits to the site or remotely with communication uplinks (discussed in the following section), allows the information to be used in-season to coordinate and prioritize other research activities and address specific management issues.

Stationary tracking can be broken down into four principal tasks: equipment selection, site selection, station installation and maintenance, and data collection and retrieval. Each task is an integral part of the process.

Station Equipment

Stationary tracking is gear intensive, and selecting equipment that will meet the research objectives and function effectively under existing field conditions is an essential first step. Tracking stations consist of a number of component parts (Figure 1) starting with a structural framework that supports and integrates the other components. Prefabricated, self-contained structures (e.g., tower) are commercially available and have the advantage of standardizing and simplifying setup procedures. Many of these structures are collapsible and designed for rapid assembly, making transportation and installation easier. These features are particularly important when working at remote or hard-to-access sites. When available, existing structures such as buildings and bridges can be used, and some researchers have been fairly innovative using trees and other natural objects. This approach can be more time-consuming since setup procedures are not standardized, but it reduces the amount of gear transported to and from the site and minimizes the impact on the surrounding area. Some man-made structures also provide additional protection from adverse environmental conditions and vandalism. However, it is important to balance the convenience and added protection gained by using a preexisting structure with the ability of the site to provide adequate coverage of the area.

Receiving Antennas

A variety of receiving antennas are available with varying capabilities (Section 5.1), and the type selected should reflect the kind of data needed. Stations generally collect two types of information: whether tagged fish are present within the area (presence-absence), and the location of the fish relative to the station (directional). Vertical dipole antennas are a good choice when collecting presence-absence data. They are easy to install, durable, and don't require precise orientation since they are omni-directional. Movements past the station can sometimes be inferred from presence-absence data, such as when fish exhibit pronounced directional movements (e.g., migrating upstream to spawning areas).

Directional antennas have greater reception range (Section 5.1), and provide more definitive information on location as well as presence-absence data. Three-element and 4-element Yagi antennas are commonly used as directional antennas for higher radio frequencies (148–172 MHz). They are easy to transport and install, and durable under harsh field conditions. Radio signals emerging from water are vertically polarized (Section 5.1), and orienting the Yagi antenna elements vertically maximize performance (Figure 1). H-antennas have similar characteristics and are more compact, but provide less reception range. Stacked arrays or Yagi antennas with more elements are more directional and increase reception range (Section 5.1). Kenward (1987) reported that 12-element Yagi antennas exhibited over twice the range of 3-element Yagi antennas. However, these larger antennas are cumbersome, less durable, and leave greater peripheral gaps in coverage, which can be problematic for stations on wide rivers.

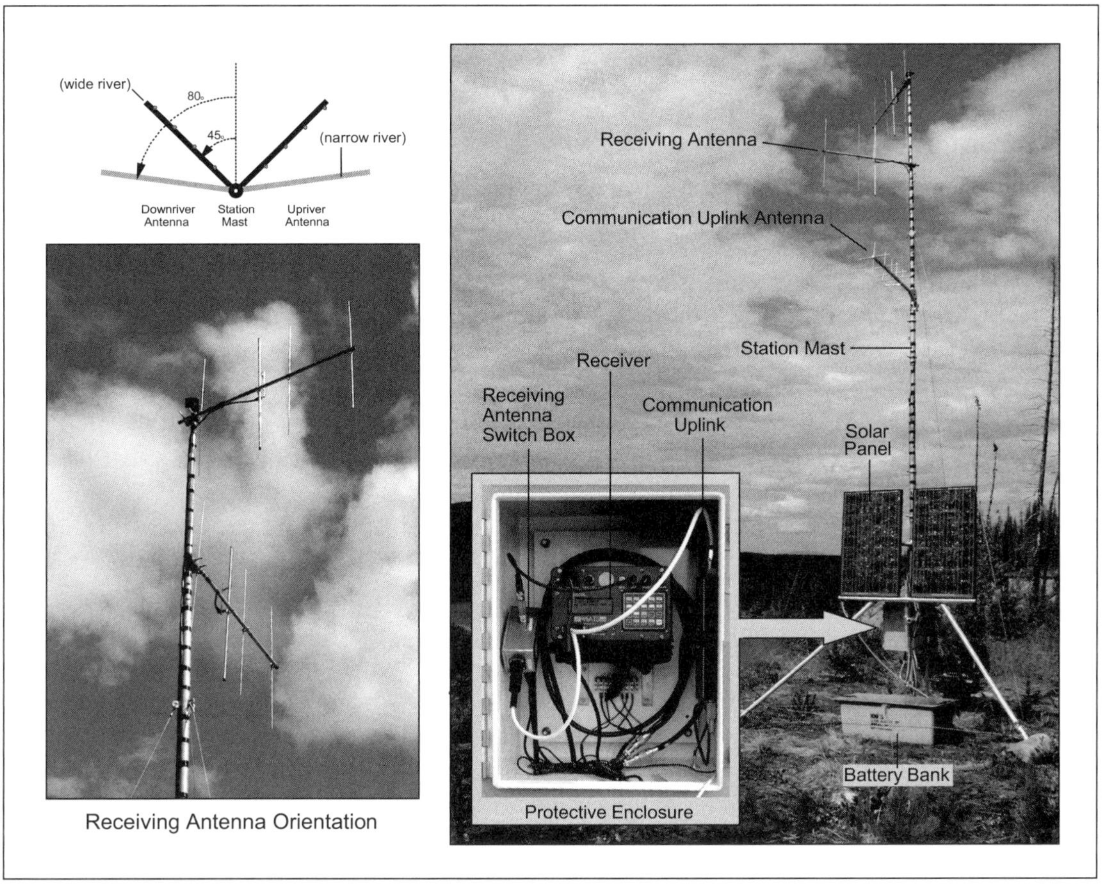

FIGURE 1. Tracking station used to detect, identify, and record radio-tagged fish passing the site. A communication uplink is used to access data collected at remote sites. The protective enclosure for the electronic components and receiving antenna orientation (for both wide and narrow sections of river) are shown.

There are benefits and tradeoffs associated with using lower radio frequencies (e.g., 30–50 MHz) to track fish. Although the signal characteristics for this frequency range reduce attenuation in water, this advantage is offset by the lack of efficient receiving antennas (Sisak and Lotimer 1998). Yagi antennas for 30–50 MHz are relatively large and cumbersome compared to those for higher frequencies, and are often unsuitable for field use (Winter 1996). Alternative solutions may therefore be necessary, such as using a dipole which limits the station to collecting presence-absence data, or loop antennas which are suitable in size, but have reduced reception range and are less directional than Yagi antennas (Section 5.1).

Underwater antennas, which typically consist of a submerged dipole or stripped coaxial cable (i.e., coax), are periodically used for stationary tracking. Although reception range is usually less than 10 m (Bunt et al. 2003; Beeman et al. 2004), this antenna type can be placed deep in the water to record signals that could not be detected by antennas on the surface. The reduced range also provides greater resolution that can be an advantage at sites where fine-scale positioning is needed, such as fish passageways (Section 6.5; Beeman et al. 2004). Underwater antennas may be impractical in less controlled environments (e.g., wide sections of river, or areas with strong and variable currents, submerged obstructions, and drift).

Antenna orientation and placement is also an important consideration, particularly when using directional antennas. Two 4-element Yagi antennas, oriented upriver and downriver from the station, were used during studies on migrating salmon (Eiler 1995). On wide stretches of river, the antennas were oriented 45° from a line perpendicular to the river (Figure 1) to maximize coverage across as well as upriver and downriver from the site. On narrower reaches, where coverage across the river was less of a concern, the angle was increased to enhance reception range in the direction of travel. Ideally, the antennas should be spaced at least one wavelength apart (e.g., 2 m when operating in the 150 MHz range) to minimize interactions that can degrade signal reception. The antennas are often mounted on a mast to provide an unobstructed view of the site, since signal reception can be impaired by topography, physical structures, and dense vegetation. Reception range is also enhanced by increasing antenna height due to the propagation and refraction patterns exhibited by radio waves at the air-water interface (Section 5.1). Steps should be taken to reinforce and guy the station mast, particularly in areas exposed to strong winds. Excessively tall masts can be unstable and are usually not practical. During salmon studies in Alaska, 10-m masts were used with few problems, although additional reinforcement and guy lines were needed at some sites.

Multiple antennas can be coupled together using a hybrid combiner to create an antenna array. The array acts like a single antenna by integrating all of the signals received (Section 5.1). Another option is to use a switchbox that alternates between individual antennas, as well as listening to the combined array. Comparing signal strength from the upriver and downriver antenna can be used to indicate where the transmitter is located relative to the station. An expanded version of this approach is used to document fish passage at dam sites. A series of inter-connected antennas is placed within different sections of the dam and associated passageways to record the position of the fish as they approach and pass through the structure (Section 6.5).

Receivers, Data Loggers, and Uplinks

The receiving antennas are connected to an integrated receiver and data logger, which is the principal component of the tracking station. There are two basic types of receivers: wide and narrow band. Wide-band receivers monitor broad frequency ranges (i.e., channels), enhancing site coverage by eliminating the need to search for transmitters on specific frequencies. Narrow-band receivers must scan through a series of frequencies to provide the same coverage, but have better reception range due to increased sensitivity and are less susceptible to problems with electronic noise (Section 5.1). Wide-band receivers are a good option when reducing scan time (i.e., the time needed to monitor all of the transmitter frequencies) is paramount. Narrow-band receivers are preferable when scan time is less critical and increasing reception range is the primary concern. Optimizing reception range and scan time are discussed later in this Section.

Although some vendors supply receivers and data loggers separately, the general trend has been to combine these into a single unit to simplify operations and improve performance. The receiver-data logger controls the operating parameters of the station (e.g., transmitter frequencies scanned, scan rate, and the antenna sequence used to detect the signals), processes the signals received, and stores the information collected. Each record typically includes the date, time, signal strength, and identity of the transmitter (Table 2). Identity is usually based on the transmitter frequency and signal pattern, often referred to as the signal code (Section

Table 2. Tracking station data for a 1-hour period, including the date (June 30), time, and identity of radio-tagged fish (based on transmitter frequency and code) within reception range of the site. Signal strength from the upriver and downriver receiving antennas is used to determined fish location in relation to the station. In this example, data is compressed into 10-minute blocks to simplify data analysis and conserve data logger memory. The shaded blocks show the movement of a fish (frequency 1840, code 46) past the site. Another fish (frequency 942, code 17) is detected downriver from the station. A reference transmitter (frequency 1183, code 12) is also recorded and used to assess station performance.

Year	Ordinal date[a]	Hr	Min	Station antenna[b]	Transmitter frequency[c]	Transmitter code	Act[d]	Signal strength
2004	182	9	0	0	1183	12	0	148
2004	182	9	0	1	1183	12	0	139
2004	182	9	1	2	1183	12	0	151
2004	182	9	2	0	1840	46	0	111
2004	182	9	2	1	1840	46	0	106
2004	182	9	10	1	1840	46	0	144
2004	182	9	18	2	1840	46	0	130
2004	182	9	18	0	1840	46	0	137
2004	182	9	22	0	1840	46	0	138
2004	182	9	22	1	1840	46	0	146
2004	182	9	23	2	1840	46	0	145
2004	182	9	30	0	942	17	0	106
2004	182	9	30	1	942	17	0	112
2004	182	9	31	0	1840	46	0	137
2004	182	9	36	1	1840	46	0	118
2004	182	9	32	2	1840	46	0	140
2004	182	9	40	0	1840	46	0	128
2004	182	9	41	1	1840	46	0	113
2004	182	9	41	2	1840	46	0	152
2004	182	9	47	0	942	17	0	110
2004	182	9	47	1	942	17	0	120
2004	182	9	51	0	1840	46	0	105
2004	182	9	51	2	1840	46	0	115
2004	182	9	59	0	942	17	0	134
2004	182	9	59	1	942	17	0	138
2004	182	9	59	2	942	17	0	102

[a] Days of year ranging from 1 (January 1) to 366 (December 31 for leap year)

[b] Downriver antenna = 1, upriver antenna = 2, and combined array (both antennas) = 0

[c] Transmitter frequency abbreviation: e.g., 942 = 150.942 and 1840 = 151.840

[d] Activity: 0 = active (transmitter motion sensors has been activated), 3 = inactive (motion sensors has not been activated within previous 24 hours.

5.1). Although the data logger can be programmed to record beeper tags (i.e., transmitters with a regular, single-pulsed signal), most stationary systems use signal processing software that only records transmitters with more complex signal patterns. This approach increases the number of tags that can be identified and minimizes problems with false positives (discussed later in this Section). It is important to make sure that the transmitters purchased are compatible with the tracking system. Some data loggers are also capable of recording additional information encoded within the transmitter signal, such as water depth or whether the fish is active. The data logger is software-driven, making it possible to select and modify functions to accommodate a variety of applications.

Accessing data stored in the data logger can be accomplished in several ways. When large amounts of information have been collected, the data files are typically downloaded with a laptop computer. Some data loggers have a flash memory card that can be removed and downloaded later. Many data loggers also have built-in mechanisms for manually viewing the data on site. Retrieving information collected at isolated locations presents a special challenge, particularly when access is limited and numerous stations are involved. Communication uplinks, equipped with transmitting antennas suitable for remote applications, can be used to access the data. This approach makes it possible to summarize study results in-season, and to monitor station performance and respond if equipment problems occur. Various types of uplinks have been used to retrieve station data, including high altitude environmental satellites (Eiler 1995), low-earth-orbit satellites (M. Sisak, Lotek Wireless, Newmarket, Ontario, personal communication), cellular modems associated with telecommunication systems (Grothues et al. 2005), and direct radio links (J. W. Beeman, U.S. Geological Survey, Cook, Washington, personal communication). Direct radio links require an unobstructed line-of-sight with the station, which often limits this method to shorter distances. In addition to one-way connections for retrieving station data, two-way uplinks are also available that make it possible to remotely operate and modify receiver-data logger functions. Although there are additional costs associated with uplinks, this approach can save substantial time and effort, as well as providing regular access to the data.

Powering the Station

Supplying power to the station is a critical issue. While most receiver-data loggers have internal batteries, few can operate for extended periods (i.e., over 8 h) without external power. The amount of power needed depends on a number of factors, including the power requirements of the electronic components, the length of time the station needs to function, and access to the site. For stations near urban areas, a direct connection to an existing power source is sometimes available and provides the easiest solution. Stations in remote areas must be self-contained and capable of operating independently for prolonged periods. An external battery is the most common solution. Although a standard 12-V "car" battery will work, a deep-cycle, sealed gel cell battery performs better and is safer to transport.

Several alternatives are available when energy demands exceed the capacity of a single external battery. When station sites are readily accessible, the existing battery can simply be replaced at regular intervals to ensure that an adequate power supply is maintained. At more isolated sites, the operational life of the station can be extended by connecting several batteries in parallel to increase power capacity, installing solar panels or small wind generators capable of recharging the external batteries, or a combination of these options. In an extreme example, stations operated in interior Alaska were powered by six 6-V batteries, connected in series and parallel, to provide 12-V current and 610 Ah of power (Eiler et al. 2006). Six-V batteries were used because they were able to operate at lower temperatures, and the battery enclosure was insulated to reduce heat loss. The batteries were charged by two 80-W solar panels, which provided adequate power to maintain the batteries even when light levels were low. A power regulator was integrated into the charging system to prevent the batteries from being overcharged during the long summer days and discharging at night. Study requirements differ, and the operational needs of the station should be discussed with knowledgeable people to ensure that the power system will supply sufficient power throughout the study and under the conditions encountered.

Supplemental Components

The electronic components (including the receiver-data logger, antenna switchbox, and communication uplink) are housed in an enclosure to protect them from the surrounding environment and vandalism (Figure 1). The power supply is typically housed separately to further protect the electronics and to isolate electronic noise produced when charging the batteries. Protective enclosures are commercially available, and should include some mechanism for locking the unit and attaching it to the station. Access ports for the station cables should be located on a sheltered surface to prevent moisture from seeping in. Enclosures can also be fabricated. During salmon studies on the Yukon River, fish totes made of thick, hardened plastic were used to house the station batteries; an aluminum plate was bolted to the lid to prevent bears from prying the enclosure open. It is usually advisable to post a sign explaining the purpose of the station. Although this may not stop vandals from disturbing the equipment, it will often dissuade people that are just curious. Camouflaging the stations can sometimes reduce vandalism, but when solar panels are used the reflective surfaces are difficult to conceal.

Cables are used to integrate the electronic components of the system, and are essential to its performance. Power cables are usually specific to the receiver and are purchased from the vendor. Coax with BNC connectors is typically used to connect the receiving antennas to the receiver, and is available from a number of sources. Signal loss increases with coax length (Belden, Richmond, IN, Product specifications and technical data), so the station design should limit the amount used. The type of coax can also be a factor. RG58 coax, typically used for mobile tracking, is often used by default for stationary applications. RG8 coax is more suitable for station use. Although thicker, heavier, and less flexible, it results in substantially less signal attenuation (Section 5.1). When exposed to the elements, coax connections should be sealed with electrical tape to prevent moisture from seeping in. In some areas, animals chewing on exposed cables can be a problem, and encasing the cables in flexible aluminum conduit can prevent damage. Conduit ends exposed to the elements should be sealed with pliable sealant putty to prevent water from running down the cable. A drip loop or "U" (with holes drilled in the bottom) should be formed with the conduit so that any water that accumulates can drain out rather than flowing into the enclosure.

Site Selection

Site selection is a key consideration since it can have a major impact on station performance and data quality. Although the general location of the station is dictated by the research objectives, the site ultimately selected is often a compromise between what is ideal (in terms of the study design) and what is technically feasible. Site characteristics, including river width, depth, and channel morphology, are important factors when selecting station sites. Narrow reaches are preferable for obtaining coverage of the entire channel. Placing additional stations on the opposite bank or strategically placed islands is an option when reception range is insufficient to cover the entire span. However, the added costs and logistics of this approach can be substantial. Fish swimming in deep water have reduced reception range (Eiler 1990; Beeman et al. 1998; David and Closs 2005), and placing stations on the bank closest to the river channel (i.e., where the water is deepest) can compensate for this limitation. Reception range increases when there are no obstructions between the fish and the station (i.e., unrestricted line-of-sight). Stations placed on an outer bend of the river provide the best unobstructed view both up and down river while providing sufficient overlap in antenna reception to minimize gaps in coverage on the far shore (Figure 2). Relatively straight stretches of river also provide adequate coverage. Islands and braided river channels can obstruct signal reception and should be avoided. Reception range can be further enhanced by increasing station height (e.g., selecting sites on bluffs or structures overlooking the river)

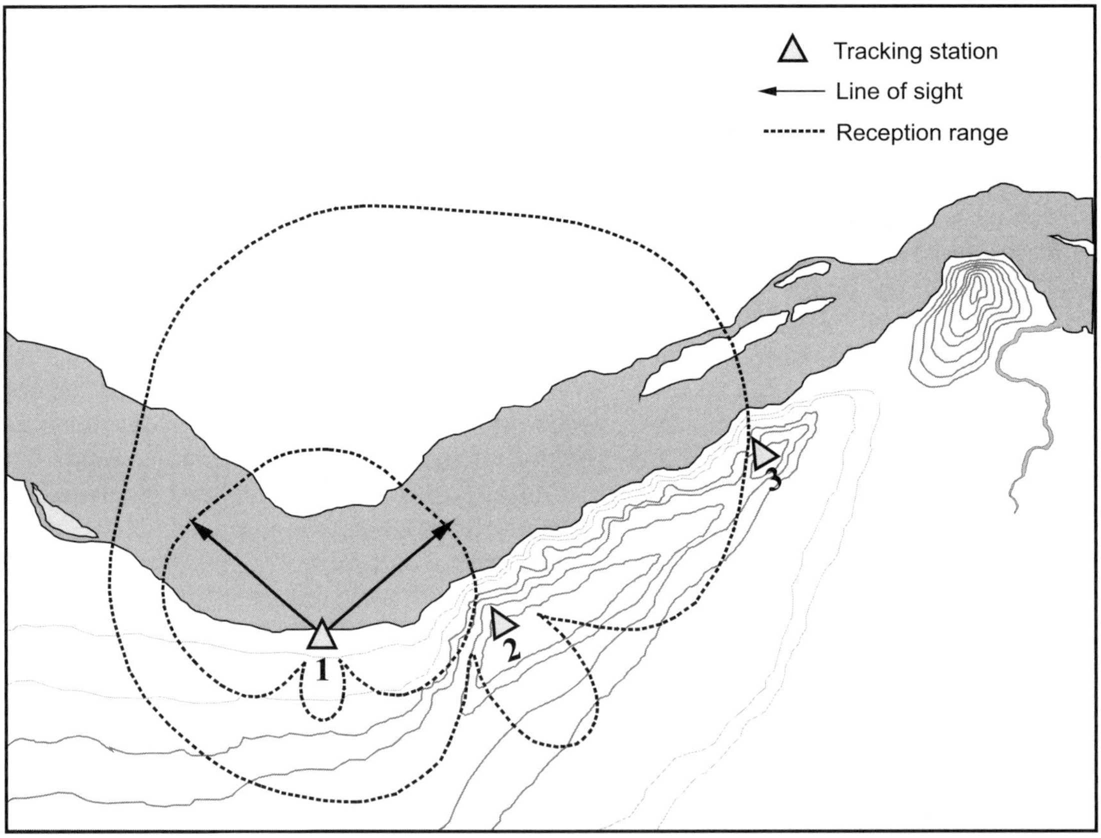

FIGURE 2. Section of river showing potential tracking station sites. Site selection is usually based on a combination of factors that maximize reception range, and facilitate station installation and site access. Placed on an outer bend, Station 1 provides the best line-of-sight upriver and downriver from the site, although the higher elevation of Station 2 substantially enhances reception range. Islands in the vicinity of Station 3 will potentially obstruct signal reception, and make this site less suitable.

due to the signal propagation properties exhibited by submerged transmitters (Section 5.1) and improved line-of-sight.

Transmitter detection can be adversely affected by ambient electronic interference, often referred to as electronic noise. Interference can occur as atmospheric noise which is relatively ubiquitous and local noise which can vary from site to site. Local noise can be constant, such as interference from power lines, or variable from sporadic radio traffic or engine noise. Excessive noise can severely impact station performance, and steps should be taken to minimize this problem. Noise may be more pronounced on certain transmitter frequencies and can be avoided simply by using quieter frequencies. Evaluating noise levels in the study area before selecting frequencies and purchasing tags can (literally) prevent unnecessary headaches. Noise levels can be determined using a spectrum analyzer or the receiving equipment used for the study. Checking with agencies that oversee telecommunications in the area (such as the Federal Communications Commission in the U.S.) can also help identify potential problems in advance. Some noise can be site specific, and efforts should be made to locate suitable alternatives. For example, what appeared to be a good station site was selected on a river near a small town. The physical characteristics of the site made it necessary to orient the downstream antenna in the general direction of the local airport. Preliminary on-site tests demonstrated unacceptably high levels of noise associated with the airport. The problem was solved simply by moving the station a short distance (less than a kilometer) upriver so that both antennas could be directed away from this source of interference.

Problems with noise can also be minimized by reducing the receiver gain or using less efficient receiving antennas. These steps reduce reception range, but may be an acceptable tradeoff when tracking slow-moving species or at sites where fish movements are restricted (e.g., velocity barriers, fish passageways, and counting weirs). The transmitters used can also mitigate problems with noise. Receivers are less discriminating than the human ear, and noise can sometimes be erroneously recorded as a legitimate signal, particularly if the transmitter signal patterns are simple and easily mimicked. Most stationary tracking systems use coded transmitters with relatively complex signal patterns so that sources of noise are less likely to duplicate the patterns recognized by the station. Excessive noise can still obscure transmitter signals and increase the time needed to process the signals detected, so locating an electronically quiet site is still advantageous.

Several other factors should be considered when selecting station sites. Locations with easy access help to facilitate station installation and maintenance, which is important when frequent site visits are necessary. Easy access can be a disadvantage in areas with a high potential for vandalism, and it may be advisable in these areas to select more isolated sites to limit public access. In the northern hemisphere, sites with an open, south-facing exposure (north-facing in the southern hemisphere) are often necessary to provide adequate sunlight when using solar panels. An unobstructed view of the horizon is needed for some communication uplinks used to transmit station data, and requirements related to the directional orientation (azimuth) and trajectory angle of the uplink antenna may dictate where stations can be placed. Some uplinks, such as cellular modems associated with telecommunication systems, are less dependent on these factors. Radio uplinks require direct line-of-sight with the station, which can influence site selection. Efforts should also be made to avoid sites with undercut banks, unstable timber, power lines, and other potential hazards. Obtaining landowner permission or land use permits (for sites on federal, state, or municipal land) often takes longer than expected, and should be initiated well in advance of field operations to avoid unnecessary delays.

A thorough evaluation of potential station sites can be time-consuming and difficult, but should be viewed as a critical part of the process due to the impact location can have on station performance and data quality. A preliminary review using topographic maps, aerial photos, or satellite imagery can simplify the initial screening by eliminating many unsuitable areas. However, locations that look good on paper (or the computer screen) may not satisfy all of the criteria, and on-site inspections are essential before making the final decision. Choosing a site almost always involves tradeoffs. There is seldom a perfect location where all of the attributes are optimal, and it is often necessary to weigh the different options in order to choose the location that provides the best overall combination of conditions. A common mistake is to base the final selection primarily on what is convenient, such as sites with easy access, needing little preparation, or requiring minimal effort to obtain the necessary land use permits. Care should be taken not to give logistical considerations undue preference over factors that impact station performance and ultimately the quality of the data collected.

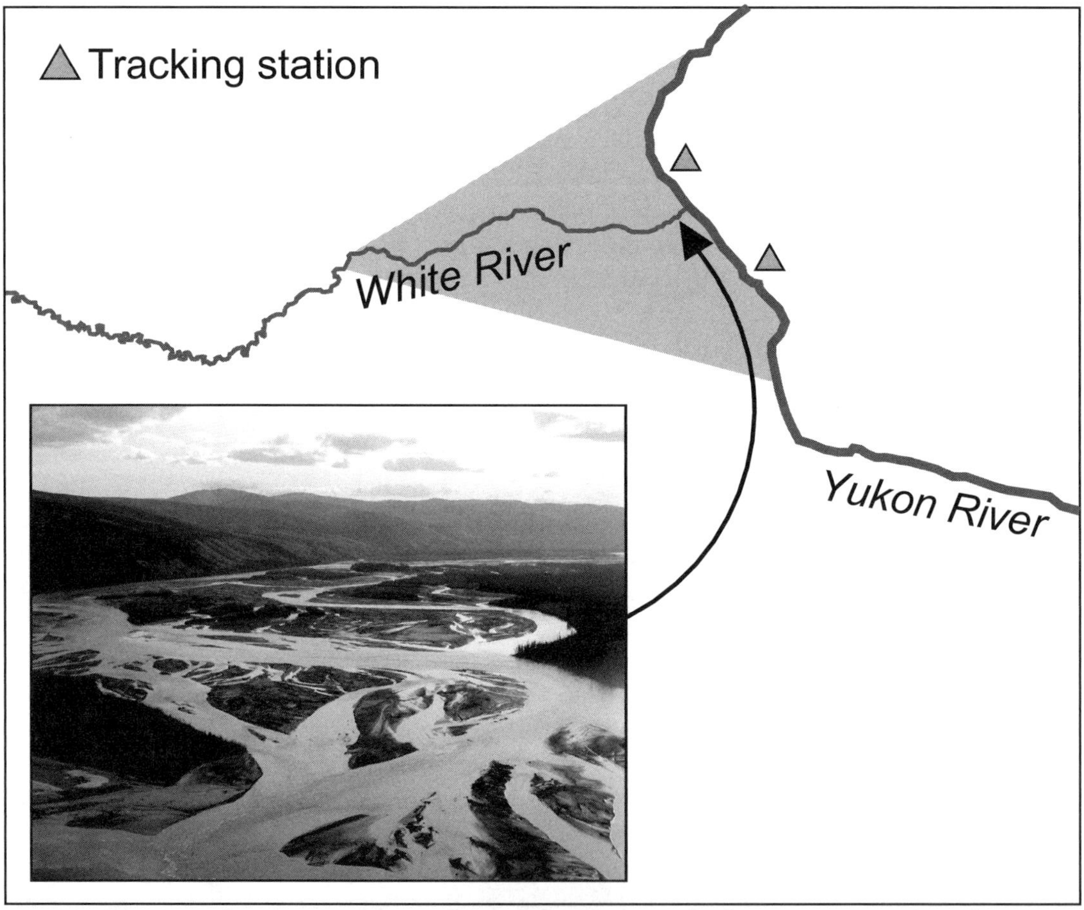

FIGURE 3. Placement of tracking stations used to record radio-tagged salmon returning to the White River, a large tributary of the Yukon River. Suitable station sites were not available in the lower reaches of the White River due to extensive braiding (see insert). In addition to effectively cordoning off the tributary, this configuration of stations also provided additional information on fish moving farther up the Yukon River main stem.

Occasionally it may be necessary to re-evaluate the overall strategy used to select station sites. For example, efforts were made to find a site on the lower White River, a large tributary of the Yukon River. The lower reaches of this river are extremely braided, and although several surveys were conducted of the area, no suitable site was found. Rather than placing stations on marginal sites within this area, two good sites were located on the Yukon River near the confluence (Figure 3). Fish that passed the lower site but not the upriver site were assumed to have moved up the White River. In addition to effectively monitoring this tributary, this arrangement also provided additional information on fish moving farther up the Yukon River main stem.

Station Installation and Maintenance

Although the specific methods used to install stations are largely dependent on the type of equipment used and the nature of the site, there are several general suggestions that should be considered, particularly when working in remote areas. Completely assembling the station before transporting the equipment into the field is strongly recommended. A trial run ensures that all of the components are present and will fit together properly during installation. Equipment can be damaged while in-transit to the site, and steps should be taken to protect items sensitive to rough handling and moisture. It is extremely IMPORTANT to have backup components and tools, particularly when site access is difficult, time consuming, or expensive. Mishaps happen, items get lost, and few things are more frustrating than having to make additional trips because the right type of wrench is missing. Use a checklist to ensure that all the gear needed for the work is on hand. Table 3 provides a general list of items that are often needed for station installation, but each situation is different and the gear needed for your application may vary.

The station should be installed and operated for an extended period before the fish arrive at the site. This precaution not only provides an opportunity to evaluate how the equipment is functioning, but ensures that there is sufficient time to identify and resolve existing problems before any data are lost. The extra lead time also avoids having to rush if there are delays due to weather or other unanticipated problems associated with field logistics.

Station maintenance is a fundamental aspect of stationary tracking. Although the quality and dependability of electronic equipment has improved over the years, malfunctions occasionally occur. Vandalism, damage by animals, and acts of nature (e.g., forest fires, floods, and falling trees) can adversely affect station operations. Undetected, these problems can result in substantial data loss, so it is important to have a way to periodically assess station performance. Regular site visits are an obvious answer. However, this solution requires additional time and effort to access the site, and visits frequent enough to provide sufficient notice when problems develop. The commonly asked question "How often should the station be checked?" is probably best answered by "How much data are you prepared to lose?" Site visits should be scheduled with the worst-case scenario in mind. An alternative solution is to equip the station with a communication uplink. Although an additional expense, communication uplinks provide a continual update on station performance that can alert the researcher when problems have occurred and help minimize data loss.

Due to the complexity of the equipment, station maintenance and repairs increasingly rely on a "black box" approach. Few researchers have the time or ability to repair electronic components in the field. The most efficient approach is to identify which item is causing the

Table 3. Basic equipment used to install and operate tracking stations. The actual gear needed will vary depending on the telemetry equipment used, the duration of the study, and the nature of the station site.

Category	Equipment item	Comment
Principal	Structural framework	
	Guy line, stakes, sandbags, etc.	To stabilize station
	Receiving antenna	Type and number dependent on study
	Receiving antenna coax[a]	Routed to protective enclosure
	Electric tape	To seal exposed coax connections
	Protective enclosure (electronics)	
	Receiver-data logger (RDL)	
	Switch box	When multiple receiving antennas used
	Coax (switch box to RDL)	
	Power system	
	External battery	Sealed gel cell recommended
	Protective enclosure	
	Battery cable (battery to RDL)[a]	
	Connectors	To join cables[a] to protective enclosures
	Wire ties/electric tape	To secure cables[a] to station structure
	Tools for station assembly	(as required)
Optional	Multiple external batteries	To increase power capacity
	Battery cables (battery to battery)	For multiple battery power system
	Solar panel and cables[a]	To recharge external battery
	Wind generator and cable[a]	To recharge external battery
	Power regulator and cable	To protect external battery
	Communication uplink	Typically placed in protective enclosure
	Communication uplink cable[a]	
	Communication uplink antenna	
	Supplemental tools	Drill, drill bits, hack saw (extra blades) utility knife, hammer, electrical tape, etc.
	Electrical meters	Volt-ohm, through-line watt, battery
Backup	RDL	
	Switch box	
	Communication uplink	
	Receiving and uplink antennas	
	Cables (coax, power, uplink, etc.)	Duplicate of all cables used in assembly
	Principal assembly tools	
	Nuts, bolts, and connectors	Duplicates of all items used in assembly
	Survival gear	For work at remote sites

[a] Encase in metal conduit for protection in areas where animal damage is a problem. Sealant putty used to prevent moisture from entering into exposed conduit openings.

problem, and replace it with one that is functioning properly. The disabled item can be returned to the vendor for repair. The black box approach needs two things to work effectively: the ability to determine which component is malfunctioning, and backup parts to replace the faulty item. The importance of backup equipment has been previously mentioned and cannot be over-emphasized. The ability to diagnose the problem is equally important. Test points should be incorporated into the station design to monitor station performance and identify faulty components when problems develop. An essential first step is to have a reference transmitter operating on-site to provide a continuous signal source that can be used to determine if the station is receiving and accurately recording transmitter signals even when no tagged fish are present. In-line monitors can be installed to determine if the switchbox and receiving antennas are functioning, and if the communication uplink is properly transmitting the information. Receiver-data loggers that have built-in mechanisms for viewing the entire data collection process in real time, from detection and identification of the transmitter signal to the data being recorded and stored, are indispensable during station installation (to ensure that the system is operating properly), and when equipment problems occur. This feature makes it possible to identify where the receiving system is malfunctioning so that the faulty components can be replaced. It is important to ask whether these types of test points are incorporated into the receiving system before selecting and purchasing equipment. Information on battery voltage is also a useful indicator; voltage dropping below prescribed levels indicates a power system problem that can adversely affect performance. Monitoring battery voltage at different points within the system (e.g., both the battery bank and receiver) not only provides more detailed information, but it also helps to isolate and identify the source of the problem. Some data loggers are able to record receiver battery voltage as part of the standard data set.

Data Collection and Retrieval

Stations collect information on the fish by having the receiver-data logger continuously scan the transmitter frequencies. Transmitters within reception range are detected, identified, and recorded. The data are combined with other relevant information, such as the date, time, and transmitter signal strength. Sequential records of individual fish are used to document site residency. For stations with directional antennas, site passage is also determined based on which antenna (e.g., downriver or upriver) detects the strongest signal. With adequate records, a distinct progression is exhibited that can provide detailed information on when the fish move past the site (Table 4). Information on station performance is also compiled and stored.

The detection zone created by the station is a function of two principal factors: reception range (spatial coverage) and scan time (temporal coverage). These two factors are influenced by a number of variables involving the type of equipment used, site characteristics, and the behavior of the fish (Figure 4). Maximizing reception range provides a distinct advantage, reducing the risk that fish will move past the site undetected. Fish detected at greater distances will also be recorded over longer periods of time, providing more detailed information on their movements. Reception range can be enhanced by the attributes of the equipment, including the sensitivity of the receiver, reception pattern of the receiving antennas, and transmitting power of the tags. The importance of the site (e.g., river width and depth, station height above the river, line of sight, and noise levels) has been previously discussed. Fish behavior can have a significant impact on reception range. Fish swimming fast and deep will be within range for less time than slower moving fish in shallow water. Range tests conducted at the site

TABLE 4. Database file showing sequential records of a radio-tagged fish (frequency 882, code 19) moving past a tracking station site (D2). Comparing the changes in signal strength between the downriver antenna (Antenna 1) and upriver antenna (Antenna 2) provides a distinctive pattern that can be used to estimate the time of passage (see shaded rows) and identify erroneous records.

Date/Time	Transmitter frequency[a]	Transmitter code	Act[b]	Station ID	Station antenna[c]	Signal strength
03-Jul-04 13:09	882	19	0	D2	1	94
03-Jul-04 13:17	882	19	0	D2	1	112
03-Jul-04 13:23	882	19	0	D2	1	115
03-Jul-04 13:26	882	19	0	D2	2	97
03-Jul-04 13:31	882	19	0	D2	1	130
03-Jul-04 13:35	882	19	0	D2	2	99
03-Jul-04 13:49	882	19	0	D2	1	143
03-Jul-04 13:44	882	19	0	D2	2	103
03-Jul-04 13:59	882	19	0	D2	1	146
03-Jul-04 13:57	882	19	0	D2	2	121
03-Jul-04 14:03	882	19	0	D2	1	149
03-Jul-04 14:06	882	19	0	D2	2	145
03-Jul-04 14:18	882	19	0	D2	1	131
03-Jul-04 14:19	882	19	0	D2	2	147
03-Jul-04 14:24	882	19	0	D2	1	107
03-Jul-04 14:29	882	19	0	D2	2	133
03-Jul-04 14:31	882	19	0	D2	1	105
03-Jul-04 14:34	882	19	0	D2	2	132
03-Jul-04 14:40	882	19	0	D2	2	124
03-Jul-04 14:57	882	19	0	D2	2	116
03-Jul-04 15:06	882	19	0	D2	2	105
03-Jul-04 15:15	882	19	0	D2	2	98

[a] Transmitter frequency abbreviation 882 = 150.882 MHz.

[b] Activity: 0 = active (transmitter motion sensors has been activated), 3 = inactive (motion sensors has not been activated within previous 24 hours.

[c] Downriver antenna = 1, upriver antenna = 2.

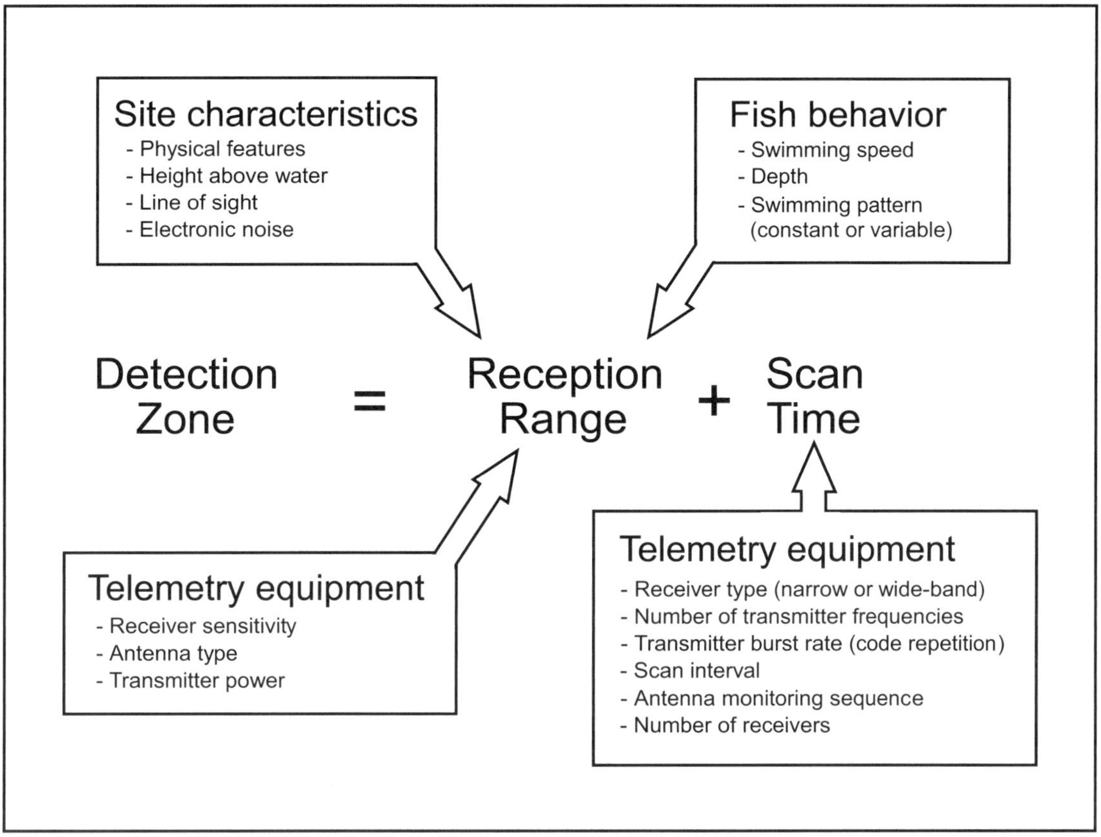

Figure 4. Factors affecting the spatial (reception range) and temporal (scan time) coverage of tracking stations used to detect and record radio-tagged fish.

with transmitters placed at different distances and depths can provide useful information on station reception. However, these results should be considered minimum estimates, since it is difficult to duplicate all of the conditions encountered during the actual study.

Reducing scan time temporally enhances the detection zone of the station. Sufficient time is needed to monitor the transmitter frequencies and process the signals received. Scan time is primarily a function of the telemetry equipment and operating parameters (Figure 4), and there are a number of ways that these factors can be adjusted to achieve suitable results. The simplest approach is to reduce the number of transmitter frequencies being scanned. As previously discussed, wide-band receivers substantially reduce scan time by monitoring broad frequency ranges. Other options are necessary when narrow-band receivers are used. Most stationary tracking systems use coded transmitters identified by two variables: frequency and signal pattern. The study design, including sample size, tagging schedule and assumed distribution and movement patterns, will dictate what combination of these variables is most appropriate. When sample size is small, individuals can be tagged with transmitters on unique frequencies. This approach becomes increasingly cumbersome when larger numbers of fish are tagged, substantially increasing the scan time needed to monitor the frequencies. Coded tags placed on the same frequency make it possible to identify individual fish while keeping the number of frequencies at a manageable level.

Transmitters with hundreds of unique codes per frequency are available. In theory, this development should make it possible for studies to use a single frequency to monitor all of the fish, eliminating problems associated with scan time. However, reducing the number of frequencies can have disadvantages. The primary issue is that it can be difficult to distinguish transmitter signals on the same frequency when the fish are grouped together in a localized area, causing what is commonly called signal "collisions." Having large numbers of transmitters on the same frequency can exacerbate this problem. Advances in pattern recognition software have increased the ability to distinguish transmitters on the same frequency. The effectiveness of these systems can vary depending on the number of transmitters present, signal pattern, burst rate, and scan time. Kuechle and Kuechle (Section 5.1) reported that 12-s scans were necessary to obtain high detection rates (99.9%) when simultaneous recording 15 transmitters (same frequency, different codes, 1 s burst rate) under laboratory conditions. Decreasing the burst rate substantially increased the time needed to obtain comparable results. An average of 5–7 transmitters were recorded during 10-s scans during field tests on the Yukon River with 10 transmitters (same frequency, different codes, 1 s burst rate) placed at the same location. Separate scans recorded different combinations of tags, and over several minutes all of the transmitters were typically recorded. Due to the impact this problem can have, it is important to discuss the capabilities of the receiving system with the vendor if there is a potential for excessive signal collisions.

Problems with signal collisions can be minimized by deploying transmitters in a way that temporally and spatially distributes tags on the same frequency. For example, when tagging fish that are migrating upriver, it is better to deploy same frequency transmitters at different times, so that they are less likely to be in close proximity. Different frequency/code combinations may be more suitable depending on the movements of the fish. When fish are widely distributed, fewer frequencies and more codes can simplify search efforts. More frequencies and fewer codes may make it easier to identify fish when large numbers are congregated in localized areas. Some receiving systems also have the ability to isolate a particular transmitter code while screening out other signals on the same frequency, making it easier to track individuals even when other fish are present.

Stations equipped with multiple receivers monitoring different transmitter frequencies have been used to decrease scan time. In an extreme case, a station was equipped with four receivers to record the passage of juvenile salmon through a narrow river canyon (N. S. Adams, U.S. Geological Survey, Cook, Washington, personal communication). The physical characteristics of the canyon severely limited reception range, making it necessary to substantially reduce scan time to ensure that fish passing the station were detected. The added cost of equipping stations with additional receivers and increased power demands are obvious tradeoffs that must be considered.

Scan time can also be reduced by decreasing the time the receiver monitors each frequency (hereafter referred to as the scan interval) to the minimum needed to detect and process the signals. The scan interval is dependent on the burst (i.e., pulse or group of pulses used to identify the signal) rate of the transmitter. Shorter intervals can be used to detect transmitters with faster burst rates. A 2-s scan interval will be sufficient to detect tags that emit a signal burst every second, but inadequate for tags transmitting every 5 s. The tradeoff is that faster burst rates either reduce transmitter life or require a larger battery. Longer scan intervals are sometimes needed to process and accurately identify the transmitter signals. To compensate, scan time can be minimized by using a variable scan interval, with the receiver-data logger switching from a short interval (used when no signals are present) to a longer interval when transmitters are detected. The antenna sequence used to detect the fish can also be modified to reduce scan time. One approach is to use the combined antenna array during the initial search phase, with the individual antennas monitored only when transmitter signals are detected.

The time available to detect fish passing the site can be calculated using the operating parameters of the station. For example, during salmon studies in Alaska, the receiver-data logger scanned 11 frequencies, had a scan interval of 2 s, and monitored the site using the combined antenna array. Scan time was less than 25 s when no fish were present. If a transmitter was detected, the receiver remained on the frequency and switched to the upriver and downriver antenna (each monitored for 10 s) before advancing to the next frequency. In the worst case with fish present on all frequencies, the time needed to complete the scan cycle was approximately 240 s (4 min). The tagged salmon were typically within reception range of the stations for several hours, so this approach provided suitable coverage of the site. In other situations, such as locations with limited reception range and faster moving fish, the detection window may be considerably shorter, and changes to the tracking system may be necessary to ensure that all the fish passing the site are recorded.

Tracking stations often collect voluminous amounts of information, and managing the data can be problematic. Most data loggers have extensive memory for data storage, but it is not unlimited. Exceeding the capacity will either result in additional data not being collected or previously recorded information being overwritten. Communication uplinks or periodic visits to the site to download the data can alleviate this problem. Another option is to compress the information so that the amount of memory used is minimized, while still providing sufficient data resolution. During the Yukon River studies, station data were grouped into 10-min segments with the "best" record (based on signal strength) recorded for each antenna (Eiler et al. 2006). This approach provided precise information on fish passage, while eliminating a substantial amount of extraneous data. However, there is a major downside to data compression, since information may be discarded that is later deemed important. If memory is not limited, the preferred approach is to store all the data collected and postprocess the information to obtain the resolution needed (Section 9.1).

Data compression is less of an issue when collecting information on station performance. The purpose of this information is usually limited to identifying equipment failures in-season and verifying the validity of the fish data during postseason analysis (i.e., no data recorded reflects the absence of fish rather than problems with the equipment). If used in-season, the data must be in an easy-to-use format and managed so that problems can be identified promptly and corrective actions taken in a timely manner. However, if not careful, data compression can obscure important details. For example, continuously recording the station reference tag will result in an overwhelming amount of data that is difficult to process and use. Only recording the tag once an hour will substantially reduce the amount of information collected, but may give a false impression of how well the station is functioning. A malfunctioning receiver may be able to sporadically detect the tag within the hour, but this level of performance would be inadequate to effectively monitor fish passage. A viable alternative is to record when the reference tag is first detected and the number of times detected within the hour. A tag detected within the first couple of minutes and recorded numerous times thereafter would indicate that the receiver-data logger is functioning properly, whereas a tag detected late in the hour or only recorded several times would suggest problems with the equipment. This approach will substantially reduce the amount of data collected, while still providing the critical information.

Data compression is also useful when utilizing communication uplinks to remotely retrieve the station data from isolated sites. Transmission windows are typically short, and the ability to distill the information to the essential details will facilitate this process. Some data loggers are able to create separate, compressed data files, which can be used to transmit the information, while the complete data set is stored in memory. Encryption programs are also available that can be used to re-format the data to further reduce message size and facilitate data transfer.

Data processing techniques can also be used to streamline the analysis of time-sensitive information, such as data on station performance. There is limited value in tracking hourly changes in battery voltage or the number of times the reference tag is recorded. The real issue is whether the station is functioning properly. Data files obtained from the station can be uploaded into an automated database with processing routines designed to flag potential problems (Eiler and Masters 2000), such as battery voltage dropping below acceptable levels or the reference tag not being regularly recorded. Brief output summaries can be produced to provide a quick overview and focus attention on potential problems. Similar approaches can also be used for processing the fish data. Fish distribution or movement rates can be determined by comparing new sightings with previous records, providing a quick overview of where groups of fish are located or the average speed that fish are traveling through the drainage.

MOBILE TRACKING

Basic Approach

Mobile tracking has been used to track fish since the inception of aquatic telemetry. Over the years, technical advances have substantially refined the techniques used. Mobile tracking provides an effective method for collecting a wide range of information, including both large and small-scale movements, macro and micro-habitat use, and activity patterns. The basic approach is to actively search for and locate the fish while traveling through the study area. Fish locations are determined either by triangulation or by following the signal back to the transmitter, sometimes referred to as "homing in" on the signal. Mobile tracking can take a number of forms, including surveys from aircraft, boats, land-based vehicles, and on foot.

Mobile tracking is dynamic and multifaceted with numerous factors to consider, but the basic techniques become easier with practice. Much of this relates to developing the ability to discern subtle changes in transmitter signal strength—a key step in the process—as well as understanding how transmission patterns and the surrounding topography impact signal reception. Conducting test trials to locate transmitters placed at a variety of locations is an excellent way to improve tracking proficiency.

Advantages of Mobile Tracking

Mobile tracking has a number of advantages (Table 1). It is an extremely flexible technique that can be adapted to numerous situations and conditions. Startup costs are substantially less than with stationary tracking, which can be particularly important when conducting feasibility, small-scale, or short-term studies. Although stationary tracking can provide detailed information at specific locations, the type and number of sites is limited by costs and logistical considerations. Mobile tracking can be used to collect data over extended areas, limited only by the ability to access the sites. Combined with advances in transmitter sensors (e.g., activity, temperature, and depth), this flexibility makes it possible to collect a remarkable variety of information that could not be obtained using more traditional research techniques.

Mobile tracking can be broken down into three basic topics: the tracking equipment used and the two principal tracking methods (triangulation and homing in). Specific techniques related to the primary methods of homing in (aerial, boat, and foot surveys) are also discussed.

Tracking Equipment

Much of the equipment used for stationary tracking is also needed for mobile applications. The receiver-data logger is again the principal component. It should be compact, lightweight, durable, and water-resistant, since mobile tracking is often conducted under a variety of settings and the equipment periodically exposed to adverse conditions. A waterproof carrying case with a neck or shoulder strap provides additional protection, and also makes it easier to handle and operate the gear. The receiver-data logger should be capable of automatically scanning through the transmitter frequencies, particularly when conducting extended surveys or monitoring more than a couple of frequencies. The receiver-data logger can be linked to a global positioning system (GPS) receiver so that spatial coordinates (e.g., latitude and longitude) are included in the data record created when a fish is identified. This approach is a major improvement over using maps to estimate and record the location of the fish, particularly when working in areas without distinct landmarks. Some tracking receivers use a separate GPS receiver, whereas others have a built-in unit which simplifies operations (fewer cables, connections, and interface issues) and improves performance. A separate tracking receiver, which is typically smaller and less expensive than units with data-logging capabilities, can be used if small numbers of fish are being tracked, but the ability to distinguish coded transmitters is diminished. In addition, the data are not recorded as an electronic file, which is a major disadvantage when it comes to processing the information.

When fully charged, most receiver-data loggers are capable of operating throughout a typical tracking survey (i.e., usually less than 8 h). However, it is advisable to have a backup unit or external power source in case the internal battery runs low. Sealed gel cell batteries provide extended power, and although cumbersome, can be used as an external power source when tracking from vehicles, boats, or aircraft. Portable battery packs are also commercially available and are smaller and much easier to handle than a standard-sized battery. They provide fewer amp hours of power but can be recharged for extended use. Power can also be accessed directly if the vehicle, boat, or aircraft is equipped with a power jack, although it is important to ensure that the receiver is adequately shielded from any noise generated by the power system, which can adversely affect signal reception.

The peripheral gear used for mobile tracking also deserves discussion. A variety of receiving antennas can be used, although the tracking method will influence the type selected from both a logistical and data quality standpoint. These options are discussed later in this Section when considering specific tracking methods. RG58 coax is typically used to connect the antenna and receiver. It is flexible, lightweight, and durable. Coax length for mobile tracking is usually short (i.e., 1–3 m), which reduces problems from signal loss. Care should be taken when using and storing coax to avoid kinks in the line which can degrade signal reception. Kinks are particularly problematic when assessing differences in signal strength (e.g., determining the direction of the strongest signal), which is an important aspect of mobile tracking.

Background noise can degrade the ability to hear the transmitter signals. Good quality, aircraft-grade earphones can enhance detections, particularly when transmitter signals are weak, and are useful for evaluating differences in signal strength. Caution should be taken when considering the use of noise-canceling headsets. Although these earphones are very efficient at screening out electronic interference, they may also reduce the ability to detect weak transmitter signals.

Triangulation

Triangulation is a common method for tracking fish. The fish are located from boats, vehicles, or on foot. The receiver is tuned to the transmitter frequency and the surrounding area swept (i.e., pan horizontally 360°) with directional antennas to determine the direction of the transmit-

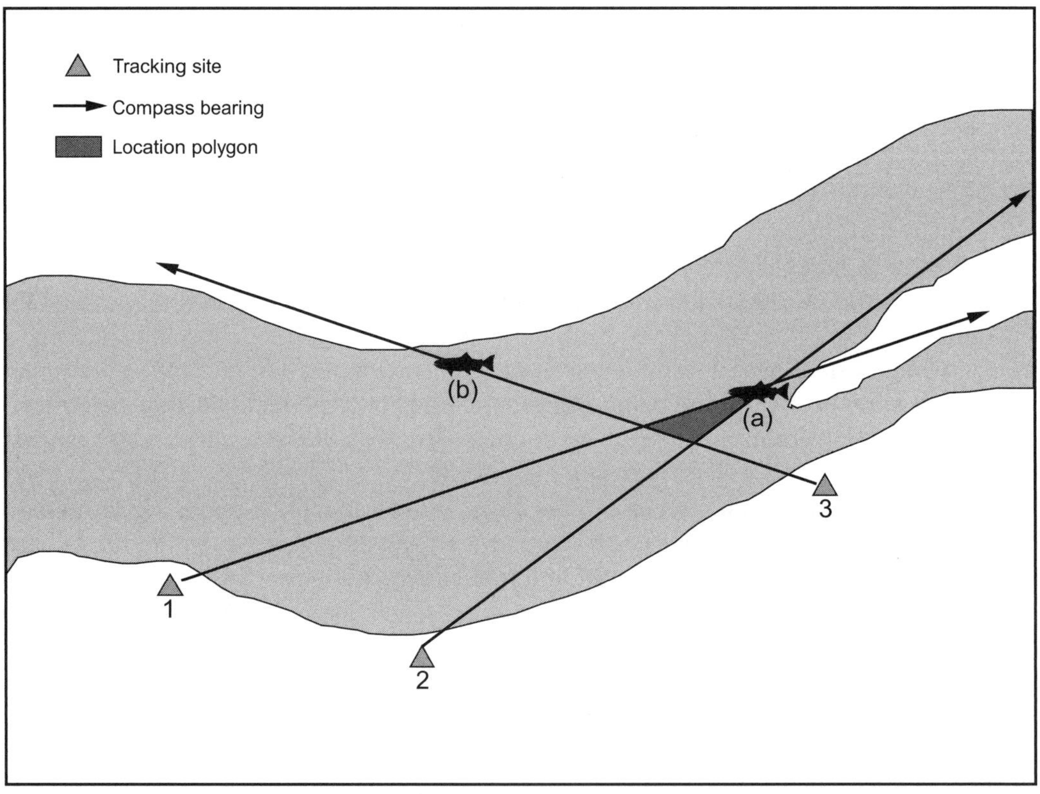

FIGURE 5. Basic method for locating radio-tagged fish with triangulation. Compass readings from tracking sites provide an estimate of the fish location. The accuracy of the estimate depends on the accuracy of the compass bearings, distance from the fish, time between readings, and the movements of the fish. The compass readings seldom converge on a single point, and the resulting polygon (darkened area) is often used to estimate the location of the fish. In this example, an erroneous estimate is obtained due to the time taken to travel from Site 1 to Site 3 and the movement of the fish from its original position (a) to its final location (b).

ter signal. Location estimates are made by taking compass bearings in the direction of the signal from several known positions and determining the intersection point (Figure 5). Although general locations can be obtained with 2 readings, accuracy is substantially increased when three or more bearings are used (Garrott et al. 1986). The most reliable estimates are obtained when the angle of the bearings approximate 90 degrees. The intersection of the bearings seldom results in an exact location (Figure 5), and numerous procedures and software programs have been developed to refine the estimates, with polygons typically used to incorporate location error (Lenth 1981; Pace and Weeks 1990; White and Garrott 1990; Withey et al. 2001). The accuracy and precision of the location estimates depend on four basic factors: the accuracy and precision of the compass bearings, distance from the fish, time between readings, and movements of the fish.

Figure 6. Common types of directional antenna use to track radio-tagged fish. Yagi antennas (top) are most often used for higher frequencies (148–172 MHz), while loop antennas (bottom) are typically used for lower frequencies (30–50 MHz). Yagi antennas have the added advantage of a more directional reception pattern than loop antennas which are bi-directional with comparable reception at both the front and back ends of the antenna (see inserts).

The receiving antenna used can have a major effect on the accuracy of the compass bearings. Four-element Yagi, 3-element Yagi, and H-antennas are commonly used for the higher frequency ranges because they are very directional and portable enough for most mobile applications. The best signals are obtained when the antenna beam is held parallel to the ground with the receiving elements oriented vertically (Figure 6) since radio signals from submerged transmitters are vertically polarized (Winter 1996; Section 5.1). Stacked arrays and Yagi antennas with more elements (e.g., 8-element Yagi) are more directional, and will improve the accuracy of the compass bearings. Stacked arrays have the added advantage of being able to produce either a peak (strongest) or null (minimum or no) signal when pointed in the direction of the transmitter (Mech 1983). Using the null can be more accurate since it is sometimes easier to distinguish when the signal can't be detected compared to when it is the strongest (Kenward 1987; Winter 1996).

Although stacked arrays and Yagi antennas with numerous elements are more accurate, the added size makes them difficult to use in small boats, vehicles, or when traveling on foot through thick vegetation or rough terrain. Placing semistationary mast antennas (i.e., guyed mast with an antenna attached at the top) at prominent sites overlooking the study area can alleviate some of the problems associated with transporting bulky equipment, making it feasible to use larger and more accurate antennas. A receiver-data logger is carried to the different sites and temporarily connected to the receiving antenna to collect data. As with tracking stations, site selection is important to ensure adequate coverage of the area, although additional readings can be taken at other sites with smaller, more portable antennas to supplement the information collected. Using multiple observers to take synchronized readings from the different sites eliminates problems associated with the time between readings and fish movements over time.

As previously mentioned, Yagi antennas for lower radio frequencies are relatively large and cumbersome, and usually impractical for mobile tracking. Loop antennas (Figure 6) are typically used even though reception range is generally a third of what would be obtained with comparable Yagi antennas (Section 5.1). Loop antennas are bi-directional with maximum signal strength obtained when either the front or back-facing edge is pointed in the direction of the transmitter, making it necessary to have a general sense of where the fish is located in order to select the correct compass bearing. A null signal is detected when the side of the antenna is perpendicular to the transmitter. Comparing the null and peak signal often helps determine the most accurate signal bearing (Winter 1996).

When sweeping the area with the antenna, the receiver gain is initially set at the maximum effective level (i.e., as high as possible to maximize reception range without recording excessive electronic noise). Once a fish is detected, the gain is reduced, which decreases the signal strength and makes it easier to differentiate the direction of the strongest signal. Reducing the gain is particularly useful when the fish is in close proximity and the signal seems equally intense over a wide range of directions. Transmitters with faster burst rates simplify efforts to obtain accurate and precise compass bearings since it is easier to compare signal strength for a signal transmitting every second than every 5 s. This consideration also applies when homing in on the transmitter (discussed in the next section). Ultimately, the type of data needed will dictate what burst rate is optimal. Burst rates of 1–2 s are ideal for fine-scale locations, whereas slower burst rates may be adequate when more general location data are acceptable. Although faster burst rates are a definite advantage when increased accuracy and precision is needed, it also impacts battery life and transmitter size.

The topography of the surrounding area can impact triangulation accuracy. Lakes generally have wide open expanses, which provide relatively unrestricted line-of-sight. Rivers are linear and convoluted, and intervening landforms can block or obscure transmitter signals. In some instances, the strongest signal detected may actually be a reflection off the surrounding landscape, commonly referred to as signal "bounce" which can bias location estimates. Multipath signals may also be detected prominently on the back or peripheral reception lobes of the antenna (described by Section 5.1). Garrott et al. (1986) reported that accurate and precise readings were obtained when 190 MHz reference transmitters were placed at locations within line-of-sight, whereas the majority of tags were inaccurately located when line-of-sight was obstructed. The best way to minimize this problem is to anticipate how the geomorphology of the area may affect signal reception and select tracking sites that are not impacted. When tracking on land, sites that are higher in elevation often provide improved line-of-sight. Increasing the number of tracking sites also helps to identify erroneous readings by providing additional information to assess potential problems associated with signal obstruction (White 1985).

Obtaining accurate compass bearings is of little value if the position of the sites used to plot the signal bearings is not accurately determined. In the past, position estimates were usually based on topographical features. Prominent landmarks along the river (e.g., river junctions, pronounced bends, and islands) and the surrounding area (e.g., hills and mountain peaks) provided points of reference. With the advent of inexpensive, handheld GPS receivers, which provide accurate position information, the problem of accurately determining site location is effectively eliminated.

The time between compass readings, and the behavior of the fish, can have a major impact on triangulation accuracy. These factors are linked since significant error can occur when the fish are moving, and the magnitude of that error increases dramatically over time. This situation is exacerbated when tracking highly mobile species that range widely over large areas. Taking simultaneous readings from different tracking sites eliminates this problem, but is labor-intensive and may not always be practical. An alternative approach is to reduce the travel time between tracking sites. Boats are often used to access the sites, and the characteristics of the river (e.g., single channel versus braided) can impact travel time. Land-based vehicles may also be an option when working in areas with good road access.

Finally, magnetic declination needs to be accounted for when taking compass readings for triangulation. Declination is the difference in azimuth degrees between true north and magnetic north, which can vary substantially depending on location. Not accounting for declination can introduce significant error when plotting fish locations. For example, the declination for some areas of the Yukon River varied over 30° from true north. Declination charts are available, and many topographic maps (e.g., U.S. Geological Survey 1:63,360 scale quadrangle series) specify the magnetic declination. Compasses designed for orienteering typically have an offset mechanism which can be used to account for declination. If not, the compass bearing obtained should be adjusted to reflect the declination of the area.

Homing in on the Signal

Locating fish by triangulation can have drawbacks. Problems associated with location accuracy can limit its use when studying fine-scale movements or habitat use patterns in areas with diverse, interspersed habitats. Triangulation is also time-consuming and labor-intensive, reducing the amount of data collected. An alternative approach is to track the transmitter signal back to its source, continually moving in the direction of the fish until the strongest signal is obtained. There are three primary methods of homing in on the signal: aerial tracking, boat tracking, and foot surveys.

Aerial Tracking

Aerial surveys are an effective mode of mobile tracking. Although more costly than other methods, this approach provides increased mobility and access, covers vast areas in a short time, and enhances reception range due to the increased height above the river and improved line-of-sight. A number of authors have described aerial tracking techniques (Kolenosky and Johnston 1967; Gilmer et al. 1981; Mech 1983), but most have focused on the terrestrial environment. Although similar in some respects, many technical considerations are different when working with aquatic species. Aerial tracking is multifaceted and several basic topics will be discussed, including the principal factors affecting tracking success, preflight preparations, and survey procedures.

Factors Affecting Tracking Success.—A fundamental objective when aerial tracking is to detect the fish present within the immediate area before moving out of reception range. Although relatively straightforward when sample sizes are small, this task becomes increasingly difficult when tracking larger numbers of fish. Tracking success can be affected by a number of factors, and being aware of these can help in planning survey flights.

Aircraft speed and altitude are major considerations. Aircraft capable of flying slower will obviously provide more time over the search area, increasing the opportunity to detect, identify, and record fish that are present. Within limits, surveying at higher altitudes provides greater reception range (Section 5.1), although at some point the transmitter signals become progressively weaker and harder to detect, and the location estimates of the fish less exact. Aerial surveys are typically flown between 150 m and 800 m above the water. During salmon studies in Alaska, tracking surveys were routinely flown at an altitude of 300 m with reception range averaging 2–4 km. The enhanced reception obtained at higher altitudes may also introduce elevated levels of electronic noise, which in some instances can be extreme. In one case tracking methods had to be radically altered due to severe, localized interference from a nearby radar station. Low-altitude surveys (<150 m) were necessary to reduce noise levels so that transmitters in the area could be detected (Holder and Eiler 2000).

Maximizing reception range and minimizing scan time is particularly important during aerial surveys due to the speed of the aircraft. This objective can be addressed in several ways. More powerful transmitters provide greater reception range, but must be balanced against increased transmitter size and the physical limitations of the fish. Using coded transmitters and conducting surveys with more than one person (i.e., dividing the frequencies) reduces the number of frequencies scanned per person, although larger aircraft and additional equipment are needed. Transmitters with faster burst rates require shorter scan intervals and reduce the time needed to scan through the frequencies.

The physical characteristics of the river can have a major impact on tracking success, and often influences how aerial surveys are conducted. Relatively straight, single channel stretches can be surveyed by flying directly over the middle of the channel, whereas highly braided reaches may require more complex flight patterns or several passes to provide adequate coverage. If the river is braided over an extended distance, it may be more efficient to survey alternate sides during the upriver and downriver portion of the flight, rather than continually doubling back to survey areas not covered during the initial pass. The surrounding topography can also impact tracking success. Mountains and hills obstruct signal reception, particularly when the fish are located immediately behind these formations, such as on pronounced river bends. Aircraft flying around the bend will have limited time to detect the transmitter signals (further demonstrating the need to reduce scan time). Making an exaggerated turn over the outer bank of the river can minimize this problem by extending the time that fish on the other side of the bend are within line-of-sight. Grid-like search patterns may be necessary in large lakes where fish are able to disperse over a wider area. Many of the tracking methods used for terrestrial species (Mech 1983; Kenward 1987) can be applied under these conditions. Considerations for tracking in lakes are further discussed in Section 6.4.

Environmental conditions can adversely affect tracking capabilities. Inclement weather can create hazardous flying conditions, which should obviously be avoided. Even when flying conditions are acceptable, low cloud cover can obscure sections of the river or make it necessary to fly at lower altitudes, significantly reducing reception range. During late fall or winter, icy snow coming into contact with the receiving antenna can create electronic interference. Hot, sunny weather can also present problems. Mid-day turbulence caused by warming air patterns sometimes creates flying conditions so physically uncomfortable that data collection is not possible. Scheduling survey flights in the early morning, late afternoon, or on overcast days with high cloud ceilings minimizes this problem.

Tracking success can also be affected by the behavior of the fish. As previously mentioned, fish that school or congregate together in localized areas complicate efforts to distinguish individuals. Fish swimming deep will be more difficult to detect due to reduced reception range, which can have a considerable effect when tracking over deep rivers and lakes. In one case, the reception range of two fish located together in a deep (25 m) section of river varied from 300–500 m to 2–3 km. Being aware of these issues and altering tracking methods accordingly (e.g., reducing aircraft speed, using flight patterns that increase the tracking time over the area) can have a major impact on survey results.

Preflight Preparations.—Aerial tracking is a dynamic process, and the situations encountered often vary in unexpected ways. Although there is no foolproof method to guarantee success, preparatory steps can be taken to minimize potential problems and enhance the quality of the data collected. Select the right type of aircraft. High-winged, single-engine airplanes are better suited for aerial tracking than those with low-winged configurations. The higher clearance provides a better view of the area being surveyed, and the wing struts are a convenient place to mount and orient the receiving antennas (Figure 7). As previously mentioned, aircraft capable of flying at slower speeds enhance tracking efficiency, particularly when the river is convoluted or the fish are grouped together. For example, a Maule aircraft has a minimum controlled air speed of around 70 km/h compared to a Cessna 185, which will stall at 100 km/h (Taylor 1989; Maule Air Inc. 2009). A Helio Courier has an even lower stall speed of around 50 km/h (Rowe 2006), although this aircraft does not have wing struts, making it more difficult to attach the receiving antennas. However, if faster air speeds don't negatively impact tracking success, the higher operating costs and limited availability of the Helio Courier or Maule may make the Cessna 185 a more practical choice, particularly when flying extended surveys. Helicopters are extremely effective for aerial tracking due to their ability to

FIGURE 7. Types of aircraft used for aerial tracking surveys. Airplanes with a high-winged configuration (left) are most commonly used and provide a convenient place to mount and orient the tracking antennas (see insert). Helicopters (right) provide an effective method for obtaining accurate fish locations due to their ability to fly low, slow, and hover.

maneuver, fly slow, hover, and if necessary, land to access remote sites. However, the substantially higher operating costs typically limit their use to special applications. Cruising range is also an important consideration, particularly when tracking in remote areas where fuel may not be readily available. It may be necessary to cache fuel at remote landing sites in order to operate effectively. The added costs and logistics associated with this process are substantial, and aircraft with an extended cruising range are an attractive option. Some aircraft generate higher levels of internal noise, usually from engine ignition, which can interfere with signal reception. Shielding engine components (e.g., installing shielded ignition cables) can minimize problems with noise. A preliminary test flight should be flown to determine whether this is an issue. The test flight also provides an opportunity to identify other unforeseen problems related to outfitting the aircraft for tracking (e.g., attaching the receiving antennas).

One of the most important decisions associated with aerial tracking is the selection of a pilot. The pilot should be experienced and familiar with the area being surveyed. Previous experience with aerial tracking is an obvious advantage. Talking with several prospective pilots, as well as with other people that have flown with them, is a good way to assess their abilities and determine if they understand (and are comfortable with) the type of flying being requested. When working with an air charter company that employs several pilots, it is important to be selective and not just settle for whoever is available. The government agency that oversees aviation practices and safety (e.g., Federal Aviation Administration in the U.S.) can provide information on company safety records and any past problems related to aircraft maintenance. Being able to talk openly with the pilot is equally important. Aerial tracking is much more involved than flying from one destination to another, and it is important to have a pilot that you can work with effectively. There should also be a clear understanding that when it comes to operational and safety issues, the pilot has the final word.

A preflight check list and inspection should be mandatory to ensure that all the necessary equipment and supplies (including backup gear) are on hand, properly installed, and in good working order (Table 5). This is particularly important when working in remote areas or flying

TABLE 5. Basic equipment used during aerial tracking surveys when a single person is tracking and receiving antennas are placed on both sides of the aircraft for improved reception coverage. The actual gear needed will vary depending on the telemetry equipment used, and the duration and nature of the survey.

Category	Equipment item	Comment
Principal	Receiver-data logger (RDL)	
	Earphone headset	
	Hybrid combiner	For joining multiple coax cables
	Receiving antenna (2)	
	Coax: RDL to hybrid combiner	
	Hybrid combiner to antenna (2)	
	Plastic wire ties	For attaching coax to aircraft
	Electric tape	For sealing exposed connections
	Reference transmitter	For pre-flight tests
	Data forms and pencils	
Optional	External power source (gel cell battery)	
	Battery cable (to receiver)	
	Power jack cables (to receiver)	When aircraft's power system is used as an external power source
	GPS receiver	If not integrated within the RDL
	Cables from RDL to GPS receiver	When using external GPS receiver
	GPS antenna and cable	
	Maps of area	
	Receiving antenna switch box	Used instead of hybrid combiner for more accurate locations
Backup	RDL	
	Earphone headset	
	Coax cables	Duplicate set of all cables
	Battery cable (to RDL)	
	Hybrid connector/switch box	
	Tools (including selected wrenches and screw drivers)	For tightening antenna elements and attachment connections
	Survival gear	For long-distance surveys in remote areas

extended distances since terminating a survey in mid-flight can be extremely costly in term of time, money, and data lost. The receiver-data logger should be checked to make sure that it is functioning properly and fully charged. The coax cables should be secured to the aircraft using electrical tape or wire ties so that they are not loose during the flight. The coax is routed into the aircraft cabin through air vents, windows, or doors sufficiently padded to prevent damage caused by crushing or abrading the cables. Electrical tape should be used to secure and waterproof the coax-antenna connections.

Special attention should be given to the receiving antennas, which are typically placed on both sides of the aircraft to minimize problems with signal shielding by the fuselage. Directional antennas, including loop, 4-element Yagi and H-antennas, are commonly used. The antennas are usually attached to the wing struts (Figure 7), which are structurally strong and provide a relatively unobstructed view of the surrounding area. Gilmer et al. (1981) reported that antennas positioned further away from the surface of the wing exhibited better reception. Reception was also enhanced by extending the antennas in front of the wing's leading edge. Antenna orientation can impact tracking success. Terrestrial surveys are typically flown with the antennas oriented to the side to cover a wider area (Gilmer et al. 1981; Mech 1983). This approach is suitable when tracking fish in large lakes, but having the antennas front facing (i.e., pointed in the direction the aircraft is flying) is preferable for fish in rivers since it maximizes the effective reception range and enhances the opportunity to detect fish farther upstream. It is a good practice to check how the antennas are mounted during the preflight inspection, particularly when flying with pilots that routinely conduct wildlife surveys and may not be aware of the difference. Front-facing antennas may be slightly off-set to the side (e.g., 20° from the front) to increase peripheral coverage when surveying rivers that are extensively braided (Figure 7). Whip antennas (also known as monopole antennas) are also used for aerial tracking. Although they are omni-directional and generally provide less range than directional antennas (Section 5.1), whip antennas are easy to mount, create less drag, and do not require special orientation. McAuley et al. (1993) reported reception ranges comparable to Yagi antennas when using whip antennas with grounded baseplates designed specifically for aircraft use.

The hardware used to mount the antennas on the aircraft should be easy to attach, but secure enough to withstand the airspeeds encountered. Mounting brackets are commercially available for some aircraft. They can also be fabricated by local metal shops if commercial models are not available. Noncommercial designs must be approved by the government agency overseeing aviation practices and safety, so it is advisable to initiate this process in advance to avoid last-minute delays. The brackets usually consist of an aluminum cuff that is clamped to the wing strut (Gilmer et al. 1981). A thin sheet of neoprene, rubber, or insulated foam is placed between the cuff and the strut to prevent damage to the aircraft. Vibrations from the aircraft can cause parts of the antenna and mounting hardware to loosen. Loose components can introduce high levels of interference, as well as being a safety hazard, and should be checked and tightened during the preflight inspection.

Once the equipment has been installed on the aircraft, the entire tracking system should be tested. One approach is to place a reference tag on both sides of the aircraft and evaluate the signals detected by the receiver. The antenna and coax on each side of the aircraft should also be tested separately, either by using a switchbox to alternate from one antenna to the other, or by separately connecting the coax from each antenna directly to the receiver. A reference tag should also be placed at a known location near the airfield, and located after takeoff to determine if the tracking system is functioning properly when airborne.

Preflight planning can significantly enhance tracking efficiency and survey results. There should be a thorough discussion with the pilot prior to takeoff detailing the objectives of the flight, areas being surveyed, distances involved, and the ideal flying characteristics needed when tracking (e.g., ground speed, altitude, and pattern of flight). Prearranged hand signals are often used to communicate while tracking, and are suitable for most basic maneuvers (e.g., circle back, increase/decrease altitude). Some aircraft are equipped with an inter-connected headphone system, making it possible to talk directly with the pilot. This arrangement has the added benefit of allowing the pilot to hear the transmitter signals, making it possible for him/

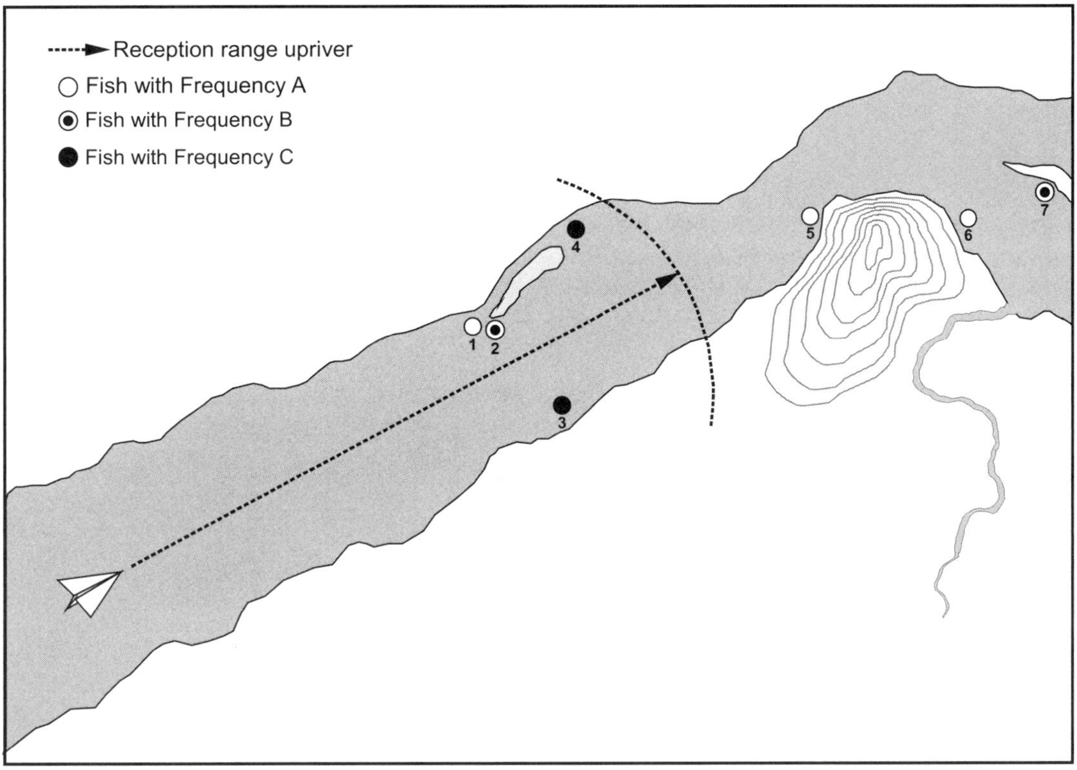

FIGURE 8. Aerial tracking survey showing a section of river, position of the aircraft, upriver reception range (i.e., distance upriver that fish can be detected, peripheral coverage not shown), and the location of radio-tagged fish within the area. Fish tagged with coded transmitters (i.e., transmitters on same frequency, but distinguished by different signal patterns) can increase detection rate when fish are concentrated in localized areas. In this example, transmitter signals are detected on three frequencies with four different fish identified.

her to anticipate what additional maneuvers might be needed. Special headsets equipped with microphones are needed, and commercial vendors can provide intercom units that will match the impedance of the aircraft's intercom system.

Tracking Procedures.—Aerial tracking is seldom straightforward. There is more than one way to track a fish, and most researchers inevitably fine-tune their methods to address specific research objectives or field conditions. However, some basic procedures are worth discussing. As an example, Figure 8 shows a stretch of river, the upriver reception range (assuming a constant aircraft speed and altitude, and fish at the same basic depth) with the gain set at the maximum effective level, and the fish located within the immediate area. The aircraft is flown over the river with the receiver-data logger scanning through the frequencies. Each time a signal is detected and decoded, the information is recorded by the data logger. This process creates a series of records that typically includes the date, time, signal strength, and the transmitter's frequency and code, which are used to identify the fish (Table 6). Receiver-data loggers equipped with GPS capabilities also record the location (e.g., latitude and longitude), which is a major advantage in areas with few distinct landmarks or when the trackers are unfamiliar with the area being surveyed. The human ear is more sensitive than the receiver, and

TABLE 6. Tracking records for radio-tagged fish located during a 2-minute period of an aerial survey with the aircraft traveling at constant altitude, direction, and speed. Transmitter frequencies are continually scanned, and signals detected are recorded. An integrated global positioning system (GPS) receiver is linked to the receiver-data logger, and spatial coordinates (e.g., latitude and longitude) included for each record. Shaded rows show the sequential records of one fish (frequency 780, code 22). The peak signal (greatest signal strength), which is used to estimate the location of the fish, is indicated.

Year	Ordinal date[a]	Hr	Min	Sec	Frequency[b]	Code	Act[c]	Signal strength	Latitude	Longitude
2004	216	16	40	7	780	22	0	98	64.3096	–145.1359
2004	216	16	40	8	780	22	0	99	64.3096	–145.1359
2004	216	16	40	9	780	22	0	107	64.3089	–145.1366
2004	216	16	40	10	780	22	0	101	64.3086	–145.1370
2004	216	16	40	23	1283	63	0	134	64.2883	–145.1538
2004	216	16	40	24	1283	63	0	133	64.2880	–145.1542
2004	216	16	40	25	1283	63	0	130	64.2876	–145.1546
2004	216	16	40	33	780	22	0	111	64.2818	–145.1613
2004	216	16	40	34	780	22	0	109	64.2818	–145.1613
2004	216	16	40	35	780	22	0	112	64.2811	–145.1616
2004	216	16	40	51	922	21	0	113	64.2697	–145.1626
2004	216	16	40	53	922	21	0	114	64.2693	–145.1627
2004	216	16	40	54	922	21	0	114	64.2690	–145.1627
2004	216	16	40	56	922	21	0	123	64.2686	–145.1628
2004	216	16	41	11	780	22	0	116	64.2595	–145.1606
2004	216	16	41	13	780	22	0	116	64.2588	–145.1599
2004	216	16	41	27	922	21	0	111	64.2561	–145.1562
2004	216	16	41	28	922	21	0	116	64.2555	–145.1553
2004	216	16	41	39	780	22	0	120	64.2493	–145.1440
2004	216	16	41	41	780	22	0	128	64.2442	–145.1319
2004	216	16	41	42	780	22	0	132	64.2440	–145.1313
2004	216	16	41	55	723	13	0	107	64.2278	–145.1331
2004	216	16	41	57	723	13	0	110	64.2265	–145.1344
2004	216	16	41	58	723	13	0	107	64.2265	–145.1344

[a] Days of year ranging from 1 (January 1) to 366 (December 31 for leap year).

[b] Transmitter frequency abbreviation: e.g., 780 = 150.780 MHz and 1283 = 151.283 MHz.

[c] Activity: 0 = active (transmitter motion sensors has been activated), 3 = inactive (motion sensors has not been activated within previous 24 hours.

it is not uncommon to hear faint transmitter signals before the receiver is able to detect them. The signal strength increases as the aircraft approaches the fish, and decreases after moving past. The peak signal is typically used to estimate the location of the fish (Table 6). The reduction in signal strength after the peak will be relatively abrupt when tracking with front-facing directional antennas, and more gradual if omni-directional whip antennas are used.

It is important to recognize that the location estimates obtained during aerial surveys reflect the position of the aircraft and not necessarily the location of the fish. Fish residing in nearby side channels, sloughs, or minor tributaries may be plotted incorrectly if only the main channel is flown. When flying over narrow or meandering rivers, it is often difficult to remain directly over the river channel, and it is not uncommon for the GPS coordinates to place the fish on land. These data obviously need to be reviewed and adjusted. Location estimates obtained when tracking multiple frequencies will also be less precise (discussed further in the next section), and therefore subject to error. Test trials, with reference transmitters placed at known locations, are useful for estimating the accuracy and precision of aerial locations. Confidence circles, developed by comparing the survey results with the actual locations, estimate the probability that the fish are within a defined area (Withey et al. 2001). This approach can also be used to evaluate reception range and location accuracy for different habitats and water depths.

Tracking fish on multiple frequencies presents special challenges. Once a transmitter is detected, remaining on that frequency until the peak signal is determined can provide an accurate estimate of its location, but fish on other frequencies within the area may be overlooked. An alternative approach is to briefly monitor each frequency, decode the signals present, and resume scanning as quickly as possible without attempting to determine the peak signal. Additional time is not spent on signals that are faint and cannot be readily decoded since they will likely be detected on subsequent scan cycles when the signal is stronger. Continually repeating this process increases the probability that all of the fish within the area are identified and recorded. Although it doesn't provide a continuous tracking record for each fish, the series of records obtained (Table 6) can provide a good approximation of the fish's location depending on the reception range and the interval between detections—once again illustrating the importance of reducing scan time. After all the fish within the area have been identified, more time can be spent to refine the accuracy of the location if such data are needed (discussed later in this Section). Figure 8 also shows the advantage of using coded transmitters. Instead of having to contend with signals on seven different frequencies, fish are present on only three frequencies, thereby simplifying data processing and reducing scan time. Software advances have enhanced the ability to distinguish multiple codes detected simultaneously on the same frequencies, although when large numbers of fish are concentrated within an area it may be necessary to make several passes to ensure that all the fish present are detected. Aircraft capable of reduced speeds are an obvious advantage in these situations.

Although the receiver-data logger records the telemetry data, a running tally (e.g., transmitter frequency, time, and general location) should be kept of the fish audibly detected. In addition to providing a backup account of the survey, this information is particularly useful for keeping track of fish that have not been decoded by the receiver. Weak signals may represent fish that are located off of the main flight path in side channels, upland sloughs, small peripheral tributaries, or anabranch streams. These signals can be misinterpreted as fish located farther upriver and may not be detected during subsequent scans. Aerial surveys can sometimes be chaotic with several things happening at once, and it is easy to lose track of specific individuals while concentrating on other transmitter signals. By keeping a running tally of the frequencies detected, it is obvious when a weak signal has not been re-acquired and decoded, making it possible to double-back to locate and identify the fish.

Location accuracy is another important consideration. In some instances only general information is needed. Determining location to the nearest kilometer may be more than adequate when tracking fish that are migrating hundreds of kilometers. In other situations, more exact information is needed due the nature of the study area and the research objectives. There are several ways to improve the accuracy of aerial locations:

- continuously monitoring the frequency provides a better indication of when the transmitter signal peaks, resulting in a more accurate location estimate
- decreasing the altitude of the plane reduces reception range, thereby isolating the signal within a smaller area and providing better resolution
- systematically decreasing the gain of the receiver reduces signal reception and restricts the area where the transmitter can be detected
- using a switchbox to compare signal strength on different sides of the aircraft and further isolate the signal to a specific area.

In practice, using a combination of these techniques provides the best approach for improving accuracy. Probably the most effective method for obtaining accurate aerial locations is with helicopters. Due to their ability to fly low, slow, and hover, it is possible to locate fish within 10 m of their actual location. Although more expensive, helicopters are also able to access the sites to recover transmitters, collect samples, and record other information (e.g., habitat characteristics and water conditions). In remote areas with limited access, this approach may not only be cost-effective, but the only way to collect the information needed.

The human factor should not be overlooked when flying aerial surveys. Although the technology is basically in place to automate aerial tracking, the human ability to assimilate and respond to changing conditions and situations (e.g., doubling back to search for a weak signal that was not identified or altering the flight plan to survey a promising area) is still a major advantage. People, however, are susceptible to fatigue, distractions, and motion sickness, and prolonged surveys can adversely impact their ability to function effectively. Gilmer et al. (1981) recommended limiting aerial surveys to under 2 h, although in remote areas this is often not practical due to the time and costs associated with accessing the sites. Simplifying tracking procedures and ensuring that conditions are comfortable can substantially extend the time that people are able to track effectively. Receiver-data loggers with easy-to-use controls and functions and that automatically record the data collected, make tracking much easier. Receivers with GPS capabilities eliminate the need to constantly determine the position of the aircraft, and make it possible to concentrate on tracking. Hockersmith and Peterson (1997) reported that location estimates obtained under adverse flying conditions were more accurate with GPS-linked receivers than those based on visual landmarks. Common sense steps, like keeping the aircraft cabin comfortable (e.g., adequate ventilation and moderate temperature) and conducting surveys on days when flying conditions are optimal, are also important. When tracking salmon in the Yukon River, tracking efficiency was maintained even though the aerial surveys typically ranged from 4 to 6 h. Similar results have been reported by other researchers (R. Brown, U.S. Fish and Wildlife Service, Fairbanks, Alaska, personal communication, T. Spencer, Alaska Department of Fish and Game, Anchorage, Alaska, personal communication).

Boat Tracking

Tracking from boats is probably the most common form of mobile tracking. Although less suitable than aerial tracking when surveying vast or remote areas, for many studies this approach offers a number of distinct advantages. It is versatile. Boats are available in a variety of shapes and sizes, provide easy access to aquatic sites, and can be readily modified and equipped with tracking gear. Considerably less expensive than aerial tracking, tracking from boats makes frequent surveys possible even when operating budgets are limited. Improved receivers and outboard motors that produce substantially less electronic noise make it easier to detect and identify the fish. When operating in small, localized areas, nonmotorized boats

FIGURE 9. Antennas used to locate radio-tagged fish from boats. General locations can be obtained with omni-directional dipole antennas based on signal strength, with the peak signal providing the best estimate of the fish's position. This estimate can be refined by reducing the receiver gain to reduce reception range. More accurate locations can be obtained with directional antennas (see insert). The area is swept horizontally with the antenna to determine the direction of the transmitter signal.

(e.g., rafts and canoes) eliminate engine-related interference, as well as being less disruptive to the fish. Accurate location estimates are also easier to obtain when in close proximity to the transmitters. In addition, it is not necessary for the vessel to travel at a high rate of speed, which simplifies tracking procedures.

Similar tracking gear (e.g., receivers, coax, power sources, and headphones) are used for both aerial and boat tracking. Increased exposure to the elements, particularly when operating from an open skiff, makes it necessary to take extra precautions to protect the equipment from moisture. A variety of receiving antenna can be used, including both omni-directional (e.g., whip, dipole) and directional (e.g., loop, Yagi) types (Figure 9). Omni-directional antennas are easier to mount on vessels since they do not have to be oriented in a particular direction, whereas directional antennas, while more cumbersome, have better reception range and provide more precise information on fish location. Tracking with large Yagi antennas (e.g., 8-element) requires a stable work platform, and small skiffs are sometimes inadequate particularly in rough water or poor weather conditions. A combination of antenna can sometimes be effective, such as using a mounted omni-directional antenna to determine when fish are present in the general area, and a hand-held directional antenna to determine the actual location.

Reception range is substantially reduced when operating from boats compared to aircraft. Winter (1996) estimated that range was two to three times less from boats. Even greater differences were observed when tracking salmon in large Alaskan rivers. Reception range from boats typical averaged 0.5 km compared to 2–3 km during aerial surveys. Other studies have reported similar tracking results, with reception typically ranging from several hundred meters to a kilometer depending on the equipment used and the behavior of the fish (Beeman et al. 1998; Babaluk et al. 2001, Monet and Soares 2001). Line-of-sight can also be limited, further complicating boat tracking efforts. Increasing the height of the receiving antenna can lessen this problem. When tracking salmon, the receiving antenna was typically mounted on a 2 m extension pole. This approach works well when operating in relatively open areas (e.g., large rivers or lakes), but it can be problematic in smaller rivers and streams due to overhanging vegetation, bridges, or other hazards. One solution is to incorporate a quick-release mechanism into the mounting bracket so that the antenna can be lowered rapidly when the need arises. Omni-directional whip antennas may be better suited when operating in more confined areas since they are more flexible than dipoles and less likely to be damaged if mishaps occur.

Differences in the frequency range used can also impact tracking capabilities. When tracking salmon tagged with low frequency transmitters (30 MHz), it was often possible to detect faint signals from fish located in side channels, anabranch streams, and small tributaries associated with braided sections of the main river, presumably due to the longer wavelength associated with this frequency range (Section 5.1). Salmon tagged in the same area with high frequency (150 MHz) transmitters, which have a shorter wavelength that is more likely to be reflected by physical obstructions, could not be detected when they were off channel, and at times were even difficult to locate in the main river when positioned behind islands or prominent gravel bars. The added coverage provided by the lower frequency transmitters is a definite advantage when boat tracking in complex habitats. However, this consideration must be weighed against the limitations inherent with low frequency antennas that have been previously discussed.

Boat tracking procedures are similar to those described for aerial surveys. In general, the boat moves through the study area with the receiver-data logger scanning through the transmitter frequencies and the gain set at the maximum effective level. The tracking process is much less demanding. The vessel's speed can be reduced to evaluate the data collected and fine-tune the direction of travel. Once the general position of the fish is determined, directional antennas are used to manually search the surrounding area to pinpoint the actual location. Similar to triangulation, the area is swept with the antenna to determine the direction of the strongest signal. As the boat moves toward the transmitter, the intensity of the signal increases until it becomes omni-directional. Reducing the gain further causes the signal to become directional again. The transmitter is relatively close if the gain is at the lowest setting and the signal is still omni-directional. Making the antenna less efficient will reduce reception further and isolate the signal to a more restricted area, resulting in fairly accurate locations (Bruno et al. 1990; Niemela et al. 1993, Koehn et al., in press) This can be accomplished in a number of ways, including (progressing in order of effectiveness):

- sweeping the area with the antenna oriented so that it is less efficient (e.g., with the antenna elements positioned horizontally when using a Yagi, or with the receiving end of the antenna directed toward the ground)
- incorporating an adjustable attenuator pad (Model 839, Kay Elemetrics Corp, Pine Brook, New Jersey) into the coax line between the antenna and receiver; the pad is used to shunt different levels of signal to ground, providing additional control on how much signal is received
- using a less efficient type of antenna (e.g., paddle antenna)
- removing the antenna, and using the coax cable as an inefficient omni-directional antenna
- removing the coax and using a small omni-directional "stub" antenna attached directly to the receiver
- listening to the signal with just the receiver (i.e., no coax or antenna)

The transmitter is extremely close (i.e., within several meters) if the signal is still audible with the antenna and coax removed, and the gain at the lowest setting. Even in this situation some directionality can still be obtained, based on differences in signal intensity, by rotating with the receiver and using your body to shield the signal. An underwater antenna can also be used to accurately determine location when in close proximity to the fish (Beeman et al. 2004). The reception range for this type of antenna is typically less than 10 m and can be substantially reduced by decreasing the gain of the receiver. Using a switchbox to alternate between an underwater antenna and a more efficient tracking antenna (e.g., 4-element Yagi) provides a convenient way to obtain general and fine-scale location information without having to physically change connections with the receiver.

As the tracking boat approaches, it may disturb the fish and cause them to move out of the immediate area or alter their behavior (e.g., diving to deeper depths or moving under cover). Disturbance is particularly problematic when collecting information on fine-scale habitat use or activity patterns. Positioning the boat upstream, turning off the engine, and drifting past the site can minimize this problem. Quietly paddling through the area may also be a viable alternative. In some cases, it may be necessary to approach close enough to get a general fix on the fish's location and then approach from the shore. Some authors have described methods used to avoid disturbing fish when approached (Leggett 1976; Katz 1986), although the actual impact is difficult to evaluate.

Foot surveys

Sometimes boats are unable to access the areas where fish are residing and tracking on foot is the only alternative. This method is ideal for small rivers and streams but can also be used along the shore or in off-channel areas associated with larger drainages. Tracking on foot is often less disruptive to the fish than approaching by boat. The tracking gear must obviously be compact, lightweight, durable, and water-resistant. Smaller antennas are essential, particularly when walking through dense brush or wading in deep water.

Many of the techniques described for boat tracking also apply to foot surveys, particularly those related to reducing reception range and antenna efficiency in order to isolate the transmitter signals within a confined area. Tracking along single-channeled stretches of river,

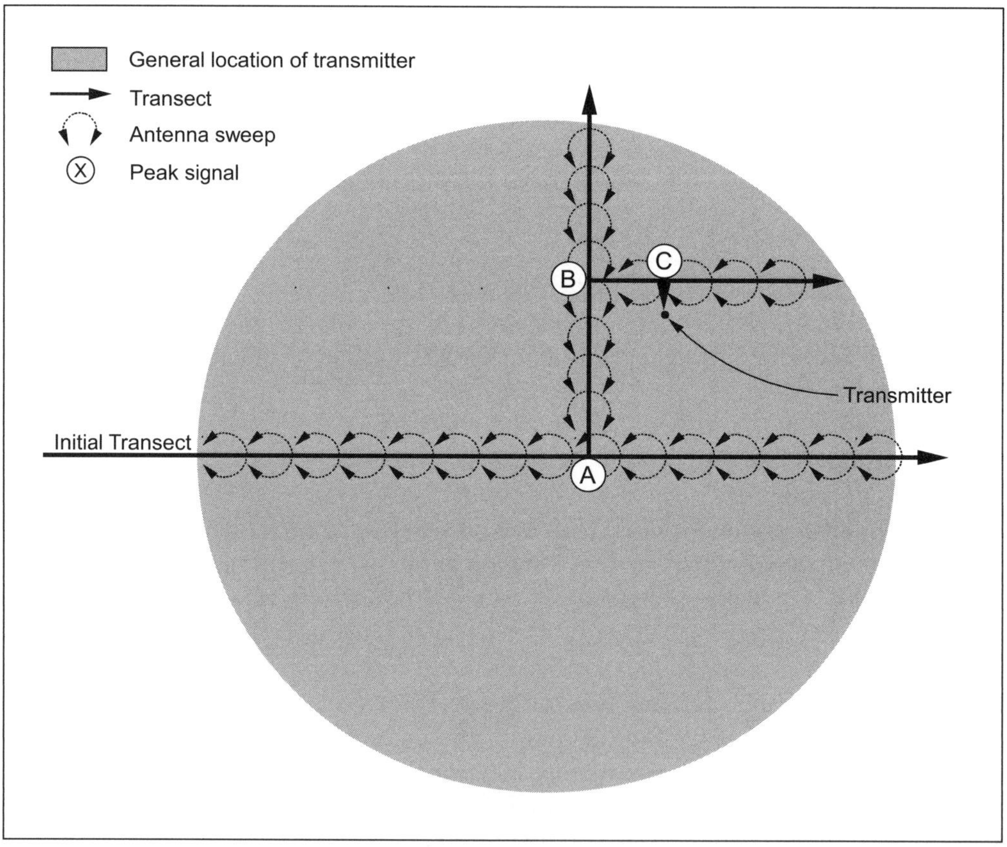

FIGURE 10. Method for locating transmitters on foot using a series of perpendicular transects. After determining the general location of the transmitter, an initial transect is made in the direction of the strongest signal. The receiving antenna is swept back and forth to determine the direction of the strongest signal. Systematically reducing the efficiency of the antenna and the receiver gain is used to maintain the directionality of the signal when in close proximity to the transmitter. Locations where the strongest signal is perpendicular to the transect (Points A, B and C) are indicated.

where it's possible to simply walk along the bank while scanning through the transmitter frequencies, is relatively straightforward. Reducing the receiver gain, combined with quick, informal triangulation, can provide a good idea of where the fish are located. This process becomes more complicated (and frustrating) when tracking in more complex habitats, such as rivers with numerous side channels and sloughs, prominent undercut banks, dense vegetation, and varied topography. Under these conditions, the direction of the signal may be difficult to determine due to signal bounce, and in some cases can mysteriously fade away as a result of signal obstruction from the surrounding landscape. To avoid these problems, it is useful to develop a general sense of where the fish is located before charging forward in the direction of the strongest signal. Doing a quick, informal triangulation to identify the approximate position, and starting the final approach from the highest point possible where line-of-sight is optimized, can avoid some of the confusion when traveling in the direction of the signal. Knowing the general location of the fish may also identify an easier way to access the site, rather than wasting time and effort by walking through waist-deep water or tangled vegetation.

Walking a series of interconnected transects is an effective way to locate the transmitter once the signal has been isolated within a confined area (Figure 10). An initial transect is made in the direction of the strongest signal while sweeping back and forth with an inefficient directional antenna. Systematically reducing the receiver gain, when the signal becomes omni-directional or too intense, helps to differentiate the direction of the transmitter. The spot where the strongest signal is perpendicular to the transect (Point A) is noted, although the search along the transect is continued until it becomes obvious that the transmitter is back in the opposite direction. A second transect is made, starting at Point A, and the process repeated until the transmitter is located. This method is particularly useful when recovering transmitters from fish that have died or shed their tags. Not surprisingly, live fish are likely to move when approached unless they are hiding in dense vegetation, large woody debris, or undercut banks. Transmitters equipped with motion sensors and activity monitors (Eiler 1990; Eiler et al. 2006) help to identify if the fish are actively moving, and are a useful option to consider when selecting tags for studies that include obtaining fine-scale locations.

COMBINING METHODS

Both stationary and mobile tracking have distinct advantages, and in many ways are complementary. Tracking stations provide continual coverage at localized sites and accurate passage information that can be used for standardized comparisons. Mobile tracking, although intermittent, provides more flexible coverage on a wider scale. In some instances, using a combination of both approaches can substantially enhance the quality of the information collected. The telemetry studies on Yukon River salmon are a case in point. Tracking stations were used to provide information on the general distribution, timing, and migration rates of the fish returning to the river to spawn; information that would have been difficult (if not impossible) to obtain with mobile tracking due to the vast size, limited access, and isolated nature of the basin. Aerial surveys were used to determine the site-specific distribution (i.e., actual locations) of the fish and identify important spawning areas; information that could not have been determined from the stations due to the localized nature of the data. Data from the stations, retrieved in-season via a communication uplink, were instrumental in coordinating the aerial surveys. By knowing in advance where the fish had traveled, it was possible to concentrate tracking efforts in areas that would provide the most information. Aerial surveys provided information on the status of the fish (e.g., unreported fishery recovery, dead from other causes, intransit, or spawning migration completed) that was essential for interpreting the data from the stations. Combining methods can also be useful in small-scale studies. The effectiveness of combining boat surveys and tracking on foot has been previously discussed. Fish missing after repeated boat tracks may be readily located from aircraft, making it possible to document unexpected behavioral patterns that might have gone unreported.

SUMMARY

Being able to effectively relocate the fish that have been tagged is essential to the success of any telemetry study, and ultimately depends on the equipment and tracking methods used, the physical characteristics of the area, behavior of the fish, and skill of the operator. The two basic methods discussed, stationary and mobile tracking, provide a wide range of options, making it possible to collect telemetry information under a variety of circumstances and conditions. There is seldom a perfect approach, and each method has advantages and disadvantages that must be weighed relative to the research objectives and logistics. Understanding and incorporating these insights into the experimental design can have a major impact on the success of the study. In some instances, using a combination of these techniques can substantially enhance the quality of the information obtained.

ACKNOWLEDGMENTS

This Section was critically reviewed by N. Adams, D. Alcorn, J. Beeman, R. Brown, T. Castro-Santos, G. Duker, N. Eiler, J. Koehn, L. Kuechle, J. Lee, M. Sisak, and R. Wildman. Many of the methods and techniques discussed have been developed and refined from years of field work, and special thanks goes to the many individuals whose dedication, insights, and imagination have helped advance our understanding and use of telemetry in fisheries research.

REFERENCES

Babaluk, J. A., H. R. Wissink, B. G. Troke, D. A. Clark, and J. D. Johnson. 2001. Summer movements of radio-tagged arctic char (*Salvelinus alpines*) in Lake Hazen, Nunavut, Canada. Arctic 54:418–424.

Beeman, J. W., P. V. Haner, and A. G. Maule. 1998. Evaluation of a new miniature pressure-sensitive radio transmitter. North American Journal of Fisheries Management 18:458–464.

Beeman, J. W., C. Grant, and P. V. Haner. 2004. Comparison of three underwater antennas for use in radiotelemetry. North American Journal of Fisheries Management 24:275–281.

Boggs, C. T., M. L. Keefer, C. A. Peery, and T. C. Bjornn. 2004. Fallback, reascension, and adjusted fishway escapement estimates for adult Chinook salmon and steelhead at Columbia and Snake River dams. Transactions of the American Fisheries Society 133:932–949.

Bunt, C. M., C. J. Cooke, and D. P. Philip. 2003. A modified downrigger for detecting radio transmitters in deep water. North American Journal of Fisheries Management 23:258–263.

Bruno, N. A., R. W. Gregory, and H. L. Schramm, Jr. 1990. Nest sites used by radio-tagged largemouth bass in Orange Lake, Florida. North American Journal of Fisherles Management 10:80–84.

Cooke, S. J., S. G. Hindh, G. T. Crossin, D. A. Patterson, K. K. English, J. M. Shrimpton, G.V. Kraak, and A. P. Farrell. 2006. Physiology of individual late-run Fraser River sockeye salmon (*Oncorhynchus nerka*) sampled in the ocean correlates with fate during spawning migration. Canadian Journal of Fisheries and Aquatic Sciences 63:1469–1480.

David, B. O., and G. P. Closs. 2005. Continuous remote monitoring of fish activity with restricted home ranges using radiotelemetry. Journal of Fish Biology 59(3):705–715.

Ebner, B. C., and J. D. Thiem. 2009. Monitoring by telemetry reveals differences in movement and survival following hatchery or wild rearing of an endangered fish. Marine and Freshwater Research 60:45–57.

Eiler, J. H. 1990. Radio transmitters used to study salmon in glacier rivers. Pages 364–369 *in* N. C. Parker, A. E. Giorgi, R. C. Heidinger, D. B. Jester, Jr., E. D. Prince, and G. A. Winans, editors. Fish-marking techniques. American Fisheries Society Symposium 7:364–369.

Eiler, J. H., B. D. Nelson, and R. F. Bradshaw. 1992. Riverine spawning by sockeye salmon in the Taku River, Alaska and British Columbia. Transactions of the American Fisheries Society 121:701–708.

Eiler, J. H. 1995. A remote satellite-linked tracking system for studying Pacific salmon with radio telemetry. Transactions of the American Fisheries Society 124(2):184–193.

Eiler, J. H., and M. A. Masters. 2000. A database-GIS mapping program for summarizing salmon telemetry data from the Yukon River basin. Pages 138–144 *in* J. H. Eiler, D. J. Alcorn, and M. R. Neuman, editors. Biotelemetry XV: Proceedings of the 15th International Symposium on Biotelemetry. Juneau, Alaska USA. International Society on Biotelemetry. Wageningen, The Netherlands.

Eiler, J. H., T. R. Spencer, J. J. Pella, and M. M. Masuda. 2006. Stock composition, run timing, and movement patterns of Chinook salmon returning to the Yukon River basin in 2004. U.S. Department of Commerce, NOAA Technical Memorandum NMFS-AFSC-165, 107 pp.

Evenson, M. J. and K. G. Wuttig. 2000. Inriver abundance, spawning distribution, and migratory timing of Copper River chinook salmon in 1999. Alaska Department of Fish and Game, Fisheries Data Series No. 00–32, Anchorage.

Garrott, R. A., G. C. White, R. M. Bartmann, and D. L. Weybright. 1986. Reflected signal bias in biotelemetry triangulation systems. Journal of Wildlife Management 50:747–752.

Gerlier, M., and P. Roche. 1998. A radio telemetry study of the migration of Atlantic salmon (*Salmo salar* L.) and sea trout (*Salmo trutta trutta* L.) in the upper Rhine. Hydrobiologia 371/ 372:283–293.

Gilmer, D. S., L. M. Cowardin, R. L. Duval, L. M. Mechlin, C. W. Shaiffer, V. B. Kuechle. 1981. Procedures for the use of aircraft in wildlife biotelemetry studies. U. S. Department of Interior, Fish and Wildlife Service Resource Publication 140. Washington, D.C.

Grothues, T. M., K. W. Able, J. McDonnell, and M. M. Sisak. 2005. An estuarine observatory for real-time telemetry of migrant macrofauna: Design, performance, and constraints. Limnology and Oceanography: Methods 3:275–289.

Hilderbrand, R. H., and J. L. Kershner. 2000. Movement patterns of stream-resident cutthroat trout in Beaver Creek, Idaho-Utah. Transactions of the American Fisheries Society 129:1160–1170.

Hinch, S. G., and P. S. Rand. 1998. Swim speeds and energy use of upriver-migrating sockeye salmon (*Oncorhynchus nerka*): role of local environment and fish characteristics. Canadian Journal of Fisheries and Aquatic Sciences 55:1821–1831.

Hinch, S. G., and J. Bratty. 2000. Effects of swim speed and activity pattern on success of adult sockeye salmon migration through an area of difficult passage. Transactions of the American Fisheries Society 129:598–606.

Hockersmith, E. E., and B. W. Peterson. 1997. Use of the global positioning system for locating radio-tagged fish from aircraft. North American Journal of Fisheries Management 17:457–460.

Holder, R. R., and J. H. Eiler. 2000. Practical aspects of aerial tracking salmon in remote Alaska rivers. Pages 154–160 *in* J. H. Eiler, D. J. Alcorn, and M. R. Neuman, editors. Biotelemetry XV: Proceedings of the 15th International Symposium on Biotelemetry. Juneau, Alaska USA. International Society on Biotelemetry. Wageningen, The Netherlands.

Katz, H. M. 1986. Migrational swimming speeds of the American shad, Alosa sapidissima, in the Connecticut River, Massachusetts, USA. Journal of Fish Biology 29:189–197.

Keefer, M. L., C. A. Peery, M. A. Jepson, K. R. Tolotti, and T. C. Bjornn. 2004. Stock-specific migration timing of adult salmon in the Columbia River Basin. North American Journal of Fisheries Management 24:1145–1162.

Kenward, R. E. 1987. Wildlife radio tracking: equipment, field techniques and data analysis. Academic Press, London.

Koehn, J. D., J. H. Eiler, J. A. McKenzie, and W. G. O'Connor. In press. An improved method for obtaining fine-scale location of radio tags. *In* J. Pepperell, editor. Advances in Tagging and Marking Technology for Fisheries Research and Management.

Kolenosky, G. B., and D. H. Johnston. 1967. Radio-tracking timber wolves in Ontario. American Zoologist 7(2):289–303.

Leggett, W. C. 1976. The American shad with special references to its migration and population dynamics in the Connecticut River. Pages 169–225 *in* D. Merriman and L. M. Thorpe, editors. The Connecticut River ecological study: the impact of a nuclear power plant. American Fisheries Society, Monograph 1, Bethesda, Maryland.

Lenth, R. V. 1981. On finding the source of a signal. Technometrics 23:149–154.

Maule Air Inc. 2009. Aircraft Model Guide: Performance Specifications. Available: http://www.mauleairinc.com/Literature/performance.pdf (March 2011).

McAuley, D. G., J. R. Goldsberry, and J. R. Longcore. 1993. Omnidirectional aircraft antennas of aerial telemetry. Wildlife Society Bulletin 21(4):487–491.

McCleave, J. D., J. H. Power, and S. A. Rommel, Jr. 2006. Use of radio telemetry for studying upriver migration of adult Atlantic salmon (*Salmon salar*). Journal of Fish Biology 12(6):54–558.

Mech, L. D. 1983. Handbook of animal radio-tracking. University of Minnesota Press, Minneapolis.

Monet, G., and I. Soares. 2001. Tracking 48 and 150 MHz radio-tagged male lake-trout during their spawning migration in a mountain regulated river. Journal of Mountain Ecology 6:7–17.

Niemela, S. L., J. B. Layzer, and J. A. Gore. 1993. An improved radiotelemetry method for determining use of microhabitats by fishes. Rivers 4(1):30–35.

Pace, R. M., and H. P. Weeks, Jr. 1990. A nonlinear weighted least squares estimator for radiotracking via triangulation. Journal of Wildlife Management 54(2):304–310.

Rowe, F. J. 2006. Helio Courier Ultra C/STOL Aircraft: an illustrated developmental history. McFarland and Company, Jefferson, North Carolina.

Sisak, M. M., and J. S. Lotimer. 1998. Frequency choice for radio telemetry: the HF vs. VHF conundrum. Hydrobiologia 371/ 372:53–59.

Taylor, M. J. H. 1989. Jane's Encyclopedia of Aviation. Portland House, New York.

Weatherley, A. H., S. C. Rogers, D. G. Pincock, and J. R. Patch. 2006. Oxygen consumption of active rainbow trout, Salmo gairdneri Richardson, derived from electromyograms obtained by radiotelemetry. Journal of Fish Biology 20(4):479–489.

White, G. C. 1985. Optimal locations of towers for triangulation studies using biotelemetry. Journal of Wildlife Management 49:190–196.

White, G. C., and R. A. Garrott. 1990. Analysis of Wildlife Radio-tracking Data. Academic Press, San Diego, California.

Winter, J. 1996. Advances in underwater biotelemetry. Pages 555–590 *in* B. R. Murphy and D. W. Willis, editors. Fisheries Techniques, 2nd edition. American Fisheries Society, Bethesda, Maryland.

Withey, J. C., T. D. Bloxton, and J. M. Marzluff. 2001. Effects of tagging and location error in wildlife radiotelemetry studies. Pages 45–75 *in* J. J. Millspaugh and J. M. Marzluff, editors. Radio Tracking and Animal Populations. Academic Press, San Diego, California.

Zigler, S. J., M. R. Dewey, B. C. Knights, A. L. Runstrom, and M. T. Steingraeber. 2003. Movement and habitat use by radio-tagged paddlefish in the upper Mississippi River and tributaries. North American Journal of Fisheries Management 23(1):189–205.

Radio Telemetry: Case Studies

Section 6.1

Two Case Studies From Washington's Olympic Peninsula: Radio Telemetry Reveals Bull Trout Anadromy and Establishes Baseline Information Prior to Dam Removal

Stephen C. Corbett and Samuel J. Brenkman

Introduction

The use of radio telemetry is a widely accepted fisheries technique being used to address numerous topics on juvenile and adult fishes in small and large river systems. A keyword search for "radio telemetry" in American Fisheries Society journals revealed a total of 265 published manuscripts on radio telemetry (accessed on 11 December 2008). For one species alone, the federally threatened bull trout *Salvelinus confluentus*, there have been 72 radio telemetry studies that involved ~3,000 tagged individuals throughout four U.S. states and two Canadian provinces (Dare 2006).

Radio telemetry studies have been conducted in a wide array of riverine systems on multiple fish species to better understand spatial and temporal movements (Curry et al. 2002; Zigler et al. 2003), identify distinct life history forms (Meka et al. 2003; Brenkman and Corbett 2005), and determine habitat preferences (Tyus et al. 1984; Muhlfeld and Marotz 2005). Radio telemetry techniques have also been used to evaluate fallback rates of fish over dams (Boggs et al. 2004), estimate sport fishing catch and release mortality (Bendock and Alexandersdottir 1993), examine diel movements (Snedden et al. 1999), and determine seasonal migrations and spawning ecology (Gray and Haynes 1979; Burger et al. 1985; Brown and Mackay 1995; Swanberg 1997; Hilderbrand and Kershner 2000; Schmetterling 2001; English et al. 2005).

Typically, the use of radio telemetry studies is confined to studies of fish movements in freshwater systems. However, radio telemetry has proven useful in understanding fish migrations between rivers and marine environments. Recent studies highlight the importance of radio telemetry in establishing the extent of anadromy, periods of marine residency, and the diverse range of freshwater, estuarine, and marine habitats used by fish that inhabit coastal watersheds [Curry et al. 2002; Brenkman and Corbett 2005; Welch et al. 2006 (acoustic and radio telemetry), Brenkman et al. 2007].

The purpose of these case studies is to discuss important technological considerations during the design and implementation phases of radio telemetry studies in wilderness river systems that drain into marine waters. We present research objectives and specific examples of information collected using radio telemetry from two case studies in small river systems (<900 km^2) on Washington's Olympic Peninsula.

In the first case study, we describe how the use of radio telemetry revealed that bull trout moved between freshwater and the Pacific Ocean and between watersheds along coastal Washington. This was the first verification of anadromy and irregular periods of freshwater, estuarine, and marine residency for bull trout (Brenkman and Corbett 2005). The second case study focuses on the use of radio telemetry to assess fish movements prior to large-scale dam removal and recolonization by anadromous Pacific salmonids in the Elwha River. Removal of the two dams in the Elwha River is scheduled to begin in 2011, and will represent one of the largest dam removals in North America (Duda et al. 2008). We conclude by highlighting some of the advantages and disadvantages of the application of radio telemetry techniques in these remote river systems and provide management implications from this work.

Radio Telemetry Studies in the Hoh and Elwha Rivers, Washington

The juxtaposition of two multiple year case studies in the Hoh (2002 to 2005) and Elwha (2005 to 2008) River systems on Washington's Olympic Peninsula offers an opportunity to examine how differing site conditions and research objectives influence project design and implementation (Table 1). The upper portions of Hoh and Elwha Rivers originate in Olympic National Park, a large wilderness area in northwestern Washington (Figure 1). These rivers flow through very remote terrain, are glacially influenced, and ultimately drain into saltwater (Figure 1). The Hoh and Elwha Rivers are particularly challenging to sample using traditional fisheries techniques because of prolonged periods of high water throughout most months of the year, low water visibility, and no road access to upper portions of each watershed where fish typically reside and spawn.

The 91 km long Hoh River is free flowing throughout its course and flows into the Pacific Ocean. The river drains 894 km^2 and descends in elevation from 1,216 m in the headwaters to its confluence with the ocean. The annual mean flow of the Hoh River is 72 m^3/s, and the highest peak flow ever recorded was 1,758 m^3/s (USGS 2007). A total of 65% of the watershed is within the boundaries of Olympic National Park with no road access along the upper 40 km of river.

Unlike the free flowing Hoh River, the Elwha River has two large hydroelectric dams, Glines Canyon Dam at river km 21.7 and Elwha Dam at river km 7.9, that have physically isolated fish populations for 97 years (Figure 1). The two impoundments created by the dams, Lake Aldwell and Lake Mills, are 4 km and 4.5 km long, respectively. The section of river upstream from Glines Canyon Dam is freeflowing, and there are two downstream river sections regulated by dams (Figure 1). The Elwha River is 73 km long, drains 833 km^2 and descends in elevation from 1,372 m in the headwaters to sea level at its confluence with the Strait of Juan de Fuca. The Strait of Juan de Fuca is the marine passage connecting the Puget Sound with the Pacific Ocean. The mean daily discharge of the Elwha River is 43 m^3/s and annual minimum flows range from 8.5 to 14 m^3/s during summer (Curran et al. 2009). Average annual rainfall ranges from 100 cm/year to over 550 cm/year in the headwaters (Duda et al. 2008). Eighty-

TABLE 1. Summary of physical and study characteristics for two radio telemetry projects that were conducted in the Hoh and Elwha River basins, Washington.

Case Study Area	Watershed Area (km^2)	Total Length of River (Rkm)	Characteristics	Mean Annual Flow (m^3)	Fish Species	Number of Fish Tagged	Tracking Infrastructure and Effort	Duration of Study
Hoh River basin	894	91	Freeflowing, drains into Pacific Ocean	72	Bull trout	128	Manual-weekly; Aerial = 90 flights; Fixed stations = 6	2002 to 2005
Elwha River basin	833	73	Two reservoirs, drains into Strait of Juan de Fuca	43	Bull trout	95	Manual-weekly; Aerial = 25 flights; Fixed stations = 5	2005 to 2008

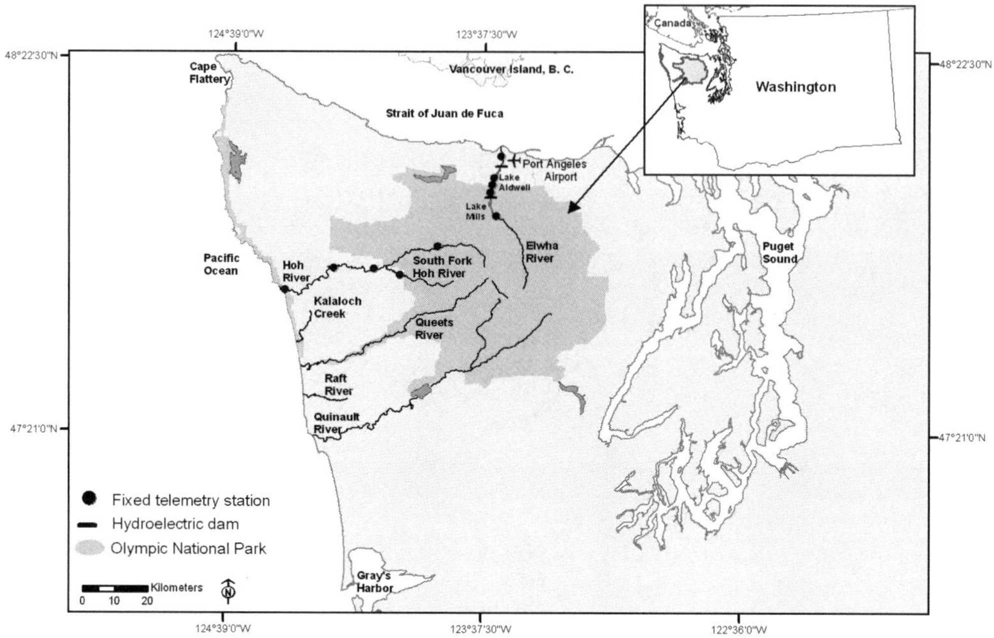

FIGURE 1. Geographic area of radio telemetry studies in the Elwha and Hoh River Basins on Washington's Olympic Peninsula. Location of fixed tracking stations, National Park boundary, and survey areas are indicated.

five percent of the watershed is within the boundaries of Olympic National Park, and there is no road access along the upper 50 km of river.

The Hoh River study, the first of the two, was designed to determine the extent that federally threatened bull trout moved to and from the Pacific Ocean. Additionally, radio telemetry was used to determine spatial and temporal movements of bull trout within the Hoh River Basin. Radio tracking quickly extended to other nearby watersheds due to extensive interbasin fish movements along coastal Washington (Brenkman and Corbett 2005 and Brenkman et al. 2007).

The Elwha radio telemetry study was designed to determine movements of bull trout in regulated and unregulated river sections of the river before the removal of two large hydroelectric dams. The Elwha River study also determined the extent that fish fall over the two hydroelectric dams, an issue that previously was not understood in that system. The following discussion chronologically outlines specific actions that we took to accomplish objectives for case studies in the Hoh and Elwha Rivers.

Literature Review and Consultation

The development of the study design and telemetry tracking infrastructure specific to our objectives and study areas benefited from review of other published telemetry studies. Many of the references provided within this case study were instrumental to the design of the two projects. We also contacted researchers within our region, and consulted experts for assistance in the study design, methods, and set-up of telemetry equipment in the Hoh and Elwha River basins.

To assist in the development and set-up of our case studies, we conducted site visits to existing projects in the Columbia River, observed data collection and radio tracking techniques, and obtained existing protocols for review. We met with telemetry manufacturers onsite and discussed study area characteristics and objectives to develop a system of tagging and tracking that could be accomplished within the project budget. Our study design and tracking infrastructure combined the applications and adaptations from many regional fisheries radio telemetry projects. It was necessary to minimize the essential equipment wherever possible in order to conduct research activities in the remote portions of each river.

Prior to the implementation of each case study, we obtained all necessary federal and state collection permits for work in the Hoh and Elwha Rivers. It was important to communicate regularly with the permitting officials during the development phases of each project, and early submission of permit applications helped to secure permits prior to the onset of each study. Similarly, early contact with manufacturers and vendors was helpful in having customized radio transmitters (also referred to as tags) (e.g., frequency, tag sizes, operational life, and labels) manufactured and shipped in a timely manner.

Study Design

The Elwha and Hoh River studies were designed to detect tagged fish movements during all months over a three-year period during highly variable riverflows and across multiple river systems. We initially considered acoustic telemetry for each study based on the likelihood of fish moving into marine waters (e.g., Hoh study) and fish moving into the reservoirs (e.g., Elwha study) where acoustic telemetry would provide some advantages. However, we selected radio telemetry over acoustic telemetry and reasoned that this was the optimal technique because: 1) fish were presumed to spend more time in freshwater where radio telemetry was deemed more useful; 2) the studies were designed to understand spatial and temporal movements within the river, and not along the Pacific coast or in reservoirs; 3) fish capture was more feasible in freshwater when compared to marine waters; 4) it was important to know the river origin of each tagged fish; and 5) there was extremely limited access along the rugged Washington coast where no structures exist to attach fixed acoustic telemetry stations.

We selected digitally coded tag technology over standard tag technology based on the assumption that fish would be highly migratory and that we would have numerous tagged fish individuals to attempt to detect over a large area. Tag technology is discussed in detail in Section 5.1. In both case studies, we limited the number of radio frequencies to monitor to maximize the likelihood of detecting tagged fish. Specifically in our case, scanning no more than three frequencies proved to be useful when detecting tagged individual fish during aerial surveys that occurred over a large geographic area and while moving at fast speeds in the airplane. Additionally, the scanning of a minimal number of frequencies allowed us to detect individual fish that typically moved rapidly past fixed telemetry sites. However, when multiple tagged fish were detected in a common location, there were collisions among radio signals within a single frequency, and telemetry receivers often failed to segregate individually coded tags. In those cases, we were able to detect the presence of multiple tagged fish, but were unable to identify some individuals. To answer specific questions, we located some of the fixed telemetry sites at terminal boundaries such as river mouths, impoundment tailraces, and the inlets to reservoirs. In anticipation of some level of tag loss, tag failure, and mortality of tagged fish, we tagged 128 fish in the Hoh River study and 95 fish in the Elwha River study so as to maximize the opportunity for data recovery. A functioning data management system was in place prior to the collection of data on tagged fish.

Fish Capture and Tagging

Fish capture was accomplished by angling and electrofishing and required extensive effort and considerably more time than was initially anticipated during the study design phase. The high levels of effort during fish capture was primarily due to the remote nature of upper sections of the Elwha and Hoh River, frequent high flow conditions during winter months, and low abundances of federally threatened bull trout.

We captured fish for tagging from several locations throughout each river so as not to target a single life history stage (e.g., anadromous or fluvial). We selected radio tag models that had the longest transmitting life (up to three years) with a total weight in air that did not exceed 2% of the body weight of the tagged fish. For adult fish (>400 mm), we selected a 150 MHz tag with 5 s burst rate that would transmit up to 761 d. For juvenile fish (>250 mm–400 mm), we selected a 150 MHz tag with a 10 s burst rate that would transmit up to 414 d. We also chose to identify the radio tagged fish with a visible T-bar anchor tag because snorkel surveys were being conducted in the basins concurrent with the radio telemetry studies.

We refined and extensively practiced our surgical and tagging methods using live surplus hatchery salmonids from a local hatchery. These surrogate fish were of similar sizes (200–600 mm) to those in the study population and handled under similar conditions.

Tagging of each fish occurred shortly after the time of capture (within 30 min) and only fish that appeared healthy and unaffected by the capture event were tagged. Captured fish were anesthetized in a buffered solution of tricaine methanesulfonate (80 mg per liter of water) for up to seven minutes. Radio transmitters (MCFT-3A, NTC-6–2, Lotek Wireless) were inserted into the abdominal cavity of bull trout such that the flexible whip antenna trailed outside the fish. Methods regarding the surgical implantation radio tags are discussed in detail in Section 4.

Prior to the actual surgical implantation of radio transmitters, fish were photographed, measured, and weighed while under anesthesia. A tissue sample was collected from the anal fin and preserved in nondenatured ethanol for genetic analysis. Scales also were collected from the posterior flank for age analysis. All radio transmitters were tested for functionality immediately before implantation into the fish. Following tagging, fish were allowed to recover in the river in a submerged holding cage for up to 45 min. When the effects of anesthesia were no longer displayed, fish were released to the river at or near the point of capture.

Radio Tracking of Fish

We used a combination of fixed telemetry stations and walking, boat, and aerial surveys to relocate radio tagged fish. Data logging fixed telemetry stations are an extremely effective and practical way to monitor fish movements, particularly in remote reaches of wilderness rivers (Eiler 1995). The operation of well functioning fixed telemetry stations required careful planning, range testing of transmitters at each site (0.8–1.3 km), and many onsite adjustments of antenna direction, gain levels, and attenuation levels. Fixed site radio telemetry is discussed in detail in Section 5.2. The fact that some of the fixed telemetry stations were located in remote areas without road access made installation and maintenance of these sites challenging and time consuming. Fixed telemetry stations were powered using 12V deep cycle batteries because of their portability and because AC power was not available at most sites. Additionally, solar panels were deemed ineffective due to extensive canopy cover and relatively low ambient light levels in each river system. We used a dual antenna array at each fixed telemetry station in an alignment to determine the directional movements of tagged fish (e.g., upstream or downstream) based on the recorded power levels at each antenna.

To maximize data storage at each fixed telemetry receiver, the data reduction function was enabled and detections were bundled in 20 minute intervals. At each site, data were frequently downloaded to reduce the possibility of losing data. The 12V batteries were replaced every 7–14 d depending on the number of tagged fish near a fixed telemetry site. For site security, radio receivers, batteries, and switchboxes were stored in locked, protective boxes. A handheld, lightweight computer with appropriate communication software was helpful for retrieving data from remote fixed sites.

The tracking and relocation of radio tagged fish were accomplished by weekly ground tracking surveys that were conducted by walking parallel to the river. We used a receiver stationed in a protective housing and a handheld collapsible three element Yagi antenna. Aerial tracking surveys were conducted up to three times per month in each case study area from a Cessna 172 that flew at altitudes that ranged from 300 to 400 m. A total of 90 flights were conducted in the Hoh River study and 25 flights were conducted in the Elwha River study. We used a telemetry receiver, an antenna switching unit, and a Yagi antenna that was mounted to each wing strut.

Ground tracking surveys were often conducted immediately following aerial tracking to verify relocation data obtained during flights. Tagged fish were relocated to the nearest river kilometer from the ocean, and relocations were based on the strongest signal at the lowest possible gain level. A regular, frequent tracking schedule was maintained throughout the active life of radio transmitters (up to three years). In the Hoh and Elwha case studies, aerial surveys were extremely important in maintaining frequent contact with radio tagged fish, especially in portions of river basins that were inaccessible or required extensive hiking times to reach. Additionally, aerial tracking surveys were an effective and economical strategy to monitor multiple watersheds from the estuary to headwaters in a relatively short period of time.

Data Management and Analysis

A data management system was designed by a data manager in the design phase of the study. Criteria for data reporting were established and definitions were used to identify inactive tags and for relocations that ultimately would be excluded from analysis based on Meka et al. (2003). We maintained relocation histories for each tagged fish that included information on: date of capture and tagging, method of relocation (e.g., fixed, mobile, aerial, or boat), date and time of relocations, name of river basin at contact, and rkm. Using relocation data over an individual's tracking history, fish movements were graphed across distance and time

(Figures 2 and 3). Tracking maps of individual fish movements and groups of fish were generated using ArcGIS (Figure 2). Migration rates were calculated for individuals that exhibited significant movements including presumed spawning migrations, post spawn migrations, juvenile emigrations to sea, and fall over dams. For example, radio telemetry data showed that juvenile bull trout emigrated from the Hoh River to the ocean at a mean rate of 3.4 km/d (SD = 2.7, range = 0.7–7.4 km/d) (Brenkman et al. 2007). Comparisons of relocation data among tagged fish revealed distinct temporal and spatial patterns from which life history information could be inferred (Figure 2).

Advantages of Radio Telemetry in Small, Wilderness Rivers

Radio telemetry was a useful technique to track salmonid movements throughout remote, roadless areas in the Hoh and Elwha River basins. With an appropriate tracking infrastructure that relies on a combination of strategically distributed fixed telemetry stations and manual and aerial tracking, radio telemetry techniques can address multiple objectives simultaneously in small rivers. One of the greatest benefits of using fixed stations is that they allow for intensive data collection seven days a week and 24 h per day at multiple sites along a river. The location of the fixed station at the Hoh River confluence with the Pacific Ocean was paramount to understanding bull trout movements to and from saltwater (Figure 1), and allowed us to determine whether fish actually entered the marine environment. From our experiences, fixed telemetry stations at a river mouth or estuary can verify movements to and from saltwater so long as range testing is conducted.

The combination of tracking methods also enabled us to obtain specific information on location of spawning areas and timing of immigration and emigration to saltwater by multiple life stages of bull trout including adults, kelts, and juveniles (Brenkman and Corbett 2005; Brenkman et al. 2007). Manual tracking of individual fish after each aerial survey provided more precise relocation data when compared to aerial flight data alone. Frequently a radio tagged fish was relocated at relatively low precision (e.g., nearest km) during an aerial survey and subsequently relocated at higher precision (e.g., nearest 0.2 km) in a manual tracking survey one to three days following.

In the Hoh River study, aerial surveys were the key to relocating fish that moved through marine waters into the lower portions of other coastal watersheds where radio transmitters could be relocated (Figure 2). In estuarine environments and coastal river mouths, consecutive and regular aerial tracking was necessary to relocate radio tagged fish. Without the use of aerial surveys, the relocation of bull trout that moved among watersheds would have been infeasible based on limited access to the lower portions of rivers along coastal Washington. A useful comprehensive report on guidelines for aerial telemetry operations and installation of equipment was summarized by Gilmer et al. (1981).

Another advantage of radio telemetry is that one can obtain meaningful data from a low sample size of tagged fish. In the Hoh River study, we observed radio tagged, anadromous bull trout moving from the Hoh River into Kalaloch Creek, an independent drainage located 17 km to the south (Figure 2a). With that information, we opportunistically tagged nine adults in Kalaloch Creek, and later relocated those individuals in several other river systems via aerial tracking along coastal Washington (Figure 2c). Another study reported important information on movements of white sturgeon *Acipenser transmontanus* by tagging only one adult

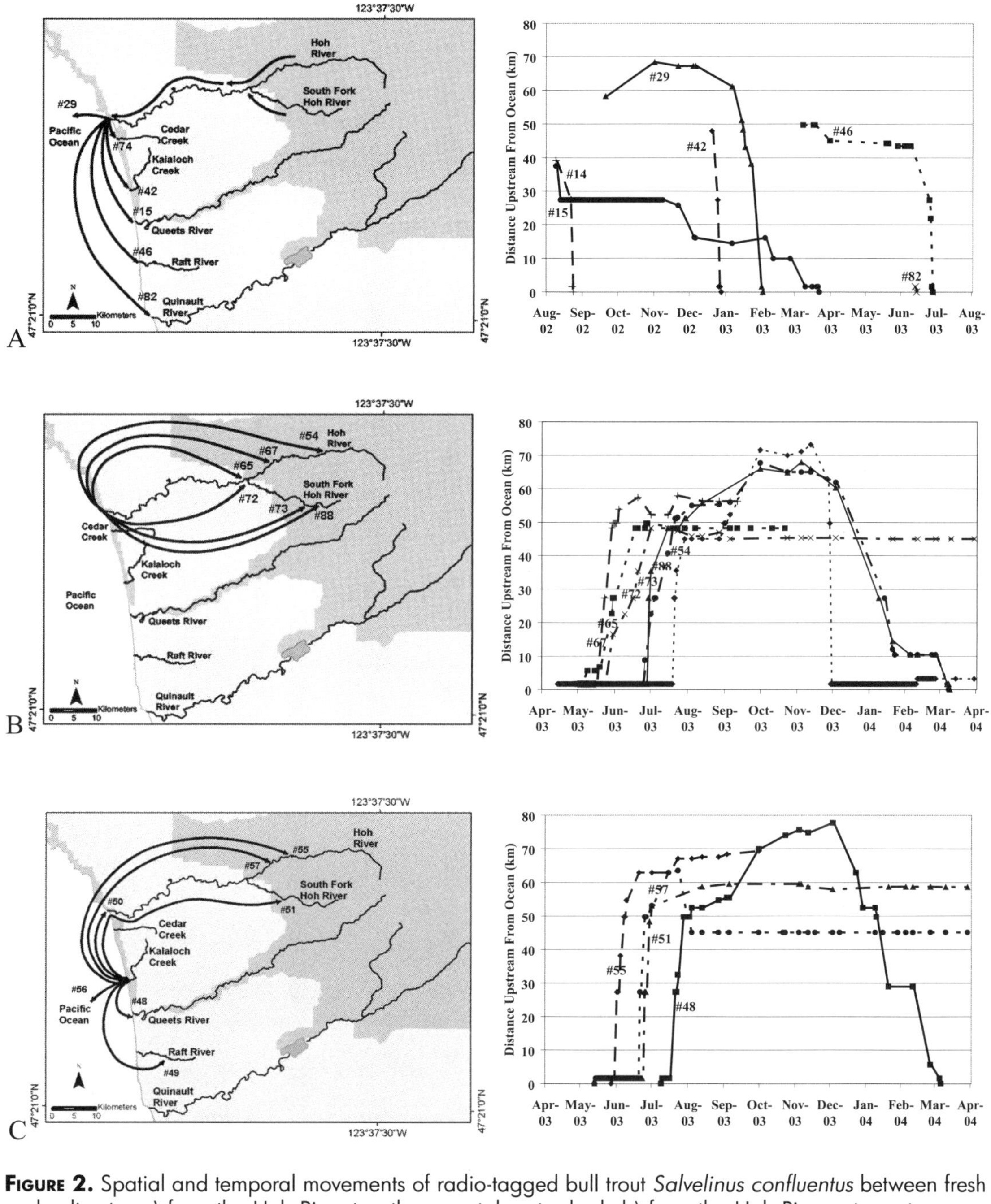

FIGURE 2. Spatial and temporal movements of radio-tagged bull trout *Salvelinus confluentus* between fresh and saltwater a) from the Hoh River to other coastal watersheds b) from the Hoh River estuary to upper reaches of the basin and c) from Kalaloch Creek to other coastal watersheds (adopted Brenkman and Corbett 2005). Maps with arrows depict individual fish movements.

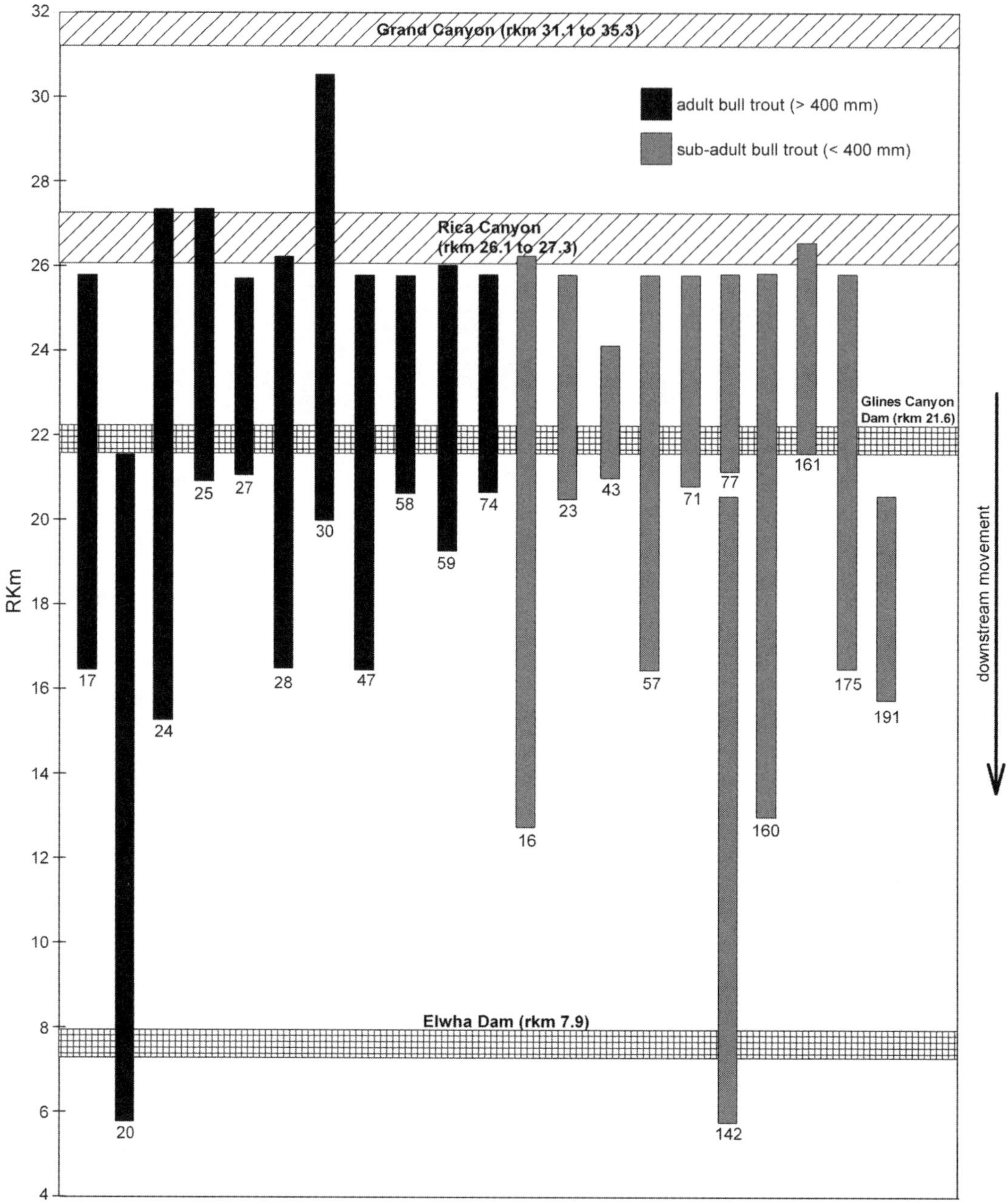

FIGURE 3. Range of downstream movements of adult and subadult radio tagged bull trout *Salvelinus confluentus* (n = 22) through canyon reaches and past hydroelectric dams on the Elwha River. Numbers associated with bars indicate radio tag codes for individual fish.

fish. Welch et al. (2006) observed white sturgeon moving 1,060 km from the initial tagging location in the Klamath River, California to the Fraser River, British Columbia, an important finding since virtually no information exists on their ocean movements.

Disadvantages of Radio Telemetry in Small, Wilderness Rivers

Although telemetry studies provide key insights from documented individual movements, there are constraints and limitations with this technique. One primary constraint of radio telemetry is that it may take several years to obtain meaningful data, and interpretations that occur too early can result in misleading findings. This is particularly the case when working with species that are highly migratory and exhibit multiple life history forms. For example, the interpretation of movements and the determination of whether a fish is anadromous or nonanadromous could easily be misinterpreted if the tracking history is incomplete and the fish had not yet entered the ocean.

Another limitation of radio telemetry is that it is difficult to obtain complete life history data for each fish. There are two parts to this problem: 1) pretagging life history information often remains unknown; and 2) each fish has a finite tracking history that is limited by tag life, tag reliability, or future loss of contact. One partial solution to obtain more complete life history information is to tag fish of various lengths and life stages and also to monitor radio tagged fish in all months of the year.

The use of radio telemetry was ineffective at relocating tagged fish in reservoirs of the Elwha River. We used radio tags that transmitted on 150 MHz and these frequencies are attenuated beyond certain depths. This limitation is discussed in detail in Section 5.1. To effectively relocate radio tagged fish that entered and exited reservoirs, it was important to install fixed stations at the inlet to each reservoir and downstream of each hydroelectric dam at impoundment tailraces. By strategically placing fixed stations in those locations, we documented the extent of fall over (22 of 96 of tagged bull trout) at the Elwha River dams (Figure 3).

One consequence of using multiple tracking methods is the differing levels of precision when relocating individual fish. For instance, aerial tracking covers the greatest area, but is the least precise. Roberts and Rahel (2005) found that aerially acquired location errors ranged from 22 to 426 m (mean 178 m). The differences in precision can influence the types of conclusions and inferences made about fish movements and corresponding habitat use. However, this may be of less concern for those studies that are designed to understand broad-scale fish movements such as the Elwha and Hoh case studies. Aerial tracking also involves much compliance, training, and scheduling. The use of airplanes in tracking fish also is greatly influenced by weather conditions, and is relatively dangerous when compared to other methods of tracking.

Perhaps the greatest challenges in radio telemetry projects are the high start-up costs and the need for year round personnel to maintain equipment and track fish. In the Hoh River, the four-year project costs for equipment and personnel were U.S. $250,000. In the Elwha River, the three-year costs for personnel and equipment were U.S. $150,000, and annual costs were less because equipment such as telemetry receivers and antennas had already been purchased. In general, radio telemetry projects require high maintenance and full-time personnel, particularly when using multiple fixed telemetry stations that require regular attention. It was important to schedule frequent visits (e.g., weekly) to fixed telemetry station sites to monitor battery power and the condition of cables, switchboxes, and antennas. Regular downloading of data were also important to minimize the potential loss of important information.

Management Implications Derived From Radio Telemetry Case Studies

From the two case studies, radio telemetry provided key insights with detailed information on individual fish movements at various spatial scales that ranged from the reach to watershed levels. In the Hoh River study, radio telemetry provided the first formal documentation of anadromy in federally threatened bull trout (Brenkman and Corbett 2005; Brenkman et al. 2007). Radio telemetry also revealed the extent of anadromy among individuals, timing of river entry from the ocean, and irregular periods of freshwater, estuarine, and marine residence times for bull trout. Radio telemetry was ultimately used to better understand broad-scale migratory patterns of bull trout that moved between multiple rivers, estuaries, and the Pacific Ocean (Figure 2). Additionally, radio telemetry methods were effective at discerning interbasin migrations by bull trout among multiple coastal watersheds up to 47 km from original tagging locations (Figure 2).

Results from the Hoh River case study were used by the U.S. Fish and Wildlife Service in the development of a federal Recovery Plan and in the designation of critical habitat for Coastal-Puget Sound bull trout (U.S. Fish and Wildlife Service 2004; Department of the Interior Fish and Wildlife Service 2005). The verification of anadromous bull trout and their use of nearshore marine waters, estuaries, and lower portions of coastal rivers ultimately resulted in the designation of 1,221 km of marine shoreline as critical habitat. The extent of anadromy and timing of movements of radio-tagged bull trout also highlighted their vulnerability to commercial bycatch in gill-net fisheries that occur in coastal rivers of Washington (Brenkman and Corbett 2005).

The results of radio telemetry in the Elwha River provided information on spatial and temporal movements of bull trout and revealed extensive fall over of fish at two hydroelectric dams (Figure 3). These findings were instrumental in the development of a plan to protect bull trout during reservoir drawdown and dam removal (Crain and Brenkman 2010). Large-scale disturbance is expected downstream of the dams when accumulated sediments (18 million m^3) erode form the reservoirs following dam removal. To minimize mortality of bull trout, fish will be translocated from the reservoir areas to portions of the upper Elwha River prior to dam removal.

Conclusions

From technological and sampling perspectives, there are few other fisheries techniques that could have provided such detailed information on the spatial and temporal movements of salmonids in remote, wilderness rivers on Washington's Olympic Peninsula. The use of other traditional techniques such as electrofishing, snorkeling, or redd surveys are particularly challenging in small rivers like the Hoh and Elwha because of prolonged periods of high riverflows, low water visibility from glacial melt, and difficult access in rugged wilderness areas. Radio telemetry provided a relatively passive technique that was highly conducive for sampling threatened fish populations in a National Park where invasive methods are less desired and additional restrictions apply. It is our hope that information gleaned from radio telemetry case studies in the Hoh and Elwha River basins will facilitate more efficient design, reduce start-up time and costs, and be used to better meet objectives of future radio telemetry studies in small rivers.

Acknowledgments

Thanks to U.S. Geological Survey biologists Noah Adams, John Beeman, Scott Evans, Marc Novick, and Rachel Reagan. We thank Brian Burke, Kathy Beirne, Jeff Chan, Judy DeLavergne, Kinsey Frick, Chris Glenney, Matt Groce, Katy Hanna, Cat Hawkins Hoffman, Barb Kelly Ringel, Phil Kennedy, David Kirner, Jay Nelson, Larry Ogg, Paul Olmstead, Roger Peters, Shelley Spalding, Larry Ward, Carmen Welch, and John Wullschleger for their assistance and support of this project. The radio telemetry projects were funded by the National Park Service and U.S. Fish and Wildlife Service. We thank Noah Adams, Carl Burger, and John Eiler for their constructive comments to an earlier version of this paper. The use of trade, firm, or corporation names in this publication is for the convenience of the reader. Such use does not constitute an official endorsement or approval by the National Park Service of any product or service to the exclusion of others that may be suitable.

References

Bendock, R., and M. Alexandersdottir. 1993. Hooking mortality of Chinook salmon released in the Kenai River, Alaska. North American Journal of Fisheries Management. 13:540–549.

Boggs, C. T., M. L. Keefer, C. A. Peery, T. C. Bjornn, and L. C. Stuehrenberg. 2004. Fallback, reascension, and adjusted fishway escapement estimates for adult Chinook salmon and steelhead at Columbia and Snake River dams. Transactions of the American Fisheries Society. 133:932–949.

Brenkman, S. J., and S. C. Corbett. 2005. Extent of anadromy in bull trout and implications for conservation of a threatened species. North American Journal of Fisheries Management. 25:1073–1081.

Brenkman, S. J., S. C. Corbett, and E. C. Volk. 2007. Use of otolith chemistry and radio telemetry to determine age-specific migratory patterns of anadromous bull trout (*Salvelinus confluentus*) in the Hoh River, Washington. Transactions of the American Fisheries Society. 136:1–11.

Brown, R. S. and W. C. Mackay. 1995. Spawning ecology of cutthroat trout (*Oncorhynchus clarki*) in the Ram River, Alberta. Canadian Journal of Fisheries and Aquatic Sciences. 52:983–992.

Burger, C. V., R. L. Wilmot, and D. B. Wangaard. 1985. Comparison of spawning areas andtimes for two runs of Chinook salmon (*Oncorhynchus tshawytscha*) in the Kenai River, Alaska. Canadian Journal of Fisheries and Aquatic Sciences 42:693–700.

Crain, P. and S. Brenkman. 2010. Elwha River Restoration Project: Bull Trout Protection and Restoration Plan. National Park Service, Olympic National Park, Port Angeles, Washington.

Curran, C. A., C. P. Konrad, J. H. Higgins, and M. K. Bryant. 2009. Estimates of sediment load prior to dam removal in the Elwha River, Clallam County, Washington. U.S. Geological Survey Scientific Investigations Report 2009–5221, Tacoma, Washington.

Curry, R. A., D. Sparks, and J. Van De Sande. 2002. Spatial and temporal movements of a riverine brook trout population. Transactions of the American Fisheries Society. 131:551–560.

Dare, M. R. 2006. Integration and application of radio telemetry data collected on a mobile fish species: a synthesis of bull trout movement research. Project Completion Report, Contract 143303G098. Available: http://www.fs.fed.us/rm/boise/publications/fisheries/rmrs_2006_dare001.pdf (February 2010).

Duda, J. J., J. E. Freilich, and E. G. Schreiner. 2008. Baseline studies in the Elwha River ecosystem prior to dam removal: introduction to the special issue. Northwest Science 82(Special Issue), 1–12.

Eiler, J. H. 1995. A remote satellite-linked tracking system for studying Pacific salmon with radio telemetry. Transactions of the American Fisheries Society. 124:184–193.

English, K. K, W. R. Koskia, C. Sliwinskia, A. Blakley, A. Cass, and J. C. Woodey. 2005. Migration timing and river survival of late-run Fraser River sockeye salmon estimated using radiotelemetry techniques. Transactions of the American Fisheries Society. 134:1342–1365.

Gilmer, D. S., L. M. Cowardin, R. L. Duval, L. M. Mechlin, and C. W. Shaiffer. 1981. Procedures for the use of aircraft in wildlife biotelemetry studies. U.S. Department of Interior, Fish and Wildlife Service. Resource Publication 140, Washington D.C. p. 1–19.

Gray, R. H. and J. M. Haynes. 1979. Spawning migration of adult Chinook salmon (*Oncorhynchus tshawytscha*) carrying external and internal radio transmitters. Journal of the Fisheries Research Board of Canada 36:1060–1064.

Hilderbrand, R. H. and J. L. Kershner. 2000. Movement patterns of stream-resident cutthroat trout in Beaver Creek, Idaho–Utah. Transactions of the American Fisheries Society 129:1160–1170.

Muhlfeld, C. C., and B. Marotz. 2005. Seasonal movement and habitat use by subadult bull trout in the upper Flathead River system, Montana. North American Journal of Fisheries Management, 25:797–810.

Meka, J. M., E. E. Knudsen, D. C. Douglas, and R. B. Benter. 2003. Variable migratory patterns of different adult rainbow trout life history types in a southwest Alaska watershed. Transactions of the American Fisheries Society 132:717–732.

Roberts, J. J. and F. J. Rahel. 2005. Accuracy of aerial telemetry in fisheries studies. North American Journal of Fisheries Management 25:660–666.

Schmetterling, D. A. 2001. Seasonal movements of fluvial westslope cutthroat trout in the Blackfoot River Drainage, Montana. North American Journal of Fisheries Management 21:507–520.

Snedden, G. A., W. E. Kelso, and D. A. Rutherford. 1999. Diel and seasonal patterns of spotted gar movement and habitat use in the lower Atchafalaya River Basin, Louisiana Transactions 128:144–154.

Swanberg, T. R. 1997. Movements of and habitat use by fluvial bull trout in the Blackfoot River, Montana. Transactions of the American Fisheries Society 126. 735–746.

Tyus, H. M., B. D. Burdick, and C. W. Mcada. 1984. Use of radiotelemetry for obtaining habitat preference data on Colorado squawfish. North American Journal of Fisheries Management. 4:177–180.

U.S. Department of the Interior Fish and Wildlife Service. Endangered and Threatened Wildlife Plants; Designation of Critical Habitat for the Bull Trout; Final Rule. Federal Register 70. 185 (September 26, 2005). (December 2010).

U.S. Fish and Wildlife Service. 2004. Draft Recovery Plan for the Coastal-Puget Sound Distinct Population Segment of Bull Trout (*Salvelinus confluentus*). Volume II (of II): Olympic Peninsula Management Unit. Portland, Oregon.

U.S. Geological Survey, 2007. USGS Real-Time Water Data for USGS 12041200, Hoh River At US Highway 101 Near Forks, Tacoma, Washington. Available: http://waterdata.usgs.gov/nwis/uv?12041200 (June 2007).

Welch, D. W., S. Turo, and S. D. Batten. 2006. Large-scale marine and freshwater movements of white sturgeon. Transactions of the American Fisheries Society. 135:386–389.

Zigler, S. J., M. R. Dewey, and B. C. Knights. 2003. Movement and habitat use by radio-tagged paddlefish in the Upper Mississippi River and tributaries. North American Journal of Fisheries Management. 23:189–205.

Section 6.2

Radio Tracking of African Cichlids in a Large Floodplain River: Challenges and Results from the Zambezi River in Namibia

Eva B. Thorstad, Clinton J. Hay, Finn Økland, and Tor F. Næsje

INTRODUCTION

Namibia is one of the driest countries in the world, and people strongly depend on the availability of open water bodies in the northern parts of the country (Cuvelai River System, Kavango River and Zambezi/Chobe River System) for fish to eat and water supplies (Mendelsohn et al. 2002). All perennial rivers are shared with neighboring countries. The Zambezi River is the fourth largest river system in Africa, flowing through Zambia, Angola, Namibia, and Zimbabwe, before emptying into the Indian Ocean in Mozambique. This drainage forms the 120 km long border between Namibia and Zambia, where the present study was performed. More than 70 fish species have been recorded in this river section (Hay et al. 2002).

The large cichlids (Cichlidae) are valuable species in small scale commercial and subsistence fisheries in the upper Zambezi and are also popular among recreational anglers. However, reports of reduced catches are a major concern for the Namibia management authorities (Anonymous 1995). To be able to develop spatial management strategies, information on seasonal fish movement and habitat use is fundamental. A series of radio telemetry studies were therefore performed in the Namibian section of the upper Zambezi River, to achieve information on movements and habitat use for a number of fish species (Thorstad et al. 2001, 2004, 2005; Økland et al. 2005, 2007).

Performing radio telemetry studies in all water systems provides a number of challenges. However, the size and complexity of a floodplain river like the Zambezi River, the general lack of knowledge on the fish resources, and the undeveloped, remoteness of the study area make studies even more challenging. The aim of this Section is therefore threefold: 1) present results on behavior and habitat use of one of the most important cichlids in the upper Zambezi River, the greenhead tilapia *Oreochromis macrochir*, 2) compare the results with previous studies on the cichlids nembwe *Serranochromis robustus*, three spotted tilapia *Oreochromis andersonii* and pink bream *Sargochromis giardi* (Thorstad et al. 2001, 2005; Økland et al. 2007), and 3) discuss some of the challenges with performing telemetry studies in such a large and complex floodplain river system.

METHODS

The Caprivi Region in Namibia, where the Zambezi River forms the border between Namibia and Zambia, is a flat area, approximately 1,000 m above sea level (Figure 1). Seasonal flooding between December and May creates extensive floodplains. The annual variation in water level is up to 7–8 m, with an annual average of 5.2 m (Van der Waal and Skelton 1984). Peak flow at Katima Mulilo in Caprivi is usually between 2000 and 4500 m^3/s, but varies extensively among years. The local human population has a rural lifestyle, depending heavily on subsistence fishery as an affordable source of protein (Abbott et al. 2007). The subsistence fisheries are also seen as a source of income needed for daily necessities. In the study area, the Zambezi River is slow flowing and consists of a wide main channel (150–500 m wide during low water), with bends, oxbows and deep pools. Small, vegetated islands, sandbanks, bays, backwaters, and narrow side streams occur frequently. There are also larger slow flowing channels and isolated pools. The turbidity in the study area is low, with recorded values in September 1997 between 0.6 and 6.8 NTU, except in a back-water section where the turbidity was 39.8 NTU (Koekemoer 2003). The conductivity is also relatively low (8.7–12.5 mS/m, total amount of dissolved salts of 58.3–83.8 mg/l, Koekemoer 2003).

Twenty-two greenhead tilapia (mean total body length 31 cm, range 25–44 cm) were captured by rod and reel in the Zambezi River; 27–60 km downstream from Katima Mulilo

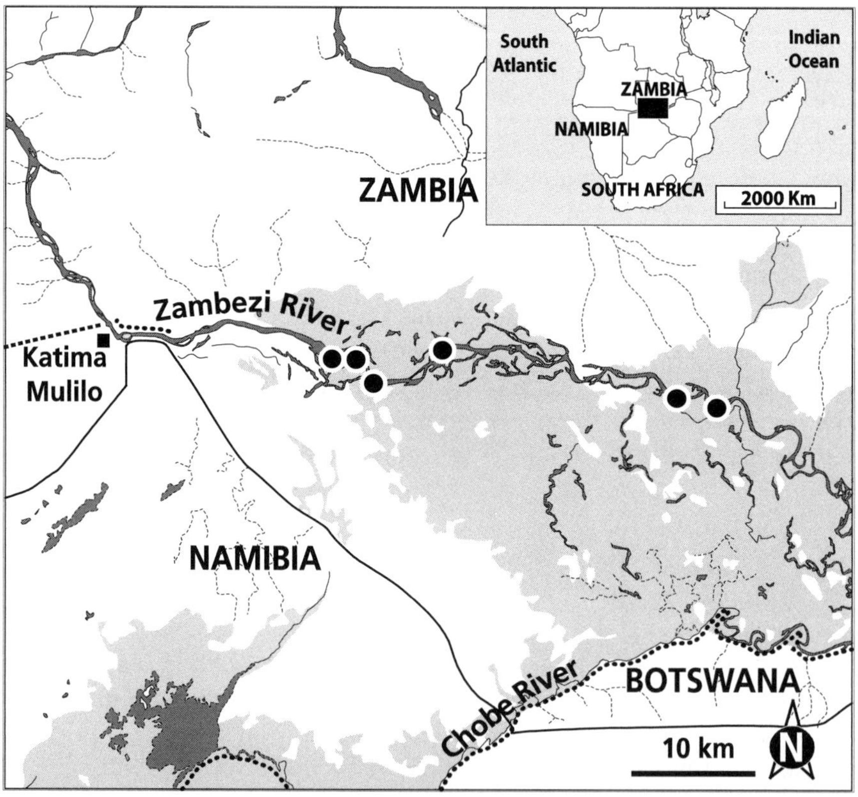

FIGURE 1. The upper part of the Zambezi River in north-eastern Namibia, bordering on Zambia. Sites where individual greenhead tilapia were radio tagged and released are indicated (•). Floodplains that are water covered during the annual flood are also indicated (gray shaded areas).

in Caprivi, Namibia, during 18–25 November 2003 (Figure 1). The fish were anesthetized (2-phenoxy-ethanol, SIGMA Chemical Co., USA, 0.7 mL/L) and radio transmitters (21 fish with model F2110, and one individual with model F1960, Advanced Telemetry Systems, Isanti, MN, USA) were externally attached to the fish, using the method described in Thorstad et al. (2001). Model F2110 were flat, rectangular transmitters with outline dimensions of 19 × 40 × 9 mm, weight in air of 12 g and guaranteed battery life of 139 d. Model F1960 were similar in shape with outline dimensions of 13 × 29 × 6 mm, weight in air of 3.6 g and guaranteed battery life of 258 d. Transmitter weight in air was average 1.9% (range 0.2–3.5%) of the body weight of the fish. The transmitters emitted signals (35 pulses per minute) within the 142.143–142.641 MHz frequency range. Transmitter frequencies were spaced at least 10 kHz apart for individual recognition of the fish. All fish were released at the capture site. The water temperature was 25–28°C during capture and tagging.

The fish were tracked by boat using a portable receiver (R2100, ATS) with 4- and 9-element Yagi antennas, using a combination of tracking the signal to the area with the loudest signal and triangulation. The fish were located with a precision of ± 10 m in the main river. Some backwaters were inaccessible by boat, and the location was estimated based on signal strength and direction. Approximately 100 km of the river were searched for tagged fish. To ensure that movements due to handling and tagging effects were not included (Thoreau and Baras 1997), positioning data before 1 December (6–13 d after capture and tagging) were not included in analyses. The fish were tracked during rising (1 December–28 March), high (29 March–2 May) and receding (3 May–5 August) water levels (Figure 2). A total of 74 tracking surveys were performed; 36 during rising water, 11 during high water, and 27 during receding water. Hence, the fish were tracked on average every 3.3 d (i.e., every 3.3 d during rising, every 3.1 d during high, and every 3.5 d during receding water).

Individual fish not recaptured or predated (see Results), were located until transmitter batteries ran out of power, which happened on different dates towards the end of the last tracking period (15 July and onwards). Some of the fish were not found during all tracking surveys. It is likely that on these occasions the fish were either in areas inaccessible by boat or at locations with reduced radio signal range (such as deep water or areas sheltered by solid structures). Hence, swim depths and proportion of time staying in narrow side channels and backwaters may potentially be underestimated.

Standardized habitat classifications were made each time a fish was located and its position determined (described in Økland et al. 2005). Data were collected on water cover (permanent water cover, or temporary water cover each year during the rain period), main habitat type, position of the fish in relation to vegetation, surface water temperature, water depth (measured by manual sounding or echo sounder), and substrate. Fish position in the water column was unknown. The distance to the nearest shore was recorded, as well as the total width of the river (determined with a laser range finder Bushnell BU Yardage 800). Due to the high water levels during peak floods, the river was frequently several kilometers wide, and the width of the river and distance to shore could not be recorded precisely.

Habitat use was described by determining the frequency of tagged individuals that were recorded in different habitats. To give a better measure of the intensity of utilization of different habitats, the mean proportion of locations recorded in different habitats was also described and compared. Descriptive statistics and statistical analyses were based on average values for individual fish. Comparisons of behavior and habitat utilization among periods of rising, high and receding water levels were made by nonparametric paired comparisons, and only fish located

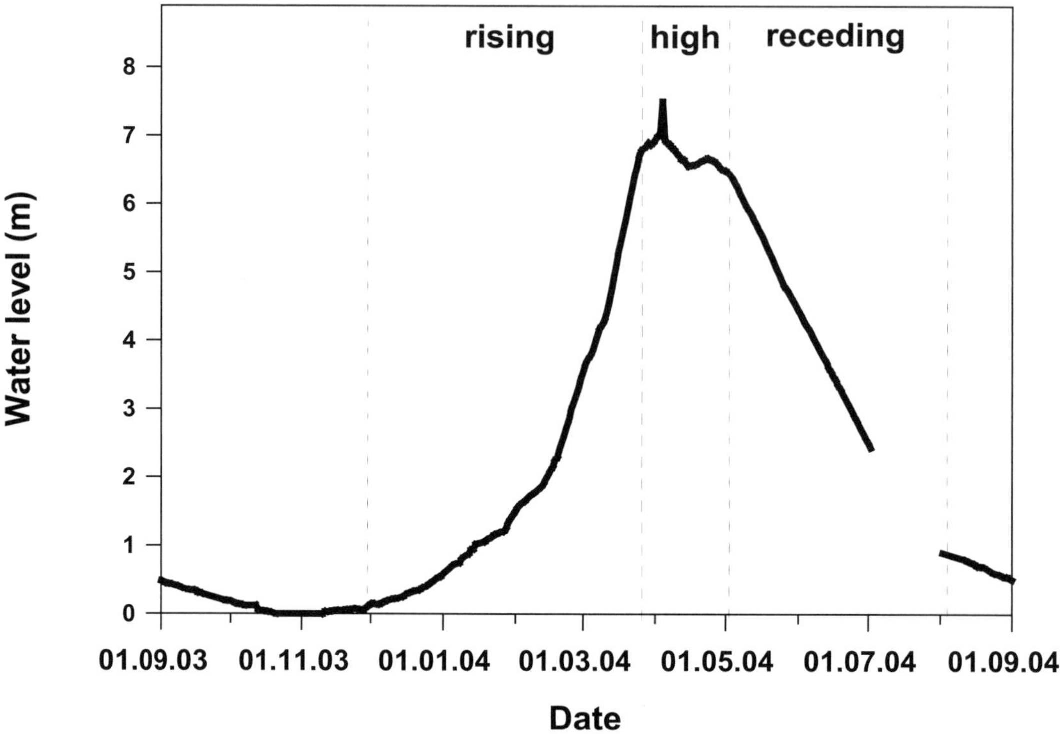

FIGURE 2. The water level in the Zambezi River at Katima Mulilo, Namibia, from 1 September 2003–1 September 2004 (solid line). The study periods during rising, high and receding water are indicated by vertical dotted lines. Data provided by the Department of Water Affairs in Windhoek.

ten times or more in each period were included in the analyses. Home ranges were calculated using the nonparametric kernel method and a probability density function (Worton 1989; Seaman and Powell 1996; Lawson and Rodgers 1997) using ArcView GIS 3.2 (Environmental Systems Research Institute, Inc.). For the kernel smoothing parameter "h," the "ad hoc" solution was rejected in favor of the least square cross-validation approach, which is more effective with multimodal distributions (Worton 1989). The utilization distribution was estimated, in terms of perimeter and area covered, at two different levels of probability (95 and 50%).

RESULTS

Two greenhead tilapia were stationary from the start of the tracking program, and were believed to be dead or the transmitters had fallen off. One fish was observed when predated by a tiger fish *Hydrocynus vittatus* immediately after release, and one fish was likely predated by a fish eagle *Haliaeetus vocifer* and two fish by otter *Aonyx capensis*, or *Lutra maculicollis*, judged by the site where the transmitters were found. Four fish were caught by local fishers shortly after tagging. Hence, 12 of the 22 tagged fish were left for long-term analyses. Results from nine fish were included in the comparisons among periods, since only those located ten times or more in each period were included.

Home ranges were relatively small, with a 50% probability of locating the fish within a mean area of 475,648 m^2 ($\approx 0.5\ km^2$) and a 95% probability of locating the fish within a mean area of 3,121,288 m² (≈ 3.1 km²) (Figure 3; Table 1). Home range size did not differ among periods (Friedman tests, 50%: $\chi^2 = 2.9$, $P = 0.24$, 95%: $\chi^2 = 0.89$, $P = 0.64$). The length of the river section used, i.e. distance between the two locations farthest from each other during the entire study period, was on average 2,554 m (SD = 3,904, range 171–13,697 m).

Total distance moved by individual greenhead tilapia during the entire study period (compilation of observations) averaged 4,076 m (SD = 5,458, range 334–19,666 m), while mean distance moved between tracking surveys was 100 m (Table 1), and did not differ among periods (Friedman test, $\chi^2 = 5.56$, $P = 0.062$). The habitat utilization varied among individuals, as 42% of the fish were only detected in permanent river sections, 50% in both permanent sections and areas temporary inundated by floods, and 8% of the fish only in temporary inundated areas. On average, 72% of the locations were in permanent river sections, whereas the remaining 28% of locations were in temporarily inundated areas (Table 1). The proportion of locations in temporarily flooded areas did not differ among periods (Friedman test, $\chi^2 = 1.6$, $P = 0.45$).

Individual fish utilized a variety of habitat types, with 67% of the tagged fish recorded in the main channel during one or more tracking surveys, 63% in side channels, 63% in

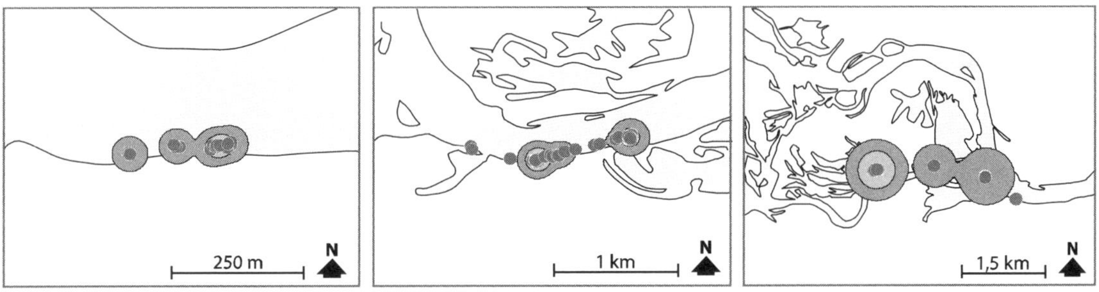

FIGURE 3. Kernel home ranges of three individual radio tagged greenhead tilapia in the Zambezi River in 2003–2004, given as examples. Each of the panels (left, middle, and right) shows the home range for an individual during the entire study period. Central dots show locations during tracking, and the gray contours of home ranges refer to two different levels of probability (dark gray shows 95% probability and light gray shows 50% probability).

TABLE 1. Home range size and distance moved between tracking surveys of radio tagged greenhead tilapia in the Zambezi River, Namibia, 2003–2004. Per cent of locations in permanent river sections *versus* areas inundated during flood are also given (mean value based on percentages for individual fish). Results are given for the entire study period (n = 12 fish). SD = standard deviation.

	Mean	Range (min–max)	SD
Home range size, 50% probability (m²)	475,648	150–4,807,453	1,369,784
Home range size, 95% probability (m²)	3,121,288	603–32,791,143	9,366,498
Distance moved between tracking surveys (m)	100	14–518	146
% locations in permanent river sections	72	0–100	35
% locations in temporarily inundated areas	28	0–100	35

backwaters, 63% in swamps, 25% on the floodplain, and 8% in the mouth of backwaters (percentages add up to more than a hundred because some fish were recorded in multiple habitat types). There was no difference in the proportion of locations (Table 2) in the main channel, side channels, backwaters, swamps or on the floodplain among periods (Friedman tests: $\chi^2 = 0.0–5.0$, $P = 0.082–1.0$). In contrast, the proportion of detections in backwaters differed among periods (Friedman test, $\chi^2 = 6.0$, $P = 0.05$), with the highest proportion during rising water level. However, no significant differences in the proportion of backwater detections were found when testing the rising level period against the high and receding water periods (Wilcoxon signed ranks tests, rising versus high: $Z = -1.6$, $P = 0.11$, rising versus receding: $Z = -1.6$, $P = 0.11$, high versus receding: $Z = 0.0$, $P = 1.0$).

The fish were recorded in different positions related to vegetation, with 83% of the fish recorded at no vegetation (>5 m away from vegetation) during one or more tracking surveys, 75% near vegetation (≤5 m away from vegetation) and 83% inside/under vegetation (percentages add up to more than a hundred because some fish were recorded in multiple habitat types). Only one fish was never recorded near or inside/under vegetation. The proportion of locations (Table 3) at no vegetation and inside/under vegetation did not differ among periods (Friedman tests, $\chi^2 = 0.86$ and 0.58, $P = 0.65$ and 0.75, respectively). However, the proportion of locations near vegetation differed among periods (Friedman tests, $\chi^2 = 7.9$, $P = 0.019$) and was higher during rising than during high water (Wilcoxon signed ranks tests, rising versus high: $Z = -2.4$, $P = 0.018$, rising versus receding: $Z = -1.5$, $P = 0.13$, high versus receding: $Z = -1.3$, $P = 0.18$).

The fish typically associated with sandy substratum. During one or more tracking surveys, 92% of the fish were recorded on sandy substratum, 58% on clay and 50% on muddy, soft bottom. Proportion of locations (Table 4) on sandy substratum differed among periods (Friedman test, $\chi^2 = 7.6$, $P = 0.023$), being lower during rising than during high water level (Wilcoxon signed ranks tests, rising versus high: $Z = -2.2$, $P = 0.028$, rising versus receding: $Z = -1.9$, $P = 0.063$, high versus receding: $Z = -0.73$, $P = 0.47$). However, proportion of locations on clay and muddy bottom did not differ among periods (Friedman tests, $\chi^2 = 2.5$ and 4.7, $P = 0.29$ and 0.097, respectively).

Water depth where the fish were recorded varied between 0.3 and 12.7 m, and averaged 4.2 m (Table 5). Water depth differed among periods (Friedman tests, $\chi^2 = 7.8$, $P = 0.21$), with water depth being larger during high water than during rising and receding water (Wilcoxon signed ranks tests, rising versus high: $Z = -2.4$, $P = 0.017$, rising versus receding: $Z = -1.0$, P

Table 2. Habitat use of radio tagged greenhead tilapia in the Zambezi River, Namibia, 2003–2004, expressed as per cent of locations in different habitats (mean value based on percentages for individual fish). Results are given for the entire study period (*n* = 12 fish). SD = standard deviation.

	Mean	Range (min–max)	SD
% locations in the main stream of the river	46	0–100	48
% locations in side channels	24	0–100	41
% locations in backwaters	14	0–100	31
% locations in swamps	4	0–24	7
% locations on the floodplain	11	0–64	24
% locations in the mouth of backwater	1	0–11	3

TABLE 3. Relation to vegetation of radio tagged greenhead tilapia in the Zambezi River, Namibia, 2003–2004, expressed as per cent of locations at no vegetation, near vegetation and inside/under vegetation (mean value based on percentages for individual fish). When recorded near or inside/under vegetation, the per cent of locations at different vegetation types is also given. Results are given for the entire study period (n = 12 fish). SD = standard deviation.

	Mean	Range (min–max)	SD
% locations at no vegetation	41	0–100	44
% locations near vegetation	19	0–67	22
% locations inside/under vegetation	40	0–100	40
% locations at marginal aquatic anchored vegetation	59	0–97	33
% locations at marginal terrestrial submerged vegetation	29	0–100	38
% locations at inner aquatic anchored vegetation	5	0–29	9
% locations at marginal terrestrial overhanging vegetation	5	0–44	14

TABLE 4. Bottom substratum used by radio tagged greenhead tilapia in the Zambezi River, Namibia, 2003–2004, expressed as per cent of locations over different substratum types (mean value based on percentages for individual fish). Results are given for the entire study period (n = 12 fish). SD = standard deviation.

	Mean	Range (min–max)	SD
% locations on sandy bottom	63	0–100	33
% locations on clay	19	0–78	25
% locations on muddy, soft bottom	18	0–89	31

TABLE 5. Water temperature and depth where radio tagged greenhead tilapia were positioned in the Zambezi River, Namibia, 2003–2004. Total width of the river where fish were positioned and distance to shore are also given (mean value based on mean values for individual fish). Results are given for the entire study period (n = 12 fish). SD = standard deviation.

	Mean	Range (min–max)	SD
Water temperature where the fish were positioned (°C)	24.8	23.8–27.2	1.3
Water depth where the fish were positioned (m)	4.2	1.8–8.0	2.1
Total width of the river where fish were positioned (m)	322	111–756	159
Distance to nearest shore (% of total river width)	16	3–32	8
Distance to nearest shore (m)	51	4–132	33

= 0.31, high versus receding: Z = –2.4, P = 0.017). Total width of the river where the fish were positioned varied between 30 and 2,000 m, and averaged 322 m (Table 5). Distance to nearest shore given as proportion of total river width averaged 16%. Distance to nearest shore varied between 1 and 1000 m, and averaged 51 m.

None of the variables recorded (distance moved, home range size, water depth, total width of river and average distance to shore) was dependent on fish body size (linear regressions, r^2 = 0.005–0.18, P = 0.17–0.83). Further, there was no difference in body length of fish utilizing temporarily flooded areas and those that did not (Mann-Whitney Test, U = 12.0, P = 0.37).

DISCUSSION

Movements and habitat use of greenhead tilapia and other cichlid species

Greenhead tilapia and most of the other large cichlids in the upper Zambezi River are valuable for subsistence fisheries and recreational angling (Van der Waal 1990; Skelton 2001). However, only limited ecological and behavioral data have been available. This study provides the first information on the seasonal movements, behavior, and habitat use of greenhead tilapia, and provides a basis of comparison with similar studies of nembwe, threespot tilapia and pink bream (Thorstad et al. 2001, 2005; Økland et al. 2007, Figure 4), which are also relatively large cichlids (total body length greater than 40–50 cm). All studies were performed in the same area, during October/November–March/August, with tracking on average every third to fourth day. However, the studies were performed in different years with variation in flood stages and the duration of study periods (Figure 4).

Greenhead tilapia were relatively stationary and stayed within a mean river stretch of 2,554 m. However, they had a larger home range and utilized larger stretches than three spotted tilapia (1999) and pink bream, which remained mainly resident within a mean river stretch of 540 and 220 m, respectively. However, annual differences in the flood cycle may affect cichlid movement patterns. Three spotted tilapia were studied in two different years and utilized a much larger stretch (mean 5,423 m) in 2000 than in 1999, when the rise in the water level was steeper (even though flood peak was at similar levels). The results from the upper Zambezi suggest that these cichlid species may be seasonally restricted to a few hundred meters up to a few kilometers, but that they do not have such a highly resident life style as previously assumed (Lucas and Baras 2001). The study of greenhead tilapia took place when the Zambezi River water level reached its highest level in more than 20 years. As suggested, the differences in flood regimes among years may affect the movements and habitat utilization, making it difficult to interpret whether differences among greenhead tilapia and other cichlids were due to species differences or annual differences in the flood cycle. Furthermore, the study period of greenhead tilapia started slightly later in the year, covering rising, high and receding water levels instead of low, rising and high water levels in previous studies.

Greenhead tilapia were recorded in the main channel of the river, in side channels, backwaters and on the floodplains. These observations were similar to those reported by Skelton (2001), that adult individuals were found in quiet waters along river margins and backwaters, in floodplain habitats and impoundments. Similarly, Winemiller and Kelso-Winemiller (2003) found that greenhead tilapia were more common in lagoons than in the river channel. In the present study, greenhead tilapia were often associated with vegetation (59% of locations near

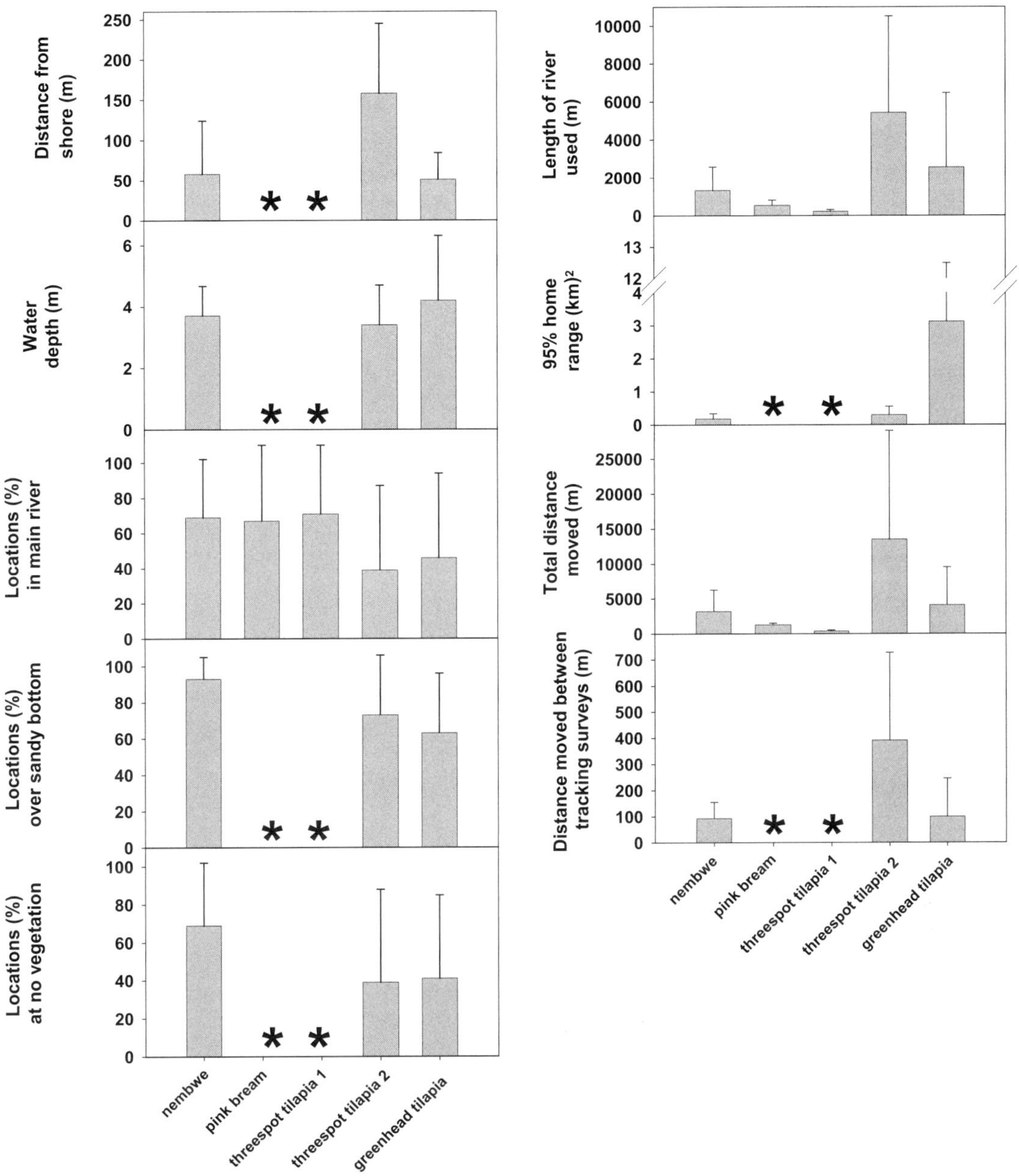

FIGURE 4. River characteristics in areas used by radio tagged nembwe, pink bream, three spotted tilapia (Thorstad et al. 2001, 2005, Økland et al. 2007) and greenhead tilapia (this study) in the Zambezi River during 1999–2003. Three spotted tilapia were studied in 1999 (three spotted tilapia 1) and 2000 (three spotted tilapia 2) (Thorstad et al. 2001, Økland et al. 2007). The duration of the study period was 5 months for pink bream and three spotted tilapia 1, 6 months for three spotted tilapia 2 and nembwe, and 8 months for greenhead tilapia. All values are mean of individual values with standard deviation. * = no data collected.

or inside/under vegetation), and were often recorded near the shore (average distance from shore 51 m). However, greenhead tilapia, nembwe and threespot tilapia showed a large individuals variation in their association with vegetation and mean distance from shore.

Water depth where greenhead tilapia were recorded varied between 0.4 m and 12.7 m during the study, similar to nembwe and threespot tilapia, but it is not known at which depths the fish actually occurred. Greenhead tilapia were less often recorded on sandy substratum and more often recorded on clay and muddy, soft bottom than nembwe. This may be linked to the more frequent utilization of other main habitats than the main river, and with their bottom feeding behavior. Greenhead tilapia mainly feed on microscopic foods such as algae, diatoms and detritus taken from the bottom (Skelton 2001; Van der Waal 1985; Winemiller and Kelso-Winemiller 2003), in contrast to nembwe that are fish predators (Skelton 2001). Greenhead tilapia and three spotted tilapia have a high dietary overlap (Winemiller and Kelso-Winemiller 2003), thus differences in habitat use between these two species might not be expected due to different feeding preferences. Nevertheless, a high proportion (average 63%) of the positional locations for greenhead tilapia were on sandy bottom, which can be linked to the widespread occurrence of sandy bottom in this area of the Zambezi River. The upper Zambezi River is a typical "sand-bank" river, with a dominance of sandy bottom (Van der Waal and Skelton 1984).

The creation of extensive floodplains during the rainy season greatly affects the habitat availability for the fish (Winemiller and Jepsen 1998; Jiménez-Segura et al. 2010). Greenhead tilapia moved on to temporarily inundated areas when the water was rising, with almost one third of the locations in temporarily inundated areas. Three fish (25%) also moved out onto the classical floodplain habitat with submerged grassland in shallow waters. Changes in behavior in connection with floods may be linked to the reproductive behavior of the fish. It has been suggested that some riverine cichlids and other fishes undertake longitudinal and lateral seasonal spawning migrations onto the inundated floodplain where their young may find favorable environments for fast growth and protection from predators, before returning to the river under receding waters (Bell-Cross 1974; Winemiller 1991; Van der Waal 1996; Fernandes 1997). Most of the greenhead tilapia in the present study had probably reached sexual maturity, according to their body lengths (Marshall 1979; Van der Waal 1985; Hay et al. 2002; Winemiller and Kelso-Winemiller 2003). Females seem to spawn at least 2–4 times annually (Bell-Cross 1974; Van der Waal 1985), and greenhead tilapia in the upper Zambezi River in Zambia breeds at least once before the floods, and at least once again during the floods (Bell-Cross 1974). In Lake Liambezi, Namibia, ripe females were found from August to March (Van der Waal 1985). Thus, the greenhead tilapia in the present study may have participated in breeding from the start of the study until March. Greenhead tilapia is a female mouth brooder, with the males attracting females to a nest in shallow water where the eggs are deposited (Van der Waal 1985; Skelton 2001). The utilization of temporarily inundated areas and floodplains during the present study may, therefore, have been in connection with spawning and nursery use. However, temporarily inundated areas may also provide abundant food, or the emigration to new areas may be triggered by fish density and social interactions (Louca et al. 2009).

In conclusion, the present and previous telemetry studies in the upper Zambezi River demonstrated that there is substantial individual variation in movement and utilization patterns among cichlids, and emphasize their mobility, association with different habitats and flexible habitat use. Basic information of annual movement, habitat preferences and habitat utilization of target species is needed to regulate the fisheries and exploitation methods used among the different countries sharing the same resources (Hocutt et al. 1994), and to evaluate the possible benefits of reserves and sanctuaries. Migration and habitat studies can provide information on which fish are most vulnerable to exploitation and when.

Based on these results, the studied cichlids may be locally vulnerable to overfishing. The relatively restricted movements of adults make them vulnerable to high local exploitation in contrast to species moving about more widely. The management and implemented regulations are, therefore, important for the local populations of these cichlids. In rivers bordering on several countries like the upper Zambezi River, multilateral management regulations are necessary even for relatively stationary species to avoid fish being protected in one country and overexploited in the neighboring country. The restricted movement of greenhead tilapia also implies that reasonably large sanctuaries will probably protect adult fish, because fish will be staying within the protected area. However, it must be emphasized that the full annual and life history cycles of the species were not studied. Juveniles, for example, may behave differently from adult fish. Such limitations must be considered when using radio telemetry data for management recommendations.

Challenges when performing telemetry studies in a large and complex river system

The studies in the upper Zambezi River demonstrated that radio telemetry was a powerful and effective technology for studying movements and habitat use of fish in a large, complex and remote river system. Radio telemetry was preferred to acoustic telemetry for several reasons. Searching for tagged fish over a 100 km long river stretch was by far more effective using radio telemetry technology. A 9-element radio antenna permanently mounted on top of the boat resulted in reception range from 700 m to over 1100 m for transmitters at 1–2 m depth, and fish could be searched for in complex habitats even when the boat was driving relatively fast. A hand-held 4-element antenna was additionally used to locate the fish once it was detected, since a small hand-held antenna is easier to move in different directions when searching for the bearing of the strongest signal. Tracking by using truck could have been an alternative if there were roads along the river, although positioning and habitat mapping would be less precise due to the reduced access and greater distances from the tagged fish. With acoustic technology, in contrast, the hydrophone receiving signals from the tagged fish has to be placed in the water, which reduces the mobility and speed of the tracking boat. When moving fast, air bubbles and turbulent water passing the hydrophone may reduce the detection range considerably (Thorstad et al. 2000). Acoustic receiving equipment is generally more expensive than radio telemetry outfit, and acoustic transmitters with a similar range and battery life are also larger and more expensive. However, acoustic telemetry has better signal reception than radio telemetry when the fish resides in deep water. The fish in this study stayed mainly in areas where the water depth was up to 8 m, but it is possible that the reason for some individuals not being recorded during some surveys was a lower range of radio transmitters when the fish stayed in deeper water. However, loosing fish at some occasions could also be linked to difficulties with fully covering the numerous backwaters and channels, and with covering the large floodplains when the river became more than 2 km wide. Gilroy et al. (2010) used a combination of acoustic and radio telemetry in a study of taimen *Hucho taimen* in a large, remote river system in Mongolia, which may be an option to utilize the advantages of both methods when sufficient resources are available.

Individuals were identified using transmitters with different frequencies, and the receiver had to search through the frequencies one at a time during tracking. Much time was needed to scan for fish at different frequencies, reducing the tracking efficiency and increasing the risk of passing fish without detecting it. An alternative would have been to use digitally coded transmitters, which have a unique numerical code transmitted at the same frequency. However, this alternative was not adopted because of the much higher equipment costs.

Only manual tracking was used during this study. Automatic listening stations (ALS) were not used, since the aim of the study was to record detailed movement and habitat use of the fish. Using ALSs may be less labor intensive, but would not have given a sufficient location resolution (i.e., where were the fish when not recorded by the ALS, and which habitats were used). Some 3D acoustic positioning systems (e.g., Cooke et al. 2005) provide more detailed information than a single- or multiple-antenna radio telemetry ALS, but they would not have provided sufficient reception range to effectively track the movements of these fish, and they are also much more expensive and technologically advanced. A problem in our studies was that some fish disappeared, either for short periods or for the remainder of the study period. Having no previous knowledge of the behavior of these fishes made it difficult to reliably anticipate the reason for fish disappearing from the study area, hence a large area had to be tracked to ensure that these fish had not left the study area. ALSs could have cordoned off the area, providing information whether individuals had left the study area or not, especially during low water (Jones and Stuart 2009). However, this was also a question of expense, and having listening stations in areas without electricity would also have demanded batteries to be changed regularly, or the purchase of solar panels. ALSs would also be difficult to deploy in an area increasing from a main river to large floodplains during the study period, increasing the risk of not recording tagged fish passing on the floodplains. When aircrafts are available, this could be a useful method to confirm the position of wide ranging fish, but there would be limitations on the exact positioning and habitat mapping (Roberts and Rahel 2005).

One of the greatest challenges was capturing a sufficient number of fish for tagging. With a high number of species present, it was difficult to target only one predetermined species as the catches were sometimes low. Due to low catchability, the decision on which species to tag was altered during the study, when it appeared that another alternative species was easier to capture within the allocated time. The difficulties in capturing fish also resulted in fish being captured and tagged over a large geographic area, extending the area that had to be tracked. Sample sizes also became lower than planned. However, when having virtually no previous knowledge on the behavior of a fish species, even results from a low number of fish will be useful for increasing the knowledge about local populations. Since annual variation in environmental variables such as flood rise, peak flooding and water temperatures may affect the fish behavior, studies should be repeated in different years to cover this variation.

Other important questions in virtually all telemetry studies are whether the capture method and site provide a representative sample of fish, whether there are inter-annual variation in parameters studied, whether the results are affected by life stage, sex and maturation, and whether a systematic bias is introduced when fish lost during the study period are excluded from analyses and only the fish found during the entire study period are included. Another challenge in our studies was the lack of basic knowledge on the species, such as life history parameters, spawning period, spawning sites and feeding. We therefore recommend collecting additional biological information on the studied species such as scale samples from live fish, and otoliths, stomach content and maturation status from additional representative samples, which can help explaining the observed behavior.

It can be concluded that radio telemetry studies using manual tracking were sufficient for obtaining critical information on the movement, habitat use, and distribution of large cichlids in the Zambezi River. The information and insights provided by these studies can be used as a knowledge base for local and regional management regulations. Considering the comprehensive and important results obtained, simple radio telemetry methods must also be regarded as relatively inexpensive, especially in countries where labor costs are comparably low. The services of a local, dedicated tracking team is invaluable and allows, as in the present study, for frequent positioning of the fish. In addition, technology transfer and capacity building are feasible using these methods (Baras et al. 2002). Manual tracking studies are also an excellent means of engaging the local community and stakeholder groups, given the easily accessible and comprehensible methods and results.

ACKNOWLEDGMENTS

The telemetry studies in the upper Zambezi River were funded by World Wildlife Fund, USAID, the Namibian Ministry of Fisheries and Marine Resources and the Norwegian Institute for Nature Research. We thank Nicoline and Rolly Thompson for extensive help during catch, tagging and tracking of the fish. We also thank Bernard Sezuni, Savage Simulebu, Garth M. Cavanagh and his family, and Kari Sivertsen for assistance during the project. We would also like to thank Steven J. Cooke and an anonymous referee for valuable comments to the manuscript.

REFERENCES

Abbott, J., L. Campbell, C. Hay, T. Næsje, A. Ndumba, and J. Purvis. 2007. Rivers as resources, rivers as borders: community versus transboundary management of fisheries in the upper Zambezi River floodplains. The Canadian Geographer 51:280–302.

Anon. 1995. White Paper on the Responsible Management of the Inland Fisheries of Namibia. Resource Management Directorate, Inland Fish Section, Namibian Ministry of Fisheries and Marine Resources, Namibia.

Baras, E., V. Benech, and G. Marmulla. 2002. Outcomes of a pilot fish telemetry workshop for developing countries. Hydrobiologia 483:9–11.

Bell-Cross, G. 1974. A fisheries survey of the Upper Zambezi River System. Occasional Papers of the National Museums and Monuments of Rhodesia B: Natural Sciences 5:278–338.

Cooke, S. J., G. H. Niezgoda, K. C. Hanson, C. D. Suski, F. J. S. Phelan, R. Tinline, and D. P. Philipp. 2005. Use of CDMA acoustic telemetry to document 3-D positions of fish: Relevance to the design and monitoring of aquatic protected areas. Marine Technology Society Journal 39:31–41.

Fernandes, C. C. 1997. Lateral migration of fishes in Amazon floodplains. Ecology of Freshwater Fish 6:36–44.

Gilroy, D. J., O. P. Jensen, B. C. Allen, S. Chandra, B. Ganzorig, Z. Hogan, J. T. Maxted, and M. J. Vander-Zanden. 2010. Home range and seasonal movement of taimen, *Hucho taimen*, in Mongolia. Ecology of Freshwater Fish 19:545–554.

Hay, C. J., T. F. Næsje, S. Kapirika, J. Koekemoer, R. Strand, E. B. Thorstad, and K. Hårsaker. 2002. Fish populations, gill net catches and gill net selectivity in the Zambezi and Chobe Rivers, Namibia, from 1997 to 2000. NINA Project Report 17:1–88.

Hocutt, C. H., S. E. Seibold, and R. V. Jesien. 1994. Potential use of biotelemetry in tropical continental waters. Revue d'hydrobiologie tropicale 27:77–95.

Jiménez-Segura, L. F., J. Palacio, and R. Leite. 2010. River flooding and reproduction of migratory fish species in the Magdalena River basin, Colombia. Ecology of Freshwater Fish 19:178–186.

Jones, M. J., and I. G. Stuart. 2009. Lateral movement of common carp (*Cyprinus carpio* L.) in a large lowland river and floodplain. Ecology of Freshwater Fish 18:72–82.

Koekemoer, J. H. 2003. A fish ecological study of rivers and floodplains in the Eastern Caprivi, Namibia. Doctoral dissertation, Rand Afrikaans University, South Africa.

Lawson, E. J. G., and A. R. Rodgers. 1997. Differences in home-range size computed in commonly used software programs. Wildlife Society Bulletin 25: 721–729.

Louca, V., S. W. Lindsay, and M. C. Lucas. 2009. Factors triggering floodplain fish emigration: importance of fish density and food availability. Ecology of Freshwater Fish 18:60–64.

Lucas, M., and E. Baras. 2001. Migration of Freshwater Fishes. Blackwell Scientific Publications, Oxford, U.K.

Marshall, B. E. 1979. Observations on the breeding biology of *Sarotherodon macrochir* (Boulenger) in Lake McIlwaine, Rhodesia. Journal of Fish Biology 14:419–424.

Mendelsohn, J., A. Jarvis, C. Roberts, and T. Robertson. 2002. Atlas of Namibia. A portrait of the land and its people. David Philips Publishers, Namibia.

Økland, F., C. J. Hay, T. F. Næsje, B. Chanda, and E. B. Thorstad. 2007. Movements of, and habitat utilisation by, threespot tilapia *Oreochromis andersonii* (Teleostei: Cichlidae) in the Upper River Zambezi, Namibia. African Journal of Aquatic Science 32:35–38.

Økland, F., E. B. Thorstad, C. J. Hay, T. F. Næsje, and B. Chanda. 2005. Movements and habitat utilisation of tigerfish (*Hydrocynus vittatus*) in the Upper River Zambezi. Ecology of Freshwater Fish 14:79–86.

Roberts, J. J., and F. J. Rahel. 2005. Accuracy of aerial telemetry in fisheries studies. North American Journal of Fisheries Management 25:660–666.

Seaman, D. E., and R. A. Powell. 1996. An evaluation of the accuracy of kernel density estimators for home range analysis. Ecology 77:2075–2085.

Skelton P. 2001. A complete guide to the freshwater fishes of Southern Africa. Struik Publishers, Cape Town, South Africa.

Thoreau, X., and E. Baras. 1997. Evaluation of surgery procedures for implanting telemetry transmitters into the body cavity of tilapia *Oreochromis aureus*. Aquatic Living Resources 10:207–211.

Thorstad, E. B., C. J. Hay, T. F. Næsje, and F. Økland. 2001. Movements and habitat utilisation of three cichlid species in the River Zambezi, Namibia. Ecology of Freshwater Fish 10:238–246.

Thorstad, E. B., F. Økland, D. Rowsell, and R. S. McKinley. 2000. A new system for automatic recording of fish tagged with coded acoustic transmitters. Fisheries Management and Ecology 7:281–294.

Thorstad, E. B., C. J. Hay, T. F. Næsje, B. Chanda, and F. Økland. 2004. Effects of catch-and-release-angling on large cichlids in the subtropical River Zambezi. Fisheries Research 69:141–144.

Thorstad, E. B., C. J. Hay, T. F. Næsje, B. Chanda, and F. Økland. 2005. Movements and habitat utilization of nembwe, *Serranochromis robustus* (Günther, 1864), in the Upper River Zambezi. African Zoology 40:253–259.

Van der Waal, B. C. W. 1985. Aspects of the biology of larger fish species of Lake Liambezi, Caprivi, South West Africa. Madoqua 14:101–1446.

Van der Waal, B. C. W. 1990. Aspects of the fishery of the Eastern Caprivi, Namibia. Madoqua 17:1–16.

Van der Waal, B. C. W. 1996. Some observations on fish migration in Caprivi, Namibia South African Journal of Aquatic Sciences 22:62–80.

Van der Waal, B. C. W., and P. H. Skelton. 1984. Check list of fishes of Caprivi. Madoqua 13:303–320.

Winemiller, K. O. 1991. Comparative ecology of Serranochromis species (Teleostei: Cichlidae) in the Upper River Zambezi floodplain. Journal of Fish Biology 39:617–639.

Winemiller, K. O., and D. B. Jepsen. 1998. Effects of seasonality and fish movements on tropical river food webs. Journal of Fish Biology 53 (Supplement A):267–296.

Winemiller, K. O., and L. C. Kelso-Winemiller. 2003. Food habits of tilapiine cichlids of the Upper River Zambezi and floodplain during the descending phase of the hydrologic cycle. Journal of Fish Biology 63:120–128.

Worton, B. J. 1989. Kernel methods for estimating the utilization distribution in home range studies. Ecology 70:164–168.

Section 6.3

The Application of Radio Telemetry to Fisheries Research in Brazil's Large Rivers

LISIANE HAHN

INTRODUCTION

Radio telemetry techniques have been widely applied to fisheries research in temperate regions since the 1970s (Baras 1991), but few studies have been performed in tropical rivers (Hocutt et al. 1994; Thorstad et al. 2001). The first projects to use radio telemetry for fish research in Brazilian freshwater rivers began in 2001 after a Brazilian researcher´s training course promoted by the NGO World Fisheries Trust and LGL Environmental Research Associates, both from Canada. Research on Brazilian fishes with radio telemetry were or are being developed in the river basins of the Amazon, São Francisco, Mucuri, Paraná, Uruguay and Sinos drainages (Figure 1), both in lakes, fishways (Itaipu, Igarapava, Volta Grande and Igarapé dams), rivers and reservoirs. The goals of these studies have been to describe the movements and behavior of migratory species (generally very poorly known), identify locations and timing of spawning, determinate fish passages selectivity and efficiency, temporal patterns of fish passage in fishways, and to assess the effectiveness of radio telemetry as a tool for research on Brazilian freshwater fish. Given the diversity of fish in Brazil and the many unknowns associated with fish biology and natural history, coupled with the difficulties of working in large rivers in relatively remote areas presented a number of challenges that had to be overcome. As such, this Section is intended to be a narrative case study that will describe how radio telemetry has been used to study riverine fish in Brazil.

METHODS

Data obtained in projects conducted from 2001 to 2008 on the south, southeast and north regions of Brazil were used in this review: Crossa et al. 2003; Schulz 2003; Silva 2004; Pompeu 2005; Godinho and Kynard 2006; Alves et al. 2007; Godinho et al. 2007; Hahn et al. 2007; Hahn et al. 2011; Pesoa and Schulz (2010); and personal communications by authors mentioned above.

Sampling and Transportation Methods

Fish were collected using gill nets, cast nets, rods with artificial and natural baits, caceia (specialized floating net) and tarrafão (specialized bottom drift net). The collected fish were

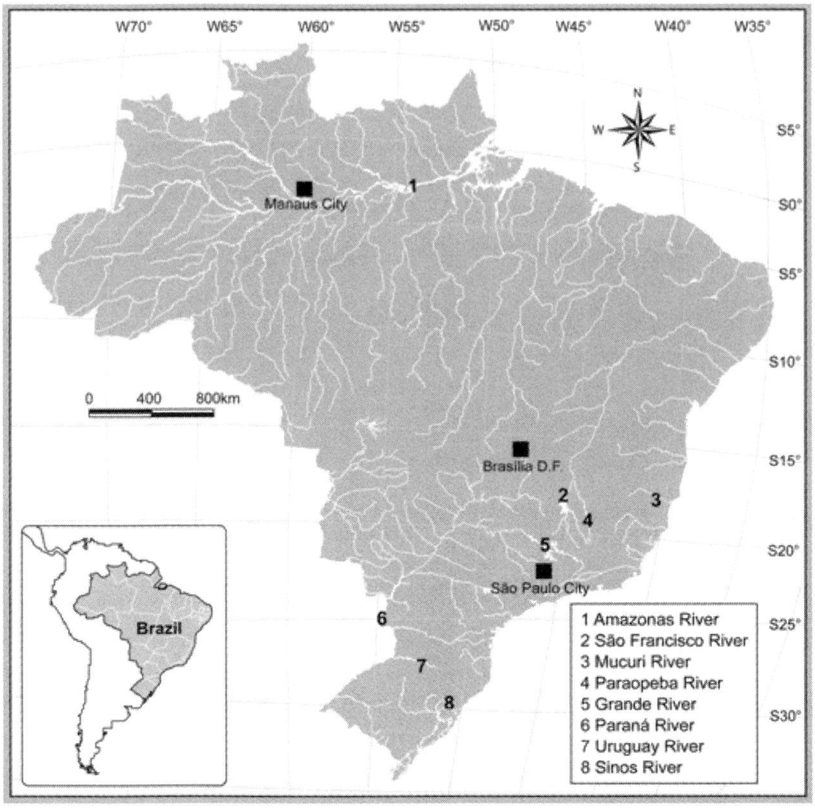

FIGURE 1. Study areas of telemetry projects in Brazil.

tagged nearby the capture site or taken to laboratories by means of boxes or holding tanks (1000–2500 L aerated) and kept either captive in laboratory tanks, sustained in the river with a rope tied around the mouth (catfish species), or even in a nearby floating cages before and after the transmitter (hereafter referred to as tag) insertion for different periods, which ranged from less than one hour up to three days.

Fish Tagging

The choice of tag specifications were based on different criteria such as compatibility between projects of different rivers of the same basin, detected background noise in specific range of frequencies, target species, and tracking protocols. All fish, except those at Santa Clara Dam and Amazonas River were radio tagged with radio tags from Lotek Wireless (Newmarket, Ontario, Canada). The various studies used tags with different frequencies, operational lives, burst rates, and models (See Table 1 for a summary of tag characteristics).

Physical immobilization (galvonarcosis, Ross and Ross 2008) and chemical sedation (clove oil and 2-phenoxyethanol) were used by Brazilian researchers to anesthetize fish. In general, researchers on fish from São Francisco River basin used galvonarcosis anesthesia. Clove oil was largely used in the Paraná River (Lateral Canal at Itaipu Dam) and Uruguay River basin projects and anesthesia with 2-phenoxyethanol was used in the Sinos River. A combination of galvonarcosis and clove oil was used on the Paraopeba River (Igarapé Dam).

TABLE 1. Telemetry equipments in Brazilian rivers.

Drainage	Citation	Tag model	Manufacture	Burst rate	Operational life
São Francisco River—Downstream Três Marias Dam	Godinho and Kynard (2006); Godinho et al. (2007)	MCFT-3FM, MCFT-3A,	Lotek Wireless	5–10 seconds	658 days to 4 years
Paraopeba River	C. B. M. Alves and A. Godinho (unpublished)	MCFT-3FM MCFT-3A,	Lotek Wireless	7–10 seconds	837 and 938 days
Mucuri River	Pompeu and Martinez (2006)	F2120	Advanced Telemetry Systems	1.7 seconds	535 days
Grande River—Funil Dam	Alves et al. (2007)	MCFT-7SP	Lotek Wireless	5 seconds	
Grande River—Igarapava and Volta Grande Dams	Silva (2004)	MCFT-3EM, MCFT-3FM	Lotek Wireless	10 seconds	730 days
Paraná River—Itaipu Dam	Hahn (2007), Hahn et al. (2007)	MCFT-3FM, MCFT-3A, MCFT-3L	Lotek Wireless	2.5–5 seconds	290 to 1010 days
Uruguay River	Hahn (2007)	MCFT-3CM, MCFT-3A, MCFT-3L	Lotek Wireless	2.5–5 seconds	150 to 812 days
Sinos River	U. H. Schulz, C. S. Anjos and M. Leal (unpublished), Pesoa and Schulz (2010)	MCFT-3FM	Lotek Wireless	5 seconds	336 days
Amazon River	Crossa (2003)		Advanced Telemetry	30/70*, 56 ppm 21–22 ms	300, 400* – 500 days (*temperature)

For immobilization using galvonarcosis variable nonpulsed 10–60-v DC was applied and fish were kept in tanks filled with 30–150 L of river water. Electric current was applied through sheet metal plate electrodes at either end of the bath. The amperage was usually increased until the fish became immobilized and unresponsive to touch, then reduced during tagging (Godinho and Kynard 2006; Alves et al. 2007; Godinho et al. 2007; Hahn et al. 2007).

The suitable concentrations of clove oil (eugenol, 4-allyl-2-methoxy-phenol) for use on Brazilian fish were first established after experimental tests for the Uruguay River fishes (unpublished data) and later applied for Paraná River fish species. A bath of anesthetic (about 1 mL eugenol/40 L water) anesthesized fish within 1–4 min and a concentration of 1 mL eugenol/80 L water kept fish anesthetized during the tagging procedures through a constant irrigation of the gills (Hahn et al. 2007; Hahn et al. 2011). For specimens of catfish *Rhamdia quelen* (jundiás) under controlled conditions, a concentration of 350 mg/L of 2-phenoxyethanol was applied and surgical implantation followed the procedures given in Adams et al. (1998) and Schulz (2003).

Tags were externally attached (Santa Clara Dam and Amazonas project), implanted through intra-peritoneal surgery, or inserted in the stomach. Tagging procedures followed Lucas and Baras (2000) with little variation between species and projects. Scales near the incision area were removed before making a 2–5 cm incision in the abdomen body wall, which could be located above or on the ventral middle line. The tag antenna exited the body wall through a catheter inserted in the incision, which was then removed. The incision was closed with separate stitches of an absorbable and nonabsorbable suture material (Vicryl Ethicon, Ethicon Ethipoint SC-20, Nylon and Silk) and finally tissue adhesives were applied right on the incision (Hahn et al. 2007). In a portion of the tagged fish, an external hydrostatic tag (LEA type tags, a small PVC tube) with a number and instructions for return were applied in the dorsum of fish. Additionally, fish were weighed and measured in all projects and in some cases sex and reproductive stage of fish were determined through macroscopic inspection (Alves et al. 2007; Godinho et al. 2007) or by means a fiber optic bioscope inserted through the incision (Godinho and Kynard 2006).

Telemetry Equipment and Tracking

Telemetry receivers were used at fixed locations and in mobile tracking. In most studies Lotek SRX-400 receivers with W7 or W32 firmware were used for both fixed station and mobile tracking applications. Fish were manually tracked on the Santa Clara Dam project using an ATS model FM 16 receiver. Every fixed station had at least two aerial 3- or 4-element Yagi antennas installed which were connected to the receiver's switcher using coaxial cables. The antennas were usually pointed upstream, downstream, or at the mouth of tributaries. Underwater antennas (stripped coaxial cable) were also installed in the Lateral Canal at Itaipu Dam. Receivers were powered by a commercial source, batteries or solar panels. The receivers, switchers and batteries were secured inside a weatherproof metal box with locks.

Preliminary tests were conducted in the Lateral Canal at Itaipu Dam on January 2004 to determinate the most suitable frequencies (such as frequencies with less interference from background noise and other sources) for future studies (Hahn et al. 2007). Range tests were also conducted for most of the projects, where a tag was immersed at different locations and depths to assess signal strength.

Manual tracking was applied in all projects, and fish positions were determined by means of an airplane, helicopter, boat, car, or on foot. In the cases where an airplane and helicopter were used (Uruguay and São Francisco Rivers), Yagi or H antennas were attached to the airplane wings or at the helicopter base. Signals sometimes could be heard when tracking fish by airplane but could not be "read" by the receiver. In these cases, the region was re-scanned until the receiver recorded the code. There are restrictions in Brazil for use of external structures attached to the airplanes and in some cases pilots did not allow bigger antennas (Yagi, for example) to be affixed to the airplane, which made it harder to track fish. Tracking by boat and by car are quite similar. One or two Yagi antennas are attached to a pole on these vehicles and pointed to the river or fish passage. In all situations the fish location was determined by a GPS or by triangulation.

Additional Data

A variety of environmental data (turbidity, temperature, discharge, flow) were also collected during the studies by measurements made directly on the river or obtained from power companies. All rivers, reservoirs and fish passages have a pronounced variety of water characteristics, but, conductivity varies from 23 to 120 μs/cm for most of them, which is not high enough to attenuate radio signals beyond useful limits. Higher values of conductivity are associated with polluted stretches of the rivers. The depth in some rivers or some stretches of the research areas, such as Uruguay River is higher than 15 m and the widths are up to 800 m.

Considering the wide array of environments covered by these projects, a brief description of the study areas is presented in Table 2.

Data Analysis

Data stored by Lotek SRX receivers were downloaded twice a week to monthly and sometimes more often during the spawning season and/or when the receiver memory filled up prior to the scheduled downloads. In the Itaipu Dam and Uruguay River projects, data logged by Lotek receivers were downloaded to a laptop computer as hex-encoded files, which were converted to standard ASCII format using software developed by LGL Limited, Sydnei, BC, Canada) (SRXW303.EXE). This software assessed several diagnostics, including the number of invalid records. If the number of invalid records was large, the receiver was downloaded a second time. The SRXW303 program also displayed the distribution of antenna noise by power level, so that problems with specific antennas could be isolated, and the appropriate troubleshooting measures could be taken. All fixed-station data received were organized into structured databases and analyzed using Telemetry Manager Version 3.0, specialized software developed by LGL Limited. The Telemetry Manager software facilitated the importing of the raw data files downloaded from the receivers and organizes these data into a database containing records for each logged data transmission from the tagged fish. The software then processes the data to remove records that did not meet the criteria specified for valid data records. Examples of invalid data included background noise, records with a signal strength below a set threshold, single records for a given frequency-code-location combination, and records that were recorded before the official release time and date. The software then compressed the data into an operational database that contained the time of arrival and departure from each zone, number of records, and maximum power for each series of detections for each radio-tagged fish. These data were immediately available for display on maps of the study area that show the location of each zone. These maps can be used to examine the numbers of fish of

Table 2. Brazilian Rivers in the telemetry projects study areas.

Drainage	Citation	Area involved	Gradient	Substrate	Channelized/Floodplain	River/Reservoir/ Fishway
São Francisco River—Downstream Três Marias Dam	Godinho and Kynard (2006); Godinho et al. (2007)	282 km	High	Mostly rocky	Channelized and narrow floodplains	River–Free flowing
Paraopeba River	C.B.M. Alves and A. Godinho (unpublished)	438 km	High and medium	Rocky and sandy	Channelized (upper portions) narrow floodplains (lower portion)	River–Free flowing
Mucuri River	Pompeu and Martinez (2006)	15 km	High and medium	Rocky and sandy		Reservoir
Grande River—Funil Dam	Alves et al. (2007)	60 km				Reservoir
Grande River—Igarapava and Volta Grande Dams	Silva (2004)	128 km	High	Rocky		Reservoirs/Fish ladder
Paraná River—Itaipu Dam	Hahn (2007)	10 km		Rocky and sandy		Lateral Channel
Uruguay River	Hahn (2007)	450 km	High	Rocky and sandy	Channelized	Free flowing
Sinos River	U. H. Schulz, C. S. Anjos and M. Leal (unpublished), Pesoa and Schulz (2010)	150 km	High and medium	Rocky and sandy	Channelized and floodplains in the medium and lower portions	Free flowing
Amazon River	Crossa (2003)	100 km	Low	Mud	Floodplain	Lake

each species seen in each zone, the first and last seen locations for all fish, or the sequence of detections for individual fish (Hahn et al. 2011). The team of the Federal University of Minas Gerais developed two new procedures to eliminate false-positive records (due to background noise, the receivers also logged signal hits of codes not used in the study—absent codes), based on the probability of a record being a false positive and checking data log records for consistency (see Godinho and Kynard 2006).

Tag Return by Fishers

Fishers from the study areas were encouraged to return the tags due to the scientific value of the information contained in the tags (rather than the value of the tag itself) (Thorsteinsson 2002), by means of advertisement in local and regional radio and newspapers, posters, informative lectures, local interviews, personal contacts, message written on the transmitter with a free telephone contact number and rewards (i.e., hats, T-shirts)

RESULTS AND DISCUSSION

A total of 464 fish from 21 species and seven families was radio tagged in river basins in Brazil (Table 3) from January 2001 to August 2008. Most (77.3%) were *Prochilodus* and *Salminus*.

Sampling and Transportation Methods

There was no report of death of any fish collected in the reviewed projects due to sampling or transportation procedures. In all gears used (except gill nets), fish were removed immediately after capture decreasing the stress level, since they spent a short time in the net. However, these methods could result in skin damage and scale losses, which facilitates the infestation by fungus, virus and bacteria (Hahn et al. 2007). Keeping fish in tanks before and after surgery allowed researchers to observe the fish behavior and ensures adequate recovery before release. However, for some species (e.g., dourado *Salminus brasiliensis*, piracanjuba *Brycon orbignyanus*) survival appears to be influenced by even short term holding so there is a need to develop strategies for reducing stress of fish associated with capture, handling and holding.

Fish Tagging

Godinho and Kynard (2006) reported that five tags implanted in *Prochilodus argenteus* (zulega) stopped transmitting signals after 224–514 d instead of specified 755 d. Three of 167 tags used in fish of the Uruguay River and Itaipu Dam projects prior to 2006 had some type of premature failure. In the Itaipu lateral channel, 16% of radio-tagged fish were not detected after release.

The tag-weight (in air) to body-weight percentages for almost all radio-tagged fish was 2% or less, except for two specimens of dourado from the Uruguay River, where the percentages were 2.6% and 8.67%, respectively. The first was not detected after release, which could also be related to the fishing activity in the region and the large dimensions of the river. The

TABLE 3. Methods used to study fish with radio telemetry in large river drainages in Brazil. [1]I = immobilization; A = anesthetic; N = none; [2]1 = boat; 2 = vehicle; 3 = foot; 4 = aircraft; 5 = helicopter

Reference	Fish species	Number of tagged fish	River	Anesthetic[1]	Tagging method	Tracking Methods: Fixed stations or Mobile[2]	Main Objectives	Main results
Crossa (2003)	*Arapaima gigas*	25	Amazon	None	External	Mobile	To identify territorial behavior, determinate daily and seasonal behavior, estimate home range size and identify lateral and longitudinal migration	Territorial behavior, longitudinal and lateral migrations (sometimes associated with schools of *Prochilodus* sp.)
C. B.M. Alves and A. Godinho (unpublished)	*Salminus franciscanus*	41	Paraopeba	I and A	Surgery	Fixed and mobile (1,4)	Determinate linear home ranges and efficiency of fish ladder	Linear home ranges of 212.8 km and 29.2 % efficiency of the fish ladder
Alves et al. (2007)	*Zungaro jahu*	01	Grande	I	Surgery	Mobile (1)	Describe migratory movements, habitat use, linear home range and diel activity	Linear home range of 31.3km, onset and daytime activity, movements between main body and the reservoir-river transition zone
C. B.M. Alves and A. Godinho (unpublished)	*Prochilodus costatus*	37	Paraopeba	I and A	Surgery	Fixed and mobile (1,4)	Determinate linear home ranges and efficiency of fish ladder	Linear home ranges of 108.3 km and 54.1% efficiency of the fish ladder

TABLE 3. Continued.

Reference	Fish species	Number of tagged fish	River	Anesthetic[1]	Tagging method	Tracking Methods: Fixed stations or Mobile[2]	Main Objectives	Main results
Godinho and Kynard (2006)	*Prochilodus argenteus*	37	São Francisco	I	Surgery	Fixed and mobile (1,4)	Identify migrations, linear home ranges, spawning and nonspawning habitats, timing of spawning	Prewpaning movements, dualistic migratory style (resident and migratory fish), linear home ranges from 26 to 127 km.
Godinho et al. (2007)	*Pseudoplatystoma corruscans*	24	São Francisco	I	Surgery	Fixed and mobile (1,5)	Migration, linear home ranges, homing, river reaches x spawning and non-spawning activies	274 km (linear home range), dualistic migration style with resident and migratory fish, ground speeds up to 31km day^{-1}
Hahn (2007)	*Salminus brasiliensis*	73	Uruguay	A	Surgery	Fixed and mobile (4)	To study seasonal movements and migratory biology	Movements upstream up to 325 km and 54.1 km day^{-1}, resident and migratory fish behavior
(author, unpublished)	*Leporinus obtusidens, Brycon orbignyanus, Prochilodus lineatus*	17	Uruguay	A	Surgery	Fixed and mobile (4)	To study seasonal movements and migratory biology	Records only from release site (except for one specimen of *B. orbignyanus*)

Table 3. Continued.

Reference	Fish species	Number of tagged fish	River	Anesthetic[1]	Tagging method	Tracking Methods: Fixed stations or Mobile[2]	Main Objectives	Main results
Hahn et al. (2007)	*Prochilodus lineatus, Cichla kelberi, Pterodoras granulosus, Pseudoplatystoma reticulate, Schizodon borelli, Leporinus friderici*	14	Paraná	A and N	Surgery and esophagic	Fixed and mobile (1,2,3,4)	Test tagging procedures, type of tags, tracking equipment and radio frequencies for studies in the bypass Channel	Telemetry protocols to be applied in further studies
(author, unpublished)	*Leporinus elongatus*	14	Paraná	A	Surgery	Fixed and mobile (1,2,3,4)	To investigate fish movements in the Itaipu bypass channel	
Hahn (2007)	*Salminus brasiliensis, Prochilodus lineatus, Piaractus mesopotamicus Pseudoplatystoma corruscans, Pseudoplatystoma reticulate, Zungaro jahu*	77	Paraná	A	Surgery	Fixed and mobile (1,2,3,4)	To evaluate the selectivity of the species that entered in the Itaipu bypass channel	High selectivity (6.5%), long-term permanence of fish in the Canal (i.e. ± 255 days to *P. lineatus*)
Pesoa and Schulz (2010)	*Prochilodus lineatus*	19	Sinos	A	Surgery	Fixed and mobile (1,4)	Diel and seasonal movement patterns, spawning migration and habitat use	Preference for river reaches surrounded by wetlands and diel activity pattern with two peaks
Pompeu (2005)	*Prochilodus vimboides, Leporinus*	15	Mucuri	None	External	Mobile (1)	Identify movements in the reservoir	Movements in the reservoir and river

Table 3. Continued.

Reference	Fish species	Number of tagged fish	River	Anesthetic[1]	Tagging method	Tracking Methods: Fixed stations or Mobile[2]	Main Objectives	Main results
U. H. Schulz, C. S. Anjos and M. Leal (unpublished)	*Salminus brasiliensis*	29	Sinos	A	Surgery	Mobile (1,4)	To obtain biological background data on the dourado movement patterns and habitat use	Movements up to 115 km upstream during spawning migration, home ranges of 2.5 km to 4 km during pre-spawning periods
L. G. M. Silva and colleagues (unpublished)	*Prochilodus lineatus*	30	Grande	I	Surgery	Fixed and mobile (1,5)	Fish behavior in and upstream fish ladder and identify key habitats for males and females	Two different movements patterns after release and no significant difference between males and females home range
Silva (2004)	*Pimelodus maculatus*	11	Grande	I	Surgery	Fixed and mobile (1,5)	Fish behavior upstream fish ladder and identify key habitats for males and females	64% remained in the reservoir, males exhibited larger home ranges than females

last was detected many times during the study period and was recorded 320 km upstream from the release site (Hahn et al. 2011).

Both for fish immobilized using galvonarcosis and for those fish anesthetized using of clove oil mortality was not reported for any project reviewed, except for one fish that died apparently due to a high concentration of clove oil wrongly applied during experimental tests at Itaipu Dam. The galvonarcosis proved to be efficient for seven different species and fish sizes with quick induction and recovery, no noted deleterious effects, and low costs. However, this procedure is controversial, since it is not completely known if a fish is only immobilized or induced to true anesthesia (Ross and Ross 2008). Also, the stress caused to fish by inappropriate voltage levels and the risks related to electric shock to researchers should be considered. Clove oil was also used to anesthetize 13 fish species in different river conditions. The induction to anesthesia was no longer than four minutes for all specimens (more than 80% were anesthetized before two minutes in the clove oil bath). At a concentration of 1 mL eugenol/40 L water differences in induction times between larger and smaller fish or between species were not detected, but further investigations are recommended. In association with galvanonarcosis, concentrations of clove oil can be much lower than when it is used alone. The advantages of the clove oil are a swift induction and recovery from anesthesia, no interference on fish behavior (swimming performance) and the low cost.

The total time to complete the surgery procedures varied from 5 to 40 min. The longer times were required especially in those cases where sex and reproductive stage were determined. The duration of the anesthesia and surgery did not seem to affect fish welfare and behavior but is strongly suggested to Brazilian researchers to investigate these topics thoroughly prior to initiating tagging studies. Records at different locations after release and recapture by fishers suggested satisfactory recoveries from surgery. Nonetheless, there are few detailed studies that evaluate tagging protocols for Brazilian fish. Researchers of the Uruguay River project recaptured with rod line and artificial bait one specimen of dourado radio- tagged and released two months before near the release site and verified the advanced process of healing and the total absorption of the sutures. Similar information was reported by fishers who captured fish and returned seven radio transmitters to the same project (Hahn et al. 2011). In Paraopeba River one specimen of curimba was recaptured by researchers one month after released with scales covering part of the incision. Healing was evident and only two of the five sutures remained (C.B.M. Alves, Federal University of Minas Gerais, unpublished data).

Schulz (2003) examined the response of 20 jundiás tagged using surgical procedures and released at high water temperatures and low water quality. Although three fish expelled the tags within 70 d of surgery, the incisions of 13 other fish had healed completely with most sutures expelled at that time. According to the author, the individual opercular strap tags (external tags attached to the pelvic fins) of the fish had more negative consequences than the dummy transmitter implanted internally and side effects are not expected to be more significant than those radio tagged fish of temperate regions. Stomach (i.e., gastric) insertion of the tags in two fish species, one specimen of *Prochilodus lineatus* (curimba or grumatã) and one of *Leporinus friderici* (piau), was successfully carried out and both fish survived until the end of study. Side effects of this approach should be better evaluated, since these fish species have feeding activity during spawning season (Hahn et al. 2007). External attachment of tags was applied in Santa Clara Dam project (Mucuri River), for 8 individuals of curimatá *Prochilodus vimboides*, 3 piaus-branco *Leporinus conirostris* and 4 piabanhas *Brycon ferox* (Pompeu 2005) and in Amazonas River, for 25 pirarucus *Arapaima gigas* (Crossa 2003).

Telemetry Equipment, Tracking and Data Analysis

The collection and data analysis in the studies were impaired for reasons such as (1) depths often exceeding 5 m and width of the river exceeding 200 m complicated signal reception (many tagged fish passing fixed stations in the Uruguay river were not detected), (2) background noise which degraded signal reception and led to increased efforts to eliminate false positives and (3) the interference from the aircraft used for tracking on the signal reception (in some older aircraft models which were used in some previous tests, the electric system is not shielded causing interference; Cesar Mozelle, Passo Fundo airport, personal communication).

In the Lateral Canal at Itaipu Dam, four underwater antennas were installed in a section with high water velocities and despite passing the cables through a robust steel pole, they were broken and had to be replaced by aerial antennas.

The procedures and software used to filter non valid codes were considered an efficient means of making the data more reliable. However, due to receiver limitations and high levels of background noise in some areas, additional reviews of the process are needed.

Fish Behavior

Based on the projects conducted so far, it was possible to identify migrations and linear home ranges for five species (Table 3): dourado (*Salminus brasiliensis*) in the upper Uruguay River; zulega and surubim (*Pseudoplatystoma corruscans*) on São Francisco River, curimatá-pioa (*Prochilodus costatus*) and dourado (*Salminus franciscanus*) on Paraopeba River and grumatã (*Prochilodus lineatus*) and dourado (*Salminus brasiliensis*) on Sinos River. For those species studied in the Lateral Canal at Itaipu Dam and at fish ladder at Paraopeba River, it was possible to determinate swimming behavior within the different sections of the fish passage route, or at least to get indications for future studies.

Advantages and Disadvantages in Use of Telemetry Techniques for Fish Research in Brazil

Fish telemetry studies in Brazil are very recent and the publications (including dissertation/thesis, papers and symposium abstracts) are less than fifteen up to 2008. These studies all tend to be descriptive and focus on very basic aspects of fish movements and behavior. Radio transmitters were used the first time in salmonids in 1956 (Stasko and Pincock 1977), 45 years before the first studies in Brazil.

In general terms, telemetry technology served the purposes of the studies in the Brazilian rivers. Results from fish telemetry projects in last eight years in Brazil are novel and remarkable. So far radio telemetry has provided unprecedented information on the movements and behavior of migratory species despite the fact that there had been decades of investigations with techniques other than telemetry used.

Most of the difficulties the researchers faced during their projects were related to (1) the lack of previous knowledge about fish natural history and behavior, (2) the dimensions of the study areas (i.e., sheer size and openness of these systems), (3) equipment failure (3) safety, (4) capture of tagged fish by fishers, (5) experience of researchers and (6) equipment costs.

For most of species, behavior data were very poorly known, making it difficult to establish efficient procedures to capture and tag them. Equipment failure was reported in some projects, such as estimated operational life of the tags and false positive records. Safety is a challenge when using telemetry in Brazil. Almost all projects had fixed stations damaged in different degrees from vandalism or theft. Receivers, antennas, solar panels and even cables were stolen or damaged periodically. Capture of tagged fish by fishers is very common in Brazilian projects. All projects had undetected fish that were likely attributable to fishing pressure and the unwillingness of fishers to report tag recoveries, vast study areas, or even to equipment failure. At Itaipu Dam, five tags were found in the bottom of the Canal and the absence of antennas suggested the fish were captured and the tags were thrown out in the water. Of the 37 zulegas tagged in the São Francisco River, signals for 19 of the fish were not detected after 19–554 d, probably because they were harvested, or the receivers did not detect them moving out of the study area (Godinho and Kynard 2006). Recovery of tags in such vast areas is a difficult task, despite the increasing efforts to advertise projects and ask for tags to be returned. In cases where tags are repeatedly detected at the same location for long periods, it is almost always impossible to find them in deep, turbid, and fast run waters. In the Uruguay River project (and probably in some others projects which started at the same time in Brazil), the failures mentioned above were related mainly to the experience level of the researchers involved (project design, surgical procedures, and data analysis), despite previous training and support from foreign researchers. It was also aggravated by the dimensions of the river and access difficulties to some portions of the river. Equipment costs limit the broad applicability of the technique for more species and other river basins in Brazil.

On the other hand, a large quantity of high quality data were collected in just 7 years (from 2001 to 2008, when most of the projects described started) for more than 20 fish species in five river basins and at different fish passages. This would probably be impossible without the application of telemetry techniques.

The next set of challenges to the application of telemetry to Brazilian fish projects involve improving project design, equipment choice, surgical procedures, (Cooke et al. 2011) and improving data analysis. The permanent interchange with foreign institutions is crucial to improving techniques and data collection in Brazil. Furthermore, it is necessary to explore other electronic tagging tools (e.g., acoustic, PIT), which could also prove useful for fish in riverine systems in Brazil.

CONCLUSIONS

Radio telemetry proved to be a suitable technique to investigate fish movements and behavior in large rivers of Brazil, despite the relatively low number of tagged fish and the small number of rivers studied. The diversity of species and environments makes it difficult to establish a standard protocol for capture, tagging and tracking fish, but the increased variety of tags and the advances in the tracking systems allowed researchers to test and define the best approach to be applied in each case. Based on experience so far, it is suggested that scientists should always incorporate a detailed planning phase, including visit to study area and checking logistic conditions (access and safety); tests for the right choice of equipment and frequencies (radio telemetry); training of all staff involved on fish tagging (especially important considering there are few Brazilian researchers with experience in it) and data analysis. As pressure to develop additional hydropower facilities in Brazil, telemetry will be even more important for providing important information on the migratory biology of riverine fish to inform management and conservation actions.

ACKNOWLEDGMENTS

I am very grateful for all Brazilian and foreign researchers and their teams who are helping to develop fisheries radio telemetry in Brazil and collaborating for the conservation of our threatened migratory fish species. I also extend my thanks to Steven Cooke for helping with language reviewing and to Carlos B. M. Alves for helping with data collection. Comments from Steven Cooke, Carlos B. M. Alves, and two anonymous reviewers improved this manuscript.

REFERENCES

Adams, N. S., D. W. Rondorf, S. D. Evans, and J. E. Kelly. 1998. Effects of surgically and gastrically implanted radio transmitters on growth and feeding behavior of juvenile Chinook salmon. Transactions of American Fisheries Society 127:128–136.

Alves, C. B. M., L. G. M. da Silva, and A. Godinho. 2007. Radiotelemetry of a female jaú, *Zungaro jahu* (Ihering, 1898) (Siluriformes: Pimelodidade), passed upstream a Funil Dam, rio Grande, Brazil. Neotropical Ichthyology 5(2): 229–232.

Baras, E. 1991. A bibliography on underwater telemetry, 1956–1990. Canadian Technical Report of Fisheries and Aquatic Sciences 1819:1–55.

Cooke, S. J., G. N. Wagner, R. S., R. S. Brown, and K. A. Deters. 2011. Training considerations for the intracoelomic implantation of electronic tags in fish with a summary of common surgical errors. Reviews in Fish Biology and Fisheries 21:11–24.

Crossa M., W. Rocha, and E.P. Sá. 2003. Investigación Participativa: Una experiencia promisora para el subsidio de programas de manejo del pirarucu (*Arapaima gigas* Cuvier) en el Bajo Amazonas. Taller Internacional de Manejo de paiche o pirarucu. WWF/IIAP Iquitos, Perú. Available: http://www.iiap.org.pe/Publicaciones/CD/documentos/L021.pdf (January 2012).

Godinho, A. L., and B. Kynard. 2006. Migration and Spawning of Radio-Tagged Zulega *Prochilodus argenteus* in a Dammed Brazilian River. Transactions of the American Fisheries Society 135:811–824.

Godinho, A. L., B. Kynard, and H. Godinho. 2007. Migration and Spawning of female surubim (*Pseudoplatystoma corruscans*, Pimelodidae) in the São Francisco River, Brazil. Environmental Biology of Fishes 80:421–433.

Hahn, L., K. K. English, J. Carosfeld, L. G. M. da Silva, J. D. Latini, A. A. Agostinho, and D. R. Fernandez. 2007. Preliminary study on the application of radio-telemetry techniques to evaluate movements of fish in the Lateral Canal at Itaipu Dam, Brazil Neotropical Ichthyology 5(2):103–108.

Hahn, L., A. A. Agostinho, K. K. English, J. Carosfeld, L. F. da Câmara, and S. J. Cooke. 2011. Use of radiotelemetry to track threatened dorados *Salminus brasiliensis* in the upper Uruguay River, Brazil. Endangered Species Research 15:103–114

Hocutt, C. H., S. E. Seibold, and R. V. Jesien. 1994. Potential use of biotelemetry in tropical continental waters. Revue d'hydrobiologie tropicale 27:77–95.

Lucas, M. C., and E. Baras. 2000. Methods for studying spatial behaviour of freshwater fishes in the natural environment. Fish and Fisheries 1:283–316.

Pesoa, N. A., and U. H. Schulz. 2010. Diel and seasonal movements of grumatã *Prochilodus lineatus* (Valenciennes 1836) (Characiformes: Prochilodontidae) in the Sinos River, Southern Brazil. Brazilian Journal of Biology 70(4s):1169–1177.

Pompeu, P. S. 2005. Estudo da regra operativa e avaliação de um mecanismo de transposição de peixes do tipo elevador com caminhão-tanque. Doctoral dissertation. University of Minas Gerais, Brazil.

Pompeu, P. S., and C. B. Martinez. 2006. Variações temporais na passagem de peixes pelo elevador da Usina Hidrelétrica de Santa Clara, rio Mucuri, leste brasileiro. Brazilian Journal of Zoology 23(2):340–349.

Ross, L. G., and B. Ross. 2008. Anaesthetic and sedative techniques for aquatic animals. 3rd edition. Blackwell, Oxford, U.K.

Schulz, U. H. 2003. Effects of surgically implanted dummy transmitters on the South American catfish jundiá (*Rhamdia quelen*). Brazilian Journal of Biology 63(2):345–348.

Silva, L. G. M. 2004. Deslocamentos de mandis-amarelos Valenciennes, 1836 em dois reservatórios do rio Grande, bacia do Alto Paraná, conectados por escada de peixes. Master's thesis. Pontifícia University of Minas Gerais, Brazil.

Stasko, A. B., and D. G. Pincock. 1977. Review of underwater biotelemetry, with emphasis on ultrasonic techniques. Journal of the Fisheries Research Board of Canada 34:1261–1285.

Thorstad, E. B., C. J. Hay, T. F. Næsje, and F. Okland. 2001. Movements and habitat utilization of three cichlid species in the Zambezi River, Namibia. Ecology of Freshwater Fish 10:238–246.

Thorsteinsson, V. 2002. Tagging Methods for Stock Assessment and Research in Fisheries. Report of Concerted Action FAIR CT. 96.1394 (CATAG). Reykjavik. Marine Research Institute Technical Report (79).

Section 6.4

Tracking Aquatic Animals in Lakes and Analyzing the Telemetry Data

JIM WINTER

INTRODUCTION

Lakes are incredible natural resources. They provide us with water for drinking, agriculture, industry, transportation, and recreation such as fishing, swimming, and boating. Lakes also provide great opportunities for scientific research, because of their diversity and complexity, and because human activities use, deplete, contaminate, and alter aquatic resources. They provide challenges and opportunities for research due to their great range in size, depth, temperature regime, salinity, water quality, ice cover, and vegetation. Telemetry is a tool that can be used to study or solve problems concerning the biology, habitat requirements, ecological interactions, and behavior of fish and other similar-sized aquatic organisms in lakes.

This Section covers radio telemetry techniques in lakes and methods to analyze the data collected. I will begin with tips on monitoring fish in lakes and will present case studies from several projects. These are projects by my colleagues, my graduate students, and myself. I hope the examples of the problems and solutions will help the reader plan studies and avoid pitfalls. Last, I will give examples of the types of telemetry studies that have been conducted in lakes and methods to analyze the data.

TIPS ON MONITORING FISH IN LAKES

Researchers locate radio-tagged fish in lakes by boat, vehicle, foot, and airplane. The equipment and tracking methods used are described by Eiler in Section 5.3; Amlaner and Macdonald 1980; Winter 1983, 1996; Kenward 1987, 2001; Samuel and Fuller 1994; Rogers 2001; and Fuller et al. 2005. Tracking by boat is the most widely used method of locating fish in lakes. Radio signals are detected from a distance by Yagi antennas on telescoping masts and pinpointed by small ("snooper") loop antennas at close range. Most fish are located by "homing-in" on the signal (Section 5.3; Winter 1996). A common error of trackers is to not get a sharp bearing before homing-in on the signal. This often is caused by assuming that the fish is in the same spot as previously determined. Because the trackers do not take an accurate path to the transmitter, they by-pass their target or even go in the opposite direction; thus, much time is lost. Other problems include the boat causing the fish to move or the boat not being maintained above the fish due to wind. Once the signal is pinpointed, the location can be recorded quickly by GPS. Also, you could use a sextant to measure the horizontal angles among three prominent landmarks that can be found on an aerial photograph or map. The sextant is slower than GPS and affected more by waves, weather, and boat drift. However, it can provide more accurate locations in small lakes that have prominent landmarks, accurate maps, and problems with GPS reception. For night tracking, you can place flashing road-construction lights at known landmarks and take sextant readings on the lights.

Investigators also can locate transmittered-fish by aligning the bearing to the fish with a landmark that is recognizable on a map. This method is useful in areas where there are many obstructions (docks, stumps, rocks, macrophytes) or breakers that prevent the boat from getting close to the fish. In heavily fished lakes, you may not want to home in on a fish, because anglers may see and target the location. Aligning fish with landmarks involves triangulation. You travel parallel to shore and obtain a bearing as perpendicular as possible to a landmark. The bearing can be made visually or with a compass through the transmitter location to the landmark. Then, you travel towards shore parallel to your first bearing, get a visual or compass bearing to the fish, and estimate the distance that the fish is from shore. Accuracy varies from <5–20 m on lakes with good landmarks and an accurate map. This method also is valuable for species that make large movements or migrations where accuracy (<20 m) is not as important as knowing the fish's path.

The advent of GPS enables one to triangulate from open water locations to a fish's location. The GPS values for the origin of each bearing are plotted on the map and the fish location is determined by the intersection of the bearings. Although this method is not as accurate as homing-in, it is quicker, less disruptive of behavior, and useful in large lakes.

Researchers also radio track fish by vehicle or by foot. Investigators use these methods when they cannot be on the lake because of bad weather, are interested in fish behavior near specific near-shore sites (river mouths), need to search many kilometers of shoreline, or must recover a transmitter near shore. Antennas, usually Yagi, are mounted on telescopic or pop-up masts in a truck (Fuller et al. 2005), or can be raised and rotated by antenna rotors (Winter 1976; Winter et al. 1978). Increasing the antenna height enhances reception range. Loop antennas or high-frequency Yagi antennas can be hand-carried. Land-tracking methods are limited by finding a sufficient number of access sites close to the shoreline. You should take bearings from landmarks that are recognizable on a map or from locations determined by GPS. Additionally, you should try to avoid power lines, large metal objects, rough terrain, and dense vegetation because they can attenuate or deflect signals. A common mistake made when tracking along the shore is to take the first bearings too close to the transmitter. The tracker may get a strong signal over the back end of the antenna and end up going along the shore in the wrong direction. The best approach is to get a good bearing away from shore and perpendicular to the transmitter and shore before moving close.

Researchers use airplanes to track highly mobile fish (e.g., salmon, trout, Salmonidae) in large lakes (e.g., Great Lakes). Equipment and techniques for airplane tracking can be found in (Section 5.3; Winter 1996; Kenward 2001; Fuller et al. 2005). Trackers usually listen for a maximum signal followed by an abrupt decrease or null as the plane passes over the transmitter (Section 5.3). Although a transmitter can be located more finely by flying circles of decreasing circumference over the transmitter and simultaneously reducing the receiver gain, trackers often do not circle because of increased time and cost. The reception range increases with flying altitude, but accuracy decreases. Because signals are not blocked on lakes by surface obstructions, altitudes of 300–400 m are good compromises. A plane should fly at an altitude where it can glide back to shore if the engine stalls; however, this may not be optimal for tracking. A good strategy for very large lakes is to use a floatplane, which provides extra safety, a lower altitude, and a slower speed to find signals.

Fish can be tracked in lakes manually from fixed-monitoring stations. Investigators often use Yagi antennas, because they have excellent directionalities, sensitivities, and ranges. Towers can be located around small lakes or bays, and the direction to the transmitter is determined by triangulation (Ross and Winter 1981; Winter 1996). Bearings cannot be plotted accurately on a map unless the locations of the towers are mapped accurately. The towers should be located so that fish cannot be near the imaginary line between two towers (low accuracy) but are likely to be near the imaginary perpendicular bisector (high accuracy) between towers (Heezen and Tester 1967) (Figure 1). The highest accuracy for triangulation with fixed or mobile systems is when bearings intersect at 90°. In some cases, fish can be located more accurately if bearings are taken from more than two towers and error polygons determined (Fuller et al. 2005). The accuracy of bearings with three-element Yagi antennas is ± 5° and increases with more elements (Fuller et al. 2005). Because fish movement can cause inaccurate triangulation, you should listen for fish activity as indicated by wavering signal strength or changing bearings. You should record notes on the quality of the signal (e.g., animal activity, signal strength) and interpret and eliminate unusual fixes later. If there is only one observer, then the time spent moving between multiple towers increases the likelihood that the fish will

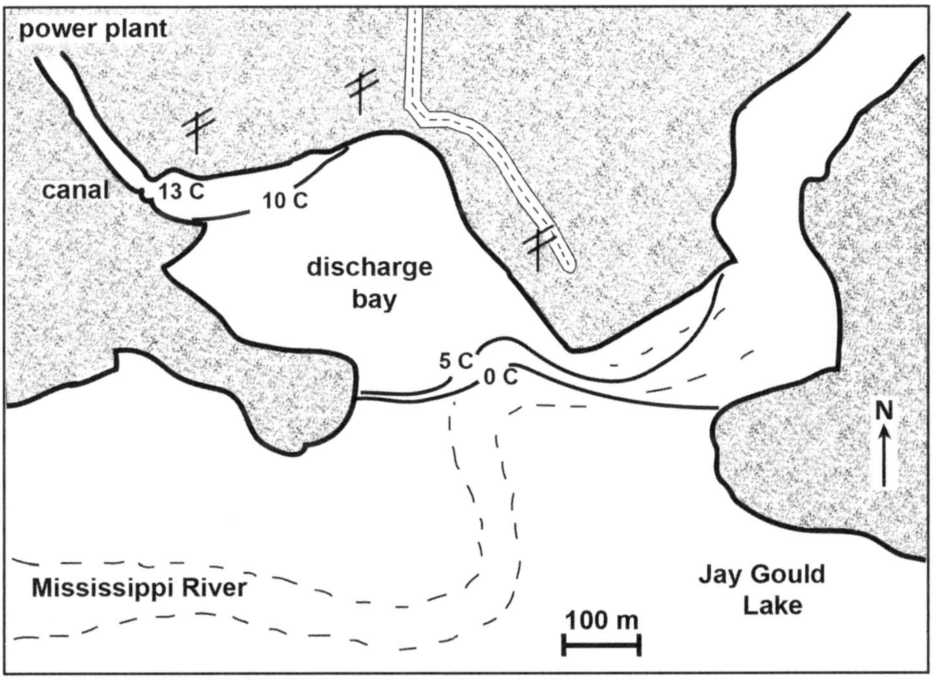

Figure 1. The positioning of tracking towers around a bay in northern Minnesota receiving a thermal discharge from a power plant (adapted from Ross and Winter 1981). The towers were located so that fish would not be on the imaginary line between two towers (low accuracy) and were likely to be between two towers (high accuracy). The north towers were positioned so the directional ends of the Yagi antennas always faced away from the power plant and received less electrical interference. The east antenna was constrained by houses, a road, and power lines. The 0°C isotherm delimits the ice covered areas.

move and its location will be inaccurate. Accuracy can be increased if two observers communicate and take bearings simultaneously from separate locations. You also should place a test transmitter at known locations and test your system regularly, determine the errors in your bearings, and calculate correction factors. Under good conditions, locations can be replicated precisely with telemetry systems; however, you should train and test each tracker so the crew has uniform precision. You should maintain the system regularly, because the antenna elements can become bent and misaligned, especially with the indicator needle for the compass rose. Inspect the coaxial cable because repeated use and stress can loosen connectors or break the wire inside. Carry spares of most components.

Methods of Searching in Lakes

The method of searching for fish equipped with transmitters is important because of cost, labor, and bias in data. Many researchers begin their searches in the place where they last located the fish because it saves time in searching. This method works well in small lakes and for species that make small movements, because you are likely to find the tagged fish all the time. However, this method often fails in large lakes with fish that are very mobile, make excursions to deep water, or have unpredictable movements (e.g., walleyes *Sander vitreus*), because they moved to a place where you did not check for them. Thus, you can have significant bias in studies on fish habitat use, if you do not find your fish regularly and sample all habitats.

The least biased method of searching for fish is to use transects. The types of transects vary with the size and depth of the lake, range of the transmitters, and the mobility of the species. The simplest and most common transect is to run parallel to the shore by boat, airplane, foot, or vehicle. Because many fish species move along the shoreline or particular depth contours, this method can be effective. If the bottom drops off slowly and you are tracking a demersal or wide-ranging species, then you should make another parallel transect farther from shore. Because outboard motors cause ambient and electrical noise, you may need to stop and listen for faint signals. The distance between monitoring points and transects should be equal to or less than the transmitter signal range to maximize transmitter detection. If the lake is broad and shallow or the species is pelagic, then you should make a series of parallel transects crisscrossing the lake.

Transects also are used with airplanes, vehicles, and by foot. Tracking by airplane is usually done parallel to the shoreline, and the signal monitoring is continuous. Tracking by vehicle along the shoreline is useful when the weather prevents a person from being on the lake in a boat or when fish are migrating near shore. It differs from boat tracking in that the monitoring points are not uniformly spaced, because access to the lake is often restricted by private property or lack of roads. Tracking by foot is useful for species with small home ranges and movements near shore. Accuracy is high if you can locate your position accurately. It also depends on access points.

Sampling Time

The time of day when you monitor a fish is important (Winter 1996). If you sample the same time every day, then you will get biased data on fish habitat preferences, activity periods, and behavior, especially of crepuscular or nocturnal species. The least biased sampling method with respect to time is to divide the 24-h day into equal periods and set up a tracking schedule where the time periods are chosen randomly. However, it is nearly impossible to maintain a random schedule from day-to-day for a long time, because of researchers' other commitments, but you might be able to sample one or two periods randomly during the week. When random tracking periods are impractical, one could systematically divide the day into 6–8 h periods, and sample each period once a week. Thus, you will sample the same time periods equally during the week and meet other commitments. Another way to spread sampling time is to start your search or transect at a different location, time, or direction each time you track. For example, you could travel clockwise around the lake on one day, and counter clockwise the next, making it unlikely to find an individual fish at exactly the same time each day.

Researchers should confirm whether their usual tracking period is representative of the species' typical activity or covers a key activity. A good method is to periodically track a few individuals over 24-h periods and discover when certain activities occur. If you know when certain behaviors likely occur, then you might proportionally allocate more monitoring time to a specific behavior (e.g., spawning) or period (e.g., feeding). This is thorough and cost-effective.

The sampling interval (i.e., time between relocations of an individual) is a potential bias. If you take locations too frequently, the fixes are not independent, because the animal has not had sufficient time to make a decision and move. The data are said to be autocorrelated and can bias estimations of home range size and habitat selection. It is tempting to take many locations on an individual in a short time period, because it is an efficient use of time and produces numerous data points. However, the sample is artificial, inflated, and biased. Kenward (2001) and Cresswell and Smith (1992) discuss methods of testing data for autocorrelation and determining a good sampling interval.

Map Considerations

Most studies in lakes require a good map to plot telemetry locations and to locate and evaluate receiving systems. Although computers process and analyze most data, it is useful to visualize the data on a map and to verify that locations are correct (Larson 2001). One should start with a good map of the lake. The most accurate depiction of the shoreline is a recent, high-resolution aerial photograph. USGS topographic maps, navigational maps, and bathymetric maps can be too old and made from lower-resolution photographs, under different water levels, and during a different season than your study. They might be fine for studies on migrations but inadequate for studies on home range and habitat use. Photos can be obtained from satellites and have resolutions that are sufficient for animals that make moderate to large movements (O'Neil et al. 2005).

A map coordinate system should be chosen so that X-Y coordinates can be assigned to each animal location and data plotted (White and Garrott 1990). Furthermore, a coordinate system will allow you to easily calculate the distances between fish locations, between fish locations and important sites, or to determine home range size (Larson 2001). The latitude and longitude system is advantageous for animals that make very long migrations, because it is continuous around the globe. The Universal Transverse Mercator (UTM) system is preferable for studies in small areas like lakes (White and Garrott 1990). It is a worldwide rectangular coordinate system that is not continuous globally, because of the curvature of the earth's surface. However, the earth is divided into 60 zones and the system is continuous within each zone. Your lake likely will fall within one zone. If not, the boundary of a zone can be extended (White and Garrott 1990). The UTM coordinates are in meters and can be easily superimposed on USGS maps by aligning your UTM grid with tick marks on the map. Distance calculations and home range analysis are easy with this system. You can record fish locations with GPS in either system or convert between systems.

You can easily make your own habitat maps. By starting with a good outline of the lake, you can make a bathymetric map of a small lake by running transects, taking a series of depth readings with an echo locator, and mapping each location with GPS. It is more difficult to make a substrate map, because you cannot see much of the substrate, and sampling and accurately defining the substrate grain size is laborious. However, an adequate substrate map can be made for general habitat use studies by using broad substrate categories (e.g., rock, silt, wood, etc., Weller and Winter 2001) or by taking a moderate number of samples and noting the relationship between depth and substrate. The locations where the substrate changes can be mapped with GPS. Paukert et al. (2002) describe a method and the precision of using visual estimates of vegetation and substrate to produce habitat maps in lakes. If you have the luxury of doing quantitative vegetation sampling, you can make a detailed vegetation map. There are methodologies being developed to measure substrate and vegetation density by depth sounders and single beam sonar (Dick et al. 2009). Furthermore, satellite photographs can measure the extent of macrophyte beds. Other types of habitat maps may be available such as the extent of plumes from power plants. Once you have a habitat map, its boundaries can be imported into GIS systems (O'Neil et al. 2005; Dick et al. 2009). The GIS maps can be overlaid on maps, aerial photographs of the lake, and fish telemetry locations (Zamora and Moreno-Amich 2002).

CASE STUDIES IN LAKES

Several case studies are presented that will illustrate how the researchers approached technical challenges and used telemetry to answer biological questions in lakes. I will divide the case studies by size of lakes. The tracking methods and numbers of fish that you can track are related to the size of the lake and the time to travel around the lake. Small lakes are ones that you can easily travel around the shoreline by foot or boat in a few hours. A medium size lake is one that you can cruise around the shoreline in a boat in a half a day to a day. A big lake is one that will take more than a day to cruise around.

Cases Studies in Small Lakes

Small lakes provide convenient laboratories to study biological processes and environmental problems with fish. You can answer experimental questions that require knowing the animal's location almost all the time such as activity patterns, temperature selection, depth selection, vertical migrations, and physiological parameters. In small lakes, you should be able to monitor the subjects at any time. The small size of the lake also helps offset physical challenges to radio telemetry such as water depth and conductivity. Small lakes range from relatively isolated to busy urban areas. They often do not have economically important fisheries.

In 1968, the Cedar Creek Bioelectronics Laboratory at the University of Minnesota developed a radio transmitter for fish and tested it in Cedar Creek. But, information was needed on the water resistance of the potting techniques and compound (Scotchcast epoxy 3M©), the transmitter range and life, the effect of depth on signals, the ability of fish to carry the tags, etc. The next step was to test the tracking system in lakes. I used several transmitters, a receiver, and loop antenna to track largemouth bass *Micropterus salmoides* in a small lake (23 ha) in northern Minnesota. This was a great laboratory because it had small size, an abundant bass population, little fishing pressure, easy access, and a variety of aquatic vegetation, substrates, and depths. Bass were caught with trap nets and angling. The transmitters were attached externally, and the fish were released where they were caught, except for one fish. This fish was transplanted from another lake to compare its behavior to the nontransplanted fish. I could not home in on the fish, because they were spooked from the noise of my canoe going through the aquatic vegetation or the sight of the canoe in the shallow, relatively clear water. Therefore, I established a grid system around each fish's area by placing flagging every 6 m along shore and floats 30 m and 61 m from shore. This simple system allowed me to take compass bearings from known locations on shore at any time or to align bearings from the water with known points on the grid system. The accuracy of the locations was within 2–4 m from distances of 75 m or less. Data from the small grids were converted to a X-Y coordinate system that was overlaid on a map of the lake. Home ranges of indigenous bass were small (<1.4 ha) and less than 3-m deep, whereas a transplanted bass roamed (Winter 1976, 1977).

Another example from a small lake concerns the effects of a thermal discharge from a power plant on fish behavior. Yellow perch *Perca flavescens* need a winter chill period of 4°C for 185 d for optimum egg viability (Jones et al. 1977). Egg viability decreases when the winter temperature is above 4°C or when the exposure to 4°C is less than 185 d. But, perch under laboratory conditions select winter water temperatures much warmer than 4°C. The U.S. Environmental Protection Agency (EPA) wanted to know whether perch in natural situations would choose warmer temperatures in the winter because of greater food availability and opportunities for growth, or would they choose temperatures near the optimum necessary for maximum egg viability.

An ideal site to test the effects of a warm thermal environment in the winter on fish behavior was the Clay Boswell power plant in northern Minnesota (Ross and Winter 1981) (Figure 1). The plant discharged 6.0–18.0°C water into a small bay (12.9 ha) of Jay Gould Lake. The bay is thermally isolated from the rest of the lake by the Mississippi River, which flows through the northwest section of the lake. The river upstream from the bay and the lake are ice-covered during the winter, whereas the bay and river downstream from the bay are ice-free. Thus, fish have a wide range of thermal habitats to select.

We chose low frequency radio signals (53 MHz), because we were uncertain about the effects of depth and conductivity. We placed three 7-m towers with 4-element Yagi antennas on two sides of the bay where access was possible. The towers were placed back from the shoreline so that the fish could not be on the baseline between the towers, and most likely would be near the perpendicular bisectors between at least two of the towers. This is the most accurate arrangement for triangulation (Heezen and Tester 1967). Each tower had a compass rose and a needle aligned with the antenna so that we could read the bearings to the fish on the compass rose. We also positioned two of the antennas so that the nondirectional ends of the Yagi always pointed towards the power plant. This minimized the heavy electrical noise from the power plant, and it is a good strategy to use in other systems with potential inference from power lines, large metal structures, and terrain. Houses and a road determined the location of the third tower. Furthermore, we planted test transmitters at known locations and took bearings regularly to verify system accuracy. Thus, we determined the correction factor to the test transmitter each tracking day, and added or subtracted that value from the bearing to transmittered-fish. This is an important step that is overlooked in telemetry studies, because the accuracy of a tracking system can change daily due to weather conditions, ambient interference, the physical receiving system, and other factors. With four-element Yagi antennas and correction factors, our bearing errors were within ±3°, or <5 m at 100 m distance and 43 m at 800 m (Winter et al. 1978). Ross (1978) suggested that a general rule from our tests and others' research is that location errors are less than 10% of the distance from the receiving antenna to the transmitters.

We took some fixes simultaneously by two observers and most with a single observer. Simultaneous fixes minimized errors due to fish movement. Because a single observer had to move 200 m between towers, the time lag increased the likelihood that a fish could move between bearings. Therefore, we also recorded information on the signal strength and fish activity as indicated by changing bearings and signal wavering. This allowed us to evaluate the quality of fixes and justifiably reject unusual fixes. The tower system also allowed us to track fish any time of the day and avoid thin ice. We found that four species of fish differed in their home range sizes, bottom depths, and distances from the thermal discharge (Ross and Winter 1981). Yellow perch preferred the areas closer to the river and moved between the open, warm water and the cold, ice-covered areas.

Other telemetry innovations were added to the Clay Boswell study. Ross and Siniff (1982) added temperature-sensing transmitters and an automatic recording system (Ross 1978) to monitor the thermal exposure or temperature selection by yellow perch. The recording system was connected to a tower and Yagi antenna. One scanning receiver monitored each fish for four minutes every 64 min, and another was set to monitor one fish continuously for 24 h. A decoder measured the pulse rate of the temperature transmitters and translated the pulse rate into milliampere current which was recorded on a strip chart recorder (Rustrack©). The system at Clay Boswell was powered by a 12-V automobile battery and was housed in a small, insulated, propane-heated shack. This system is crude by today's standards, but it worked. Today, you can readily buy scanning receivers with built-in signal decoders and data loggers. The biological conclusion is that yellow perch selected relatively cool winter temperatures of 5.4°C for all locations and 6.3°C while in the discharge affected areas (Ross and Siniff 1982). Thus, their thermal exposure was close to the optimum necessary for maximum egg viability.

Other telemetry studies have used small lakes as convenient natural laboratories. Small quarry lakes were used to investigate the thermal niche of striped bass *Morone saxatilis* (Coutant and Carroll 1980). Savitz et al. (1983) manipulated the food supply in feeders in a small lake and demonstrated that home range size in largemouth bass was inversely related to food supply. Mackenzie-Grieve and Post (2006) compared the thermal habitat use of lake trout *Salvelinus namayacush* in two lakes that differed greatly in depth, available temperatures, and food supplies. Dick et al. (2009) used fixed-recording stations to compare the use of substrates and zones within a small lake by different size classes of Arctic char *S. alpinus*. Fixed-recording stations in small lakes have been used to determine the swimming speed and activity patterns of yellow perch (Zamora and Moreno-Amich 2002) and the three-dimensional activity of largemouth bass (Hanson et al. 2007). These studies demonstrate the potential of small lakes in telemetry studies.

Take-home messages from small lakes.—There are several take-home messages from these case studies on small lakes. First, small lakes make simple laboratories where the investigator can be in contact with the transmittered-fish almost continuously and answer many important questions on fish behavior and ecology. Second, studies on small lakes can be done with simple equipment and done relatively inexpensively. Graduate students, managers, and beginning researchers can be intimidated by telemetry projects, because of the elaborate, expensive equipment available and the large numbers of fish or sophisticated data-collecting techniques used in studies on economically important and endangered species. Fortunately, there are many opportunities available for small studies with well-designed research questions and simple telemetry equipment. Third, a simple flagging and float system can be used to locate fish with localized movements often more accurately than sextants and GPS. Fourth, you can possibly locate over 50 individuals per day by boat in a small lake because the search time is low. More could be tracked by using fixed stations and coded tags. Fifth, towers and fixed-recording systems on small lakes can answer a variety of interesting biological questions. Sixth, it can be difficult to find suitable physical sites for fixed radio-recording stations, because of vegetation, power lines, rough terrain, and human development. Because the stations are conspicuous, they should be located on private property where they will be safe from vandalism.

Case Study in a Medium-Size Lake

Medium-sized lakes are common and support much sport fishing. They can be rather homogeneous to incredibly diverse, have few to many fish species, and have one to many sport fish. Telemetry studies are often conducted on medium-sized lakes, because there is an economically important species that requires management information.

Chautauqua Lake, New York is a good example of a medium-sized lake with a diverse fish community (Einhouse 1981). Chautauqua Lake is 27 km long, about 2 km wide, and 5300 ha in area. It has a shallow (mean depth = 3.5 m), eutrophic southern basin separated by a narrows from a deep (mean = 7.8 m), mesotrophic northern basin. The conductivity of the water is about 200 μS/cm. The lake supports a rich fish fauna and excellent sport fisheries. Most of the shoreline has cottages and small communities, and there is heavy recreational use from boating, sailing, fishing, and swimming. There are abundant macrophytes that are controlled by mechanical harvesters and annual herbicide spraying.

Chautauqua Lake was an interesting place to study walleye movements and presented challenges to telemetry, because of its size, conductivity, productivity, diversity of depths and substrates, and human impacts. Our objectives were to determine the walleye's home range size, habitat use, activity patterns, schooling/aggregating behavior, spawning migrations and sites, and fidelity to spawning sites or summer areas. We implanted radio transmitters into the walleyes (Einhouse 1981; Winter 1996). We chose low frequency (53 MHz) signals because of the moderately high conductivity and depths. We tracked walleye with Yagi antennas mounted on a pontoon boat, a fast 5-m long boat, and a snowmobile. Fish were located mostly by homing-in, and the locations were recorded with sextant readings on excellent landmarks. We also located walleye by aligning their positions with landmarks that were recognizable on a map and taking bearings. This method was necessary because of the enormous numbers of docks on Chautauqua Lake, dense macrophyte beds, and numerous anglers.

Little tracking was done from vehicles because of insufficient road access due to the dense housing development. We did not use towers, because the walleye did not have localized movements, access sites for towers were limited, and the towers could not be protected from vandals. Furthermore, tracking from an airplane was not accurate enough and too expensive.

Tracking a mobile species like walleye in a moderate-size lake is very time consuming and potentially expensive. You can use the amount of time that it took us to track walleye in Chautauqua Lake as a useful planning tool for projects on similar-size lakes with a mobile species. It took 8–12 h per day to travel around the shoreline and stop, detect, and home in on signals. It can take one half hour to detect, home in, and record a position, and longer if limnological measurements are taken, too. Tracking efforts are complicated further, because walleye have large home ranges and make long movements. Consequently, the maximum number of mobile fish that can be located in a day is about 24–36 in a medium-size lake. More fish can be tracked if a species (e.g., largemouth bass) shows strong site fidelity or if efforts are alternated daily between one area or basin. Tracking from a boat also requires much labor and hence cost—someone to take bearings on the signal and someone to drive the boat towards the signal. You also should have more than one person from a safety standpoint. Many projects today are using fixed-recording stations, especially submerged ultrasonic systems, and choosing many, less-detailed locations on large numbers of fish instead of detailed locations and habitat use of small numbers of fish. You can buy several fixed-recording units for the cost of an additional tracker.

Take-Home Messages from Medium-Sized Lakes.—There are several take-home messages on medium-size lakes. First, you probably cannot track as many fish as you would like, because of the time that it takes to search the lake. Second, it takes more time to home in on a fish than you think because of moving fish and interference by boat motors and boat wakes. You can decrease the time to locate a fish by using GPS instead of sextants or compasses. Third, species with small home ranges (largemouth bass) will be easy to find in a medium-size lake. However, mobile species like walleye can be elusive in a medium-size lake because they can move long distances in a short period of time (>0.75 km/h) and they can occupy depths that significantly decrease the signal range. This coupled with moderately high conductivity means that you need a good, systematic searching technique.

Case Studies in Large Lakes

Large lakes present the greatest challenges to telemetry studies. Their great size makes it difficult and expensive to monitor the entire shoreline or lake. Tracking from a boat may not cover sufficient area within the available sampling period. Large lakes also can have frequent bad weather that prevents tracking from boats. Furthermore, they typically have deep areas that can inhibit radio signals. On the other hand, large lakes often have economically important fisheries and interesting biological problems to study. I will discuss tracking salmon (*Oncorhynchus* spp.) and trout (*Salmo* spp.) in the Great Lakes as an example of radio telemetry in large lakes and a difficult challenge.

Stocking of salmon and trout in the Great Lakes has created valuable sport fisheries. Initially, many anglers did not know enough about the behavior of the fish to catch them. Managers still do not know enough about where their stocked fish travel and who catches them. Several telemetry studies have been conducted on salmon and trout in the Great Lakes (Winter 1976; Wenger et al. 1984, 1985; Haynes et al. 1986; Haynes and Gerber 1989; Nettles et al. 1987). These investigators chose radio telemetry because of the long distances that salmon and trout can travel (easily 16 km per day) and because the water conductivity (<100 μS/cm to 300 μS/cm) was not high. During calm weather, investigators used boats with 53-MHz Yagi antennas mounted on telescoping masts. Although a 53-MHz Yagi antenna is bulky, we chose that frequency to alleviate problems of signal attenuation with depth. Trackers could search at least 40 km along the shore. Usually, they stopped the boat every kilometer where fish were likely, turned off the engine, and listened for signals. If trackers knew that a fish was in the area, they often listened for signals and operated the boat at a slow speed. Tracking by boat was moderately successful when the fish were close to shore during spawning migrations and postspawning dispersal. Trackers often located fish by aligning the signal with a landmark and making a parallel transect to estimate the fish's distance from shore. The accuracy of this method was often within 25 m and sufficient to document the migration route of highly mobile species. The main reasons that trackers did not home in on the fish were because it was time consuming, the fish often traveled very close to shore and could be spooked, and boats were difficult to handle near rocks and breakers.

Bad weather was a huge obstacle to tracking in the Great Lakes and reduced boat use to a few days a week. If investigators missed a few days of tracking, transmittered-salmon and trout were difficult to find, because they moved far from their last location (>48 km) and even reversed direction (Haynes and Gerber 1989). Alternative tracking methods were used.

Trucks with Yagi antennas on telescoping masts were useful for tracking. In some areas, roads followed the shoreline closely and there were good places to pull off. In general, good access was patchy. Triangulation depended on finding sites close together and recognizable on the map or easily mapped. Bearings were made by compass or visually aligning with a landmark. Investigators covered long distances (>75 km) with a truck and maintained contact with some rapidly-migrating fish. Trucks also were useful for monitoring stream mouths for fish. During the spawning season, salmon and trout often stopped near stream mouths as part of their pre- and postspawning behavior (Winter 1976; Wenger et al. 1984).

Tracking fish by airplane was a valuable search method on the Great Lakes. The equipment is described in Section 5.3, Winter 1996, Kenward 2001, Fuller et al. 2005. An airplane easily covered over 175 km of shoreline per day and found some missing fish. In general, a slow floatplane was safest and best. Investigators needed tightly-fitting headsets to block airplane noise and to detect faint signals. A scanning receiver made searching for frequencies easier, but using coded tags on fewer frequencies would have increased listening time per frequency. Researchers usually flew transects parallel to shore and within 3 km of shore. Fish locations were determined by listening for the maximum signal and a sharp drop in the signal passing over the fish (Winter 1996; Section 5.3). This location method is time efficient and less expensive. Most locations were made visually with respect to landmarks, but GPS would be optimal. The accuracy of aerial locations are ±100–200 m (Fuller et al. 2005) and is sufficient to determine the route of highly mobile or migrating species, because of their large movements between each radio location. At river mouths, we sometimes flew in circles to locate the transmitter closer (<100 m). This approach was time consuming.

Steelhead trout *O. mykiss* and salmon moved offshore in the Great Lakes in the spring possibly as the nearshore waters warm above 10°C and thermal fronts moved offshore, especially the 10°C isotherm (Haynes et al. 1986). Trackers often lost contact with transmittered-fish due to attenuation of radio signals with increased depths and to the vast amount of space to search. The signal range can decrease 50% in <5 m depth and 90% in <10 m depth (Winter 1996). Summer movements test the limits of telemetry with respect to signal range and cost of big vessels and crew time. Therefore, we supplemented telemetry data with other types of sampling. Olson et al. (1988) used vertical gill nets to determine distribution of salmon and trout from shore, their depth distribution, and their preys' depth distribution. Salmon and trout were most abundant <4 km offshore and partitioned their resources around the thermocline. Chinook salmon *O. tshawytscha* were above the thermocline, brown trout *S. trutta* within the thermocline, and lake trout below the thermocline. The species ate the most abundant prey within their habitat. Few steelhead trout were caught suggesting that they were more widely distributed. Wenger et al. (1985) found that steelhead and brown trout returned to near shore waters when the water temperatures dropped below 20–21°C. The techniques described for large lakes were successful for tracking prespawning movements near shore, stream testing behavior, false runs, and finally spawning runs.

Take-Home Messages from Large Lakes.—The take-home messages from these case studies are that the tracking techniques described can be applied to many species in large lakes. Second, tracking in large lakes and with highly mobile species is difficult and expensive, because it requires a large boat (>7 m), and airplane time (at least once a week). If contact with a migrating fish is lost for several days, the probability of contacting it again is low. Although you are likely to have few locations per individual for most of your subjects, the small number of locations can be sufficient to determine migration routes. The number of individuals tracked concurrently can be and may need to be large (>50 per season) to accumulate sufficient data. Time spent searching is long and determining locations may be short. Little information will be accumulated on fish responses to the habitat or environmental variables. You also could use a series of fixed-radio stations or submerged hydrophone stations along the shoreline to record whether transmittered-fish pass by. The savings from not using airplanes and using less tracking from boats could pay for the stations. However, suitable station sites can be scarce, because of insufficient access, human activity, and wind swept shorelines. To obtain information on the behavior of highly mobile fish and their responses to environmental variables, you could choose one individual or release a fish in open water and follow the individual continuously for as long as weather and logistics permit. A combined acoustic-radio transmitter (CART) could alleviate the depth-attenuation problems, but the large lake area and expensive vessel time will limit tracking in the open water during the summer. A large vessel capable of withstanding sudden storms and a crew is essential. Because telemetry cannot answer all questions, other methods must be used to fill in the gaps in knowledge: e.g., netting, creel census, hydroacoustic, etc.

PURPOSES OF TELEMETRY PROJECTS IN LAKES

Lakes have many environmental problems affecting fish and aquatic organisms and provide many opportunities for researchers to investigate and solve (Priede and Swift 1992; Thorstad et al. 2002). They also can serve as great natural laboratories to answer biological questions. Telemetry can be used to answer many biological questions. The most common types of telemetry studies and data analyses will be discussed: home range, habitat use, response to environmental variables, migration and dispersal, survival and exploitation, population estimation and assumptions, and animal interactions.

Home Range

Investigators, especially wildlife biologists, have studied locations, sizes, shapes, and structure of animal home ranges, and animals' use of space for over half a century. The history and methodology of the concept are documented well (Macdonald et al. 1980; Kenward 1987, 2001; Millspaugh and Marzluff 2001; Kernohan et al. 2001; Fuller et al. 2005). The simplest and most common definition of home range is the area covered by an animal in its normal activities of finding food, mating, and caring for young (Burt 1943). The definition was developed for mammals. It is applicable to fish except that spawning migrations should be excluded because fish are leaving the area that they usually occupy. Furthermore, seeking cover, finding suitable environmental variables, and resting are normal activities in fish, whereas parental care often is not. Although the term "normal activities" is criticized as subjective, it does stimulate researchers to determine where animals spend time and why. Investigators assume that the shape and size of a home range is related to an animal's need for resources, response to environmental variables and terrain, and interactions with other animals (conspecifics, predators, or prey). Most often home range studies are coupled with information on the habitat and resources and used to determine the animal's habitat use or selection (Fuller et al. 2005; McDonald et al. 2005). Gerking (1953) applied the concept of home range to stream fishes and Winter (1976, 1977) applied telemetry and home range techniques to largemouth bass in a small lake. Rogers and White (2007) discuss home range methods with respect to fish.

There are many assumptions with home ranges and analytical methods. First, excursions and migrations are not considered part of the home range. Many approaches have been proposed to exclude these locations. Second, the sizes and locations of home ranges may change with seasons. Therefore, you should separate data and define home ranges by season. Because animals do not use our calendar, I suggest that you use the dates when animals shift their locations or of environmental changes as the beginning or end of the seasonal home range. Weller and Winter (2001) used changes in lake stratification or water temperature to define seasonal categories. Third, many methods for measuring home range size and shape are dependent on the number of locations taken. Therefore, you need to compare home ranges of animals or seasons that have similar numbers of telemetry locations, or plot the cumulative increase in home range sizes versus number of locations (area-observation curve) to determine when home range size is stable, or use analysis methods that are less sensitive to sample size (Odum and Kuenzler 1955; Winter 1977; Fuller et al. 2005). Although Odum and Kuenzler (1955) recommended determining the home range when the observation-area curve changes less than 1%, a change of 5% or 10% is more reasonable (Winter and Ross 1982). Home ranges are dynamic, and animals likely make minor changes in response to food supply or environmental factors.

There are many methods of determining an animal's home range (Macdonald et al. 1980; White and Garrott 1990; Kenward 2001; Fuller et al. 2005; Rogers and White 2007). Most methods are available in software at little or no cost (Larson 2001). Kernohan et al. (2001) classified the methods as polygon, grid-cells, and probabilistic methods. The minimum convex polygon is the simplest and most common method (Mohr 1947) (Figure 2). The extreme outermost animal locations are connected so that the angles open to the inside and form the smallest convex polygon. Only one polygon fits this description so the method is not arbitrary. It does provide some picture of an animal's behavior with respect to the physical habitat and

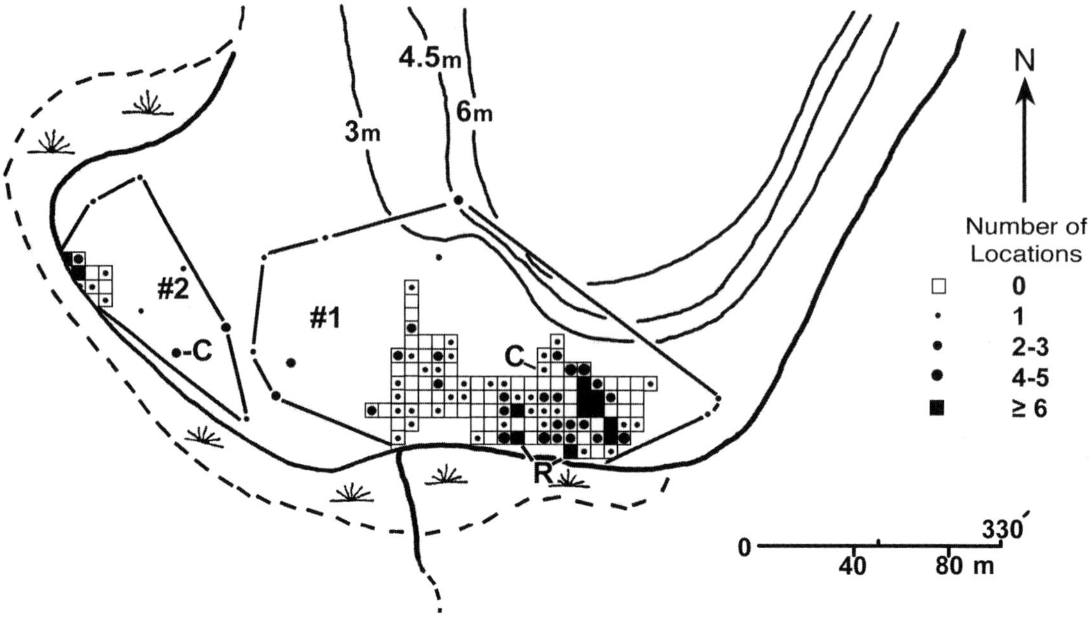

Figure 2. The summer home ranges and intensity of habitat use of two largemouth bass in a Minnesota lake (Winter 1976, 1977). Convex polygons enclose the maximum home range area. The grid-squares represent the primary or core areas and show intensity of habitat use. C indicates capture locations and R shows midday resting areas.

perhaps to other transmitter-tagged animals. However, this method is very sensitive to different numbers of telemetry locations and to excursions and wanderings. It is probably more representative of an animal's maximum or total range. Kernohan et al. (2001) suggest that convex polygons require more than 100 fixes to obtain reliable estimates of home range size; however, few studies achieve this number. Additionally, the polygon can include area that is not habitat, especially cutting across irregularities in the shoreline. You can use the shoreline as one boundary, determine the area that includes land, and subtract it from the home range area (Winter 1977). These biases make it difficult to statistically compare areas of convex polygons.

Methods such as peeled, concaved, and cluster analysis polygons have been developed to remove extreme outlying points, eliminate extraneous areas from convex polygons, and measure core areas (Kenward 2001). Peeled polygons strip away the outliers according to some criteria such as a specified distance from a center of activity. Concave polygons can be drawn according to some criterion, which is often subjective and not applicable to other studies. Cluster analysis polygons are formed by using nearest-neighbor distances between locations to link the locations in clumps (activity cores) (Kenward 2001). The clusters are delimited by convex polygons and the areas of the multiple or merged polygons are summed.

Grid-cell methods are a second category of home range methods (Siniff and Tester 1965; Winter 1977; Voigt and Tinline 1980) (Figure 2). These methods create a grid over the map of the study area with the cell (square) size not smaller than the accuracy of the telemetry location method. The home range is determined by summing the areas of cells containing an animal's locations. Because there are usually empty cells that an animal must have traveled through to get between locations, the investigator usually devises a ("fill") system to include these empty cells in the home range estimate. The criterion for the fill is often a certain distance that an animal can travel in a given time. The shape of the grid-cell home range does not have much biological significance. An advantage of the grid-cell method is that it can be combined with habitat maps to measure intensity of habitat use. It provides an estimate of the animal's core area. The number of locations needed to estimate grid-cell areas is probably less than convex polygons.

Probabilistic and contour approaches are the third category of home range methods (Kernohan et al. 2001). These methods examine the distribution of locations and often produce better statistical comparisons of home range sizes. Although they are more difficult to understand and visualize, they are easily calculated by readily-available computer programs (Larson 2001). Probabilistic methods determine an animal's home range by assuming a certain probability distribution (e.g., bivariate normal) or characterizing the distribution of locations (e.g., harmonic mean and kernel methods) (Kernohan et al. 2001). An advantage of these methods is that you need fewer locations (30–50) to get valid home range estimates. The tolerance to smaller numbers of locations makes these methods useful for comparing home ranges between seasons, sexes, and size classes, or for studies where it is difficult to obtain large data sets (e.g., large lakes and mobile species).

The first probabilistic method, the bivariate-normal distribution, assumes that an animal's activity is concentrated in the center of the home range and that the probability of finding an animal decreases from the center of activity according to a normal distribution (Jenrich and Turner 1969; Fuller et al. 2005) (Figure 3). It is not greatly influenced by the number of locations and can be used to identify the number of independent locations needed to get a specific reliability estimate of home range size. The disadvantages of the bivariate-normal method are that animals often have more than one center of activity, no activity center in the center of the range, and a pattern of locations that is not elliptical (Fuller et al. 2005). Peripheral locations cause the ellipse to extend in the opposite direction to compensate for their effect on the normal distribution (Rogers and White 2007). Furthermore, the ellipse can include nonhabitat, such as irregular shorelines in fish tracking studies.

The harmonic-mean method is widely used. It estimates home range size by using the distances between all the animal's locations and all intersections of a grid lattice (Dixon and Chapman 1980; Fuller et al. 2005). Harmonic means are calculated for each grid intersection (node) and contours (e.g., 95%) enclosing the locations are calculated. The method calculates the home range boundary using the complete utilization distribution of locations and makes few statistical assumptions (i.e., nonparametric) (Kernohan et al. 2001). It provides poor estimates of home ranges for animals with linear use patterns (e.g., fish along narrow depth contours) or traditional travel corridors.

Another category of probabilistic home range methods is the Fourier-transformation method (Anderson 1982) (Figure 4). This method represents the density of locations as three-dimensional stacks on a map grid or plane. The abrupt transitions between peaks are smoothed with a Fourier transformation that decomposes the function into a series of sine and cosine components. The home range size is calculated as the smallest area encompassing a percentage chosen by the investigator (e.g., 95%) of the volume of the utilized distribution or 3-dimensional surface (White and Garrott 1990; Fuller et al. 2005). The advantages of the Fourier method are that it is nonparametric and is not sensitive to grid cell size and using a fill factor for cells.

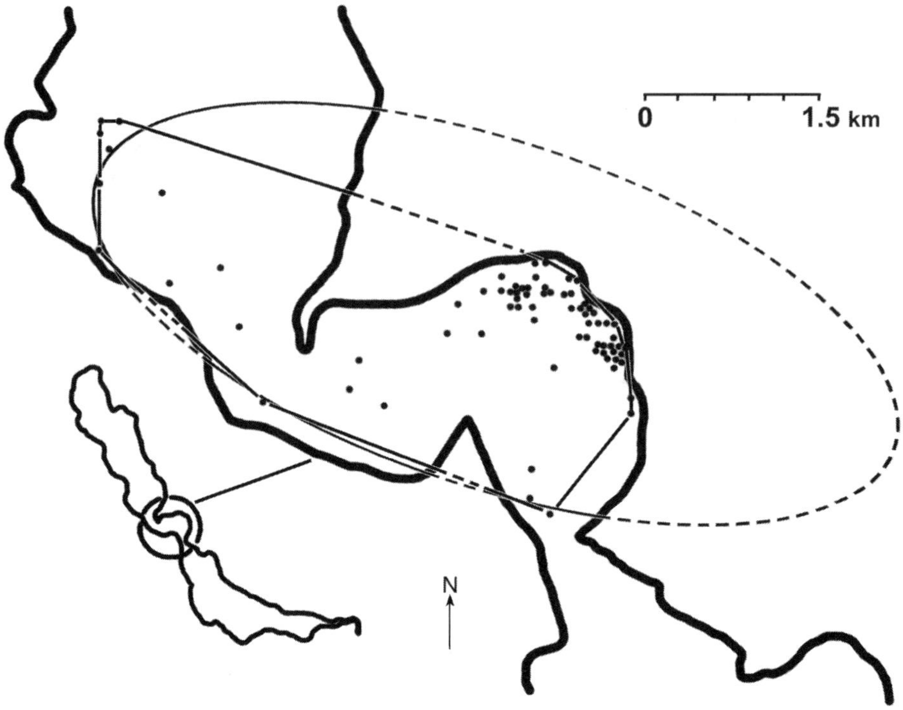

Figure 3. A comparison of the modified convex polygon and the 95% probability ellipse methods for determining the home range of a walleye (#1369) in Chautauqua Lake, New York (adapted from Winter and Ross 1982). Irregularities in the shoreline cause the boundaries of both methods to cross land. The dashed lines show the areas of land that need to be subtracted from the home range areas.

The last group of probabilistic methods is the kernel methods (Figure 4). These are similar to the harmonic mean in that they examine the location density at intersections (nodes) of cells in a grid system and calculate a density function (Kenward 2001). A density function, called a kernel, is calculated for each data point. These methods determine a minimum area that includes a fixed percentage (e.g., 95%) of the density distribution or three-dimensional volume of locations (Fuller et al. 2005). The kernel methods use a smoothing parameter, which is called the bandwidth, that controls the width of the individual kernels. There are two types of kernel methods—fixed and adaptive (Kenward 2001). The fixed-kernel method uses the same bandwidth over the evaluation area whereas the adaptive kernel method uses a local bandwidth for each observation. The cell size and spacing do not affect the distribution. The kernel methods are nonparametric, but assume that locations are independent (White and Garrott 1990). They have a strong statistical basis. However, the choice of bandwidth and smoothing is subjective and a limitation, and kernel methods overestimate home range area for small samples (Seaman and Powell 1996). The fixed kernel usually has less bias and a better fit, but the adaptive kernel may work better when peripheral locations cause uncertainty (Seaman and Powell 1996; Kernohan et al. 2001). The choice of kernel method is less important than choice of bandwidth (Kernohan et al. 2001). Lastly, fixed and adaptive kernel methods produce different estimates and different computer programs produce different estimates (Kenward 2001).

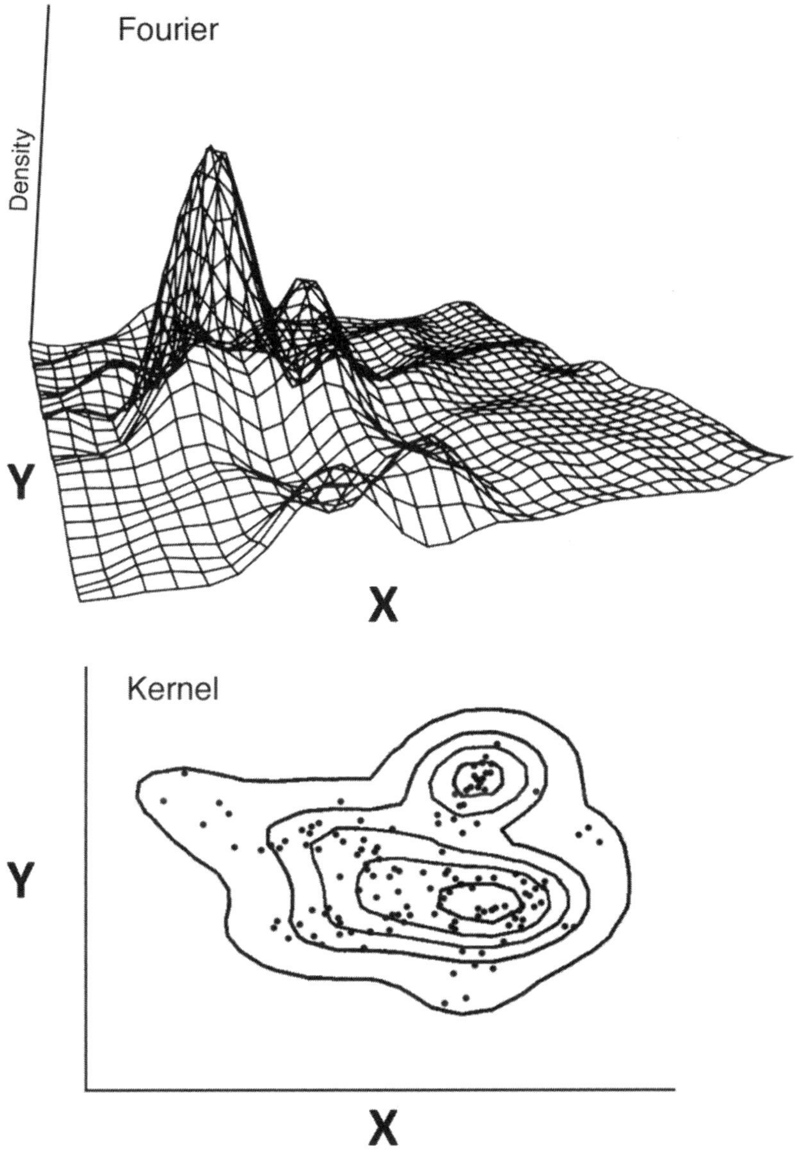

FIGURE 4. Graphical representations of home ranges from two probabilistic or contour home range methods. The home range size is calculated from a boundary enclosing a chosen percentage (often 95%) of the volume of the utilized distribution. The Fourier transformation figure is adapted from Anderson (1982). The fixed-kernel figure is adapted from Seaman and Powell (1996) and shows the 95, 72.5, 50, 27.5, and 5% contours. The data sets are different.

Kernohan et al. (2001) evaluated and ranked home range methods. They examined the performance with respect to seven criteria: sample size needed, effects of autocorrelation of data, using the complete utilization distribution, being nonparametric, handling multiple centers of activity, tolerance to outliers, and comparability to other estimators using the same dataset. Fixed and adaptive kernel methods ranked the highest (6 of 7 criteria) of the home range methods. They scored well with all criteria except comparability. Cluster analysis was next highest (5), and was rated well for everything except utilization distribution and outliers. The harmonic mean and Fourier transformation tied at 4 and scored well for utilization distribution, nonparametric, multiple centers of activity, and outliers. The grid-cell method also scored 4, and scored well for autocorrelation, nonparametric, multiple activity centers, and comparability. The minimum convex polygon method scored 3, and was rated well in autocorrelation, nonparametric, and comparability. There were three other methods that scored 2, including the weighted bivariate normal.

All home range methods have limitations and the choice of method depends on the investigator's purpose. Harris et al. (1990) suggested that one should use at least two home range estimators with each dataset, and one estimator should be the minimum convex polygon method for comparison to other studies. Kernohan et al. (2001) disagreed and state that comparisons with the convex polygon method are unreliable due to sample size influences and handling of multiple activity centers and outliers. They recommend kernel methods and cluster analysis. Despite the sampling and statistical shortcomings of the convex polygon method, it continues to be widely used, because it is simple, easily understood, and easily visualized. I think its greatest value is that it provides a picture or concept that the public or decision-makers can comprehend. Its measure of core areas can often be improved by peeled, concaved, and cluster analysis polygons. The grid-cell method is valuable when it is combined with habitat maps and used to measure intensity of habitat use. Probabilistic methods are recommended for making fine comparisons of home range and statistical analyses between seasons, sexes, and size classes. However, they are more difficult to calculate, comprehend, and visualize. Probabilistic methods often enclose 95% of the utilized distribution; however, the value is arbitrary and is not related to $\alpha = 0.05$ in statistical tests (White and Garrott 1990). The home range estimates for a particular method differ among software programs so valid comparison of home range sizes between studies can only be made with the same program, grid systems, user-defined options, and sample sizes (Kenward 2001; Larson 2001). Although home range sizes are difficult to compare between studies, behavior of fishes and conclusions probably are comparable. Rogers and White (2007) suggest that raw location data or movement rates can be used to test hypotheses and avoid problems associated with home range analysis. Therefore, the choice of analytical method will depend the purpose of the study and the fineness of desired comparisons.

Many researchers have used telemetry to study home range behavior of fish in lakes. Winter (1977) found that resident largemouth bass in a small lake had definite home ranges as opposed to a transplanted bass. He related home range theory to fish and used the convex polygon and grid-cell home range methods. To examine the influence of a thermal discharge into a bay on fish species, Ross and Winter (1981) determined home ranges using the convex polygon method. Home ranges of the four species differed in size and location to the thermal plume and ice cover. Weller and Winter (2001) determined the seasonal variation in home range size by flathead catfish *Pylodictus olivarus* in a productive, warm-water reservoir. The size and locations of the home ranges were greatly influenced by temperature and structure

(Weller and Winter 1999). Knight et al. (2008) used cluster analysis to examine the relationship between long excursions or dispersal distances and home range size in northern pike *Esox lucius*. The excursion distance was positively correlated with winter and spring home ranges, but not to summer and autumn ranges. Fixed-kernel methods have been used to compare the seasonal home ranges of largemouth and spotted bass *M. punctulatus* (Hunter and Maceina 2008a) and lake trout (Morbey et al. 2006) with respect to water temperature. Ahrenstorff et al. (2009) used an adaptive-kernel method to examine home ranges of largemouth bass with respect to coarse, woody habitat. Largemouth bass in high coarse, woody habitat had smaller home ranges.

Habitat Use and Selection

A second purpose of telemetry studies in lakes is to determine habitat use or selection by fish. Terrestrial ecologists have much experience in measuring habitat types and producing maps largely because terrestrial vegetation is more persistent, easier to measure, and easier to map than aquatic vegetation. Thus, aquatic ecologists can borrow much theory and methodology for analyzing habitats from terrestrial ecologists (Fuller et al. 2005; McDonald et al. 2005; O'Neil et al. 2005). When analyzing data from telemetry, you should use the animals as the experimental or sampling units for comparisons and avoid pooling data from all individuals (White and Garrott 1990; Winter 2000).

Knowledge of a species' habitat use and selection can reveal important aspects about a species' physical, biotic, and environmental requirements. Habitat use or selection can provide biologists with information on a species' needs or on the effects of human activities on environmental components. Managers can protect habitat types or perhaps manipulate habitat components to increase a species' population. Habitat in its broadest context includes all the abiotic and biotic features of the environment (McDonald et al. 2005). Habitat use is the number of telemetry locations or amount of time that a species spends in a particular habitat type, or the association of an animal with the habitat type. Habitat selection is the disproportionate number of telemetry locations or time spent in a habitat relative to the abundance or availability of that habitat type (Johnson 1980). Habitat availability is the quantity of a habitat type accessible to the animal or population of animals, whereas abundance is the quantity of the resource in the environment (McDonald et al. 2005). The term "preference" is often used interchangeably with selection, but technically involves the probability that a habitat type would be chosen if available equally with other habitat types (Fuller et al. 2005). It is measured under controlled conditions such as enclosures. Some people use the term "avoided" for habitat types that are used disproportionately less than their availability. Fuller et al. (2005) discourage this use, because a telemetry location in the habitat type indicates that the habitat is used, not "avoided."

Habitat selection is studied on three levels (Fuller et al. 2005; McDonald et al. 2005). First-order selection occurs at the physical or geographic ranges of the species. Second-order selection involves the selection of a home range within a study area or the geographical range of the species. Third-order selection involves the importance of habitat types within an animal's home range. It does not answer why the home range was selected (Fuller et al. 2005). The available habitat types might be defined for the whole study area or lake. This could include habitat used rarely by individuals in a population (e.g., deep water, anoxic water). Inclusion of unavailable habitat on a lake-wide basis will not greatly obscure the relative importance of other habitat types (McDonald et al. 2005). Another method of examining habitat selection is to consider only the habitat types within a home range as available to the animal. A person could compare habitat types within home ranges to habitat found outside.

There are many methods of statistically analyzing habitat selection. McDonald et al. (2005) list characteristics associated with twelve methods. Some of the older methods use chi-square analyses to compare habitat use versus availability (Neu et al. 1974). Other tests are Friedman's test, Quade method, and the PREFER method. These techniques are being replaced by compositional analysis and discrete-choice modeling such as log-linear modeling and logistic regression (McDonald et al. 2005). McDonald et al. (2005) present examples using chi-square analysis, selection ratios, Bonferroni adjusted confidence limits, and Resource Selection Function (RSF). A selection ratio for a habitat type is the ratio of the proportion used to the proportion available. If the ratio is close to one, there is little selection. The selection ratios for individual animals can be averaged so that the number of animals is the effective sample size and pseudoreplication is eliminated. Mean selection ratios, standard errors, and Bonferroni 95% confidence limits can be calculated for each habitat type (Byers and Steinhorst 1984). The selection ratios can be standardized; i.e., divided by the summation of the mean selection ratios. The standardized selection ratio for each habitat type can be compared to the Bonferroni confidence limit to determine the statistical significance for each habitat type. McDonald et al. (2005) discuss resource selection functions for categorical data and use logistic regression to analyze habitat selection. The logistic regression is a specialized regression technique for multiple continuous and discrete variables, and can be used to analyze habitat categories on selection of locations. Their description is lengthy and detailed, and they indicate that you can use any computer software package that fits a logistic regression function. The functions can estimate the probabilities that habitat types were selected and to rank the habitat types according to their use by animals. In conclusion, the chi-square methods, Bonferroni confidence limits, and other methods are adequate for general habitat selection studies. However, if you need fine comparisons among seasons, sexes, and size classes, or are interested in modeling habitat selection, then you should consider the resource selection functions described by McDonald et al. (2005).

Many studies have used telemetry to study habitat use or selection by fish in lakes. Often, studies look at the distribution of locations with a habitat variable, and compare the means or distributions with common statistical tests. Hubert and Lackey (1980) compared smallmouth bass *M. dolomieu* locations to bottom temperatures, flow rates, and water levels using chi-square tests and Kendall's coefficient of rank correlation, and Schaffler and Isely (2002) used chi-square to test for differences in habitats used and time of day by striped bass. Both studies used manual tracking. Hanson et al. (2008) and Suski and Ridgway (2009) used fixed-hydrophone arrays and pressure and/or temperature-sensing transmitters to monitor depth selection of fish. Habitat use data were analyzed with repeated-measure analysis of variance (ANOVA). Kobler et al. (2008) used selection ratios and Bonferroni confidence intervals to analyze physical habitat selection by northern pike whereas Weller and Winter (2001) used chi-square and Bonferroni confidence intervals to determine substrate and depth selection by flathead catfish. Bettoli (2005) use mortality-temperature transmitters to describe the fundamental thermal niche of landlocked striped bass. Because telemetry provides many observations on individual fish, it is possible to determine habitat use and selection of many variables.

Migration and Dispersal

Many telemetry studies are concerned with fish migration and dispersal in lakes. We study migrations and dispersal to better understand the biology of the species. We want to know what triggers these events, what directions do fish go, how fast do they travel, where do they go to, do they return, and what happens to them? From a management standpoint, we want to know where different fish stocks go, who harvests them, where do they spawn, and what do they need for spawning or survival. In general, migration and dispersal studies are conducted on highly mobile species and in moderate-size to huge lakes, lakes that are interconnected, or reservoirs that often have multiple bays and inlets. The logistics are different than for home range and habitat studies. The search areas are very large, the fish may cover long distances in a short period of time, and transmittered-fish may be scattered over a large areas. Searches are conducted with methods (see section on large lakes) that cover large areas relatively quickly and make less accurate locations. The error in locating a fish is small compared to the fish's daily movements and does not obscure the migration route or pattern. Many studies now use a series of fixed-recording stations (Section 5.3) or submerged-hydrophones (Section 7.1) with data-logging receivers, because it is less time consuming and potentially less costly than mobile search. Furthermore, you can monitor more transmitters with recording stations than mobile searches. People usually determine total distance traveled and compute movement rates. These are usually compared with other factors such as size-class, sex, season, or stock (Fuller et al. 2005). Movement rates computed from frequent intervals are usually different than movement rates calculated between longer intervals, because fish do not travel long in a straight line and may even double back. The rate of movement should be expressed in units appropriate to the accuracy of the distance measured and the tracking interval (km/d or m/h not cm/s). Analysis of an animal's direction of movement may involve circular statistical tests because of compass directions (Fuller et al. 2005), although fish directions are usually truncated on at least one side by the shore.

Many studies examine migrations and dispersal of fish in large lakes or reservoirs. Winter (1976) radio tracked from an airplane, boat, and truck to examine prespawning migrations and postspawning dispersal of rainbow trout *O. mykiss* in Lake Superior. The dispersal direction after spawning was 13.8% SW, 20.7% NE, and 65.5% SE offshore. Some individuals traveled long distances close to shore, but most moved out of tracking range quickly. Wenger et al. (1984) followed prespawning behavior of salmon in eastern Lake Erie. The prespawning movements were predominantly southwestward from the capture locations and within 400 m of shore. Salmon tested other streams and made false runs in the spawning steam before the final run (Winter 1976; Wenger et al. 1984; Haynes and Gerber 1989). Radio-tagged lake sturgeon *Acipenser fulvescens* quickly leave the Sturgeon River after spawning and remain near the river confluence from 3 to 53 d before likely traveling long distances in Lake Superior (Auer 1999). Brown trout migrate out of Box Canyon Reservoir when the summer water temperatures reach 19–20°C and seek refuge in cooler tributaries (Garrett and Bennett 1995). After fall spawning, half the fish return to the reservoir and the rest returned during high flows in March. Similarly, striped bass migrated from the main part of a southeast U.S. reservoir when it stratified to the cooler, upper section of the reservoir (Young and Isely 2002). The striped bass returned to the main area in the fall. These studies demonstrate that telemetry can reveal patterns and timing of migrations and correlations to environmental variables.

Survival and Exploitation

Telemetry can provide valuable information on survival and exploitation of fish (Section 9.2; White and Garrott 1990; Winterstein et al. 2001). It provides better information on when death occurs and the likely causes of mortality than with conventional tags. Furthermore, anglers return most transmitters with a reward label compared to a traditional fish tag. For example, angler exploitation of radio-tagged salmon captured in Lake Erie in the fall was substantial, ranging from 47.1% in one year to 54.5% in the next (Wenger et al. 1984). Baker and Jennings (2005) transplanted adult Gulf-strain striped bass with transmitters into a Georgia reservoir. Because of a drought during the study, some cool-water spring refuges dried up and survival was poor. The management implications are that stocking and re-introduction of striped bass into southern rivers and reservoirs depends on maintaining water levels and cool-water springs. Hightower et al. (2001) followed a cohort of striped bass with telemetry to determine exploitation, survival rates, and time of mortality. They used this method to verify assumptions in fisheries survival models and get better estimations of parameters. The survival and dispersal of tournament-caught largemouth bass (Wilde and Paulson 2003; Hunter and Maceina 2008b) and smallmouth bass (Kaintz and Bettoli 2010) have been monitored with telemetry. Largemouth bass seem to disperse slower from the release site than smallmouth bass, and transporting largemouth by boat after tournaments could reduce their susceptibility to angling and competition.

Response to Environmental Variables

Telemetry can be used in lakes to determine the response of fish to environmental variables. Behavior of fish is most often studied in response to temperature, depth, light, oxygen, food supply, or water discharge, but telemetry can be used to assess responses to many other variables. Fixed towers and temperature-sensing transmitters were used in a thermally enriched bay and revealed that four fish species differed in temperature and depth utilization, distance from the thermal plume and ice cover, and home range sizes (Ross and Winter 1981; Ross and Siniff 1982). Mackenzie-Grieve and Post (2006) used temperature-sensing transmitters and found that lake trout in two dissimilar Yukon lakes selected temperatures cooler than the optimum thermal range (8–12°C) predicted by other studies. Morbey et al. (2006) examined lake trout behavior in a lake without pelagic schooling prey. Lake trout made frequent trips into lake temperatures that exceeded their optimum for physiological performance. Pressure-sensing transmitters and three submerged-hydrophone recording stations were used to monitor the depth selection and activity patterns of brown bullheads *Ameiurus nebulosus* in a bay of a New Zealand lake (Dedual 2002). Brown bullheads were shallower in the summer than winter and more active at night. They preferred the cover of rooted plants and might have moved deeper in the winter, because of decreasing cover and increasing light penetration. Cooke and Schreer (2003) used electromyogram radio transmitters to monitor the activity of carp *Cyprinus carpio* in a thermal canal from a power plant. The transmitters sensed electromyographic activity in the axial muscle of carp and the signals were recorded with a fixed-receiving station. Locations from tracking suggested that carp were relatively immobile while in the canal. However, the electromyogram data provided finer information than location telemetry and indicated that carp were significantly more active when the temperatures were changing. Könignson et al. (2002) studied the response of transmittered-whitefish

(*Coregonus lavaretus*) to strobe lights in an enclosure. The light directed at the front or side of the fish caused the fish to turn and swim away from the light. They suggest that strobe lights might be used to prevent fish from swimming into specific areas such as water intakes of power stations. The response of largemouth bass to highly variable oxygen levels under the ice was assessed with a 13-hydrophone array system (Hasler et al. 2009). Dissolved oxygen content of 2.0 mg/L might be a critical threshold that induces behavior responses by bass. In conclusion, telemetry is a valuable tool to study the response of fish to environmental variables, and the types of variables that can be assessed are enormous.

Population Estimation and Assumptions

Telemetry can be used to estimate population size or density and to evaluate assumptions of population indices (White and Shenk 2001). Transmitters can be used as the primary markers in mark–recapture indices or to correct density estimates from other surveys (Kenward 2001). An advantage of using transmitters for population estimates is that you know the number of transmittered-animals in an area better than with passive tags and can get better estimates. You also can evaluate assumptions of indices such as mixing of the population, emigration, mortality, and likelihood of capture. The use of telemetry to estimate population size in aquatic systems is relatively new. A recent model uses telemetry mark–recapture to estimate abundances of populations of fish migrating through a river system to lake tributaries (Bromaghin et al. 2010). The stratified model handles capture probabilities and migratory timing that vary among time periods.

Animal Interactions

Telemetry can be used to study animal interactions (Macdonald et al. 1980; Kenward 2001). It is commonly used in terrestrial systems to study territoriality, search for mates, or predator–prey behavior. It has not been used much in aquatic systems. Basically, researchers look at the amount of overlap in home ranges among individuals, the mean distances between transmittered-animals, and indices of association. Einhouse (1981) examined whether walleyes traveled together in schools as anglers commonly believe. Although transmittered-individuals were sometimes close together, most of the time they were far apart and moved independent of each other.

CONCLUSIONS

Telemetry is a powerful technique for studying the basic biology and ecology of fish in lakes and problems impacting fish due to human activities. You should start by identifying a problem or research question and reviewing the literature. You should consult the literature (Winter 2000; Garton et al. 2005) and Sections in this book (Section 3) for help on planning the project. Refine your research idea into well-formulated questions, especially ones that yield "yes" or "no" answers, because they will yield unambiguous results (Winter 2000). Because statisticians do not like to band-aid results, you should think about your experimental design and methods for data analysis before collecting data. Next, you need to choose a telemetry system and equipment. There is a wealth of information that you should consider in this book on equipment and techniques for tracking aquatic animals and on methods to study problems and analyze data. Furthermore, you should consult researchers who have conducted studies similar to your proposed study, because they can provide good advice on equipment, methods, and mistakes to avoid. Do not be intimidated by sophisticated equipment, cost, or size of other projects. There is still much opportunity for projects with well-formulated hypotheses, simple equipment, or small lakes as natural laboratories. You need to decide whether your experimental question requires many, frequent locations on a small number of fish or a large number of fish with few, less frequent locations. People probably overestimate the number of fish that can be tracked manually. On a medium-size lake, the maximum number of mobile fish that can be located in a day is 24–36. Over 50 fish can be located per day in a small lake because search time is low and in a large lake because location time is low and aircraft might be used. Fixed-recording stations can monitor the most fish, because location time is minimal (presence/absence) and scanning receivers, data loggers, and coded transmitters are used. This Section provides information and valuable citations on telemetry and data analysis methods to study fish biology in lakes. If you do careful research and planning, you likely with have a successful and, hopefully, enjoyable project.

REFERENCES

Ahrenstorff, T. D., G. G. Sass, and M. R. Helmus. 2009. The influence of littoral zone coarse woody habitat on home range size, spatial distribution, and feeding ecology of largemouth bass (*Micropterus salmoides*). Hydrobiologia 623:223–233.

Amlaner, C. J., Jr., and D. W. Macdonald, editors. 1980. A handbook of biotelemetry and radio tracking. Pergamon Press, New York.

Anderson, J. D. 1982. The home range: a new nonparametric estimation technique. Ecology 63:103–112.

Auer, N. A. 1999. Population characteristics and movements of lake sturgeon in the Sturgeon River and Lake Superior. Journal of Great Lakes Research 25:282–293.

Baker, T. L., and C. A. Jennings. 2005. Striped bass survival in Lake Blackshear, Georgia during drought conditions: implications for restoration efforts in Gulf of Mexico drainages. Environmental Biology of Fishes 72:584–596.

Bettoli, P. W. 2005. The fundamental thermal niche of adult landlocked striped bass. Transactions of the American Fisheries Society 134:305–314.

Bromaghin, J. F., K. S. Gates, and D. E. Palmer. 2010. A likelihood framework for joint estimation of salmon abundance and migratory timing using telemetric mark-recapture. North American Journal of Fisheries Management 30:1385–1394.

Burt, W. H. 1943. Territoriality and home range concepts as applied to mammals. Journal of Mammalogy 24:346–352.

Byers, C. R., and R. K. Steinhorst. 1984. Clarification of a technique for analysis of utilization-availability data. Journal of Wildlife Management 48:1050–1053.

Cooke, S. J., and J. Schreer. 2003. Environmental monitoring using physiological telemetry—a case study examining common carp responses to thermal pollution in a coal-fried generating station effluent. Water, Air, and Soil Pollution 142:113–136.

Coutant, C. C., and D. S. Carroll. 1980. Temperatures occupied by ten ultrasonic-tagged striped bass in freshwater lakes. Transactions of the American Fisheries Society 109:195–202.

Cresswell, W. J., and G. C. Smith. 1992. The effects of temporally autocorrelated data on methods of home range analysis. Pages 272–284 *in* I. G. Priede and S. M. Swift, editors. Wildlife telemetry: remote monitoring and tracking of animals. Ellis Horwood, New York.

Dedual, M. 2002. Vertical distribution and movements of brown bullhead (*Ameiurus nebulosus* Lesueur 1819) in Motuoapa Bay, southern Lake Taupo, New Zealand. Hydrobiologia 483:129–135.

Dixon, K. R., and J. A. Chapman. 1980. Harmonic mean measure of animal activity areas. Ecology 61:1040–1044.

Dick, T. A., C. P. Gallagher, and A. Yang. 2009. Summer habitat use of Arctic char (*Salvelinus alpinus*) in a small Arctic lake, monitored by acoustic telemetry. Ecology of Freshwater Fish 18:117–125.

Einhouse, D. W. 1981. Summer-fall movements, habitat utilization, diel activity and feeding behavior of walleyes in Chautauqua Lake, New York. Master's thesis. State University College, Fredonia, New York.

Fuller, M. R., J. J. Millspaugh, K. E. Church, and R. E. Kenward. 2005. Wildlife radiotelemetry. Pages 377–417 *in* C. E. Braun, editor. Techniques for wildlife investigations and management, 6th edition. The Wildlife Society, Bethesda, Maryland.

Garrett, J. W., and D. H. Bennett. 1995. Seasonal movements of adult brown trout relative to temperature in a coolwater reservoir. North American Journal of Fisheries Management 15:480–487.

Garton, E. O., J. T. Ratti, and J. H. Giudice. 2005. Research and experimental design. Pages 43–71 *in* C. E. Braun, editor. Techniques for wildlife investigations and management, 6th edition. The Wildlife Society, Bethesda, Maryland.

Gerking, S. H. 1953. Evidence for the concepts of home range and territory in stream fishes. Ecology 34:347–365.

Hanson, K. C., S. J. Cooke, C. D. Suski, G. Niezgoda, F. J. S. Phelan, R. Tinline, and D. P. Philipp. 2007. Assessment of largemouth bass (*Micropterus salmoides*) behaviour and activity at multiple spatial and temporal scales utilizing a whole-lake telemetry array. Hydrobiologia 582:243–256.

Hanson, K. C., C. T. Hasler, S. J. Cooke, C. D. Suski, and D. P. Philipp. 2008. Intersexual variation in the seasonal behaviour and depth distribution of a freshwater temperate fish, the largemouth bass. Canadian Journal of Zoology 86:801–811.

Harris, S., W. J. Cresswell, P. G. Forde, W. J. Trewhella, T. Woolard, and S. Wray. 1990. Home-range analysis using radio-tracking data—a review of problems and techniques particularly as applied to the study of mammals. Mammal Review 20:97–123.

Hasler, C. T., C. D. Suski, K. C. Hanson, S. J. Cooke, and B. L. Tufts. 2009. The influence of dissolved oxygen on winter habitat selection by largemouth bass: an integration of field biotelemetry studies and laboratory experiments. Physiological and Biochemical Zoology 82:143–152.

Haynes, J. M., D. C. Nettles, K. M. Parnell, M. P. Voiland, R. A. Olson, and J. D. Winter. 1986. Movements of rainbow steelhead trout (*Salmo gairdneri*) in Lake Ontario and a hypothesis for the influence of spring thermal structure. Journal of Great Lakes Research 12:304–313.

Haynes, J. M., and G. P. Gerber. 1989. Movements and temperatures of radiotagged salmonines in Lake Ontario and comparisons with other large aquatic systems. Journal of Freshwater Ecology 5:197–204.

Heezen, K. L., and J. R. Tester. 1967. Evaluation of radio-tracking by triangulation with special reference to deer movements. Journal of Wildlife Management 51:399–401.

Hightower, J. E., J. R. Jackson, and K. H. Pollock. 2001. Use of telemetry methods to estimate natural and fishing mortality of striped bass in Lake Gaston, North Carolina. Transactions of the American Fisheries Society 130:557–567.

Hubert, W. A., and R. T. Lackey. 1980. Habitat of adult smallmouth bass in a Tennessee River reservoir. Transactions of the American Fisheries Society 109:364–370.

Hunter, R. W., and M. J. Maceina. 2008a. Movements and home ranges of largemouth bass and Alabama spotted bass in Lake Martin, Alabama. Journal of Freshwater Ecology 23:599–606.

Hunter, R. W., and M. J. Maceina. 2008b. Dispersal of tournament-displaced largemouth bass and spotted bass in Lake Martin, Alabama. North American Journal of Fisheries Management 28:678–687.

Jenrich, R. I., and F. B. Turner. 1969. Measurement of non-circular home range. Journal of Theoretical Biology 22:227–237.

Johnson, D. H. 1980. The comparison of usage and availability measurements for evaluating resource preference. Ecology 61:65–71.

Jones, B. R., K. E. F. Hokanson, and J. H. McCormick. 1977. Winter temperature requirements for maturation and spawning of yellow perch *Perca flavescens* (Mitchell). Pages 189–192 *in* M. Marios, editor. Towards a plan of actions for mankind. Proceedings of the World Conference, Volume 3, biological balance and thermal modifications. Pergamon Press, New York.

Kaintz, M. A., and P. W. Bettoli. 2010. Dispersal of smallmouth bass from a simulated tournament weigh-in site. North American Journal of Fisheries Management 30:976–982.

Kenward, R. E. 1987. Wildlife radio tagging: equipment, field techniques and data analysis. Academic Press, San Diego, California.

Kenward, R. E. 2001. A manual for wildlife radio tagging. Academic Press, San Diego, California.

Kernohan, B. J., R. A. Gitzen, and J. J. Millspaugh. 2001. Analysis of animal space use and movements. Pages 125–166 *in* J. J. Millspaugh and J. M. Marzluff, editors. Radio tracking and animal populations. Academic Press, San Diego, California.

Knight, C. M., R. E. Gozlan, and M. C. Lucas. 2008. Can seasonal home-range size in pike *Esox lucius* predict excursion distance? Journal of Fish Biology 73:1058–1064.

Kobler, A., T. Klefoth, C. Wolter, F. Fredrich, and R. Arlinghaus. 2008. Contrasting pike (*Esox lucius* L.) movement and habitat choice between summer and winter in a small lake. Hydrobiologia 601:17–27.

Königeson, S., A. Fjalling, and S. Gunnar Lunneryd. 2002. Reactions in individual fish to strobe light. Field and aquarium experiments performed on whitefish (*Coregonus lavaretus*). Pages 39–44 *in* E. B. Thorstad, I. A., Fleming, and T. F. Naesje, editors. Aquatic telemetry. Proceedings of the Fourth Conference on Fish Telemetry in Europe. Kluwer Academic Publishers, Norwell, Massachusetts.

Larson, M. A. 2001. A catalog of software to analyze radiotelemetry data. Pages 397–421 *in* J. J. Millspaugh and J. M. Marzluff, editors. Radio tracking and animal populations. Academic Press, San Diego, California.

Mackenzie-Grieve, J. L., and J. R. Post. 2006. Thermal habitat use by lake trout in two contrasting Yukon Territory lakes. Transactions of the American Fisheries Society 135:727–738.

Macdonald, D. W., F. G. Ball, and N. G. Hough. 1980. The evaluation of home range size and configuration using radio tracking data. Pages 405–424 *in* C. J. Amlaner and D. W. Macdonald, editors. A handbook on biotelemetry and radio tracking. Pergamon Oxford and New York.

McDonald, L. L., J. R. Alldredge, M. S. Boyce, and W. P. Erickson. 2005. Measuring availability and vertebrate use of terrestrial habitats and foods. Pages 465–487 *in* C. E. Braun, editor. Techniques for wildlife investigations and management, 6th edition. The Wildlife Society, Bethesda, Maryland.

Millspaugh, J. J., and J. M. Marzluff, editors. 2001. Radio tracking and animal populations. Academic Press, San Diego, California.

Mohr, C. O. 1947. Table of equivalent populations of North American small mammals. American Midland Naturalist 37:223–249.

Morbey, Y. E., P. Addison, B. J. Shuter, and K. Vascotto. 2006. Within-population heterogeneity of habitat use by lake trout *Salvelinus namaycush*. Journal of Fish Biology 69:1675–1696.

Nettles, D. C., J. M. Haynes, R. A. Olson, and J. D. Winter. 1987. Seasonal movements and habitats of brown trout (*Salmon trutta*) in southcentral Lake Ontario. Journal of Great Lakes Research 13:168–177.

Neu, C. W., C. R. Byers, and J. M. Peek. 1974. A technique for analysis of utilization-availability data. Journal of Wildlife Management 38:541–545.

Odum, E. P., and E. J. Kuenzler. 1955. Measurement of territory and home range in birds. Auk 72:128–137.

Olson, R. A., J. D. Winter, D. C. Nettles, and J. M. Haynes. 1988. Resource partitioning in summer by salmonids in south-central Lake Ontario. Transactions of the American Fisheries Society 117:552–559.

O'Neil, T. A., P. Bettinger, B. G. Marcot, B. W. Luscombe, G. T. Koeln, H. J. Bruner, C. Barrett, J. A. Pollock, and S. Bernatas. 2005. Application of spatial technologies in wildlife biology. Pages 418–447 *in* C. E. Braun, editor. Techniques for wildlife investigations and management, 6th edition. The Wildlife Society, Bethesda, Maryland.

Paukert, C. P., D. W. Willis, and R. S. Holland. 2002. Sample size requirements for *in situ* vegetation and substrate classifications in shallow, natural Nebraska lakes. North American Journal of Fisheries Management 22:1329–1333.

Priede, I. G. and S. M. Swift. 1992. Wildlife telemetry: remote monitoring and tracking of animals. Ellis Horwood, New York.

Rogers, A. R. 2001. Recent telemetry technology. Pages 79–121 *in* J. J. Millspaugh and J. M. Marzluff, editors. Radio tracking and animal populations. Academic Press, San Diego, California.

Rogers, K. B., and G. C. White. 2007. Analysis of movement and habitat use from telemetry data. Pages 625–676 *in* C. S. Guy, and M. L. Brown, editors. Analysis and interpretation of freshwater fisheries data. American Fisheries Society, Bethesda, Maryland.

Ross, M. J. 1978. Distribution of fish and temperature preference of yellow perch in the thermal plume of a power plant as determine by radio telemetry. Master's thesis. University of Minnesota, Minneapolis.

Ross, M. J., and D. B. Siniff. 1982. Temperatures selected in a power plant thermal effluent by adult yellow perch (*Perca flavescens*) in winter. Canadian Journal of Fisheries and Aquatic Sciences 39:346–349.

Ross, M. J., and J. D. Winter. 1981. Winter movements of four fish species near a thermal plume in northern Minnesota. Transactions of the American Fisheries Society 110:14–18.

Samuel, M. D., and M. R. Fuller. 1994. Wildlife radiotelemetry. Pages 370–418 *in* T. A. Bookhout, editor. Research and management techniques for wildlife and habitats. The Wildlife Society, Bethesda, Maryland.

Savitz, J., P. A. Fish, and R. Weszely. 1983. Effects of forage on home-range size of largemouth bass. Transactions of the American Fisheries Society 112:772–776.

Schaffler, J. J., and J. J. Isely. 2002. Habitat use by striped bass in relation to seasonal changes in water quality in a southern reservoir. Transactions of the American Fisheries Society 131:817–827.

Seaman, D. E., and R. A. Powell. 1996. An evaluation of the accuracy of kernel density estimators for home range analysis. Ecology 77:2075–2085.

Siniff, D. B., and J. R. Tester. 1965. Computer analysis of animal movement data obtained by telemetry. Bioscience 15:104–108.

Suski, C. D., and M. S. Ridgway. 2009. Seasonal pattern of depth selection in smallmouth bass. Journal of Zoology 279:119–128.

Thorstad, E. B., I. A. Fleming, and T. F. Naesje, editors. 2002. Aquatic telemetry. Proceedings of the Fourth Conference on Fish Telemetry in Europe. Kluwer Academic Publishers, Norwell, Massachusetts.

Voigt, D. R., and R. R. Tinline. 1980. Strategies for analyzing radio tracking data. Pages 387–404 *in* C. J. Amlaner and D. W. Macdonald, editors. A handbook on biotelemetry and radio tracking. Pergamon Oxford and New York.

Weller, R. R., and J. D. Winter. 1999. Temperature selection by flathead catfish in a west Texas reservoir. Proceedings of the Annual Conference of the Association of Fish and Wildlife Agencies 53:42–49.

Weller, R. R., and J. D. Winter. 2001. Seasonal variation in home range size and habitat use of flathead catfish in Buffalo Springs Lake, Texas. North American Journal of Fisheries Management 21:792–800.

Wenger, M. N., R. M. Lichorat, and J. D. Winter. 1984. Pre-spawning and spawning behavior of coho salmon and chinook salmon in eastern Lake Erie. New York Fish and Game Journal 31:146–164.

Wenger, M. N., R. M. Lichorat, and J. D. Winter. 1985. Fall movements and behavior of radio-tagged brown trout and rainbow trout in eastern Lake Erie in 1979–1980. New York Fish and Game Journal 32:176–188.

White, G. C., and R. A. Garrott. 1990. Analysis of wildlife radio-tracking data. Academic Press, San Diego, California.

White, G. C., and T. M. Shenk. 2001. Population estimation with radio-marked animals Pages 329–350 *in* J. J. Millspaugh and J. M. Marzluff, editors. Radio tracking and animal populations. Academic Press, San Diego, California.

Wilde, G. R., and L. J. Paulson. 2003. Movement and dispersal of tournament-caught largemouth bass in Lake Mead, Arizona-Nevada. Journal of Freshwater Ecology 18:339–342.

Winter, J. D. 1976. Movements and behavior of largemouth bass (*Micropterus salmoides*) and steelhead (*Salmo gairdneri*) determined by radio telemetry. Doctoral dissertation. University of Minnesota, Minneapolis.

Winter, J. D. 1977. Summer and home range movements and habitat use by four largemouth bass in Mary Lake, Minnesota. Transactions of the American Fisheries Society 106:323–330.

Winter, J. D., V. B. Kuechle, D. B. Siniff, and J. R. Tester. 1978. Equipment and methods for radio tracking freshwater fish. University of Minnesota Agricultural Extension Service Miscellaneous Report 152.

Winter, J. D., and M. J. Ross. 1982. Methods in analyzing fish habitat utilization from telemetry data. Pages 273–279 *in* N. Armantrout, editor. Proceedings of the symposium on acquisition and utilization of aquatic habitat inventory information. American Fisheries Society, Western Division.

Winter, J. D. 1983. Underwater biotelemetry. Pages 371–395 *in* L. A. Nielsen and D. L. Johnson, editors. Fisheries Techniques. American Fisheries Society, Bethesda, Maryland.

Winter, J. D. 1996. Advances in underwater biotelemetry. Pages 555–590 *in* B. R. Murphy and D. W. Willis, editors. Fisheries Techniques, 2nd edition. American Fisheries Society, Bethesda, Maryland.

Winter, J. D. 2000. Designing telemetry studies and other technical and analytical considerations. Pages 229–247 *in* J. H. Eiler, D. Alcorn, and M. Neuman, editors. Proceedings of the 15th International Society of Biotelemetry. National Marine Fisheries Service, Juneau, Alaska.

Winterstein, S. R., K. H. Pollock, and C. M. Bunck. 2001. Analysis of survival data from radiotelemetry studies. Pages 351–380 *in* J. J. Millspaugh and J. M. Marzluff, editors. Radio tracking and animal populations. Academic Press, San Diego, California.

Young, S. P., and J. J. Isely. 2002. Striped bass annual site fidelity and habitat utilization in J. Strom Thurmond Reservoir, South Carolina-Georgia. Transactions of the American Fisheries Society 131:828–837.

Zamora, L., and R. Moreno-Amich. 2002. Quantifying the activity and movement of perch in a temperate lake by integrating acoustic telemetry and a geographic information system. Hydrobiologia 483:209–218.

Section 6.5

Design and Performance of Radio Telemetry Systems for Assessing Juvenile Fish Passage at Three Hydroelectric Dams

John W. Beeman, Eric E. Hockersmith,
and John R. Stevenson

INTRODUCTION

Studies of the effects of hydroelectric dams on fish populations are common (Williams 2008). Dams block passage of migratory and resident fish, alter habitats from free-flowing to lacustrine, and can alter water temperatures both upstream and downstream. At some dams, structures or operations are modified to reduce their effects on fish populations. In these cases, it is recommended that a series of studies be conducted before and after the alterations to help assess the effectiveness of the actions. We will describe three studies at hydroelectric dams on the Columbia and Snake rivers in the Pacific Northwest of the United States prompted by a need to reduce their effects on fishes, primarily salmonids, listed under the Endangered Species Act (ESA 1973).

Hydroelectric development on the Columbia and Snake rivers occurred chiefly between the early 1930s and the late 1970s. Fish originating in the upper portions of the Columbia and Snake rivers must pass as many as eight dams on these rivers during their seaward migration and again on their trip back to their natal waters. Small changes in passage survival at each dam can be important, due to the multiplicative effects of the series of dams. For example, if downstream passage survival at each of eight dams and reservoirs was 90% and it was increased by only 3% per dam, the numbers of fish surviving through the entire hydro system would increase from 43% to 56%. Thus, precisely measuring small changes in passage survival are important to the overall program. This has been achieved by designing efficient telemetry systems and releasing large numbers of tagged fish.

Initially, efforts to estimate smolt survival through reaches of the Snake and lower-Columbia rivers relied on thermal branded fish (Groves and Novotny 1965; Bentley and Raymond 1968; Raymond 1968, 1974, and 1979). However, those efforts required large numbers of fish to be marked, and only provided survival estimates for extended reaches of river spanning a number of dams. More recently, PIT-tag technology has been utilized to assess in-river survival (Muir et al. 2001a, 2001b; Bickford et al. 2000a, 2000b; Smith et al. 2003). However, this method also requires a large number of fish. Furthermore, since not all routes of passage can be monitored for PIT-tagged fish, route-specific survival cannot be ascertained for volitional passage. Another method to estimate survival uses balloon tags (Heisey et al. 1992). While relatively small numbers of fish need to be tagged in this type of evaluation, estimates of survival through a specific route (e.g., turbines, spillways, bypass systems, etc.) provide only estimates of direct survival or mortality, and excludes indirect effects such as delayed mortality due to injury, or predation due to injury or disorientation after passing the project. Studies of juvenile salmonids based on balloon-tags are also typically limited to uses where the natural behavior of the animals are not important (e.g., direct passage through a turbine), due to the added burden of the balloon. Because of these limitations, investigations into the feasibility of using radio telemetry as a tool to evaluate project, pool, dam and route-specific survival at hydroelectric projects began in 1999 (Lady et al. 2000; Stevenson et al. 2000; Skalski et al. 2001).

The purpose of this Section is to describe methods used to collect data on the behavior, passage, and survival of juvenile salmonids at a hydroelectric dam. Three examples will be used to describe methods to achieve this goal. Three different research groups operating independently of one another conducted these studies.

METHODS

Study sites

Hydroelectric dams in the Columbia and Snake rivers have many features in common. The waters immediately upstream from the dam are called the forebay and those immediately downstream the tailrace (Figure 1). Most dams include an earthen dam, spillway, powerhouse, and some type of juvenile bypass system. Dams located in the lower Columbia River, and the four lower Snake River dams all have a navigation lock. Some also have floating structures ("trash booms") in the forebay to divert debris from the powerhouse. Spillways are composed of spill bays regulating water releases with gates separated by concrete piers. Powerhouses are fitted with a series of turbines with up to three intakes each. Main piers separate turbines and intermediate piers, narrower than main piers, separate intakes. Vertical slots in the powerhouse at each turbine intake, called headgate slots, allow dam operators to lower barriers to separate turbines from the forebay water for maintenance. At many dams on the Columbia and Snake rivers, screens are lowered into the headgate slots to divert fish from turbine passage and into various orifices, pipes, and flumes comprising a juvenile bypass system (Raymond 1988; Figure 2). Some dams also use a surface-flow collector located within the forebay of the dam at a location where fish naturally congregate to pass fish downstream. These devices collect the juvenile fish and pass them directly to the tailrace, or divert them into a bypass flume that empties into the tailrace. Pool-and weir fish ladders provide upstream passage at these dams.

The studies were conducted at one hydroelectric dam on the Columbia River and two on the Snake River (Figure 3). Rock Island Dam is owned and operated by Public Utility District No. 1 of Chelan County, and is located on the Columbia River at river kilometer 729.7, about 21 km downstream from Wenatchee, Washington. The dam consists of two powerhouses, separated by a spillway. The original powerhouse (Powerhouse 1) was completed in 1933 and has 10 vertical axis turbines, each with 3 intakes, and is situated near the left bank of the

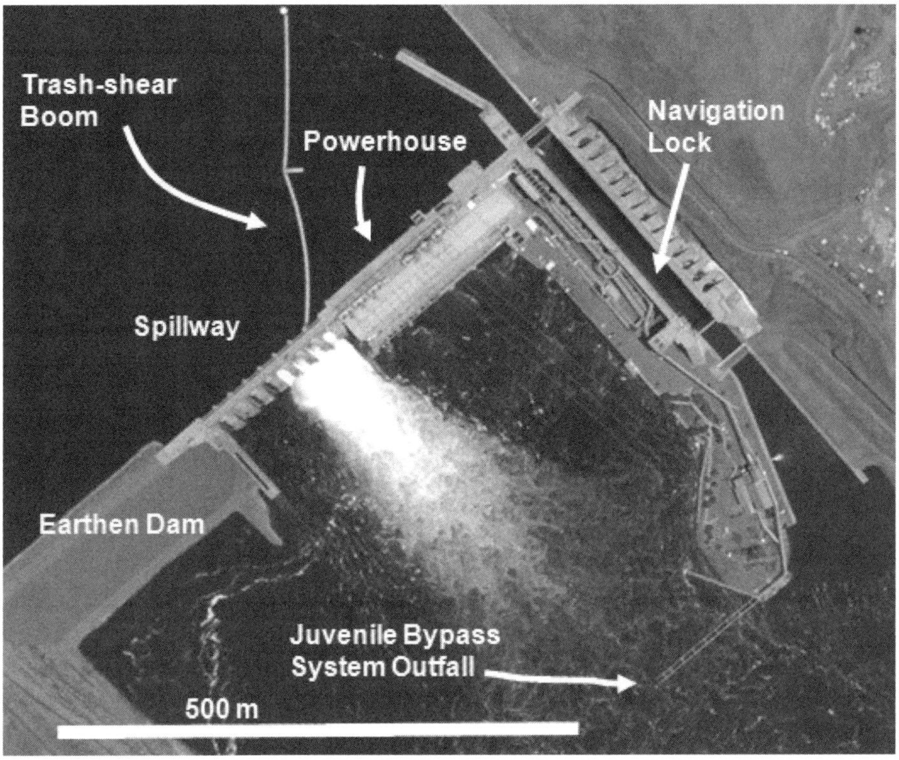

FIGURE 1. Photo of Little Goose Dam showing important hydroelectric dam features.

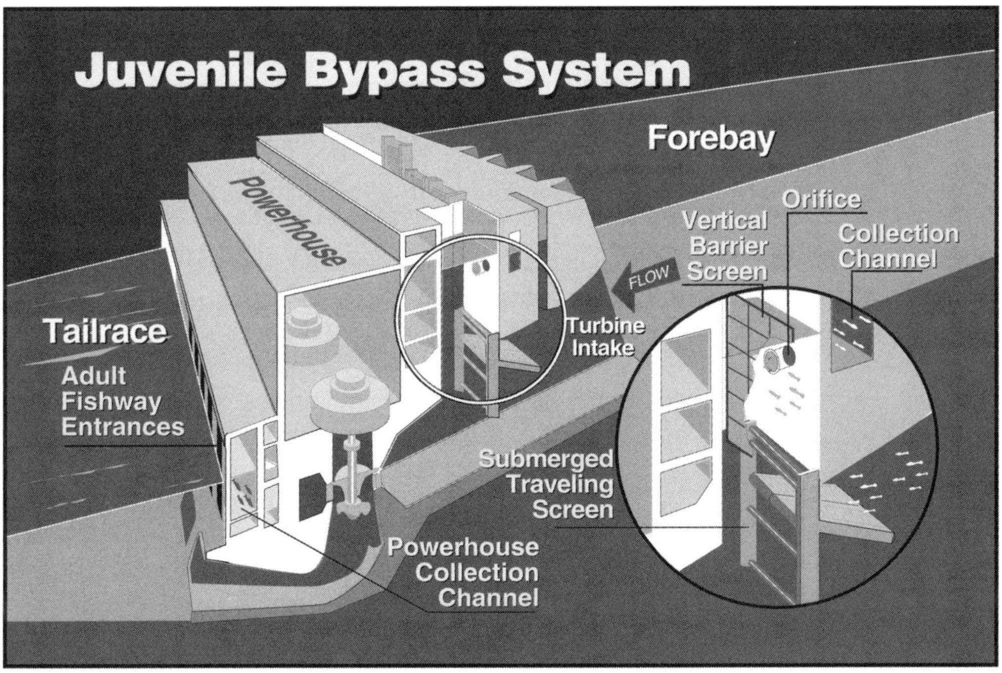

FIGURE 2. Schematic of a juvenile bypass system similar to that at Little Goose and Ice Harbor dams. Submerged screens are 6.1 m wide x 12.1 m long. Schematic from U.S. Army Corps of Engineers.

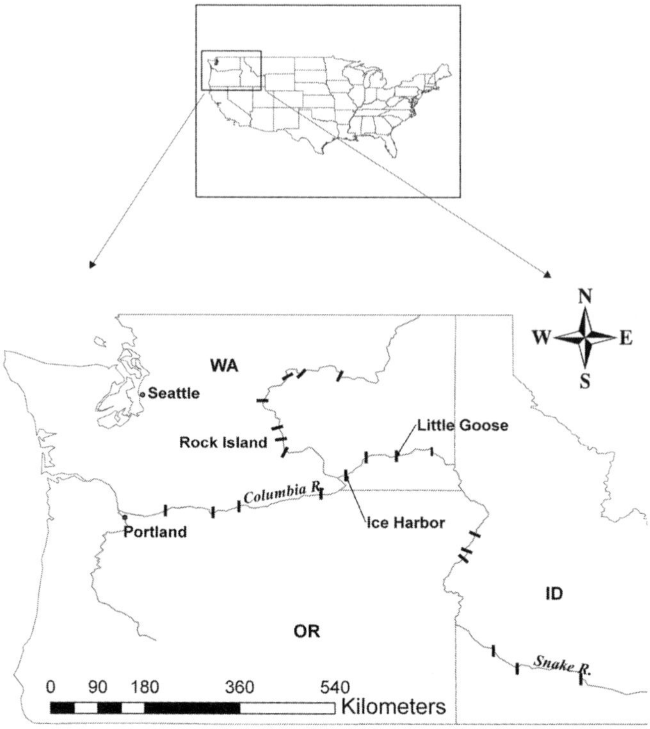

FIGURE 3. Map of study area in the northwestern United States showing Rock Island, Little Goose, and Ice Harbor dams relative to the states of Oregon (OR), Washington (WA), and Idaho (ID).

river; the generating capacity of Powerhouse 1 is approximately 213 MW. Powerhouse 2 was completed in 1979 and has 8 horizontal bulb-type units, each with two intakes and is situated near the right bank. Powerhouse 2 is capable of generating approximately 410 MW. The spillway consists of 31 spill bays that are 9 m wide and range from 7 to 17 m deep. The spillway is separated between bays 14 and 16 by an adult fishway. The project spans 1,022 m of the Columbia River, has a hydraulic head of about 12 m, and has an overall hydraulic capacity of 6,230 m^3/s. Little Goose Dam, located at Snake River kilometer 113, was completed in 1978 and has a generating capacity of 810 MW. It is 809 m long and creates a hydraulic head of 30 m. It has an eight-bay spillway operated with 15 × 18 m tainter gates, a six-unit powerhouse, an earthen dam and a navigation lock, and has a hydraulic capacity of 3,680 m^3/s. Inclined screens within each turbine intake are used to divert juvenile fish from turbine passage and into a juvenile bypass system. Ice Harbor Dam at Snake River kilometer 16 was completed in 1976. It is similar in design to Little Goose Dam, except the generating capacity is 604 MW, it has a 10-bay spillway, and is slightly longer at 860 m. All three dams include fishways for upstream fish passage.

Study Periods and Experimental Groups

The study at Rock Island Dam was conducted during the spring migration of yearling Chinook salmon *Oncorhynchus tshawytscha*. The tagging period spanned a total of 42 d, beginning April 21, 2000 and concluding June 1, 2000, with 10 replicates of each group being released at each paired-release location. As a run-of-river project, flows were variable and a function of the spring freshet. However, spill (water released through the spillway) at the project was relatively consistent during the study period, with an average of 866 m^3/s being spilled daily, which amounted to an average of 18.7% of the overall flow at the project. Spill was provided as a protective measure for juvenile salmonids as they migrated downstream of the dam.

The study at Little Goose Dam occurred during the spring and summer migration periods of juvenile salmonids. The spring study period at Little Goose Dam was from April 18 to June 14, 2007 and the summer period was from June 6 to July 24, 2007. The two periods were necessary to study juvenile steelhead (*O. mykiss*, anadromous rainbow trout) and yearling Chinook salmon that migrate seaward primarily during the spring and subyearling Chinook salmon whose migration is primarily during the summer. There were three distinct periods of operations during the spring and one during the summer. A condition with 30% of the river discharge passed through the spillway 24-h per day was bisected with a 2-week period of a condition consisting 30% spill during the day and 38% spill at night from April 29 to May 13, 2007. The increased spill in mid-season was intended to coincide with the peak of the juvenile salmon migration to increase their passage survival. A single operation of 30% spill 24-h per day was used during the summer study period.

The study at Ice Harbor Dam also occurred during the spring and summer. The spring study period at Ice Harbor Dam was from May 2 to June 10, 2007 and the summer period was from June 12 to July 14, 2007. The same species were studied at Little Goose and Ice Harbor Dams during the spring and summer time periods. Project operations at Ice Harbor Dam followed a 2-d random block design with two spill treatments. The first treatment operation was 1,274 m^3/s spill during the day and spill to the gas cap (up to 100% of river flow or to dissolved gas limits defined by state or local law) at night (hereafter referred to as high spill). The second treatment operation was a reduced volume of spill (between 30 and 40% of river flow volume) 24-h per day (hereafter referred to as reduced spill).

Fish collection, tagging, and releasing

Each study used commercially available radio transmitters with external antennas. Lotek Wireless[1] (Newmarket, Ontario Canada) manufactured transmitters used at Rock Island and Little Goose dams. Model MCFT-3GM was used at Rock Island Dam. At Little Goose Dam, model NTC-3–1 was used for yearling Chinook salmon and juvenile steelhead *O. mykiss* (anadromous rainbow trout), and model NTC-M-2 was used for subyearling Chinook salmon. The transmitters used at Ice Harbor Dam were custom-built by Advanced Telemetry Systems (Isanti, Minnesota, USA). Each study used transmitters emitting coded bursts enabling unique identification of several hundred transmitters per frequency. The transmitters chosen for each study were a compromise between tag dimensions, signal strength, and life (Table 1). For example, at Little Goose Dam the transmitters chosen for the yearling Chinook salmon enabled 99% of all fish available to be tagged with a maximum tag weight-to-body weight ratio of about 5% (including the PIT tag). Recommendations of tag sizes suitable for use in juvenile salmonids vary widely, ranging from 2 to over 10% (Winter 1996; Adams et al. 1998; Brown

[1]Reference to trade names does not imply endorsement by the U. S. Government.

TABLE 1. Specifications of radio transmitters used in the three studies. Cohorts are juvenile steelhead *Oncorhynchus mykiss* (STH), yearling Chinook salmon *O. tshawytscha* (CH1), and subyearling Chinook salmon (CH0). Size is by length, width, and height except for Rocky Reach Dam, which is listed as length and diameter.

Cohort	Number of Frequencies	Size (mm) L, W, H	Antenna Length (cm)	Weight (g) in		Burst Rate (s)	Expected Life (d)
				Air	Water		
			Little Goose Dam				
STH	8[a]	14.5, 6.3, 4.5	16	0.87	0.53	2.0	20
CH1	8[a]	14.5, 6.3, 4.5	16	0.87	0.53	2.0	20
CH0	8	13.5, 5.3, 3.0	16	0.44	0.30	2.5	20
			Rocky Reach Dam				
CH1[b]	19	18.9, 8.2	30	1.75	1.40	2.2	22
CH1[c]	19	18.9, 8.2	30	1.75	1.40	2.2	22
			Ice Harbor Dam				
STH	9	14.0, 6.3, 4.4	30	0.98	--	2.0	10[d]
CH1	9	14.0, 6.3, 4.4	30	0.98	--	2.0	10[d]
CH0	9	12.9, 5.4, 3.1	30	0.71	--	2.0	10[d]

[a] identical frequencies
[b] hatchery origin
[c] run-of-river origin
[d] user-defined shut off at 10 d

et al. 1999). We used the same transmitter in juvenile steelhead (mean weight 84.8 g, range 23.8–185.8 g) and yearling Chinook salmon (mean weight 25.6 g, range 14.3–49.7 g) so we could compare results without the potential bias from transmitters with different output powers. The transmitter chosen for use in subyearling Chinook salmon (mean weight of tagged fish 14.1 g, range 10.0–29.6 g) was simply the smallest and lightest transmitter meeting the life requirements given the desired burst rate; 89% of the available fish were suitable for tagging with this transmitter based on a minimum fish weight of 10 g. Burst rates of the tags in each study were chosen to enable detection within the passage routes with the highest water velocities. Sizes of the fish in each study were generally similar and will not be described further.

Each study relied on fish releases at one site upstream from the dam and another site downstream from the dam to enable survival estimation using methods to be described shortly. At Rock Island Dam, fish were released in 10 replicate groups of 35 fish at each of the two sites over a 36-d period from April 21 to June 1, 2000, for a total of 700 fish. The study was based on in-river migrants from various hatchery and wild sources upstream and those taken directly from a hatchery. Migrating fish were captured within the surface collector at Rocky Reach Dam, which is 32.6 km upstream of Rock Island Dam. Hatchery fish were taken from the Turtle Rock Hatchery, which is located immediately upstream of Rocky Reach Dam and operated by Chelan PUD. The upstream release (treatment group) was at the Rocky Reach tailrace, and as such, a tagging and holding facility was established at that site. The downstream release (control group) occurred in the tailrace of Rock Island Dam. For that release group, a separate tagging and holding facility was established downstream of that dam. Both control and treatment fish were treated similarly to minimize potential bias. That is, transport from the originating site to the tagging facility was standardized in respect to the length of transport time. Also, water temperatures were monitored, and supplemental oxygen was used. Fish were surgically tagged using similar procedures as those described in Adams et al. (1998), with modifications as described in Stevenson et al. (2000). Fish were allowed to recovery for approximately 40–48 h prior to release, with releases occurring mid-morning. Release of fish at both sites occurred mid-channel in relatively high velocity areas to minimize predation. Collectively, 693 hatchery yearling Chinook and 698 run-of-river yearling Chinook were released at the two locations. Since the survival models used assume that all fish detected are alive, it was necessary to test the hypothesis that a fish killed passing the dam would not drift downriver and be detected at a downstream site. Therefore, 20 dead fish implanted with transmitters were released within the Rocky Reach tailrace and 18 dead fish with transmitters were released within the Rock Island tailrace.

Fish tagged at Little Goose Dam were released daily at each site from April 18 to June 25 during the spring (yearling Chinook salmon and juvenile steelhead) and June 6 to July 2 during the summer (subyearling Chinook salmon). Individuals of hatchery origin were selected from the conglomerate of hatchery and wild fish collected in the juvenile bypass system at the dam using existing fin clips or other marks. Fish captured in the bypass system were removed each morning, held in tanks with flowing river water for about 24-h prior to tagging, and were held again about 30-h after tagging prior to release. Transmitters were surgically implanted using procedures similar to those in Adams et al. (1998), except a passive integrated transponder (PIT) tag weighing 0.07 g in air was inserted in the body cavity prior to closing the incision. The PIT tag enabled detection in the juvenile bypass system as part of the system-wide monitoring program, aiding detection within this route. For yearling Chinook salmon and juvenile steelhead, an average of 32 treatment and 22 control fish were released each day. For subyearling Chinook salmon, an average of 57 treatment and 33 control fish were released daily. Release times were alternated daily between approximately 1800–2100 hours and 1000–1300 hours, with the order of which group was released first (treatment or control) alternated as well. This was to ensure fish would be passing the dam in roughly similar proportions throughout the day and night. We released the treatment fish 21 km upstream from the dam and the control fish 500 m downstream from the dam by gently pouring them out of their 19-L holding buckets into a mid-river location after transportation to access sites by auto and boat. As in the Rock Island study, the distances from the tagging facility and release sites differed for treatment and control fish, so the control fish were transported longer than necessary to equalize the handling between the groups. The transport times were about 60 min. We released about 2,000 of each cohort with 60% apportioned to the treatment group. More treatment than control fish were released each day to account for potential mortality of treatment fish between release upstream from the dam and mixing with the control fish downstream from the dam (Burnham et al. 1987). About 50 fish of each cohort were also tagged, euthanatized, and released at the control site to test the assumption of detection of dead fish with live tags.

For the Ice Harbor Dam study, fish collection, tagging, and handling were similar to those previously described at Little Goose Dam except fish were collected from the juvenile bypass system at Lower Monumental Dam (32 km upstream from Ice Harbor Dam) in the afternoon prior to tagging. Treatment fish were released 50–60 km upstream from Ice Harbor Dam during the spring and 5 km upstream during the summer study. Control fish were released 2 km downstream of the dam during both the spring and summer studies. All releases were from a boat anchored in the thalweg of the river. To facilitate temporal mixing of the treatment and control fish, releases were distributed over a period of 4–5 h during both daytime and nighttime. About 2,000 of each cohort were released with 60% apportioned to the treatment group. Mortalities from tagging or posttag holding (usually about 1% of the number tagged) were released in the tailrace to test the assumption of detecting dead fish with live tags.

TELEMETRY SYSTEMS

The physical environment, the biology of the study animals, the research questions, and the available technologies dictated the type and design of the telemetry systems. The environments near these dams are characterized by freshwater with conductivities of about 85–300 μS/cm, localized electronic noise from transformers and power lines, turbulent tailrace areas, and abundant electricity and water. Oddly, the last two are often the hardest to acquire at dams and their use requires coordination in advance of the study. The populations of interest were juvenile salmonids during their seaward migration, some of which weigh less than 10 g. An actively transmitting tag was used to describe fish approach and route-specific passage, due to the requirement to detect fish at distances of many meters from the dam and in rivers several hundred meters wide. At the time of these studies, radio telemetry offered smaller transmitters with longer lives than established acoustic telemetry systems, which were attractive attributes given the small sizes of the fish. In addition, the low operating frequencies of acoustic telemetry are not conducive to monitoring fish in the turbulent tailrace areas, which was an area of interest in some studies. The primary disadvantage of radio telemetry relative to acoustic telemetry was the poor transmission distance through water (Winter 1996) and presence of an external antenna (Murchie et al. 2004). Radio telemetry was chosen for the studies, though each group used slightly different systems.

Rock Island Dam

The receiving system at Rock Island Dam was based on commercially available equipment (Lotek Wireless Inc.) operating at a frequency near 150 MHz as described by Skalski et al. (2002). The aerial systems generally had a range of about 100–300 m, depending on the water conductivity, the transmitter, its depth, and other factors (Sisak and Lotimer 1998; Johnson et al. 2000; Peters et al. 2008). The expected ranges of underwater antennas for the transmitters used and the conductivities encountered were about 5–9 m (Johnson et al. 2000; Bunt et al. 2003; Beeman et al. 2004). Underwater antennas included stripped coaxial cable and standard dipole types described by Beeman et al. (2004).

Aerial detection arrays at the dam consisted of 6-element Yagi antennas and Lotek SRX-400 receivers spaced about equally across the dam. These receivers scan individual frequencies for a user-specified time, usually slightly longer than the burst rate of the transmitter, and cycle among them repeatedly. This type of receiver, termed narrow-band, is less susceptible to ambient nose than some other types, but only listens to one frequency at a time. As such, it will fail to detect a tagged fish if it passes the detection area when listening to the other frequencies. The dam arrays were installed primarily along the upstream side of the dam and were configured with similar gain settings, antennas, and transmission line losses to enable the antenna nearest the fish to be assigned based on received signal strength and to result in a consistent detection range across the dam. This enabled the areas of approach to the dam to be described based on a series of independent systems with similar ranges and sensitivities. Aerial antennas in the tailrace were used to describe fish presence or to confirm a passage location, but were not used to determine passage survival due to the potential for fish from different routes to mix in the tailrace and detecting dead fish with live tags near the dam. Aerial antennas in the forebay were used as part of the system to determine passage route and time at the spillway. Those in the tailrace were used to confirm passage when only one of the two arrays in the forebay detected a fish passing.

Underwater antenna arrays at the dam were used to identify movements within a few meters from the dam and fish passage locations. Each array consisted of a series of antennas connected to Lotek SRX-400 and Lotek DSP-500 receiver combinations. The addition of the DSP-500 enables scanning of 25 frequencies simultaneously. This type of receiver, called wide-band, is generally more susceptible to ambient noise than narrow-band types, but may have a higher probability of detecting tagged fish due to the lack of a scan time between frequencies. At the spillway, some bays were capable of spilling at two elevations, so it was necessary to monitor both locations. At these bays dipole antennas were placed on the upstream apex of the pier noses (structural divisions between spill bays or turbine intakes) 2 m above the spillway crest, and 2 m below the water surface minimum operating pool elevation. All other spill bays (top-spill only) had antennas on each pier nose 2 m below the water surface. At both powerhouses, primary arrays of dipole antennas were mounted at the center of each turbine intake at a depth of 9.1 m below the water surface, approximately halfway between the normal water surface elevation and the roof of the turbine intake. Secondary arrays at both powerhouses were located within the head-gate slots of each turbine intake, with dipole antennas located mid-way between the intake floor and roof. At Powerhouse 1, the antennas were mounted to removable sleds on each side of the intakes at a depth of 15.2 m. At Powerhouse 2, the dipole antennas were bolted to the intake walls on each side of the intake 24.4 m below the surface. Passage through the three adult fishways was monitored using a pair of dipole antennas within and near their exits, but no tagged fish were detected there and it will not be included in the results.

Aerial arrays were installed at several locations along the river downstream from the dam to estimate survival and recapture probabilities of fish that passed the dam. These arrays used 6-element Yagi antennas and SRX-400 receivers mounted near shorelines or on islands creating three lines of detection located 27, 31, and 36 kilometers downstream from Rock Island Dam. The receivers were powered with solar panels and batteries.

The low power of the small transmitters demanded careful receiving system design to enable high recapture probabilities. Transmission losses in the coaxial cable and combiners connecting antennas and receivers were abated to maximize received signal strength and detection distance. A loss of 6 dB will reduce received power by 50% (see Section 5.1), so abating transmission losses can be important. This is especially true in areas where a large volume of water must be monitored (e.g., turbine intakes), and when using small transmitters with their inherently low output powers. In these studies, transmissions transited up to about 100 m of cable. Aerial antennas were connected to receivers with RG-58 coaxial cable with a signal loss at 150 MHz of 4.8 dB per 30 m. This cable is flexible, relatively inexpensive, and easy to work with. Signal losses in RG-58 cables were mitigated by amplifying the signals near the antennas (after a short piece from the antenna) and attenuating them near the receiver to a net loss of ±1 dB. Underwater antennas used RG-58 coaxial cable except at the second powerhouse secondary array. At that location, an armored cable with similar transmission characteristics to RG-58 was used due to high velocities associated with the horizontal bulb-type turbine units (approximately 3 m/s). At locations where nonarmored cable was used and abrasion was an issue, coaxial cable was routed through protective conduit. At both Rock Island powerhouses, individual antennas were amplified near the water surface to minimize additional signal degradation, and we abated transmission losses from this point to the receiver as described previously.

All telemetry systems at Rock Island Dam were checked daily. During these inspections, all coaxial cables to the receivers were visually inspected, as were individual antennas where possible. Also, all telemetry receivers were checked to ascertain the status of the internal memory, the voltage of the backup battery, to ensure that the charging system was functioning normally, and to confirm that the receiver's internal clock was within 1 s of satellite GPS time. More detailed inspections and testing occurred about three times per week. During these tests, voltage to all amplifiers was tested, and all antennas that were easily accessible were tested with a live transmitter or a stripped coaxial cable antenna attached to a signal generator. On a weekly basis, antennas that had not been tested throughout the week were checked. These antennas were typically within turbine intakes and the secondary spillway array. Due to head-gate covers and spill activity that created safety issues, these antennas were difficult to access and test. Downloading receiver data at Rock Island Dam occurred weekly, unless high detection volume created memory issues. This occurred at receivers monitoring locations where high fish passage occurred, or where tagged fish congregated and milled about prior to passing the dam. For most receivers, it was necessary to download data two to three times per week. For receivers where little detection occurred, we typically downloaded the receivers to simply inspect the data and ensure that all systems were functioning properly. All data were archived on external media, a desktop computer hard drive, and the laptop computer hard drive used to download the data. Data were then uploaded to a central FTP site where it was available to the analytical team.

Little Goose Dam

The receiving systems at Little Goose Dam were 150 MHz equipment similar to that used at Rock Island Dam. The transmitters used at Little Goose Dam were smaller and weaker than those at Rock Island Dam, so the detection ranges were shorter. The aerial systems at Little Goose Dam generally had a range of about 100–300 m, but the ranges of underwater antennas were about 4–7 m (unpublished data). Underwater antennas included stripped coax, standard dipole, and armored dipole types described by Beeman et al. (2004). Aerial and underwater antennas in the tailrace were used to describe fish presence and movements.

Aerial detection arrays at the dam consisted of 3-element Yagi antennas and Lotek SRX-400 receivers spaced about equally across the dam. The dam arrays were installed along the upstream sides of the dam and, as at Rock Island Dam, were configured consistently among coverage areas to enable the antenna nearest the fish to be assigned based on received signal strength and to result in a consistent detection range across the dam. Aerial antennas were used to describe fish entry into the forebay (2 km upstream from the dam) and the area of approach to the dam. Detections in the tailrace were used to confirm presence and movements.

Underwater antenna arrays each consisted of a series of antennas connected to one of two Multi-protocol Integrated Telemetry Acquisition System (MITAS; Sigma8, King City, Ontario, Canada) telemetry receivers. The independent arrays consisted of a pair of standard dipole antennas on the sides of each pier nose at elevations of 6.2 m (primary array) and 5.6 m (secondary array) above the spillway crest, resulting in four antennas within each spill bay at elevations of 9.8 m and 10.4 m below the water surface. The antennas were mounted to a trolley that was slid to the desired elevation within a 30 cm diameter steel pipe bolted to the spillway pier nose as shown in Figure 4. At the powerhouse, pairs of armored dipole antennas were mounted equidistant from each side of the 5.8 m wide fish screen frame near its bottom, at the tips of the inclined screen, and near its top for six antennas per intake slot. The inputs from the antennas were combined by turbine unit near the powerhouse deck such that the antennas at the screen tip were the primary array and the others were the secondary array. Armored dipole antennas were also installed in the Little Goose Dam juvenile bypass collection channel near its outlet. The PIT tag interrogation system in the juvenile bypass, part of the separate system-wide monitoring program, was used as the second detection array in this route. Passage through the adult fishway was monitored using four stripped coaxial cable antennas, but very few fish were detected in this route and it will not be included in the results. Fish passage route and time was assigned based on detections at underwater antennas within the routes.

Transmission losses through coaxial cables were abated at Little Goose Dam, though slightly differently than at Rock Island Dam. Losses in RG-58 cables from all aerial antennas and underwater antennas at the spillway were abated as at Rock Island Dam. The underwater antennas at the powerhouse at Little Goose Dam were connected to the receiving systems via equal lengths of RG-8 cable (signal loss at 150 MHz = 2.1 dB per 30 m) armored with braided stainless steel wire and additional insulation. This cable has a larger diameter, is heavier, and

FIGURE 4. Photo of pipes on pier noses of the Little Goose Dam spillway used to locate underwater dipole antennas. The inset shows a trolley and dipole antennas being placed in one of the pipes at Ice Harbor Dam prior to the use of dual plane dipole antennas there.

is less flexible than RG-58 cable, but has less signal loss and greater resistance to abrasion. We chose this cable because the chance of abrasion in this application was high and the distances from the underwater antennas to the nearest access point above water were often nearly 30 m, increasing the potential for signal loss. Losses between the access point near the top of the head gate slot and the receiving system were abated with amplifiers and attenuators as previously described.

Aerial arrays were installed at three locations along the river downstream from the dam as in the Rock Island study. Arrays on each bank of the river used 3- and 6-element Yagi antennas, SRX-400 receivers, and a 90 amp-hour deep-cycle battery connected to a regulator and a 75-W solar panel. These arrays were located 12, 27, and 45 km downstream from the dam.

The telemetry systems at Little Goose Dam were downloaded and inspected daily. Data from underwater antennas was automatically written to a computer hard drive near midnight each day via the MITAS system. That system also identified antennas that had not had a detection for a user-specified time, we used 12-h, and these were checked daily by visually inspecting the antenna and cabling as well as sending a radio signal along the cabling from the antenna insertion point. Aerial arrays were checked manually for proper receiver settings, the remaining memory, and the presence of detections identified as possible collisions or noise. Some arrays were downloaded manually and others were downloaded using 900 MHz radio modems. Transmitters with replaceable batteries ("beacon tags") were placed at the distal ends of the underwater arrays as a check of overall system components. Beacon tags had a specific frequency and code and sent signals through the system several times per hour.

Ice Harbor Dam

The receiving system at Ice Harbor Dam was based on wide-band equipment developed by NOAA Fisheries operating at a frequency of 30 MHz. Receivers and antennas were installed similar to that described for the other studies with some minor differences related to project-specific objectives, physical attributes of study areas, and equipment used. The 30 MHz aerial systems generally had a range of about 400–600 m for a transmitter 3 m below the surface and the underwater antennas had a range of 15–18 m (authors' unpublished data). Aerial antennas installed along the shorelines and on a floating platform near the middle of the river 500 m upstream from the dam were used to determine time of entry into the forebay. Aerial and underwater antennas in the tailrace were used to describe fish presence and movements, and confirm passage but not to assign passage route. Only fish detected in the tailrace were assumed to have passed the dam.

Underwater antenna arrays at Ice Harbor Dam were used to identify movements within a few meters of the dam and fish passage locations. Individual spill bays were monitored by an independent array consisting of four shielded dual plane dipole antennas on the medial side of each pier nose. The antennas were placed at elevations of 6 m and 12 m below the forebay water surface using a trolley and pipe arrangement like that shown in Figure 4 (note that standard dipole antennas are shown in Figure 4 rather than dual plane dipole antennas). Aluminum plate shielding between antennas minimized detection on the back side of the antenna and was used to limit detection of fish in adjacent bays. The dual plane dipole antenna is similar to dipole antennas described by Beeman et al. (2004) except that it has two sets of dipoles, one each in the horizontal and vertical planes. The dual plane dipole antenna has a larger field than standard dipole antennas because it maximizes both the horizontal and vertical fields. Underwater antennas at the powerhouse were pairs of striped armored RG-8 coaxial cable with braided stainless steel wire and additional insulation attached to each side of the 5.8 m wide fish screen frame near its bottom. Inputs from the three intake slots per turbine were combined providing six antennas per turbine. Aerial loop antennas were installed in the juvenile bypass collection channel near the primary dewatering screens. The PIT interrogation system in the juvenile bypass as part of a separate system-wide monitoring program was also used for detection in this route. Passage through the adult fishway was monitored by a PIT interrogation system, but very few fish were detected in this route and it will not be included in the results.

The combination of amplifiers and attenuators to balance the range between arrays monitoring individual routes was not used at Ice Harbor Dam. At this dam an approach that included a combination of a) equal cable length runs and b) detection field mapping and range testing was used. The attenuation of radio signals through cable at 30 MHz is less than half that at 150 MHz. Individual receiver attenuation settings were adjusted during range testing to balance the detection field for receivers used to describe approach and passage behavior.

Aerial arrays were installed at four locations downstream from the dam to estimate survival and recapture probabilities of fish passing the dam. The arrays were 15, 20, 61, and 66 km downstream from the dam and each was powered with two 110 amp-hour batteries connected to a regulator and pair of 58-W solar panels.

At Ice Harbor Dam telemetry receivers were automatically downloaded four times daily using polling software that communicated with each receiver using 2.4 GHz wireless radios. Clocks in individual receivers were adjusted to within 1 s of satellite GPS time as part of the automated downloading process. After downloading, individual data files were compressed by recording the first time a radio-tagged fish was detected and counting the number of detections where the time difference between adjacent detections was less than or equal to 5 min. When the difference between adjacent detections became greater than 5 min, a new line of data was created. All compressed data were combined and loaded into a database where automated queries and algorithms were used to remove erroneous data. Beacon transmitters on different frequencies were placed in the detection field of every receiver location to check system performance. These tags transmitted 20 times every hour. The automated download process also included a series of performance monitoring checks which included successful communication with receivers, successful downloading, and verification of beacon tag records to verify functionality. E-mail notifications were automatically sent to electronics technicians if the performance monitoring system identified potential problems.

DATA PROOFING

Each study used automated methods to convert raw data records to final data sets prior to analysis. The methods used in each study were generally similar, consisting of a series of logical steps to remove false positive records from the data. Records were eliminated as known false positives if they were not from transmitters released as part of the study, based on radio frequency and transmission code; were from a time prior to when that particular fish was released; or passed through a route unavailable at the time (i.e., a closed spill bay). The data were further reduced by identifying potential false positives by examining the geographical order of detections, e.g., a valid detection could not occur at a downstream site before it occurred at an upstream site. A minimum received power level, determined from know false-positive records, was also used at Rock Island Dam. Each study also used some form of a minimum number of records from a fish per unit time (e.g., at least two records within 30 s, at least 3 records in 5 min) and none accepted single detections of a fish. Records that did not fit the criteria were marked for manual inspection by at least two people and were discarded unless there was agreement between them as to the validity of the record. The Rock Island study took this further, and used two independent private companies to proof all the data and compared results to arrive at a final data set for analysis. Each study also manually proofed a subset of the records to verify the automated process was working as intended.

STATISTICAL ANALYSES

Each study was designed primarily to estimate the proportions or probabilities of passage through each route at the dam as well as the route-specific survival probabilities. The studies at Little Goose and Ice Harbor dams were also designed to estimate time and general location of fish arrival to the forebay, survival through it, the initial approach areas at the dam, times to pass through the forebay and tailrace, and measures of guidance success into the fish bypass systems. These measures included the fish guidance efficiency (FGE), defined as the proportion or probability of fish entering the fish bypass system divided by that of fish detected entering turbine intake slots; the fish passage efficiency (FPE), defined as the total proportion or probability of passage at all nonturbine passage routes at the dam; and the spillway passage effectiveness (SPY), defined as the proportion or probability of fish passing via the spillway divided by the percentage of total dam discharge passing via the spillway.

Estimates of survival were made using extensions of the mark-release–recapture methods of Cormack (1964), Jolly (1965), and Seber (1965). Inasmuch as the detection of fish at any site is a product of the probability of survival to the site and the probability of detection at the site, these probabilities must be known or estimated separately. This entails a minimum of two detection arrays at each site, or one detection site downstream from the area of interest so that recapture probabilities can be estimated for the sites upstream. Estimates of survival in areas upstream from the dam of interest were made using single-release models as described by Burnham et al. (1987) and are based on the fish released at a single location upstream. Estimates of dam passage survival were based on paired-released models of Burnham et al. (1987), and represent the ratio of survivals of a treatment group released upstream from the dam to that of a control group released downstream. The purpose of this model is primarily to restrict the space over which survival is estimated to the distance from the dam itself to the site of control group release. In these studies, the distances of the control group release sites from Rock Island, Little Goose, and Ice Harbor dams were 0.5, 0.5, and 1.5 km, respectively. Studies at Little Goose and Ice Harbor dams also estimated survival through the dam forebays, which were defined as the waters beginning 2.0 km and 0.5 km, respectively, upstream from the dams.

Route-specific and overall dam passage survivals in the studies at Rock Island Little Goose dams were estimated using the route-specific survival model (RSSM) described by Skalski et al. (2002). This model requires two independent detection arrays within each route of passage for the estimation of route-specific recapture probabilities via a Lincoln/Peterson-type estimator, and results in route-specific passage probabilities and route-specific survival probabilities as well as overall dam passage survival weighted by the passage probabilities of the routes. The primary and secondary arrays used for this purpose were identified previously. Time and route of passage at Rock Island Dam was based on underwater antennas at the powerhouses, aerial and underwater antennas at the spillway, and at times aerial antennas in the tailrace. All passage routes and times were assigned based on underwater antennas within the routes of passage at Little Goose Dam.

Route-specific and overall dam passage survivals at Ice Harbor Dam were estimated using the paired-release model of Burnham et al. (1987). In this study, the recapture probabilities were assumed equal among the routes of passage and thus differences in recapture probabilities and their uncertainties were not incorporated in the estimates of survival probabilities. This assumption was based on past studies where high detection percentages (approaching 100%) for all passage routes combined were observed. As such, route-specific passage proportions were estimated directly from the numbers of tagged fish detected passing each route. Route of passage through Ice Harbor Dam was based on the last time a fish was detected on an underwater passage-route antenna and was assigned only to fish that were subsequently detected in the tailrace on stilling-basin, turbine draft tube, or tailrace-exit telemetry receivers. Tailrace detections were used to confirm passage because fish could be detected on a passage-route receiver while still in the forebay. Overall dam survival, and route-specific survival were independently estimated by regrouping treatment fish into various 'virtual release groups' based on their detections. The dam survival estimate included fish which did not pass the dam as well as fish that passed the dam with or without an assigned passage route. Route-specific survival included only those fish for which a passage route could be determined.

RESULTS AND DISCUSSION

The methods employed in each study were successful in the goal of estimating survival probabilities with small standard errors (Table 2). Data from yearling Chinook salmon of hatchery origin are used as examples in the tables, but results from other groups were also described as part of each study. Meeting the goals of the studies resulted from the combination of proper design, high recapture probabilities (except as noted below), and large sample sizes. Proper design elements included adequate spatial coverage of passage routes, use of transmitters appropriate for the sizes of the fish to be studied, and spacing downstream arrays far enough away from the dams to prevent detections of dead fish with live transmitters. The choices each research group made along these topics were based on subject-matter knowledge of fish behavior, radio telemetry, and the study areas. Several euthanatized fish with live transmitters were detected at the first downstream array in the Little Goose Dam study, so data from this array were omitted from analyses. This was possible because that study used three downstream sites, rather than the minimum of two, in case one had technical problems or became unsuitable for analysis. No euthanatized fish were detected at downstream arrays in the other studies.

Each of the studies resulted in high recapture (i.e., detection) probabilities at most sites. At Rock Island and Little Goose dams the estimated recapture probabilities at each of the routes were typically near 1.0 (Table 3). The exceptions were the spillway arrays at Little Goose Dam, and the powerhouse 2 arrays at Rock Island Dam. This shows that high recapture probabilities can be achieved at dams with careful deployment of radio telemetry equipment even when the study sites are large and the transmitters are weak. High recapture probabilities are not required to estimate survival, but they increase the information used from each fish and contribute to smaller errors in estimation. The key to high recapture probabilities is designing and implementing properly sized and placed detection fields. The sizes of detection fields can be estimated from the literature (Bunt et al. 2003; Beeman et al. 2004), or determined empirically at the study site with the transmitters to be used. The dams in these studies were large, yet the passage routes themselves were often suitable for efficient radio detection systems. For example, the turbines had intakes as deep as 30 m, which is well beyond the range of small (weak) transmitters and aerial antennas, but the area within each turbine intake was small enough to monitor with strategically placed underwater antennas. At Little Goose Dam, the PIT tag interrogation system was part of the detection array within the juvenile bypass system, but this route is relatively easy to monitor with aerial or underwater radio systems and the high recapture probability there ($\geq$0.999) could probably be achieved without the PIT system.

The double array requirement of the RSSM provides a useful method of estimating recapture probabilities, but only when the arrays are properly designed. To determine passage, a proper detection array is one that detects fish where they are committed to passage and one that covers the entire passage route. Perry and Skalski (2008) showed that when using a double array system at least one array must cover the entire passage route to result in unbiased estimates of recapture probabilities. In most cases these criteria were met, but not at the Little Goose Dam spillway. Each array at the Little Goose Dam spillway was near the deepest part of the tainter gate (the passage area), but the coverage area was apparently insufficient. The data from that double array indicated the recapture probability of yearling Chinook salmon was 0.998 (SE 0.001), but there were many fish detected downstream that were not assigned

TABLE 2. Estimates (SE) of survival probabilities (S) and passage (Pa) results for yearling Chinook salmon *Oncorhynchus tshawytscha* of hatchery origin from the three studies. Results from Little Goose and Ice Harbor dams are listed by treatment. Passage results are probabilities at Rock Island and Little Goose dams and proportions at Ice Harbor dam. The λ parameter represents the joint probability of survival and detection at sites downstream from the dam (see Skalski et al. 2002). ph1 and ph2 indicate first and second powerhouses at Rock Island Dam.

	Little Goose Dam			Ice Harbor Dam		Rock Island
	Bulk1	GasCap	Bulk2	High Spill	Low Spill	Dam
Parameters	Estimate (SE)	Estimate (SE)	Estimate (SE)	Estimate (SE)	Estimate (SE)	Estimate (SE)
S pool[1]	0.957 (0.012)	0.998 (0.003)	0.947 (0.011)	--	--	0.969 (0.012)
S forebay	0.998 (0.004)	0.981(0.007)	0.996 (0.004)	--	--	--
S dam[2]	0.994 (0.018)	0.989 (0.010)	1.005 (0.018)	0.921 (0.019)	0.940 (0.019)	0.978 (0.021)
S spill	0.995 (0.018)	0.991 (0.010)	1.011 (0.018)	0.955 (0.019)	0.968 (0.019)	1.015 (0.022)
S turbine	0.995 (0.498)	0.845 (0.109)	0.880 (0.114)	0.722 (0.176)	0.834 (0.066)	--
S ph1	--	--	--	--	--	0.912 (0.071)
S ph2	--	--	--	--	--	0.972 (0.024)
S bypass	0.989 (0.033)	1.000 (0.015)	0.998 (0.035)	0.927 (0.072)	0.949 (0.037)	--
λ	0.968 (0.012)	0.987 (0.007)	0.947 (0.013)	0.825 (0.018)	0.839 (0.018)	0.960 (0.010)
Pa spill	0.826 (0.022)	0.814 (0.018)	0.834 (0.018)	0.933	0.750	0.245 (0.025)
Pa turbine	0.015 (0.007)	0.026 (0.007)	0.030 (0.008)	0.021	0.163	--
Pa ph1	--	--	--	--	--	0.074 (0.014)
Pa ph2	--	--	--	--	--	0.681 (0.027)
Pa bypass	0.159 (0.022)	0.160 (0.017)	0.136 (0.017)	0.046	0.086	--
FPE	0.985 (0.007)	0.974 (0.007)	0.970 (0.008)	0.979 (0.005)	0.914 (0.008)	--
FGE	0.916 (0.040)	0.859 (0.038)	0.821 (0.047)	0.686 (0.032)	0.654 (0.030)	--
SPY	2.72	2.37	2.75	1.36	2.43	--

[1]Represents survival from release to detection at the forebay entrance at Little Goose and Ice Harbor dams and release to detection at the dam at Rock Island Dam.

[2]Represents survival from the forebay entrance to the control group release point at Little Goose and Ice Harbor dams and detection at the dam to the control release point at Rock Island Dam.

Table 3. Estimates (SE) of route-specific recapture probabilities (*P*) at Rock Island, Little Goose, and Ice Harbor dams.

Route	Array1 *P (SE)*	Array2 *P (SE)*	Overall Route *P (SE)*
Rock Island Dam			
Powerhouse 1	0.864 (0.073)	0.905 (0.064)	0.987 (0.086)
Powerhouse 2	0.428 (0.034)	0.420 (0.034)	0.668 (0.02)
Spillway	0.893 (0.041)	0.682 (0.055)	0.966 (0.056)
Downstream[2]	--	--	0.960 (0.011)
Little Goose Dam			
Turbine	1.000 (na)[3]	0.893 (0.058)	1 (na)[3]
Bypass	0.994 (0.006)	0.988 (0.008)	0.999 (na)
Spillway[1]	--	--	0.956 (0.007)
Downstream[2]	--	--	0.980 (0.003)
Ice Harbor Dam			
Turbine[3]	1 (na)	--	1 (na)
Bypass[3]	1 (na)	--	1 (na)
Spillway[3]	1 (na)	--	1 (na)
Downstream[2]	--	--	0.991 (0.003)

[1] Determined using downstream detections, not the double arrays.
[2] Represents pooling of data from sites downstream from the dam.
[3] Not measured using a double array and assumed to be 1.000

a passage route. A closer inspection of the data suggested fish passing the spillway undetected was the most likely cause, so a variant of the RSSM was used to estimate recapture probability for this route. In that model, the recapture probability of the bypass route was estimated using the double arrays, that of the turbine route was set manually to 1.0 (based on previous information), and the recaptures downstream were used to estimate the recapture probability of the spillway route. The manual setting for the turbine route was required because very few fish were detected in that route and the data available to estimate recapture probability was sparse, resulting in failure of some models to converge. The sparse detections could also be due to very poor recapture probability, but infrequent turbine passage is consistent with studies at this dam based on active hydroacoustics (Ham et al. 2007), so this was not strongly suspected. The results of this model agreed with results of the overall recapture probability at the dam based on recaptures downstream, and indicated the recapture probability at the spillway was 0.956 (SE 0.007) for yearling Chinook salmon. The disparity was greatest for juvenile steelhead (data not shown), where the double array system indicated a recapture probability of 0.994 (SE 0.002) and the revised method indicated 0.768 (SE 0.017). Additional underwater antennas such as shallow and deep ones like those used at Ice Harbor Dam may have decreased or eliminated this bias. We have used this type of array at several Columbia River dams with success and it was implemented at Little Goose Dam for a study in 2009 without evidence of bias (unpublished data).

The low recapture probability at the Rock Island Dam Powerhouse 2 exemplifies the value of estimating route-specific recapture probabilities. Assuming constant and high recapture probabilities among all routes would have resulted in biased estimates of passage and survival probabilities. The low recapture probabilities were believed to be due to the high water velocities present there. Velocities near the antennas in the turbine intakes at this powerhouse were about 3 m/s. Given the large area to monitor within these intakes, it is probable that some tagged fish passed through the detection field before a radio transmission could be recorded. Likewise, the primary array at this location also had a low recapture probability relative to other locations. It may well be that tagged fish approached the turbine intakes at a depth that precluded detection at this underwater array.

At Ice Harbor Dam the recapture probabilities were assumed to be 1.0 at all routes. Only the major passage routes were monitored and fish could pass the dam via the unmonitored navigation lock. Of 1,040 yearling Chinook salmon and 978 juvenile steelhead passing Ice Harbor Dam, the passage route was determined for 99.4%. Only six yearling Chinook salmon and five juvenile steelhead passed the dam without a passage route determined. The passage route was determined for 99.6% of the 1,191 subyearling Chinook salmon passing the dam. In situations like the study at Ice Harbor Dam, where we were able to assign a passage route to over 99% and the majority of the fish passed through a single route (the spillway), the RSSM model would produce similar results. Variances of estimates increase as parameters are added to models, due to the error associated with additional estimation, so including the extra parameters for recapture probabilities will therefore inflate variances. On the other hand, incorporating recapture probabilities into the estimates of survival can prevent biased results, such as those that would have occurred at Rock Island Dam due to the differences in recapture probabilities among routes. The RSSM model also provides maximum-likelihood estimates of route-specific passage probabilities, which we see as an additional advantage over some other methods. However, the large number of parameters estimated by the model often requires data from large numbers of fish to converge, which may not always be available.

Two of these studies were based on the premise of estimating route-specific passage and survival probabilities using estimates of recapture probabilities at each route. As such, they required more equipment than the study at Ice Harbor Dam where recapture probability within each route was assumed to be 1.0. However, where recapture probabilities are assumed to be 1.0, researchers need to incorporate system diagnostics and more redundancy to mitigate for equipment failure which could reduce the recapture probability. In practice, few studies of fish achieve recapture probabilities of 1.0. In studies where route specific survival is not required, the amount of equipment, complexity of the study, and ultimately study cost can be reduced. For instance, fewer receivers would be needed to describe just forebay residence times, or migration timing between dams, if the need to adjust the data for the potential of missed fish was not present.

Part of the success of the detection systems was due to ongoing system diagnostics provided by beacon tags, detailed and thorough preseason testing, adequate back up equipment, and timely response to problems to minimize data losses. Abating losses in the transmission cables preserved received signal strength and resulted in optimal system range. For example, some cables at Little Goose Dam were up to 100 m long and would have reduced received signal strength by 16 dB if not mitigated. A loss of this magnitude will result in a reduction of range by a factor of a 6. In some cases, receivers might be moved closer to the antennas or receivers and antennas could have been added to the system to reduce cable lengths, but telemetry receivers are expensive and simply adding more is not always possible. Mitigating for losses in transmission lines is relatively inexpensive and can reduce the number of receivers required. This affects overall costs by reducing receiver numbers and potentially downloading and maintenance times, and simplifies creating consistency among receiving systems by reducing their number.

The choice of operating frequency during the studies (i.e., 30 MHz versus 150 MHz) was not a significant obstacle, but required frequency-specific designs and implementation. For example, the size and spacing of aerial and underwater antennas is dependent on operating frequency. The U.S. Government recently adopted the 162–174 MHz range for its radio telemetry studies and other groups have made similar choices. However, these choices were not based on the technical merits of these frequencies for animal telemetry, but rather convenience in allocating private and government uses of the radio spectrum. The Bureau of Commercial Fisheries (now NOAA Fisheries Service) purchased the licenses for the frequencies used at Ice Harbor Dam from the Federal Communications Commission in the 1970s. At that time, 30 MHz was chosen because of its advantage in transmission through water compared to higher frequencies. Frequency selection is a tradeoff among several factors, including size of transmitting and receiving antennas and transmission losses through water and other media.

The studies we described achieved similar goals with different applications of radio telemetry. Each system was chosen based on a variety of factors including the study animals, the study environment, the available technologies, and each researcher's perceptions about how these interacted. There are certainly conditions under which other radio or acoustic telemetry systems could meet these study objectives. Each study was successful in their goals to develop systems with high recapture probabilities by adhering to the basic tenets of the chosen technology.

ACKNOWLEDGMENTS

Many people were responsible for the completion of the research studies we describe. These include staff within our respective organizations, those caring for fish passing through the dams studied, dam operators and their staff, and funding agencies. The study at Rock Island Dam was funded by the Public Utility District No. 1 of Chelan County and those at Little Goose and Ice Harbor dams were funded by the Walla Walla District of the U.S. Army Corps of Engineers.

REFERENCES

Adams, N. S., D. W. Rondorf, S. D. Evans, J. E. Kelly, and R. W. Perry. 1998. Effects of surgically and gastrically implanted radio transmitters on growth and feeding behavior of juvenile Chinook salmon. Transactions of the American Fisheries Society 127:128–136.

Beeman, J. W., C. Grant, and P. V. Haner. 2004. Comparison of three underwater antennas for use in radio telemetry. North American Journal of Fisheries Management 24:275–281.

Bentley, W. W., and H. L. Raymond. 1968. Collection of juvenile salmonids from turbine intake gatewells of major dams in the Columbia River System. Transactions of American Fisheries Society 97:124–126.

Bickford, S. A., J. R. Skalski, R. Townsend, D. Parks, S. McCutcheon, and R. Frith. 2000a. Project survival estimates for yearling summer steelhead migrating through the Wells Hydroelectric Facility, 1999. Douglas County Public Utility District No. 1, Wenatchee, Washington.

Bickford, S. A., J. R. Skalski, R. Townsend, R. Frith, R. Alexander, D. Park, and S. McCutcheon. 2000b. Survival estimates for radio-tagged and PIT-tagged yearling summer steelhead migrating through the Mid-Columbia River, 1999. Douglas County Public Utility District No. 1, Wenatchee, Washington.

Brown, R. S., S. J. Cooke, W. G. Anderson, and R. S. McKinley. 1999. Evidence to challenge the "2% rule" for biotelemetry. North American Journal of Fisheries Management 19:867–871.

Bunt, C. M., C. J. Cooke, and D. P. Philip. 2003. A modified downrigger for detecting radio transmitters in deep water. North American Journal of Fisheries Management 23:258–263.

Burnham, K. P., D. R. Anderson, G. C. White, C. Brownie, and K. H. Pollock. 1987. Design and analysis methods for fish survival experiments based on release-recapture. American Fisheries Society, Monograph 5, Bethesda, Maryland.

Cormack, R. M. 1964. Estimates of survival from the sighting of marked animals. Biometrika 51:429–438.

Endangered Species Act (ESA). 1973. U.S. Code, Title 16. pp. 1531–1544.

Groves, A. B., and A. J. Novotny. 1965. A Thermal-Marking Technique for Juvenile Salmonids. Transactions of American Fisheries Society 94:386–389.

Ham, K. D., P. S. Tizler, and D. J. Klos. 2007. Hydroacoustic evaluation of fish passage at Little Goose Dam in 2006. Report to U. S. Army Corps of Engineers, Walla Walla District, contract DACW68–020D-0001, Walla Walla, Washington.

Heisey, P. G., D. Mathur, and T. Rineer. 1992. A reliable tag-recapture technique for estimating turbine passage survival: application to young-of-theyear American shad (*Alosa sapidissima*). Canadian Journal of Fisheries and Aquatic Sciences 49:1826–1834.

Johnson, G. E., N. A. Adams, R. L. Johnson, D. W. Rondorf, D. D. Dauble, and T. L. Barila. 2000. Evaluation of the prototype surface bypass for salmonid smolts in spring 1996 and 1997 at Lower Granite Dam on the Snake River, Washington. Transactions of the American Fisheries Society 129:381–397.

Jolly, G. M. 1965. Explicit estimates from capture-recapture data with both death and immigration-stochastic model. Biometrika 52:225–247.

Lady, J., J. R. Stevenson, J. R. Skalski, and A. E. Giorgi. 2000. A pilot study to estimate route-specific survival and passage probabilities of steelhead smolts at Rocky Reach and Rock Island dams, 1999. Final report prepared for Public Utility District No. 1 of Chelan County, Wenatchee, Washington.

Muir, W. D., S. G. Smith, J. G. Williams, and B. P. Sandford. 2001a. Survival of juvenile salmonids passing through bypass systems, turbines, and spillways with and without flow deflectors at Snake River Dams. North American Journal of Fisheries Management 21:135–146.

Muir, W. D., S. G. Smith, J. G. Williams, E. E. Hockersmith, and J. R. Skalski. 2001b. Survival estimates for migrant yearling Chinook salmon and steelhead tagged with passive integrated transponders in the lower Snake and lower Columbia Rivers, 1993–1998. North American Journal of Fisheries Management 21:269–282.

Murchie, K. J., S. J. Cooke, and J. F. Schreer. 2004. Effects of radio-transmitter antenna length on swimming performance of juvenile rainbow trout. Ecology of Freshwater Fish 3:312–316.

Perry, R. W., and J. R. Skalski. 2008. The design and analysis of salmonid tagging studies in the Columbia Basin Volume XXIII: Effects of array configuration on statistical independence of replicated telemetry arrays used in smolt survival studies. Report to Bonneville Power Administration, contract 00025091, Portland, Oregon.

Peters, L. M., U. G. Reinhardt , and M. A. Pegg. 2008. Factors influencing radio wave transmission and reception: use of radiotelemetry in large river systems. North American Journal of Fisheries Management 28:301–307.

Raymond, H. L. 1968. Migration rates of yearling Chinook salmon in relation to flows and impoundments in the Columbia and Snake Rivers. Transactions of American Fisheries Society 97:356–359.

Raymond, H. L. 1974. Marking fishes and invertebrates: state of the art of fish branding. United States National Marine Fisheries Service. Marine Fisheries Review 36(7):1–6.

Raymond, H. L. 1979. Effects of dams and impoundments on migrations of juvenile Chinook salmon and steelhead from the Snake River, 1966 to 1975. Transactions of American Fisheries Society 108:505–529.

Raymond, H. L. 1988. Effects of hydroelectric development and fisheries on spring and summer Chinook salmon and steelhead in the Columbia River Basin. North American Journal of Fisheries Management 8:1–24.

Seber, G. A. F. 1965. A note on the multiple recapture census. Biometrika 52:249–259.

Sisak, M. M., and J. S. Lotimer. 1998. Frequency choice for radio telemetry: the HF vs. VHF conundrum. Hydrobiologia 371/372:53–59.

Skalski, J. R., J. Lady, R. Townsend, A. E. Giorgi, J. R. Stevenson, C. M. Peven, and R. D. McDonald. 2001. Estimating in-river survival of migrating salmonid smolts using radio-telemetry. Canadian Journal of Fisheries and Aquatic Sciences 58:1887–1997.

Skalski, J. R., R. Townsend, J. Lady, A. E. Giorgi, J. R. Stevenson, and R. S. McDonald. 2002. Estimating route-specific passage and survival probabilities at a hydroelectric project from smolt radiotelemetry studies. Canadian Journal of Fisheries and Aquatic Sciences 59:1385–1393.

Smith, S. G., W. D. Muir, E. E. Hockersmith, R. W. Zabel, R. J. Graves, C. V. Ross, W. P. Connor, and B. D. Arnsberg. 2003. Influence of river conditions on survival and travel time of Snake River subyearling fall Chinook salmon. North American Journal of Fisheries Management 23:939–961.

Stevenson, J. R., J. R. Skalski, J. Lady, R. Townsend, A. E. Giorgi, and R. McDonald. 2000. A pilot study assessing the feasibility of using radiotelemetry and PIT-tag techniques to estimate project, pool, and dam survival of steelhead smolts at Rocky Reach and Rock Island dams, 1999. Report to Public Utility District No. 1 of Chelan County, Wenatchee, Washington.

Williams, J. G. 2008. Mitigating the effects of high-head dams on the Columbia River, USA: experiences from the trenches. Hydrobiologia 609:241–251.

Winter, J. D. 1996. Underwater biotelemetry. Pages 555–590 *in* L. A. Nielsen and D. L. Johnson, editors. Fisheries Techniques. American Fisheries Society, Bethesda, Maryland.

Acoustic Telemetry: Principles and Practice

Section 7.1

Acoustic Telemetry Overview

Douglas G. Pincock and Samuel V. Johnston

INTRODUCTION

Acoustic telemetry is the use of an acoustic transmitter (hereafter referred to as tag) attached to or implanted in an aquatic animal to locate and gather information about its presence, movements, and behavior in the underwater environment. Fish are the most commonly tagged animals in acoustic telemetry studies, but other aquatic animals have been studied (Sauer et al. 1997; Niezgoda et al. 2002; Cooke et al. 2004). Early acoustic telemetry technology (Trefethen 1956) relied on relatively large tags that were only suited for use in larger fish such as adult salmon (Johnson 1960; Hallock et al.; 1970). As technology evolved, smaller tags became available (Figure 1) allowing smaller fish to be tagged (Voegeli et al. 1998; Steig 2000), often by gastric or surgical implantation (Moore et al. 1990; Adams et al. 1998). Tags and receivers evolved together to allow for greater range and reliability (Stasko and Pincock 1977; Ehrenberg and Steig 2003; Grothues 2009). Through the use of more sophisticated signal coding methods, it has become possible to simultaneously monitor much larger numbers of fish (Cole et al. 1998; Bach et al. 2003; Ransom et al. 2008; McMichael et al. 2010) and to monitor the movements of fish over large geographic areas (Hubley et al. 2008; Evans et al. 2010).

An acoustic monitoring system consists of the tags and the receivers. The fundamental components of an acoustic tag are the acoustic transducer, the battery, and the electronics, which is everything else that makes the tag function. The acoustic transducer (universally a lead zirconate titanate (PZT) cylinder as discussed later) converts electrical energy to acoustic energy that in turn propagates through the water and is detected by the receiver. The electronics facilitates and controls the conversion of the electricity from the battery to acoustic energy. The end result is a tag that emits a signal consisting of short bursts of sound waves. As sound propagates relatively well in water, tags can often be detected at significant ranges (up to 1000s of m) with appropriate signal characteristics, receiver design, and environmental conditions. Receivers can be used on a mobile platform, such as a boat, to follow or track (often referred to as mobile tracking) a tagged fish or receivers can be placed at a fixed location to detect tagged fish that move near enough to the receiver to be detected (often referred to as passive tracking). Multiple receivers may be employed to cover larger areas, and/or to position tagged fish near or within the receiver array. This Section will focus on the underlying principles, methodology, and application of acoustic telemetry.

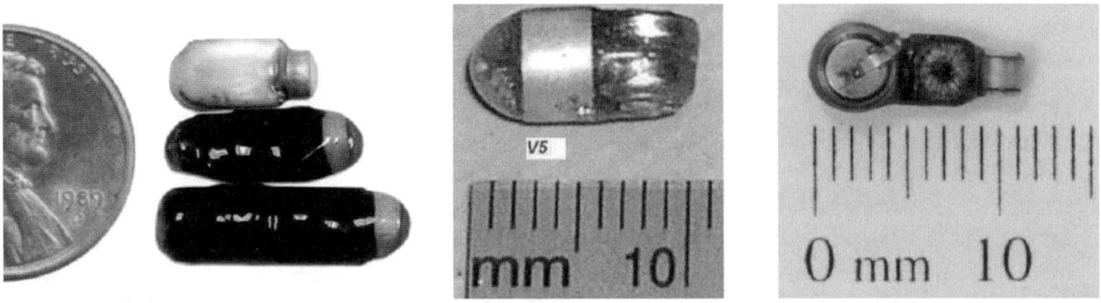

(a) (b) (c)

FIGURE 1. Representative miniature commercial transmitters: (a) HTI Model 795 (The US One cent coin shown has a diameter of 19.5 mm.) (b) Vemco V5 Prototype (c) JSATS AMT Transmitter (With permission of USGS-Columbia River Research Laboratory).

UNDERWATER ACOUSTICS BASICS

The Basic Signal

Because acoustic signals can be distorted as they travel through the water, acoustic telemetry systems use pulsed communication, where each tag transmission consists of a short burst of sound containing a number of pressure wave cycles (Figure 2). Depending on the application and the system design used, the transmission frequency generally falls in the range of 30–400kHz, and the duration of the transmission, often referred to as the pulse width or pulse length, falls in the range of less than 1 ms to 10 ms or more. The strength of the signal emitted by the tag is usually stated as a measure of acoustic pressure and is expressed in decibels (dB) with respect to 1 micro Pascal measured at a distance of 1 m from the tag (i.e., dB re 1 μPa @ 1 m). In more familiar terms, 141 dB re 1 μPa @ 1 m is equal to 1 milliwatt of acoustic power, 151 dB is equal to 10 milliwatts, 161 dB equals 100 milliwatts, and so on.

Frequency Considerations

In acoustic tags, the size of the tag is a function of the frequency because lower frequencies require larger diameter resonant elements (acoustic transducers) and higher frequencies require smaller resonant elements. For example, while they can to some extent be used at lower frequencies, a 30 mm diameter element is most efficient with a transmission frequency of 30 kHz and a 3 mm diameter element is most efficient at 300 kHz. Table 1 shows the diameter of a resonant element for some of the commonly used transmission frequencies. This dimension clearly places constraints on the achievable tag size at the lower frequencies. Since the frequency has a significant effect on the maximum range, the researcher must select a frequency that is suitable for the size of the species but will also yield the desired detection range. In other words, the frequency used for a study is influenced by the size of the tag that can be carried by the study animal without significantly affecting the health and behavior of the animal and the reception range requirements. For this reason, tags for smaller fish are usually restricted to higher frequencies with correspondingly shorter maximum ranges.

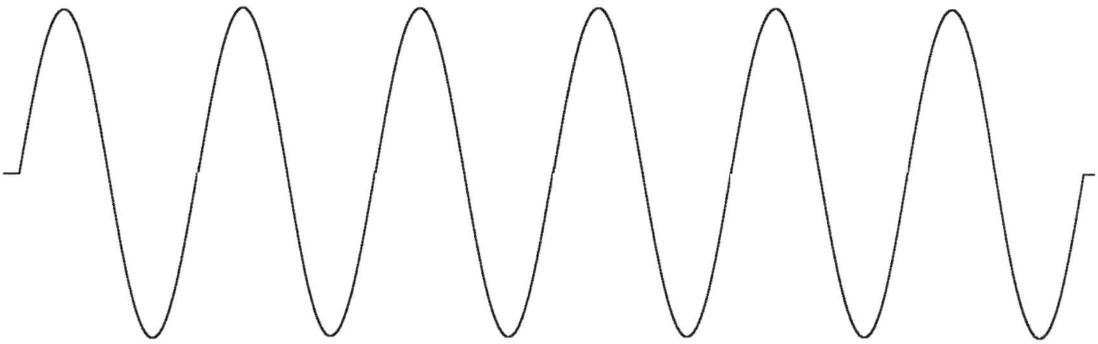

FIGURE 2. A single frequency tone burst. The diagram is not to scale as there would normally be many more cycles within the burst (For example, a 10 millisecond burst at 75 KHz would contain 750 cycles).

Detection Range under Ideal Conditions

Detection range is of primary interest in most acoustic telemetry studies, and range, for a given signal strength is dependent on three major factors, namely spreading loss as distance between transmitter and receiver increases, distortion suffered by the signal as it travels through the water and noise levels at the receiver. To understand how signal distortion effects and noise sources can impact detection range, it is first useful to have some knowledge of the predicted range under ideal conditions. This can be calculated using the passive sonar equation (Urick 1983):

$$SNR = SL - 20\log(R) - \alpha R - NL \tag{1}$$

SNR = Signal to Noise Ratio (dB) at Range R
SL = TagTag Signal Level (dB re 1 μPa @ 1 m), R = Range (m)
α = Absorption (dB/m)
NL = Noise Level (dB re 1 μPa).

In the equation, absorption (α) depends on water properties such as temperature, salinity, or conductivity as well as the frequency of the transmission (Tucker and Gazey 1966) and this is the factor that creates range variations with frequency and between salt and fresh water.

Note that, due to the form of the equation, one cannot calculate detection range directly; rather, one iterates values of R until the point is reached that the right hand side of equation (1) is equal to the minimum SNR at which the receiver will detect a signal (a specification the manufacturer should provide). Table 1 shows the results of typical maximum range calculations made this way (See Table caption for assumptions.). From this, one can see that, in general, range is greater in freshwater than salt water at any given frequency, and range is greater at lower frequencies and then decreases as the frequency increases. These results also assume that the propagation of the signal occurs in a normal spherical pattern.

Table 1. Maximum detection ranges for tags of various frequencies in salt and fresh water where tag transmission level = 150 dB re 1 μPa @ 1 m, minimum signal to noise ratio for detection = 6 db, receiver bandwidth = 1 kHz, depth = 10 m, temperature = 20°C, wind speed = 0. Spreading is assumed to be spherical and the receiver is assumed to have characteristics representative of a reasonably sensitive active tracking receiver (i.e., bandwidth = 1 kHz; Minimum SNR for detection = 6 dB).

Frequency (kHz)	Diameter of Resonant Transducer (mm)	Range in Fresh Water (m)	Range in Salt Water (m)
24	37	27,421	5,997
30	30	25,813	4,654
70	13	10,515	1,432
200	6	1,966	417
300	3	983	298
400	2.5	616	242

Factors Causing Reduction in Detection Range from Ideal

Conditions in practice, however, are seldom ideal and, hence, the theoretical maximum range calculations shown in Table 1 are seldom observed. In particular, conditions in both fresh and salt water environments can degrade the signal to the point that detection range is so poor that acoustic telemetry is not feasible. Because environmental conditions vary greatly within and between study sites, the most reliable way to estimate the detection range of a particular system in a given environment is through in situ range testing. Additional information on conducting range tests can be found in Section 7.2 and in the section "Study and Application Considerations" below.

Fresh water is not pure, and among other chemical properties, includes some level of salinity. The effect of salinity in freshwater is often not accounted for when identifying factors that could affect detection range. The results shown in Table 1 assume a salinity of zero when, in fact, most freshwater has a salinity of at least a few ppt. This discrepancy has little impact at lower frequencies but, as the frequency increases to around 300 kHz and above, absorption losses due to even low salinity levels can result in significant decreases in the detection range.

The characteristics and capabilities of the receiver used in an acoustic telemetry system can also significantly impact achievable detection ranges. The receiver characteristics used to calculate the results in Table 1 are representative of a reasonably sensitive receiver. While it is sometimes sufficient to simply detect the signal in active tracking applications, passive tracking often requires the receiver to detect information encoded in the transmission. Depending on the coding scheme employed, the range at which a tag signal can be decoded may be considerably less than the range at which its presence is merely detected (Ehrenberg and Steig 2009). This is exacerbated by the fact that, in order to keep the size of the receiver down and to extend the length of time the receiver can function using the internal battery, many passive tracking receivers are less sensitive than the one assumed in Table 1.

Acoustic Noise.—Acoustic noise in the underwater environment is often the major limitation to achieving maximum tag detection range, especially when lower frequencies are used. Natural sources of underwater acoustic noise include wave action (Dahl 1994), flow noise, riverine bed-load movement (Belleudy et al. 2010), biological noise such as produced by snapping shrimp (Readhead 1997), rain (Bjorno 1994), and even snow (Crum et al. 1999). As an example of the significance of some of these noise sources, range in the ocean for an acoustic system operating at 70 kHz could be reduced from almost 1.5 km under ideal conditions to a few hundred meters in high winds accompanied by heavy rain (Hobday and Pincock 2012). With frequencies above 100 kHz, the dominant noise source is inherent thermal noise (i.e., that caused by the random vibration of water molecules) with, except in a few unusual cases, other environmental noise sources having a negligible effect.

Man-made noise from any device or machine that produces high frequency vibrations can also interfere with the ability to detect the signal. Examples of man-made noise sources include sonars or fish finders, boat motors and shipping traffic, hydroelectric turbines, and pumps to name a few. With the exception of high frequency sonars operating at frequencies near that of the tags, man-made sources will cause some interference, but this can usually be mitigated and is generally not a significant factor when using frequency above 200 kHz.

Electrical Noise.—Ambient acoustic noises can interfere with the detection of the signal, but electrical noise can also affect the performance of the acoustic system. Once the signal is detected and converted to an electrical signal by the receiving hydrophone, electrical noise can degrade the signal and affect the detection range. To minimize unwanted electrical noise in the receiving system, proper grounding and shielding procedures should be implemented. Most manufactures of acoustic telemetry systems have addressed this issue when they designed and built the equipment. However, the researcher needs to understand the basic principles of properly grounding and shielding electrical systems, or should consult the manufactures to ensure the system is properly installed. For example, a significant amount of electrical noise will be present in the system if it is plugged directly into the electrical system on a boat or directly into a gas powered generator. In these instances, a device (such as a power inverter or battery backup unit) that aids in removing the electrical noise should be used between the acoustic system and the main source of electricity.

Signal Distortion Effects.—Physical structure in the environment can interfere with the propagation and/or integrity of the signal as it travels from the tag to the receiver, thereby affecting the detection range as well as the ability to decode information that might be embedded in it. High levels of entrained air may entirely block the acoustic tag signal because the sound is scattered by the air bubbles and the signal is degraded to a point that it never arrives at the hydrophone. Because woody debris and aquatic plants usually contain some gas bubbles, they can have a similar effect on the acoustic signal. High concentrations of large suspended particles, such as sand or silt, can have a similar effect, causing the signal to be scattered (Richards et al. 1998). Similarly, the interface between bodies of water that have different densities, like haloclines or thermoclines, can potentially reflect and/or refract the signal in such a way that it never reaches the receiver (Tucker and Gazey 1966). Further distortion of the signal can be caused by Doppler shift (Urick 1983), water currents, and temperature and salinity gradients.

Echoes due to reflections from the water surface, bottom, or any other object can be a problem if they reach the receiver with sufficient strength to be detected. Such echoes, often referred to as multipath signals, can either distort the signal seen by the receiver by extending its apparent duration or produce what appear to be additional transmissions or in more extreme cases, a replica of the transmitted signal. These affects will depend heavily on the relative positions of tags to the hydrophones, and the presence of reflective surfaces such as the bottom of the river or lake, the surface of the water, and other underwater structure.

While distortion effects such as multipath and Doppler shifts do not usually impact the ability to detect the presence of a tag, they can have a significant impact on the range at which the information contained in the signal can be decoded. As well, because the speed of sound in water is many orders of magnitude less than the speed of radio waves, the impact of such distortion is much greater than with radio telemetry. Furthermore, these effects make it impossible to use many modulation schemes that work well with radio telemetry.

Velocity of Sound

The speed of an acoustic signal as it travels through the water is not particularly important from the standpoint of simple tag detection, but it is an important variable for systems that use triangulation to determine the position of tagged animals near or within a hydrophone array. Mackenzie (1981) provides a simple equation for finding the speed of sound in seawater of a given temperature, salinity, and depth. Del Grosso and Mader (1972) provide a similar calculation for freshwater. Both methods produce estimates that are generally acceptable for most acoustic telemetry applications. More precise, but more complex equations for the speed of sound in seawater may be found in http://www.comm-tec.com/Library/Technical_Papers/speedsw.pdf.

ACOUSTIC TELEMETRY PRINCIPLES

Coding Information into the Signal

In studies where the objectives are relatively simple and only require tracking the movements of a few individuals, the signal transmitted by the tag can be relatively simple. However, when objectives require using tags with sensors to collect additional information such as temperature and depth, the signal must be encoded to enable this information to be transmitted to the receiver. In the case of monitoring large numbers of fish, encoding the signal enables each tag to emit a unique signal that can be used to identify individual fish. This unique signal is called the tag identification code, or ID code. In the early days, the number of fish that were tagged and released as part of any study was small; so the number of ID codes required was small. In the past ten years, however, the amount of data encoded in the signal as well as the number of fish tagged and released has increased dramatically. Many experiments now tag hundreds or thousands of fish, and thus require comparable numbers of ID codes and systems able to detect and decode information from numerous tags simultaneously. Also, the amount of data transmitted by each tag is greater due to the increased use of sensors. The need for unique ID codes is also the result of the worldwide increase in the use of acoustic telemetry (O'Dor et al. 2009). With more researchers using the same monitor system, more unique ID codes are needed to eliminate the risk that redundant ID codes are detected within multiple study areas, which would result in the collection of erroneous data and could put scientific conclusions at risk. While encoding the signal seems like a simple way to increase the capability of the monitoring system, it does not come without a cost. In particular, encoding the signal can, as mentioned above, reduce the maximum range at which tag signals can be decoded and may limit the number of tags that can be simultaneously detected by receivers, and reduce the precision of the estimates of the position of the tags.

Coding Method Taxonomy

Because of the distortion of amplitude, pulse length and phase described above, the only realistic building block of any coding scheme is a tone burst (Figure 2) which transmits on a single frequency. With a tone burst (which for simplicity, we will refer to as a pulse) as the fundamental building block, the methods of coding data can be grouped into one of four categories based on two criteria.

The first criterion is the number of pulses used to transmit the signal. In what we call a Pulse Interval Scheme (Figure 3), each transmission consists of a single pulse and the data are encoded in the period between successive pulses. In what we call a Coded Scheme, each transmission consists of a series of pulses with the intervals between successive pulses representing a binary number. Figure 4 shows a typical Coded Scheme transmission where the length of time needed to transmit the data are called the Code Length and the time between successive transmissions is called the Transmission Interval.

The second criterion deals with the nature of the pulses themselves. When all transmitted pulses are the same, we will refer the method as a Single Color Signaling (or simply Single Color) scheme. On the other hand, a Multiple Color Signaling (or simply Multiple Color) scheme is one in which there is more than one type of transmitted signal. Multiple color signaling systems can be achieved in a number of ways including using a different frequency for individual pulses or by changing the phase (Phase Shift Keying) or frequency (Frequency

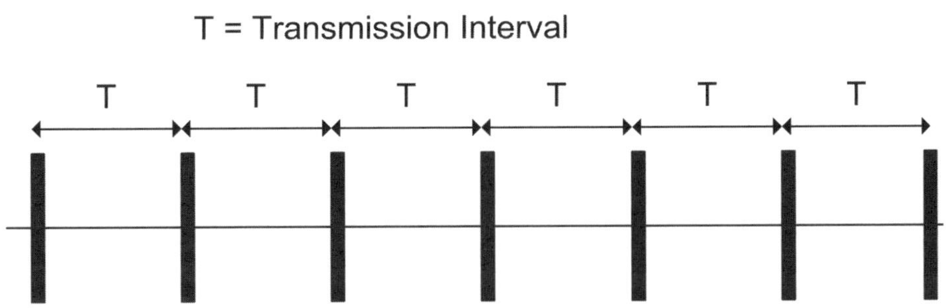

FIGURE 3. Example single frequency continuous coding. Each pulse shown represents a tone burst as shown in Figure 2.

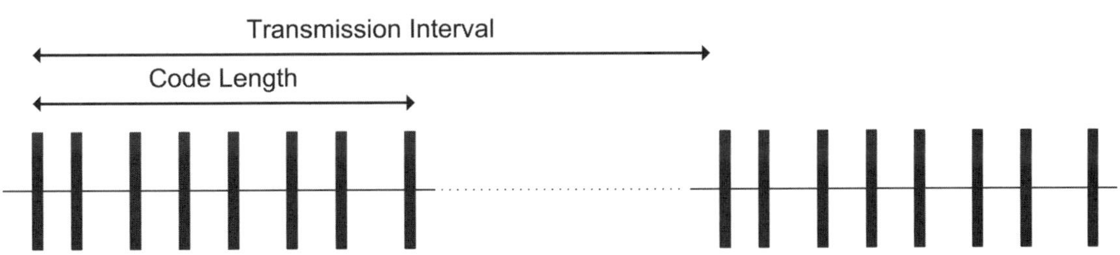

FIGURE 4. Example single frequency multiple pulse coded tag scheme.

Shift Keying) of the signal within each pulse to represent a binary number. The motivation for any multiple color scheme is to increase the amount of data that can be transmitted. The tradeoffs to achieve this are discussed below.

Table 2 classifies the most common coding methods used today according to the above criteria and in the following section we summarize the features, advantages and disadvantages of each approach to coding.

Coded with Single Color Signaling.—The advantages of using a coded approach are the ability to encode hundreds, thousand or even millions of uniquely identifiable codes, the ability to encode sensor value(s) and the ability to easily automate the detection algorithms. As well, only a single detection is necessary to identify individual fish when multiple fish are within the detection range of the system. On the other hand, as will be discussed below, coded signals can collide with one another and prevent either signal from being decoded by the receiver. In summary, single color schemes are very effective when individual receivers are not required to simultaneously detect a large number of tagged fish in a short period of time. It also offers the advantage that relatively simple tag and receiver implementations are possible. As a consequence, receivers can be relatively small and be deployed without servicing for a year or longer. This feature has led to the emergence of a number of large collaborative projects used to monitor the movements of fish over vast geographic areas. (e.g., The Ocean Tracking Network, the Australian Acoustic Tracking and Monitoring System, the Atlantic Cooperative Tracking Network and the Pacific Ocean Shelf Tracking project)

Pulse Interval with Single Color Signaling.—The advantages of pulse interval coding with single color signaling include near immunity from the potential of multiple signals colliding when many tagged fish are within the detection range of the receivers (Ehrenberg and Steig 2003), all the energy transmitted by the tag is used for both detection and decoding (i.e., the maximum detection range and maximum decoding range are the same as is the case with any single color scheme) and the ability to retrieve movement and behavior information from a single hydrophone and receiver. Because of the ability to predict the expected arrival of the next pulse with high precision it becomes possible to graphically display time ordered pulses such as those in Figure 5. Disadvantages include the requirement that multiple pulses must be

TABLE 2. Categorization of commonly used coding schemes.

Coded	Pulse Interval
Single Color Signaling	**Single Color Signaling**
Vemco Coded	HTI
Lotek SA	Vemco Continuous
Multiple Color Signaling	**Multiple Color Signaling**
JSATS	Vemco VRAP with
Lotek MAP	Continuous Tags[1]

[1]Although each tag uses only a single color, each uses a different frequency; so the receiver has to deal with multiple colors.

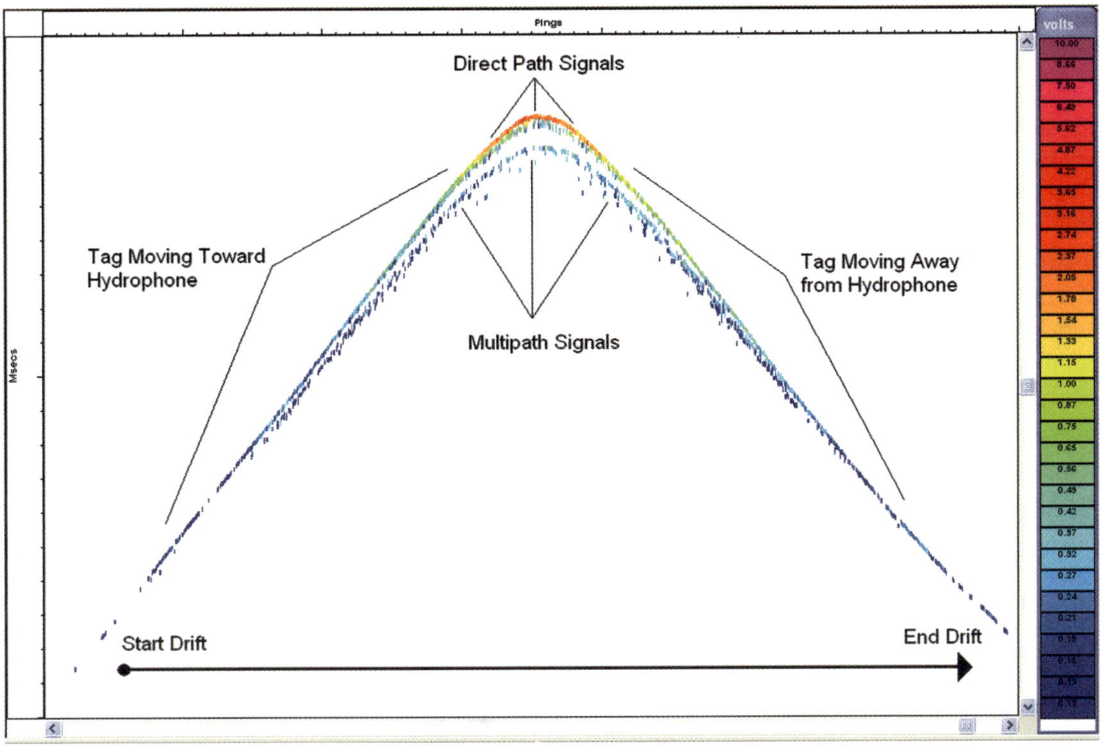

FIGURE 5. Acoustic tag signals from a constant period tag (3,028 ms) drifting past a fixed receiver deployed in a river. The total time (horizontal axis) displayed is 35 minutes, total change in period (vertical axis) is 200 ms. Note multipath signals below direct path tag receptions.

detected to ensure positive identification of the tag signal, and that more sophisticated tag and receiver designs are required to precisely control transmission intervals in the tag and allow for precise time stamping of tag receptions.

Using a single pulse interval to distinguish individual tags limits the number of potential tag codes to only several thousand unique ID codes. When a second pulse is transmitted at the same overall period, but at a fixed time interval from the first transmission, then the total available number of unique ID codes increases to hundreds of thousands. The spacing of the second pulse can also be used to transmit sensor data. Regularly spaced pulses allow for automating the process of tracking the tag receptions from individual hydrophones (through time), or from multiple hydrophones in a positioning array. Short, regular pulse intervals also provide the best detection scenario for mobile tracking.

Coded with Multiple Color Signaling.—This approach has the advantages of single color signaling with the added advantage of increasing the available data bandwidth. Higher bandwidth allows more tags to be simultaneously present at a receiver and still be reliably detected or, for a given number of tags, more frequent updates from each tag. More rapid updates can offer a number of advantages—for example, greater certainty if fish are moving rapidly past a receiver, more frequent updates from any tag sensors and, in fine positioning applications (see below), more detail on movements.

Pulse Interval with Multiple Color Signaling.—While other possibilities may exist, the only technique of this type used to date involves the use of a number of Pulse Interval coded tags, each operating on a different frequency. Because the individual fish can be identified by their frequency, the interval between successive pulses can be used to encode sensor data. While the individual tags use single color signaling in such a scenario, the receiver must be able to identify the transmitted frequency for each detected signal. This is the simplest example of using multiple color signaling to increase bandwidth. For example, if each tag were to use a transmission interval of one second, data would be updated for all tags simultaneously at this rate, provided the receiver is able to detect pulses on any of the frequencies simultaneously. This approach can be very effective in a situation with a requirement to monitor detailed interactions between individual fish (Sauer et al. 1997). However, its applicability is limited as it cannot be used with a large number of tagged fish since, because of the low frequencies used in acoustic telemetry and the need to maintain sufficient separation between the frequencies used, only a small number of frequencies can be used. Therefore, we will not consider this approach in subsequent discussion of coding system performance issues where the focus is on large numbers of tagged fish.

Data Rate in Coded Systems

Although this Section deals with coded systems which are digital in nature, transmitted signals in Pulse Interval coded systems are also subject to distortion and hence one must consider the proportion of transmitted data that is successfully recorded by the receiver. However, since tag data (ID code or sensor data) is inherently analog (i.e., the time between pulses), the issues are different and are addressed separately below.

Received Data Rate

Central to any discussion of acoustic telemetry system performance are the related issues of data rate and communication errors of various types. In a telemetry application where one or more tags are transmitting data to be detected at a receiving site, we can define the Transmitted Data Rate as the total amount of data transmitted by all tags within range of a receiver (e.g., in the case of n tags, n times the amount of data transmitted by each tag). Except in the simplest of situations, not all the signals reaching the receiver will be successfully detected which leads to the definition of Received Data Rate as the amount that is correctly recovered at the receiver.

There are three factors to consider in determining the received data rate in a particular situation:

1. Distortion Errors: These arise from distortion of the signal between the time it leaves the tag and arrives at the receiver, and the distortion is sufficient enough that the receiver fails to correctly detect the signal.

2. Collision Losses: These occur when transmissions from two or more tags overlap at the receiver and the receiver fails to decode at least one of them.

3. False Positives: These occur when a decoding failure from one of the two mechanisms above is not detected by the receiver and appears to be valid data but is not. These errors have a far more serious implication than reduction of received data rate as they can introduce false data. For example, recording detections as true when they are not can lead to the false conclusion that a fish was present when it was not. See Sections 7.2 and 9.5 for further discussion on this issue.

Taking these errors into account, the effective received data rate of a system is given by:

$$\text{Received Data Rate} = \text{TR} - \text{DL} - \text{CL} - \text{FP} \tag{2}$$

TR = Transmitted Data Rate
DL = Distortion Error Rate
CL = Collision Loss Rate
FP = False Positive Rate

Error Detection in Coded Systems

Like any digital communication system, coded acoustic telemetry systems will exhibit transmission errors of one or more of the types described above. Furthermore, because transmission errors occur frequently in an underwater acoustic environment, acoustic telemetry systems are particularly vulnerable.

Most coded tag schemes use the standard approach of an Error Detection Code (EDC) as a basis for the rejection of erroneous transmissions. This involves adding extra bits to the transmission (the EDC). At the receiver, the EDC is recalculated from the decoded data and compared to the received EDC. Transmissions in which the calculated and received EDCs differ are rejected as this could only occur as a result of a transmission error in either the EDC bits or the data bits in the received signal or both.

The use of EDCs to detect errors can never be perfect. Regardless of the scheme used, there is a nonzero probability that a transmission error will slip through the comparison test and result in creating a false positive. The best one can do is increase the "robustness" of the EDC to the point that the probability of such false positives becomes as small as possible. Of course, the cost of more robust error detection is that the portion of transmitted bits used for error detection rather than the actual data are higher, hence reducing the decoded data rate. This tradeoff is severe in acoustic systems because the limited bandwidth only permits a small number of total bits to be transmitted.

The actual statistics on the probability of a false positive depends on the coding method, nature of the EDC, number of tags present, and the characteristics of the transmission path. A rigorous discussion of false positive probability is beyond the scope of this paper. However, as an example, Pincock (2008) addresses false positive probability and identified methods to identify them in a typical single color signaling coding method. The author also showed that without the use of EDCs, the false positive rate would be unacceptably high in all but the least demanding situations.

Performance Considerations for Coded, Single Color Systems

Estimating Received Data Rate.—The major factor that reduces received data rate (equation (2)) when a number of single color signaling tags are present in the monitoring array is the loss of data due to collisions. How the system performs when multiple tags are present depends on the coding method used and, to some extent, the algorithm used in the software of the receiver. Most vendors will provide some level of detail about how the system is designed to deal with potential collisions of the signals from multiple tags. If they do not, the researcher should ask for this information because it can have a significant effect on the performance of the system. At a minimum, the vendor needs to provide a general idea about the number of tags that can be monitored simultaneously by the receiver (often referred to as the number of resident tags). With this information, the researcher can use the following equation to estimate the performance of the system (Pincock et al. 2010).

$$TD = (RT \times CL)/TI \tag{3}$$

TD = Tag Density
RT = Resident Tags
CL = Average Code Length (Figure 4)
TI = Transmission Interval (Figure 4)

Using this formula, and assuming the option exists for the researcher to specify the rate at which the tag emits the signal (often referred to as the transmission interval), the transmission interval can be chosen to be sufficiently long enough to decrease tag density to an acceptable level. Table 3 shows the estimated numbers of detections that will be recorded for various tag

TABLE 3. Percentage of transmissions that will be detected as a function of tag density.

Tag Density	Percent Detected
0.05	91%
0.1	83%
0.25	63%
0.37	50%
0.5	39%
0.75	23%
1.0	13%
1.25[1]	8%
1.5	4%
2.0	1%

[1]Some tags are detected even if tag density is greater than 1 as the randomness of the process will create some empty space

TABLE 4. Average time in minutes (unless otherwise stated) between detections of each resident tag in a representative 16 bit coded single color signaling system with Transmission Intervals (TI) of 60 and 120 s and two examples of coding parameters (conservative: minimum interval = 300 ms and timing sensitivity = 10 ms; aggressive: minimum interval = 100 ms and timing sensitivity = 2 ms).

Resident Tags	Conservative		Aggressive	
	TI = 60 sec.	TI = 120 sec.	TI = 60 sec.	TI = 120 sec.
5	1.5	2.4	1.1	2.1
10	2.4	3.1	1.3	2.2
15	3.9	3.9	1.4	2.4
25	10	6.2	1.8	2.7
50	118	20	3.4	3.7
75	> 20 hours	67	6.4	5.0
100		220	12	6.8

density values. For example, a tag density of 0.05 will yield a detection efficiency of about 91% and a tag density of 2.0 will yield a detection efficiency of about 1%. Clearly, longer transmission rates are good for system performance and can reduce the tag density values (Table 4; Pincock 2008). However, reducing the transmission rate decreases the frequency at which information is obtained about the study animal. Therefore, a balance must be reached between increasing the transmission interval to minimize the potential for collisions and having a transmission rate that is so long that the frequency of data collected does not meet the objectives of the study.

Effect of Coding Parameters on Performance.—Code length also determines the overall performance of the system when multiple tags are simultaneously present in the monitoring array. Equation (3) shows that for a given number of resident tags and transmission interval, the percentage of transmissions detected by the receiver increases as the length of the code (code length) is decreased. For a particular scheme with a given number of bits, code length is determined by the two parameters as defined below:

1. Minimum Interval: The smallest interval between successive pulses used in the coding scheme. This essentially determines how immune the scheme is to multipath. Multipath echoes which die out in less than this time interval will not interfere with the decoding process.

2. Timing Sensitivity. This is the accuracy to which the difference between successive pulses must be measured at the receiver for the code to be interpreted.

Traditionally, single color coded systems have used conservative coding parameters to ensure satisfactory operation over a wide range of conditions and, in particular, for large scale applications in the ocean. For example, a minimum interval of 300 ms and timing sensitivity of 10 ms have been shown to be reliable in virtually all ocean environments to a range of 1 km or more. However, schemes that have shorter intervals and increased time sensitivity requirements can be used to increase the rate at which data are transmitted and decoded to allow for a larger number of resident tags. Table 4 illustrates the points discussed through a presentation of received data rate calculations with a representative scheme encoding 16 bits per transmission (including a 30% allowance for an error detection code). The results illustrate the performance improvement that can be obtained in high residency situations through the use of a longer transmission interval as well as the performance improvement that can be achieved through the use of more aggressive coding parameters. To illustrate the point, the table shows two extremes of coding parameter choices and, of course, less extreme values are possible.

Performance Considerations for Coded, Multiple Color Systems

With multiple color systems, it is not realistic to present a similar section to the above because the wide variety of potential approaches makes it impossible to generalize. However, it would be useful if users with experience, or the vendors, were to publish such information for the currently available systems. In the absence of this, we present some general comments with respect to the approach taken by the two existing commercial systems.

The two existing commercial schemes are very aggressive in that they use a large number of signal colors to create the potential for very high data rates at the cost of significant reduction in the maximum range (compared to that of a single color system) at which signals can be reliably detected. It should be noted that this range reduction is the result of the aggressive coding (i.e., the use of a very large number of signal colors). It is certainly possible to take a more conservative approach which uses a small number of signal colors to provide a modest increase in bandwidth over that of a single color scheme with little or no reduction in range; however, to date, there are no such products on the market.

Although the details are different, the two currently available commercial products using multiple color schemes (Table 2) are similar in that they use thousands of signaling colors with each color representing the identification code for a particular tag. These transmissions are short and repeated at regular intervals, typically every few seconds. As a result of the short percentage of time that each transmission occupies, loss of data due to collision with transmissions from other tags is significantly less, two or more orders of magnitude less than those with a single color signaling system. This, along with the relatively rapid rate at which data are transmitted by each tag provides the potential for very high data bandwidth. However, the potential for collecting data at relatively high rates and the dramatic reduction in lost data due to signal collisions will, as range increases, be accompanied by increases in distortion error rate and false positive rate (equation 2). The reason for this is that the receiver is required to accurately determine which one of the thousands of possible colors corresponds to the actual transmission and, as range increases, so does distortion making this task more and more difficult.

To date, there is little to no published information on the magnitude of these potential errors. However, the results reported by Weiland et al. (2011) for one of the systems currently using multiple color signals (JSATS system) has provided some insight into the performance of this type of system. The authors showed received data rate (for which they use the term Decoding Efficiency) for a study conducted at John Day Dam on the Columbia River, Oregon. About half of the data were lost at a range of 120 m from the hydrophones, and nearly all the data were lost at a distance of 140 m. While the inherently noisy environment upstream of a hydroelectric dam may contribute to this relatively low performance, the major issue is the small amount of distortion (due to multipath effects and Doppler in particular) that can be tolerated before signals, while detectable, will be incorrectly decoded. On the other hand, with single color coding, as discussed above, detected signals are highly likely to be correctly decoded with resulting greater useful range for equivalent conditions.

Decoding Data in Pulse Interval, Single Color Systems

Pulse interval tags use the interval between tag transmissions to code a unique tag ID. To work properly, pulse interval systems using a large number of tags require very high precision in the timing of both the tag transmissions and the reception of tag signals at the receiver. This level of precision is relatively easy to accomplish with modern digital electronics and processors.

Pulse interval systems differ from coded systems in that they only require that tag signals arrive repeatedly within a known time window to be included as tag receptions. Multipath or noise signals arriving outside that window can be rejected. However, reception of a single tag transmission does not allow for recovery of tag ID. Multiple transmissions from each transmitting tag are required for positive ID. For this reason, it is useful that all potential tag transmissions received at the hydrophone be collected. This allows for review of all potential tag transmissions (either graphically as described in Ehrenberg and Steig 2003; or using automated processing methods described in Steig and Johnston 2010) to sort out any possible false positive tag IDs, and to assess general data quality with respect to changing environmental conditions.

Single color pulse interval systems (as currently implemented in commercially available systems) do not suffer from lost detections due to overlapping transmission intervals. For instance, if a tag is transmitting at 5 s intervals, with a 1 ms pulse length, then there are 4.999 s out of 5.000 s of time over which a pulse from a second tag may arrive without interfering with the first tag. As more tags are detected by a single receiver, the probability increases that two tag signals will arrive simultaneously. In the rare case that tag signals from two tags do arrive at the hydrophone at the same time, then the combined received signal level from both tags would be anywhere between the sum of both signals, and no signal at all. The percentage of time that a combined signal would be detected is very high, but would depend on the signal strength of both arriving pulses, the relative phase of the arriving signals, the signal level threshold of the receiver, and the pulse width acceptance criteria. However, most simultaneously arriving pulses would be detected. Because the arrival time history of detections for each tag is known, the combined detection signal can be assigned to both transmitting tags.

For pulse interval systems, each tag is transmitting with a different interval between pulses, so if a combined detection does occur, it will be guaranteed not to occur on the next transmission of each tag.

The structure of the signal and the rate it is transmitted makes single color pulse interval systems particularly useful in high residency situations. Because there is only a single (or in some cases two) short tag transmission(s), the interval between transmissions can be very short relative to the interval between groups of pulses in a coded scheme. For this reason, single color pulse interval systems are also useful in situations where tags are within range of the receiver for a limited time. Without tag detection lost due to collisions, and with short transmission intervals, high numbers of tags transmitting simultaneously can be detected in short periods of time.

ACOUSTIC TELEMETRY EQUIPMENT

Tag Components

Acoustic Transducer.—Because it is the smallest shape that produces roughly omnidirectional radiation at resonance, cylinders are universally used for this purpose. As well the material universally used in transducer manufacture is lead zirconate titanate (PZT) as this provides the highest electric to acoustic efficiency (Herman 1975).

We have previously discussed in connection with Table 1 the tradeoff between the desire to use a low transmission frequency to achieve good transmission range and the fact that small transducers are only efficient at higher frequencies. This trade off is particularly important in salt water applications. For example, if a 3 mm diameter transducer were chosen in an effort to produce a small tag, the optimum electric to acoustic efficiency is obtained at mechanical resonance (approximately 300 kHz). However, at this frequency in salt water (20°C, 35 ppt salinity), acoustic absorption losses are much higher (10.2 dB per100 m) than would be the case if a lower frequency could be used (e.g., absorption loss is 3.8 dB per 100 m at 100 kHz). The effect is less in fresh water but can be significant at very high frequencies. While it is possible to use a small transducer and drive it at a frequency below its resonance frequency to achieve the benefits of a lower transmission frequency, the resulting loss in electric to acoustic efficiency will lead to shorter and shorter range the further one moves away from the optimal resonance for the transducer.

Batteries.—The most important battery requirement for acoustic tags is high energy per unit of weight or volume which would suggest the use of lithium batteries. Lithium batteries are used in some cases but, because of their explosion risk, inadequate peak current capacity for high power tags and unavailability of small versions suitable for use in miniature tags, other formulations are more prevalent. In particular, silver oxide batteries are often used for miniature tags and alkaline batteries in larger tags requiring higher power output. A more detailed description of the battery types used in tags (including types in addition to those discussed above) and the associated tradeoffs can be found in Section 5.1 of this book.

Electronics

The electronics of the tag, which is essentially everything besides the transducer and the battery, perform the functions of creating the acoustic signal, transmission sequence encoding ID, and delivery of power to the acoustic transducer. If an environmental sensor is included in the tag, then the electronics must also determine the sensor readings and encode these into the tag transmissions.

Over the past decade, advances in low power electronics have made it feasible to use a single microcontroller chip with firmware running in self contained flash memory to realize the above functions. In addition, microcontroller firmware can also support functions such as turning the tags off for periods of time to save battery life (often call "sleep" periods). Similarly, if study objectives require getting different information during different times over the duration the tag is in the fish, the microcontroller can be used to alter the coding of the data to meet these needs. In addition to the microprocessor, other components are usually needed to control the tuning of the tag to the proper frequency (usually with a separate crystal), to provide a time reference, and to provide a way to activate the tag before (or potentially after) it is implanted in the fish. As an alternative to the microprocessor option, tags with simple functionality (e.g., single color pulse interval tags) can be constructed with circuitry which is potentially smaller than a microcontroller chip.

Activation of the tag can be accomplished in several ways. It can be as simple as two wires that the user solders together and then protects in some way. As this process is potentially time consuming and subject to failure (particularly with miniature tags using very fine wire), internal magnetic switches are commonly used with various forms of activation, the simplest being the removal of a magnet attached to the tag. At least one manufacturer provides the ability to activate the tag without using a magnet as well as program the operational parameters of the tag in the field.

Some tags, particularly larger ones, will have additional components such as a power amplifier (possibly including inductive components to help maximize efficiency), signal condition components, and circuitry required to integrate some sensors. Table 5 summarizes sensors that are commonly used in commercially available acoustic tags. Various other sensors including dissolved oxygen, pH, electromyography (EMG), and electrocardiogram (ECG) have been used but are not generally available in commercial products.

Assembly and Encapsulation.—Electronic assembly routinely uses small surface mount parts in preference to bulkier through-hole parts that were used in the past. For the smallest tags these are often un-encapsulated chips (Figure 6a) which are connected directly to the printed circuit board. After the transducer, battery, and electronics are assembled they must be encapsulated in some way to protect them from damage. Tags are normally encapsulated in one of two ways. The first encloses the components in an epoxy-filled cylindrical tube (Figure 6b) which has the advantages of being rugged and protecting the electronics from damage

TABLE 5. Sensors commonly used in acoustic telemetry tags.

Sensor	Type	Comments
Pressure	Strain gauge(s) mounted on back of diaphragm exposed to water	Use to determine swimming depth. Major source of errors is temperature dependence of strain gauge resistance; errors can be reduced by arranging strain gauges in a bridge
Temperature	Thermistor	Low cost, small and reliable; accuracies up to 0.1°C are readily available
Activity and Motion	MEMS (Micro Electrical Mechanical System) integrated circuits, motion sensitive mercury switch and various custom mechanical devices	Usually three axis sensors are used. Normally acceleration changes are at too high a rate to be coded into transmissions. The normal practice is for on board processing to calculate desired information (e.g., average acceleration) and transmit that. Non MEMs implementations are often bulky

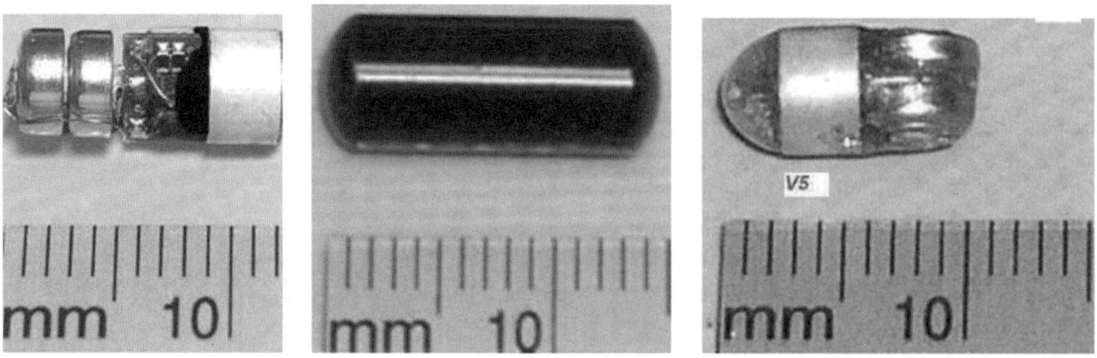

(a) (b) (c)

FIGURE 6. Representative commercial transmitters (Vemco) showing the traditional cylindrical arrangement before (a) and (b) after encapsulation. From left to right the components are battery cells, electronics board and acoustic transducer. The V5 prototype (c) shows the potential for size reduction with a rearrangement of components and conformal coating.

due to increased pressures at depth. The second widely used way to encapsulate the components of the tag is called conformal encapsulation. This process uses a thin layer of protective coating to encapsulate and protect the components (Figures 1a, 1c and 6c). Depending on the mechanical configuration of the components, this approach can lead to significant size and weight reduction over using the cylindrical tube encapsulation process. However, conformal coating is inherently less robust, thereby creating a higher risk of tag failure due to physical damage to the components.

Tag Miniaturization.—One of the challenges in using any type of active tracking technology to study small animals (for example fish that are less than 100 mm in fork length) is the need for extremely small tags that can be carried by the study animal with minimal affect on its behavior. In recent years, there has been considerable interest in making tags smaller than those currently available (Figure 1). While this is feasible, there are some constraints that factor into how small a tag can be, and miniaturization often results in an increase in the cost of the tag, reduction in the battery life, and can decrease the detection range of the tag. Currently, the size and shape of the battery is a major factor limiting the miniaturization of the tag. The flat arrangement of currently available batteries shown in Figure 7 provides one example of a configuration that facilitates miniaturization of tags. The configuration shown is based on a single 4.8 mm diameter battery (the smallest silver oxide cells available at the time of writing) and an electronics board sufficient to contain a single integrated circuit (in the form of a bare chip) and tuning elements. The acoustic transducer shown is less than 3 mm in diameter, indicating a transmission frequency in the vicinity of 300 kHz. Using a single cell (unlike most current miniature products which typically use two) in such a design limits the demands that can be put on the battery. Furthermore, a single cell provides relatively low battery voltage and reduces the potential power output by a factor of four from what would be expected if two cells were used. This configuration also uses a 300 kHz transducer which can reduce detection range compared to tags that use lower frequencies.

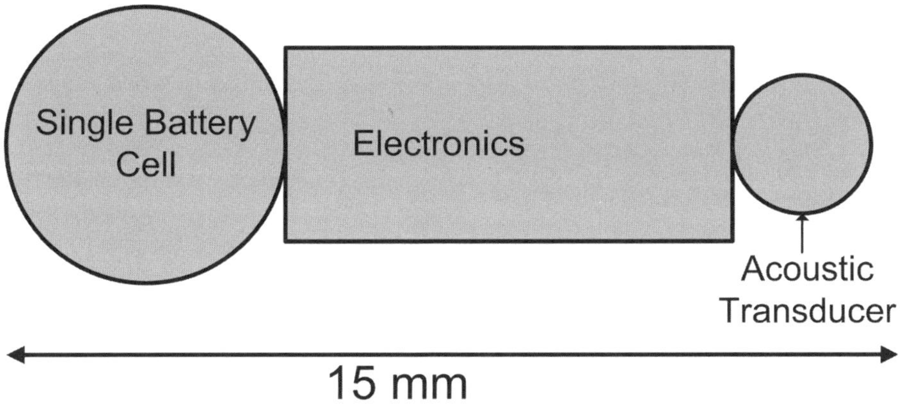

FIGURE 7. Flat arrangement of transmitter components.

To achieve further miniaturization of the tag, significant reductions must be made to the size of the battery or transducer. While it is unlikely that the size of the transducer can be reduced, a breakthrough in battery technology could have a significant impact on tag size and available voltage, however, little has changed in the last few decades in battery technologies. The focus area for the battery industry has been elsewhere and smaller batteries for miniature tags is a relatively small part of the market with the result that battery companies are reluctant to invest research and development costs that would likely not be recuperated.

Life, Size and Range Tradeoffs for Larger Tags.—While the objectives and parameters of each individual acoustic telemetry study will dictate the acceptable tag to body mass ratio (Jepsen et al. 2005), in general, larger fish can accept larger tags with larger batteries so the available energy allows for greater latitude in what can be achieved with respect to range, sensors, and programming. For example, Table 6 demonstrates tradeoffs between size and life for a typical tag operating at 80 kHz (i.e., using a 12 mm transducer). Assumptions behind the specifications shown are outlined in the table caption. Freshwater range at 80 kHz is not shown in the table because the best configuration for a larger transmitter at this frequency would be relatively large with a range capability far beyond what would normally be required in freshwater.

Tag life is of course impacted by the addition of sensors. However, since sensors are typically activated briefly before each transmission, the effect is usually minimal. For example, a sampled sensor consuming 0.5 ma would reduce battery life from those shown in Table 6 by about 10%; but this would rise about 95% without sampling (e.g., 11 d instead of 145 d for a tag that does not have a sensor, as shown in the first row of the table).

Finally, for some tags that are activated using a magnetic switch, battery energy is consumed at a low level even when they are stored in the deactivated state. This is often referred to as quiescent drain. Therefore, it is important for users to be aware of the potential for the tag to use battery power even when the tag is turned off. If this is the case, the vendor should provide data to determine the length of time between manufacture and activation which can be tolerated without significant battery life deterioration.

Table 6. Representative specifications for larger tag operating at 80 kHz. Specification shown are based on the following assumptions: (1) electric to acoustic efficiency is 75%, (2) electronics consume 1 uA when not transmitting (conservative with today's technology) and (3) pulses being transmitted 1% of the time (battery life would adjust proportionally if percentage is different). Range calculations are based on the use of a broadband receiver.

Battery	Typical Tag Dimensions (mm)	Maximum Theoretical Range in Ocean (Broadband Receiver)	Useful Life
Silver Oxide (3 V, 150 ma-hrs)	60 × 15	950 m	145 days
Silver Oxide (9 V, 150 ma-hrs)	80 × 15	1,175 m	145 days
Lithium (3.6 V, 2,100 ma-hrs)	100 × 15	950 m	5.6 years

Hydrophones

Hydrophones are the acoustic equivalent of radio antennas in that they convert acoustic signals to electrical signals. Usually, a transducer similar to the type used in the tags is used to accomplish this function in the hydrophone. In some submerged passive telemetry receivers, the hydrophone is integral to the unit while, in other cases, where separation between the hydrophone and receiver components is required, cabled connections are used. Mobile tracking receivers would normally use a cabled connection.

Hydrophones are generally categorized as omidirectional or directional. Most passive tracking applications use omnidirectional hydrophones to increase the probability of detecting tags regardless of the direction the fish approaches the monitoring area. While size considerations don't preclude the use of spherical transducers in such hydrophones, a single PZT cylinder is usually used for cost and ease of manufacturing considerations. While, as would be expected, these transducers are most sensitive at their resonant frequency, satisfactory performance can be achieved below the resonant frequency and they are often used this way.

In the simplest implementation of mobile tracking, the researcher follows a fish by continuously steering towards the strongest signal. This requires that the hydrophone be directional i.e., producing a stronger signal when pointed in the direction of the tag, thereby giving an indication of the direction the signal is coming from. There are different ways to make a hydrophone directional. If the length of the hydrophone in any direction exceeds a few wavelengths, it becomes directional on that plane. The Vemco VH110 (Figure 8a) is an example of this approach using a number of interconnected cylinders in a line with sound absorbent material at the back of the hydrophone so there is not a 180 degree ambiguity as to where the signals is coming from. Since the cylindrical shape is maintained vertically, looking forward, the unit is directional horizontally but omnidirectional vertically. An alternative to this approach is the use of a parabolic reflector (e.g., Sonotronics DH-4 shown in Figure 8b). However, a hydrophone like this cannot be used when the tracking vessel is moving at higher speeds.

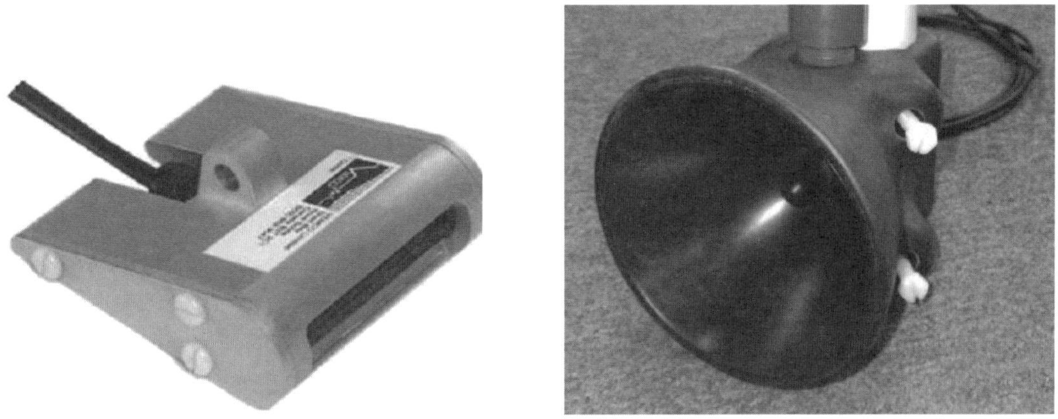

(a) (b)

Figure 8. Representative directional hydrophones: (a) Vemco VH110 which is directional horizontally due to length of transducer array and (b) Sonotronics DH4 which uses a reflector to provide directionality (Photos taken from company websites with permission).

A parabolic reflector directional hydrophone is directional vertically as well as horizontally so that one would need to search vertically as well as horizontally to find a tag signal while operating on deep water.

Another approach that can be used in a mobile tracking application is described by Vickery (1998). They define this approach as a Short Baseline Assembly and can be an attractive option since it does not require large separation between the hydrophones and does not create much drag when on a moving boat. The basic setup for this system involves having multiple hydrophones (3–4) that are slightly offset from one another. This approach allows the researcher to monitor a number of tags simultaneously in a number of different directions from the boat without the need to steer the boat towards each tag. Unfortunately, the complexity of this approach (installation issues, requirement for a multichannel receiver, and specialized training for the users) has limited its use. The only commercial available product that we are aware of that was capable of achieving this was the now-discontinued Vemco VH4 series.

Noise Considerations and Preamplifiers.—Ideally, the hydrophone converts the acoustic signal to an electrical one with no degradation of the tag signal. However, the quality and specification of the hydrophone can degrade the signal, so it is important for the researcher to know what they are getting when they acquire hydrophones. One of the important specifications is referred to as the noise floor of the hydrophone/preamplifier combination. The noise floor is the acoustic pressure which corresponds to the electrical noise at the hydrophones output when there is no signal. In short, it is a measure of how much background noise is generated by the hydrophone itself when there is nothing the hydrophone is listening to. Ideally, this should be less than approximately 26, 30, 36 and 40 dB re 1 $\mu Pa/\sqrt{Hz}$ at 75, 150, 300 and 400 kHz respectively (Wenz 1962). If the noise floor is less than these values, it is safe to say that the hydrophone will not cause further degradation of the signal. The manufacturer should have accounted for this when they designed the hydrophone and well designed hydrophones operating up to about 100 kHz usually meet these specifications. However, above 100 kHz where acoustic noise approaches the thermal limit as discussed above, added noise contributed by the preamplifier electronics will cause the noise floor to increase above the acoustic level seen by the hydrophone.

Receiving Systems

Receivers typically perform two basic functions. The first is the detection of the actual transmission as distinct from other acoustic sources and noise. The second involves decoding information within the transmission, i.e., the individual ID of the tag and sensor data if any is present in the signal. With the exception of some basic mobile tracking receivers, receivers today store detection data for subsequent retrieval, transmission, and/or analysis. In some systems, this information includes all potential tag signals, whether or not the signal can be decoded. Some receiving systems will also record specific aspects of potential tag signals such as amplitude, pulse length or shape, and signal to noise ratio. These additional measurements may assist in distinguishing true tag signals from multipath or noise. Most systems also time stamp the incoming tag signals with various degrees of accuracy. The accuracy of the time stamp depends on the objectives of the study. If simple presences/absence data are all that is required, time stamping to the nearest second would be quite adequate. However, if the data are being used to triangulate the position of the signal in 3 dimensions based on differences in the time the signal arrives at four or more hydrophones in the detection array, the accuracy of the time stamp is much more critical.

Broadband Receivers.—Broadband receivers are basic systems consisting of minimal filtering and simple circuitry to detect the presence of a transmitted pulse. The motivation for this simplicity is to minimize size and/or power consumption to facilitate easy deployment and autonomous operation for long periods. These are important requirements in many passive tracking applications. The disadvantage of such receivers is that because of their greater bandwidth, sensitivity of the receiver is sacrificed.

Narrowband Analog Receivers.—As the name implies, the heart of such a receiver is a narrow band filter matched to the bandwidth of the transmitted pulses (taking into account frequency distortion created by the underwater medium). Usually, such receivers shift the incoming signal to a higher frequency (often a standard intermediate frequency such as 455 kHz) before filtering.

Not surprisingly a narrow band receiver will outperform a broadband receiver in most situations. For example, the range of 950 m shown in the first row of Table 6 would rise to almost 1500 m if a good quality narrow band receiver were used. Of course, with the additional circuitry, power consumption will be significantly higher. Mobile tracking has traditionally used narrowband analog receivers (e.g., Vemco VR60, Sonotronics USR-08 and Lotek SRX 400). Some passive tracking receivers (e.g., Sonotronics SUR-1 and Vemco VRAP) are also narrowband receivers.

Digital Receivers.—The major motivation for digital receivers is to overcome limitations of analog receivers; these limitations include:

- The dynamic range of analog receivers is limited (i.e., if the gain is not properly set, the signal may either be too low to be detected, or so high that the receiver is saturated with the strength of the tag signal so no detections are recorded)

- It is not realistic to construct very complex filters for analog receivers and the accuracy of their coefficients is limited by the accuracy of components used. Therefore, potential detection range is lower than theoretically possible.

- Detection range can be increased by using multiple filtering strategies depending on acoustic conditions. The potential to do this with an analog receiver is very limited.

The underlying concept of a digital receiver is straightforward: conversion of the amplified hydrophone output to a digital data stream followed by digital processing to implement the detection algorithm. Ongoing improvements in the amount of processing that can be achieved by the receiver at a given power consumption rate makes the use of digital receivers more and more practical. Of course, power consumption is still significantly higher than for an analog receiver with similar performance (typically on the order of ten times as high at the time of writing). Examples of systems that use digital receivers are Vemco VR100, Lotek MAP 600, and HTI 290, 291, and 295.

Fine Positioning Systems

The Basic Principle.—Where there is an array of receivers and tags are simultaneously identified by at least three of them, it is possible to determine two dimensional fish position to (depending on factors discussed below) a resolution of a meter or even better (Ehrenberg and Steig 2002). Positioning is accomplished by measuring the difference in arrival time of the same transmitted signal at different hydrophones/receivers. When detection is by four or more hydrophones, with hydrophones separated vertically as well as horizontally, three-dimensional positioning is possible (Evans et al. 2010). However, in shallow-water studies where tag transmission range is several times that of water depth, wide spacing of receivers vertically isn't possible. One approach to three dimensional positioning in shallow water is to position tags horizontally using 3 (or more) hydrophones deployed in a horizontal plane, and then use tags that have depth sensors to provide the third dimension (Sauer et al. 1997).

Error Sources.—The accuracy of any fine-scale positioning system is limited by three factors: array geometry, the accuracy to which hydrophone positions are known, and timing synchronization between detections at the various hydrophones. These are discussed briefly below.

Ideally, the tag should be within the volume defined by the hydrophones in question. The position of tags can be determined outside of this volume, but the position accuracy degrades rapidly as the distance from the array increases. The hydrophones should be reasonably uniformly spaced as well; otherwise, variability in the shorter dimension(s) will be greater.

Uncertainty or variability in hydrophone positions will directly impact the calculated tag positions. The locations of hydrophones can be determined at the time of installation using a GPS receiver, and/or can be adjusted or fine tuned with 'time of flight' measurements of signals transmitted by the hydrophones themselves. For this analysis, sound velocity measurements are important to convert the 'time of flight' measurements to distances. If positioning accuracy requirements are demanding, significant effort is required to ensure that hydrophones are positioned or moored in such a way that they will not move with wind or current. If hydrophones move, then some method of recording their movement and applying the changes in position to the received tag data are required.

Positioning algorithms are based on differences between times of arrival of the signal at each hydrophone and, therefore, any discrepancies between the time references for detections at various hydrophones will increase positioning error. For systems with cabled hydrophones and a single receiving site (e.g., HTI model 290), this issue doesn't arise. An alternate approach used by the Vemco VRAP System transmits detection information by radio in real time to a central receiving site. A disadvantage of the two approaches above is that the infrastructure (cables, radio links, etc.) required limits the geographic scope over which fish can be monitored to approximately 1 km^2 or less. These systems also tend to be costly.

The Vemco VPS system (Espinosa et al. 2011) takes a different approach by using an extendable array of autonomous receivers and synchronization or reference tags in known positions. Post processing algorithms are then used to correct for drift between the time references of the individual receivers. If array geometry is such that each group of three receivers reliably detects at least one reference tag, this approach can be as accurate as the traditional approaches described above (Vemco 2010). On the other hand, if less accuracy is required, one can use fewer receivers to cover a given area. In addition, because of the simplicity of the autonomous receivers, these systems can be deployed unattended for long periods of time.

STUDY AND APPLICATION CONSIDERATIONS

Mobile versus Passive Tracking

Early acoustic telemetry studies almost always involved mobile tracking—typically a single tagged fish (with or without a sensor) was followed from a boat. While this provided some previously unknown knowledge of fish behavior, studies were limited by the restriction to only tracking one fish at a time, and the relatively short duration of tracking activities due to the need to keep in continual contact with the fish. As a result of these limitations, the vast majority of today's activities involve passive tracking in which fish presence (and possibly sensor data) is logged by stationary receivers.

Passive tracking studies can be broadly divided into two classes: Habitat Usage and Migration. Migration studies most commonly use lines of receivers to monitor passage of fish down a river (Perry et al. 2010) or along a coast (Chittenden et al. 2009). Habitat usage studies, on the other hand, are concerned with the behavior of a group of fish in a relatively restricted area—for example, in a lake or bay (Semmens 2008; Bacheler et al. 2009; Suski and Ridgway 2009), near a dam (Steig 2000; Ransom et al. 2008), near a coral reef (Chapman et al. 2005), near a fish aggregation device (Bach et al. 2003), or in a river (Bowen et al. 2009). In such studies fish position may be very rough (i.e., presence or absence of a tagged fish at a particular receiver) or may involve fine positioning.

Recently, the availability of the marine gliders (e.g., Webb Research's Slocum Glider and Liquid Robotics' Wave Glider) has led a number of researchers to investigate the potential for mobile tracking mission which would survey large expanses of the ocean for the presence of tagged fish.

Hydrophone Placement for Passive Tracking Applications

In most cases, researchers have choices as to where to place hydrophones. This is particularly true for migration studies where there is seldom a precise requirement of where lines should be placed. With the above in mind, one should try to adhere to the following guidelines as much as possible:

- Avoid hydrophone placements where underwater sound sources (e.g., fish finders, acoustic water level measurement devices, Acoustic Doppler Current Profilers) might be present

- Keep hydrophones/receivers and cables (if receivers are wired) away from electrical equipment, pumps, motors, power lines, etc.

- Avoid hydrophone placements in areas where air bubbles might be present.

- Be aware that woody debris, aquatic plants and the like in the path between the tag and receiver will seriously degrade performance.

- Be aware that noise from boat engines and propellers can cause temporary degradation of the ability to detect signals.

Range Testing

While it is possible to calculate the detection range of a given receiver and tag combination for assumed ocean or freshwater conditions, the noise and propagation characteristics of the local environment for a particular study are often unpredictable, particularly in freshwater environments. For this reason, in situ range tests are recommended to verify that tags are detected throughout the range of interest (Celedonia et al. 2008; Webber 2009; Loher et al. 2010).

While the details of range testing will vary somewhat depending on the nature of the study and the equipment being used, the following guidelines should be kept in mind:

- Tags used in the range tests should have a similar power level to those planned for use in the study.

- In situations where seasonal (e.g., in many riverine environments) or weather-related (e.g., in the ocean) conditions vary significantly during the study period, one should try to take range measurements in the worst case situation. If this is not realistic, then the results should be derated to compensate for this.

- If a boat is used as part of the test procedure (e.g., to position tags), ensure that engine noise, boat electronics, bubbles from the wake of the boat, or the boat hull itself, do not obscure or interfere with tag signals. Often this possibility can be eliminated by suspending the tag on an independent drifting float.

In determining the range requirement (or receiver placement) in studies involving the detection of fish passing a given area, one needs to take into account the interval between tag transmissions and the potential migration rate of passing fish. If the swimming speed of the fish, and/or the water currents in the study area are extremely fast and the transmission rate of the tag is relatively slow it may be possible for the fish to pass through the study area in less time than the transmission interval. See Hobday and Pincock (2012) for a discussion of this issue.

Fine Scale Positioning

Studies incorporating two or three dimensional positioning can address objectives such as determining where within a body of water fish spend the most or least amount of time (Semmens 2008), or whether or not environmental variables are correlated with fish behavior. While such behavioral studies can be and have been used in a wide variety of situations, they are particularly important for determining the effects of man-made structures on fish populations (Steig 2000; Brown et al. 2011). Summarizing data can sometimes be challenging, as modern positioning systems can produce two or three-dimensional positions for large numbers of fish at intervals sometimes down to a few seconds. One approach to summarizing such data are to divide the study area into cells, count the number of individuals that enter the cell, and develop a density plot to describe areas utilized by more or less individuals (Steig and Johnston 2010). Steig et al. (2009) used this method to summarize the density of sockeye salmon smolts (*Oncorhynchus nerka*) by color, as they passed through the forebay of Rocky Reach Dam on the Columbia River, near Wenatchee Washington, USA (Figure 9).

Visualization tools can be a useful alternative to the analytic methods referred to above. For example, the VPS system described above produces KML files which produce a two dimensional animation in Google Earth (see, for example Vemco 2011). When fine scale positioning data are overlaid on geo-referenced aerial images (see Figure 10, for example.), behavior of fish around specific features of a water body can be studied (Bowen et al. 2009).

APPLICATION ENVIRONMENTS

While no two studies are exactly alike, broad categories can be defined based on the environment where the tags and receivers are deployed. Most environments can be represented in one of four categories: marine; riverine; lakes and reservoirs; and dams, power plant intakes and other man-made underwater structures. Some considerations with respect to each of these are discussed below.

Marine Environments

Most marine environments allow unrestricted movement of the tagged fish, so whether the study involves habitat usage or migration, receivers need to be installed with prior knowledge of the behavior patterns of the tagged fish. This can be particularly challenging in the case of studies involving fish which might move large distances. For example, although migrations close to coastlines involving paths of thousands of kilometers are often monitored today (Welch et al. 2003), detections away from the coast are unlikely due to the sheer area

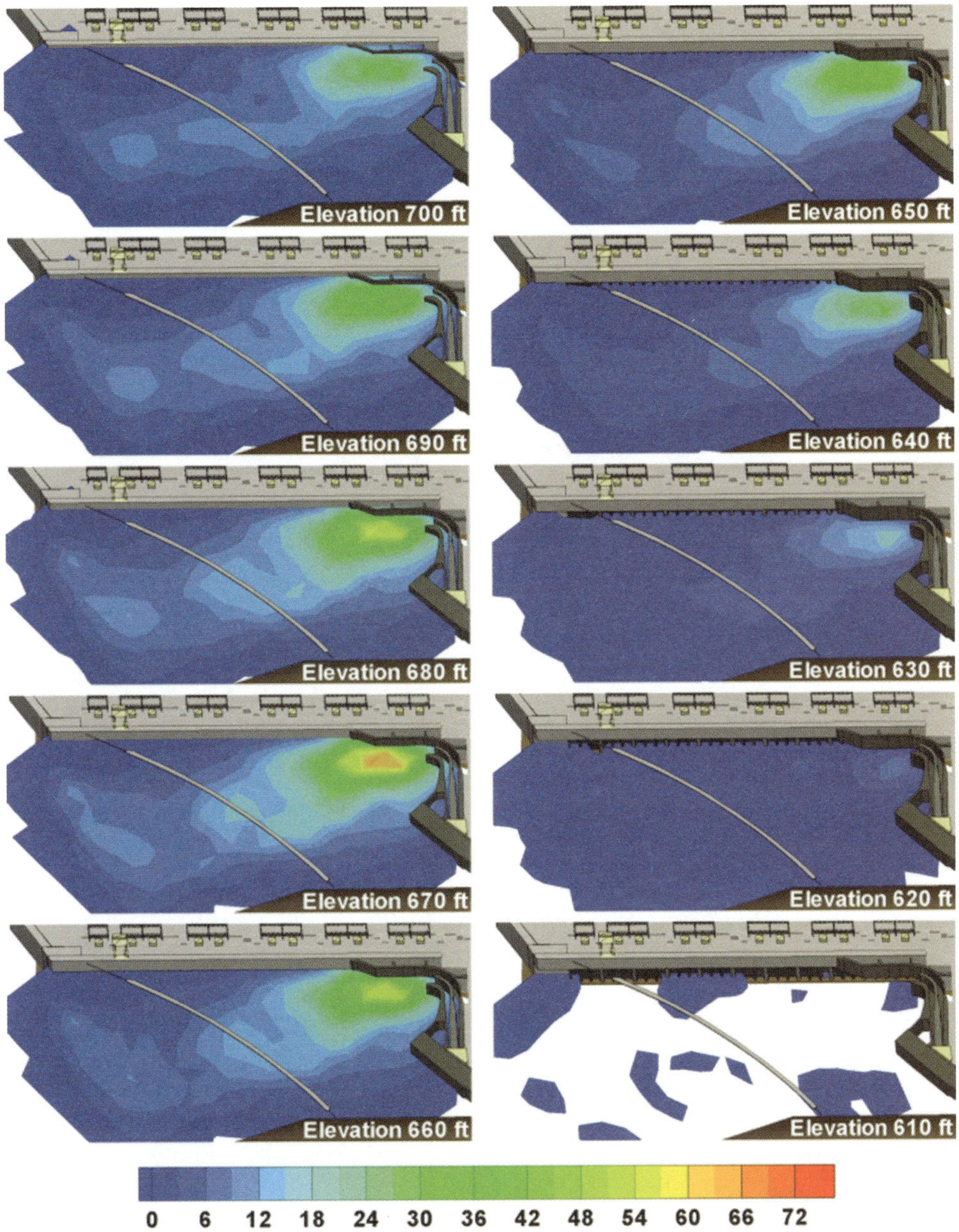

FIGURE 9. Binary densities of sockeye passage through the Surface Collector at Rocky Reach Dam during the spring of 2008. Density plan view figures are shown in 10 ft slices from elevation 700 ft (top left) to elevation 610 ft (bottom right). (Reproduced from Steig et al. 2009 with permission).

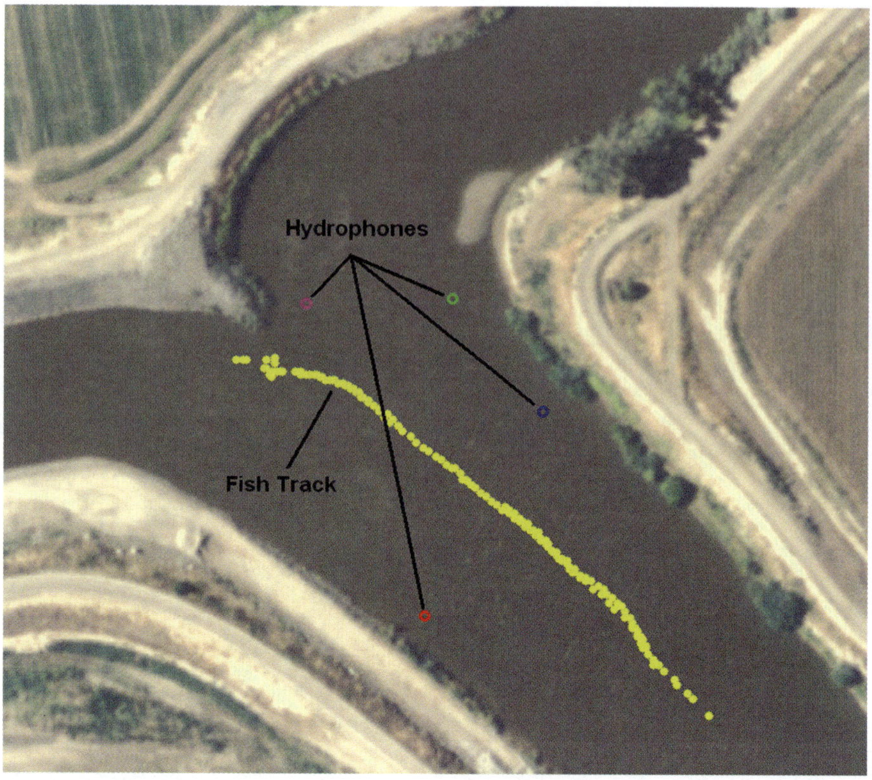

Figure 10. Two dimensional track of a tagged juvenile Chinook salmon migrating downstream through an array of four hydrophones overlaid on a geo-referenced image of the study site. (from Bowen et al. 2009, with permission).

involved. However, various approaches to address this are now being investigated. Exploratory missions using marine gliders have already been mentioned. Others include using ocean observation buoys as receiving platforms of opportunity (e.g., NOAA 2011) and using so called Business Card Tags (e.g., Vemco Mobile Transceivers) which add an acoustic receiver and data storage to acoustic tags carried by large mobile animals (e.g., sharks and seals) which might interact with fish of interest and hence record detections. The business card tag approach is described in more detail with some early results by Holland et al. (2009).

Tag size is also an issue in the ocean. As discussed above, high frequency transmissions suffer from significant absorption loss in the ocean, and the majority of studies require tags to function for long time (months or years). As a result, very small tags which must operate at high frequency and use very small batteries are only suitable in a small subset of applications in the marine environment. For example, they could be used in studies designed to characterize the habitat usage of fish in relatively small, shallow areas along the coast or in estuaries.

Riverine Environments

Hydrophone deployment in rivers can be difficult, since the environment is often very dynamic, with changing water velocities, high debris loads, boat traffic, and even changing bottom profiles. Many areas (e.g., near rapids or waterfalls) are simply not suitable for acoustic tag studies because of large amounts of entrained air. Other areas may appear ideal at some times of the year but rapidly change to unsuitable locations with changes in flow and debris load.

Basic information such as migration rates or survival is often estimated using fixed hydrophones or hydrophone arrays placed at intervals along the river (Ransom et al. 2008; Evans et al. 2010). However, as more and more natural watercourses are impounded or altered for irrigation, hydroelectric generation, and/or flood control man-made structures are becoming increasingly important as subjects of fisheries research in rivers. Often, such studies require fine positioning capability to provide detailed information about movements such as preferred paths and responses to the structures in question. Typically, a significant amount of planning and testing is required for this type of study as the structures often create artificial aquatic environments which present challenges due to unusual electrical and/or acoustic noise as well as multipath signals created by confined spaces, often with good reflecting surfaces.

Lakes and Reservoirs

Freshwater lakes and reservoirs usually represent the most trouble free environments for acoustic tag studies. Noise levels are often low and the absorption of the acoustic signal is low so maximum detection ranges tend to be high. This environmental condition can simplify implementation of acoustic studies. For example, in smaller lakes, it is often possible to position receivers to allow complete coverage of the entire lake (Hanson et al. 2007). On the other hand, in cases where there are good reflective surfaces (e.g., rocky cliffs or bottom), echoes can arrive at the receiver long after the direct transmission arrives causing a great deal of lost data for systems using coding tags. Where this might be an issue, extensive range testing designed to determine suitable receiver locations is required to optimize data collection. These types of test could indicate that a tag with a weaker signal may be better than a tag with a very strong signal that bounces off of structures and generates a significant amount of multipath.

SUMMARY

While acoustic telemetry may not be the desired technology for all research applications, the flexibility of this technology allows the researcher to meet many common study objectives. The ability to collect relatively coarse information about the movement of fish over large geographic areas and simultaneously obtain 2 dimensional and 3-dimentional behavior information with an accuracy of less than 1 m makes this a very versatile technology. With this versatility comes system complexity and advanced knowledge and familiarity with the technology is critical to designing and implementing a successful study that delivers robust results. It is our hope that the material presented in this Section, together with the information in other Sections of this book, will provide the researcher ample knowledge to successfully apply acoustic telemetry technology to address research and management questions.

REFERENCES

Adams, N. S., D. W. Rondorf, S. D. Evans, and J. E. Kelly. 1998. Effects of surgically and gastrically implanted radio tags on the growth and feeding behavior of juvenile chinook salmon. Transactions of the American Fisheries Society 127:128–136.

Bach P., L. Dagorn, A. Bertrand, E. Josse, and C. Misselis. 2003. Acoustic telemetry versus monitored longline fishing for studying the vertical distribution of pelagic fish: bigeye tuna (*Thunnus obesus*) in French Polynesia. Fisheries Research 60(2–3):281–292.

Bacheler, N. M., L. M. Paramore, J. A. Buckel, and J. E. Hightower. 2009. Abiotic and biotic factors influence the habitat use of an estuarine fish, Marine Ecology Progress Series 377:263–277.

Belleudy, P., A. Valette, and B. Graff. 2010. Passive hydrophone monitoring of bedload in river beds: first trials of signal spectral analysis. USGS, Scientific Investigations Report 2010-5091.

Bjorno, L. 1994. Underwater rain noise: sources, spectra and interpretations. J. Phys. IV France 04 (C5) C5–1023-C5–1030.

Bowen, M. D., S. Hiebert, C. Hueth and V. Maisonneuve. 2009. Effectiveness of a Non-Physical Fish Barrier at the Divergence of the Old and San Joaquin Rivers (CA). U.S. Department of the Interior, Bureau of Reclamation. Technical Memorandum 86–68290-11.

Brown, L. M., T. F. Savoy, J. P. Manderson and D. A. Fox. 2011. The Atlantic Cooperative Telemetry Network (ACT). Available: http://www.nero.noaa.gov/prot_res/atlsturgeon/wsdoc/day2/Research%20Updates/Brown_ACT_sturgeon_workshop.pdf (March 2012)

Brown L. S., A. Haro, and T. Castro-Santos. 2009. Three-dimensional movements of silver-phase American eels in the forebay of a small hydroelectric facility. Pages 277–291 *in* J. M. Casselman and D. K. Cairns, editors. Eels at the edge: science, status and conservation concerns. American Fisheries Society, Symposium 58, Bethesda, Maryland, .

Celedonia M. T., R. A. Tabor, S. Sanders, D. W. Lantz, and I. Grettenberger. 2008. Movement and habitat use of Chinook Salmon smolts and two predatory fishes in Lake Washington and the Lake Washington Ship Canal, 2004–2005 Acoustic Tracking Studies. U.S. Fish and Wildlife Service, Western Washington Fish and Wildlife Office, Lacey, Washington.

Chapman D. D., E. K. Pikitch, E. Babcock , and M. S. Shivji. 2005. Marine reserve design and evaluation using automated acoustic telemetry:a case-study involving coral reef-associated sharks in the Mesoamerican Caribbean. Marine Technology Society Journal 39:42–55.

Chittenden, C. M., R. J. Beamish, C. M. Neville, R. M. Sweeting, and R. S. McKinley. 2009. The Use of Acoustic Tags to Determine the Timing and Location of the Juvenile Coho Salmon Migration out of the Strait of Georgia, Canada. Transactions of the American Fisheries Society 138(6):220–1225.

Cole, D., D. A. Scruton, G. H. Niezgoda, R. S. McKinley, D. F. Rowsell, T. Lindstrom, M. N. Ollerhead. and C. Whitt. 1998. A coded acoustic telemetry system for high precision monitoring of fish location and movement: application to the study of nearshore nursery habitat of juvenile Atlantic cod (*Gadus morhua*). Marine Technology Society Journal 32(1):54–62.

Cooke, S. J., S. G. Hinch, M. Wikelski, R. D. Andrews, T. G. Wolcott, and P. J. Butler. 2004. Biotelemetry: a mechanistic approach to ecology. Trends in Ecology and Evolution 19:334–343.

Crum L. S., R. Roy, H. Pumphrey and A. Prosperetti. 1999. The Underwater Sounds Produced by Impacting Snowflakes. Journal of the Acoustical Society of America 106(4): 1765–1770.

Dahl, P. H. 1994. High frequency noise emitted from ocean breaking waves. in Sea Surface Sound '94: Proceedings of the 3rd International Meeting on Natural Physical Processes Related to Sea Surface Sound. M. J. Buckingham and J. R. Potter, editors. World Scientific Publishing Company, Singapore, University of California, Lake Arrowhead, California.

Del Grosso, V.A. and C. W. Mader. 1972. Speed of sound in pure water, Journal of the Acoustical Society of America 52:1442–1446.

Ehrenberg, J. E., and T. W. Steig. 2009. A study of the relationship between tag-signal characteristics and achievable performances in acoustic fish-tag studies. ICES Journal of Marine Science 66:1278–1283.

Ehrenberg, J. E., and T. W. Steig. 2002. A method for estimating the "position accuracy" of fish tags. ICES Journal of Marine Science 59:140–149.

Ehrenberg, J. E., and T. W. Steig. 2003. Improved techniques for studying the temporal and spatial behavior of fish in a fixed location. ICES Journal of Marine Science 60:700–706.

Espinosa, M., T. J. Farrugia., D. M. Webber, F. and C. G. Lowe. 2011. Testing a new acoustic telemetry technique to quantify long-term, fine-scale movements of aquatic animals. Fisheries Research 108(2–3):364–371.

Evans, S. D., C. E. Walker, S. J. Brewer, and N. S. Adams. 2010. Assessing survival of Mid-Columbia River released juvenile salmonids at McNary Dam, Washington, 2008–09. U.S. Geological Survey Open-File Report 2010–1237.

Grothues, T. M. 2009. A review of acoustic telemetry technology and a perspective on its diversification relative to coastal tracking arrays. Tagging and tracking of marine animals with electronic devices. Reviews: Methods and Technologies in Fish Biology and Fisheries 9(2):77–90.

Hallock, R. J., R.F. Elwell and D.H. Fry. 1970. Migrations of adult King Salmon (*Oncorhynchus tshawytscha*) in the San Joaquin Delta as demonstrated by the use of sonic tags. California Department of Fish and Game, Anadromous Fisheries Branch, California Department Fish and Game. Fish Bulletin (151):1–92.

Hanson K. C., S. J. Cooke, C. D. Suski, G. Neizgoda, F. J. S. Phelan, R. Tinline and D. P. Philipp. 2007. Assessment of largemouth bass (*Micropterus salmoides*) behavior and activity at multiple spatial and temporal scales utilizing a 3-D whole-lake telemetry array. Hydrobiologia 582:243–256.

Herman, C. H. 1975. Underwater Sound Transducers. IEEE Transactions on Sonics and Ultrasonics, SU-22(5):281–290.

Hobday, A. J. and D. Pincock. 2012. Estimating detection probabilities for linear acoustic monitoring arrays. Pages 325–346 *in* J. R. McKenzie, B. Parsons, A. C. Seitz, R. K. Kopf, M. Mesa, and Q. Phelps, editors. Advances in fish tagging and marking technology. American Fisheries Society, Symposium 76, Bethesda, Maryland.

Holland, K. M., C. G. Meyer, and L. C. Dagorn. 2009. Inter-animal telemetry: results from first deployment of acoustic 'business card' tags. Endangered Species Research 10:287–293.

Hubley, P. B., P. G. Amiro, J. F. Gibson, G. L. Lacroix, and A. M. Redden. 2008. Survival and behaviour of migrating Atlantic salmon (*Salmo salar* L.) kelts in river, estuarine, and coastal habitat ICES Journal of Marine Science (2008) 65(9):1626–1634.

Jepsen N., C. Schreck, S. Clements and E. B. Thorstad. 2005. A brief discussion of the 2% tag/bodymass rule of thumb. Pages 225–260 *in* M. T. Spedicato, G. Lembo, and G. Marmulla, editors. Aquatic telemetry: advances and applications. FAO/COISPA, Rome, Italy.

Johnson, J. H. 1960. Sonic tracking of adult salmon at Bonneville Dam, 1957. U.S. Fish and Wildlife Service, Fishery Bulletin 60(176):471–485.

Loher, T., B. M. Leaman, S. R. Hare, D. Carlile and C. R. Lunsford. 2010. Range test for a deepwater acoustic tag listening array, IPHC Report of Assessment and Research Activities. Available: www.iphc.washington.edu/publications/rara/2010/2010.363.Rangetestforadeepwateracoustictaglisteningarray.pdf (March 2012).

Mackenzie, K.V. 1981. Nine-term equation for the sound speed in the oceans, Journal of Acoustical Society of America 70(3):807–812.

McMichael, G. A., M. B. Eppard, T. J. Carlson, J. A. Carter, B. D. Ebberts; R. S. Brown, M. Weiland, G. R. Ploskey, R. A. Harnish and Z. D. Deng. 2010. The juvenile salmon acoustic telemetry system: a new tool. Fisheries 35(1):9–22.

Moore, A., I. C. Russell and E. C. E Potter. 1990. The effects of intraperitoneally implanted dummy acoustic tags on the behaviour and physiology of juvenile Atlantic salmon, Salmo salar. Journal of Fish Biology 37:713–721.

Niezgoda, G., M. Benfield, M. Sisak, and P. Anson. 2002. Tracking acoustic tags by code division multiple access (CDMA)-based telemetry. Hydrobiologia 483:275–286.

O'Dor R., L. Dagorn, K. Holland, I. Jonsen, J. Payne (5), W. Sauer, J. Semmens, M. Stokesbury, P. Smith and F. Whoriskey. 2009. The Ocean Tracking Network, OceanObs' 2009. Available: abstracts.congrex.com/scripts/jmevent/abstracts/fcxnl-09a02a-1662001-1-cwp4a09.pdf (March 2012).

NOAA. 2011. National Ocean Observing System to See Marine Animal Migration, Adaptation Strategies, Available: www.noaanews.noaa.gov/stories2011/20110304_ioostagging.html (March 2012).

Perry, R. W., J. R. Sklaski, P. L. Brandes, P. T. Sanstrom, A. P. Klimley, A. Ammann and B. MacFarlane. 2010. Estimating survival and migration route probabilities of juvenile Chinook salmon in the Sacramento-San Joaquin River Delta. North American Journal of Fisheries Management 30(1):142–156.

Pincock, D. G. 2008. False detections: what they are and how to remove them from detection data. Available: www.vemco.com/pdf/false_detections.pdf (March 2012).

Pincock D, D. W. Welch, S. G. McKinley, and G. Jackson. 2010. Acoustic Telemetry for Studying Migration Movements of Small Fish in Rivers and the Ocean—Current Capabilities and Future Possibilities. Pages 107–119 *in* K. S. Wolf and J. S. O'Neal, editors. PNAMP Special Publication: Tagging, Telemetry and Marking Measures for Monitoring Fish Populations—A compendium of new and recent science for use in informing technique and decision modalities: Pacific Northwest Aquatic Monitoring Partnership Special Publication 2010–002. Available: http://www.pnamp.org/node/2892 (April 2012)

Ransom, B. H., T. W. Steig, M. A. Timko, and P. A. Nealson. 2008. Basin-wide monitoring of salmon smolts at US dams. The International Journal on Hydropower and Dams 15(3):43–49.

Readhead, M. L. 1997. Snapping shrimp noise near Gladstone, Queenland. Journal of the Acoustical Society of America 101:1718–1722.

Richards, S. D., N. R. Brown, and T. G. Leighton. 1998. Turbidity in future high frequency sonar performance models, Proceedings of the joint meeting of the 16th International Congress on Acoustics and the 135th Meeting of the Acoustical Society of America. Pages 1349–1350 *in* P. K. Kuhl, L. A. Crum, editors. Acoustical Society of America, Melville, New York.

Sauer, W. H. H., M.J. Roberts., M.R. Lipinski, M.J. Smale, R.T. Hanlon, D.M. Webber and R.K. O'Dor. 1997. Choreography of the squid's "nuptial dance". Biological Bulletin 192:203–207.

Semmens, B. 2008. Acoustically derived fine-scale behaviors of juvenile Chinook salmon (*Oncorhynchus tshawytscha*) associated with intertidal benthic habitats in an estuary. Canadian Journal of Fisheries and Aquatic Sciences 65:2053–2062.

Stasko, A. B., and D. G. Pincock. 1977. Review of underwater biotelemetry with emphasis on ultrasonic techniques. Journal of the Fisheries Research Board of Canada 34:1261–1285.

Steig, T. W. 2000. The use of acoustic tags to monitor the movement of juvenile salmonids approaching a dam on the Columbia River. Pages 296–304 *in* J. H. Eiler, D. J. Alcorn, and M. R. Neuman editors. Biotelemetry 15: Proceeding of the 15th International Symposium on Biotelemetry. Juneau, Alaska USA, 9–14 May 1999.

Steig, T. W. and S. V. Johnston. 2010. Behavioral results from acoustically tagged fish using innovative techniques for analyzing three-dimensional data. Proceedings of the Oceans 2010 MTS/IEEE Seattle Conference and Exhibition, Seattle, Washington.

Steig, T. W., P. A. Nealson, K. K. Kumagai, B. J. Rowdon, J. R. Selleck and B. D. McFadden. 2009. Route specific passage of juvenile Chinook, sockeye and steelhead salmon using acoustic tag methodologies at Rocky Reach and Rock Island Dams in 2008. Final report for Chelan County Public Utility District No. 1, Wenatchee, Washington.

Stokesbury, M. J., M. J. Dadswell, K.N. Holland, G.D. Jackson, D.W. Bowen, and R. K. O'Dor. 2009. Tracking diadromous fishes at sea: the electronic future using hybrid acoustic and archival tags. Pages 311–320 *in* A. J. Haro, K. L. Smith, R. A. Rulifson, C. M. Moffitt, R. J. Klauda, M. J. Dadswell, R. A. Cunjak, J. E. Cooper, K. L. Beal, and T. S. Avery, editors. Challenges for diadromous fishes in a dynamic global environment. American Fisheries Society, Symposium 69, Bethesda, Maryland.

Suski, C. D., and M. S. Ridgway. 2009. Seasonal patterns of depth selection in smallmouth bass. Journal of Zoology 279:119–128.

Trefethen, P. S. 1956. Sonic equipment for tracking individual fish. U.S. Fish and Wildlife Service, Special Technical Report—Fisheries 179:1–11.

Tucker, D. G. and B. K. Gazey. 1966. Applied underwater acoustics. Pergamon Press, California

Urick, R. J. 1983. Principles of underwater sound. McGraw-Hill, New York.

Vemco. 2010. High Accuracy Achieved with VPS in Local Lake. Available: www.vemco.com/news/vps_banook.php (March 2012).

Vemco. 2011. VPS Sample 2. Available: www.youtube.com/watch?v=62FkpmQbVxw (March 2012).

Vickery, K. 1998, Acoustic Positioning Systems. Available: www.dynamic-positioning.com/dp1998/SVickery.pdf (March 2012).

Voegeli, F. A., G. L. Lacroix, and J. M. Anderson. 1998. Development of miniature pingers for tracking Atlantic salmon smolts at sea. Hydrobiologia 371/372:35–46.

Webber, D. M. 2009. VEMCO Acoustic Telemetry New User Guide pp 16–17. Available: www.vemco.com/pdf/acoustic_telemetry.pdf (March 2012).

Weiland, M. A., Z. D. Deng, T. A. Seim, B. L. LaMarche, E. Y. Choi, T. Fu, T. J. Carlson, A. L. Thronas, and M. B. Eppard. 2011. A cabled acoustic telemetry system for detecting and tracking juvenile salmon: Part 1. Engineering design and instrumentation. Sensors 11:5645–5660.

Welch, D. W., G. W. Boehlert, and B. R. Ward. 2003. POST–the Pacific Ocean salmon tracking project. Oceanologica Acta 25:243–253.

Wenz, G. M. 1962. Acoustic ambient noise in the ocean: Spectra and sources. Journal of the Acoustical Society of America 34:1936–1956.

Section 7.2

Detection Efficiency in Telemetry Studies: Definitions and Evaluation Methods

MICHAEL C. MELNYCHUK

INTRODUCTION

Acoustic and radio telemetry are valuable tools for studying the movements and dynamics of animal populations. Some studies combine telemetry methods with mark–recapture or tag-recovery modeling to estimate population abundance or demographic rates such as survival, individual growth, recruitment, and transition among habitats (reviewed in Pine et al. 2003). Other studies use telemetry techniques for assessing movement patterns as they relate to habitat use, foraging behavior, or interactions with other species. Many kinds of biological questions can be posed, and answering them often involves estimating how many tagged animals are still alive and/or still present in an area at some time.

Since some tagged animals may not be detected during the sampling process, estimates of numbers alive and biological rates (survival, movement) are confounded and therefore depend on the probability of transmitter (hereafter referred to as tag) detection. This detection probability (or equivalently, detection efficiency, *DE*) nuisance parameter must be estimated to obtain unbiased estimates of the survival or movement parameters of interest. Despite the widespread and increasing use of telemetry, there has generally been a paucity of studies that evaluate different methods for estimating *DE* (Clements et al. 2005).

Biological hypotheses and study objectives should drive the design of telemetry arrays and sampling methods (Heupel et al. 2006). Similarly, the meaning of *DE* and methods to quantify it also differ depending on study objectives. This Section is written to present the diverse usage of "detection efficiency" in telemetry studies and to describe the variety of evaluation methods available. Several environmental and technological factors which can affect *DE* are also outlined. The text generally refers to fish, but the ideas are transferable to telemetry studies involving other aquatic taxa and/or terrestrial animals. Examples given in this Section generally involve acoustic tracking studies, but the definitions, topics, and methods presented have application to radio telemetry as well.

OBJECTIVES FOR ESTIMATING DETECTION EFFICIENCY

Life history strategies, behaviors, and movement patterns are diverse among aquatic taxa. Some species (or individuals, or stages within a species) are relatively sessile, with restricted home ranges and forays of only short distances or duration. Others are far-ranging, spending little time in a particular area during dispersal or migration events. As a result of this diversity, definitions of *DE* have varied among telemetry studies depending on study objectives, and are categorized as probabilities of detecting: (1) individual tag transmissions, (2) tagged fish residing in an area, (3) tagged fish migrating past a specific location, and (4) tags present during mobile transect surveys.

Probability of detecting individual tag transmissions with stationary receivers

A common measure of *DE* is the proportion of known tag transmissions that are detected by a receiver or an array of multiple receivers over some time period. This gives the average *probability of detecting a single transmission of a tag,* DE_{single}, typically specified at some distance from tag to receiver. Systematic range testing with test tags makes extensive use of this measure, often for designing tracking studies (Clements et al. 2005; Hobday and Pincock 2012). A common objective is to determine an appropriate spacing between individual receivers deployed in a line (i.e., 'gate' or 'curtain') or grid pattern. At increasing distance from tag to receiver, the proportion of transmissions successfully decoded decreases, and the researcher can determine from this relationship what the necessary maximum spacing should be to maintain some minimum DE_{single} (Lindholm et al. 2007) or to evaluate the distance beyond which tags are no longer detected (Topping et al. 2006). A similar metric is the number of detections recorded over a fixed time period (Simpfendorfer et al. 2002; Grothues et al. 2005; Heupel et al. 2006). This can be used to compare the effect of distance on tag reception but can only be transformed to a proportion if the number of signals transmitted during this period is known.

The relationship between DE_{single} and tag-to-receiver distance is also used in positioning algorithms (Simpfendorfer et al. 2002). To evaluate fine-scale movement patterns within a grid array of receivers, it is necessary to quantify the proportion of detections received on multiple receivers in proximity of a tag over some short time period. The differential proportion of transmissions detected among receivers provides information on the most likely position of the tag within the array. The temporal sequences of detections or centers of activity are used in turn to quantify movement patterns and speeds within the array (Simpfendorfer et al. 2002; Ohta and Kakuma 2005; Hedger et al. 2008). An alternative method is to use ≥ 3 synchronized receivers and measure differential arrival times of a tag signal to triangulate position e.g., HTI (Seattle, Washington) 2D and 3D tracking systems, or VEMCO (Halifax, Nova Scotia) VRAP system.

Arguably, there is little direct biological interest in DE_{single}. Compared to the definitions that follow, DE_{single} rarely has any direct relevance to vital population rates or movement parameters. As such, this measure is more of an indicator of how a receiver or array of receivers perform technically, in addition to being useful for planning telemetry studies.

Probability of detecting resident fish with stationary receivers

Telemetry is often used for studying movement patterns of species within or away from some defined area. Some species show high degrees of site fidelity with small home ranges. Others may use some particular area repeatedly, but move to different areas intermittently for spawning, foraging, or searching for new territories (Klimley and Holloway 1999; Lindholm et al. 2007). Quantifying the time over which these various patterns are exhibited, or the timing of removal from an area due to fishing (Hightower et al. 2001) are often of interest. These can only be reliably estimated if there is a high *probability of detecting the tagged fish within the study area over some period of time, given that it is present there,* DE_{res}. If this probability were instead low (e.g., Hedger et al. 2008), the accuracy of any estimate would be confounded by the increased or unknown probability that the fish was absent from the area due to foray behaviors (Meyer et al. 2007) or removal.

The DE_{res} measure differs from DE_{single} in that determining presence in an area may require only one or two detections within some time period. It can be estimated either on a per-receiver basis or over an entire grid; multiple receivers within the area of interest will usually provide a higher DE_{res} for establishing presence. If receivers in stationary grid arrays are deployed densely enough that complete, continuous detection coverage is achieved, DE_{res} can be assumed to be 100% (Heupel and Simpfendorfer 2002). As a result, biological attributes of interest like movement or mortality rates can be estimated more precisely than if DE_{res} were also estimated. Incorrect assumptions of 100% DE_{res}, however, will cause underestimates of the proportion of time that tagged fish spend within the array of receivers.

Probability of detecting migrating fish with stationary receivers

Receivers can be deployed along migratory routes utilized by fish to estimate migration and survival rates (Rechisky et al. 2009; Welch et al. 2009; Perry et al. 2010) or passage by existing structures such as dams and velocity barriers (Section 8.1). Multiple receivers, arranged in lines or grid patterns, can be used to effectively "seal off" possible migration routes of tagged fish. In constricted areas such as rivers, estuaries, fjords, straits, and channels, a line of receivers can be positioned over the full width of such passageways. However, migrating fish may travel rapidly between adjacent receivers and at different depths, and may therefore be within detection range for only a short or variable period of time. This can result in a relatively low *probability of detecting a tag moving past a specific location,* DE_{mig}. This probability can change over time due to variation in background noise; typically, an average probability over the migratory period is quantified.

The measure DE_{mig} is equivalent to p in the Cormack-Jolly–Seber (CJS) open-population mark–recapture model for estimating survival (Cormack 1964; Jolly 1965; Seber 1965; Lebreton et al. 1992), where instead of the classic temporal form, a spatial form is considered for migrating fish (Burnham et al. 1987). Survival estimates in each segment of the migration between successive detection stations depend on the probability of tagged fish being detected at these stations as they cross (Section 9.2). When DE_{mig} is assumed to be 1, the modeling approach becomes that of known-fate models (Kaplan and Meier 1958). When there are additional receiver stations after the one of interest, DE_{mig} is estimable either in CJS models, or independently as the ratio of fish detected at and after the line of interest to those detected after the line of interest (Manly and Parr 1968). The CJS framework is often used to study salmonids that migrate in rivers and coastal areas, either adults returning upstream to spawn (Eiler 1995) or juveniles migrating downriver towards the ocean (Skalski et al. 2001; Melnychuk 2009a). Comeau et al. (2002) used a typical mark–recapture calculation to estimate DE_{mig} on a line where, instead of later lines providing information on the proportion of fish not detected at the line of interest, it was the same line at different times of year. Cod moved offshore past the line in the fall, and then returned onshore past the same line in the spring.

Generally, DE_{single} is a reasonable index of DE_{mig}, but there may be exceptions. If DE_{single} is low (e.g., <0.5, Simpfendorfer et al. 2008), that does not necessarily imply that DE_{mig} will be low. Only one detection (or two, to rule out the possibility that a single detection is a false positive) is required to substantiate the presence of a fish as it crosses a receiver line. If receivers are deployed with a short enough spacing between them to ensure that a minimum of several tag signals are transmitted during the time a tagged fish crosses the line (Welch et al. 2003), the probability that all transmissions will be missed will usually be small. Conversely, if DE_{single} is relatively high, DE_{mig} could still be low. If the programmed pulse rate of tags is low enough relative to the maximum swimming speed of the fish (or travel speed assisted by river currents), the fish could cross the line with an expectation of <1 signal transmitted while within detection range. At typical tag pulse rates of every 120 s or less, this last possibility would be rare for acoustic tags except perhaps for fish moving downstream in very fast rivers (>1 $m \cdot s^{-1}$ water velocity). Receivers can be deployed close enough together to reduce this possibility.

Probability of detecting tags during mobile surveys

The previous three definitions primarily relate to studies that use stationary receivers. Detections at fixed locations depend on tagged fish moving close enough to be detected. If a tagged fish dies early in its migration, ceases its migration, or has a limited range of movement, it may never come within detection range of a receiver. In these cases, mobile tracking surveys can be used to sample in areas surrounding an array or between stationary receivers (Ng et al. 2007; Lowe et al. 2009; Melnychuk and Christensen 2009). Researchers can conduct continual transects with a hull-mounted or towed hydrophone (e.g., Holland et al. 1985) or periodically stop a boat at sampling stations to monitor for tags. Similar to the case of migrating fish crossing over a stationary array, however, searching for relatively sedentary fish can result in a low *probability of detecting tags present within the area sampled by mobile surveys,* DE_{mobile}. Understanding detection ranges of mobile sampling gear and how these change with environmental conditions and boat speed is crucial for designing mobile sampling patterns.

Mobile tracking studies have different objectives. Many studies involve continuously following tagged individuals to quantify movement patterns, habitat use, and environmental correlates with these patterns (Holland et al. 1985; Block et al. 1997; Topping et al. 2005). In these types of studies, there is little need for estimating DE_{mobile}. Other studies involve sampling for an unknown number of tags currently present in some area, where the location of each tag is initially unknown (Hightower et al. 2001; Melnychuk and Christensen 2009). Tags may or may not be easily detected. In transect surveys, it is crucial to estimate DE_{mobile} if any inferences are to be made about either the number of tagged fish present in the area during the sampling period or survival rates of tagged populations. The number of tags detected by periodic mobile searches is sometimes reported (Voegeli et al. 1998), but if DE_{mobile} associated with this sampling process is not quantified, it is not possible to estimate how many tagged individuals were in the area but *not* detected during mobile surveys.

Similar to the preceding definitions, achieving a higher DE_{mobile} results in less ambiguous inferences about biological movement patterns and population vital rates. As DE_{mobile} (or relocation probability, where the same area is repeatedly sampled for the same group of tags over the course of a study) decreases, so does the precision of other model parameters like natural and fishing mortality rates (Pollock et al. 2004). As DE_{mobile} increases, the failure to detect a particular tag increasingly implies that it is not present within the sampling area rather than the alternative of it being present but not detected.

Other definitions of detection efficiency

Comeau et al. (2002) presented two definitions of *DE*. One of these was a typical mark–recapture definition for tagged fish migrating across a fixed receiver line, but the other did not involve detection data. This second metric was calculated as a ratio of the average detection range based on test tag measurements to the average distance between adjacent receivers on the line. Although this estimate was close (72%) to the mark–recapture DE_{mig} (80%), using it involves risk of relying too heavily on range tests. If oceanographic conditions, current velocity, or tag depth vary between the period of range testing and when fish actually cross the receiver line, range testing results may not reflect actual detection processes at the time of fish crossing events. Using data from tagged fish to determine *DE* (as in their other estimate) eliminates the need for detection range assumptions.

FACTORS AFFECTING DETECTION EFFICIENCY

Several factors have been shown to affect detection probabilities. Many have been recently summarized (Heupel et al. 2006; Simpfendorfer et al. 2008) and were alluded to earlier in this Section, so let us now focus only on the principal factors. These factors can be controlled by the choice of equipment or study designs, affected by environmental conditions, and affected by the behavior of tagged fish in the area. They likely have similar influence for all definitions of *DE* identified previously.

Tag-to-receiver distance

The distance between a transmitting tag and receiver is the most commonly assessed factor affecting *DE*, forming the basis for range tests. Two types of losses of acoustic strength occur as a result of increased distance: geometrical spreading of the wavefront through water, and absorption of sound which is affected by environmental factors (Voegeli and Pincock 1996). Over the long term (as environmental acoustic conditions vary), the decline of *DE* with increasing distance has been suggested to be linear (Simpfendorfer et al. 2002; Heupel et al. 2006), near-Gaussian (Hobday and Pincock 2012), logistic (Szedlmayer and Schroepfer 2005), or exponential beyond some threshold distance (R. Hedger, Norwegian Institute for Nature Research, personal communication). At any given moment (during a short period of constant acoustic conditions), the relationship between *DE* and distance has been suggested to be near-uniform (Hobday and Pincock 2012) or logistic with variable rates of decrease (Melnychuk and Walters 2010). For acoustic tags in rivers, turbulence is an especially severe problem; detection distance can drop from an order of 100 m to <10 m quite abruptly, with a shift from smooth (laminar) to turbulent river flow.

User-controlled factors

Researchers have direct control over how receivers are deployed as well as the programming characteristics of tags. Some hydrophones (those with spherical transducer elements) are truly omni-directional, but others are most sensitive to signals in a particular plane (see Section 7.1). For a stationary receiver with a cylindrical hydrophone element oriented vertically, this plane is typically perpendicular and radial to the vertically-oriented receiver, so sounds are detected at greater distances when they are within a broad horizontal plane of the hydrophone (Clements et al. 2005). Preliminary studies suggest that in some environments this plane of greatest sensitivity may be narrow, with *DE* decreasing when tags are more than a few meters above or below the hydrophone plane (Farmer 2009). Users cannot control the depth of tagged fish, but receivers can be deployed at a depth where tagged fish are most likely to occur. Receivers can also be deployed to maintain direct line-of-sight between tags and receiver, by not having topographical or vegetative obstructions near the hydrophone (which may block the signal and prevent detection). Mounting equipment for the receiver such as submerged floats, lines, or even the receiver itself (Clements et al. 2005) may also obstruct sound signals reaching the hydrophone regardless of the type of transducer element used.

Transmitter output and frequency directly affect transmission distance and therefore *DE* (Simpfendorfer et al. 2002). Many manufacturers provide a range of acoustic frequencies or power levels among tag models. Lower frequencies generally transmit further in water, but require a tag transducer with greater diameter to emit the sound, so are typically associated with larger-sized tags. Even within the same tag model, minor variation in strength among individual tags may translate into heterogeneity of *DE* among tags (Melnychuk 2012). Variation in hydrophone sensitivity may also exist among receivers, and can be assessed prior to deployment using a test tag. When several tags transmit within an area, some signals may collide (i.e., overlap) and prevent detection. The frequency with which this occurs is a function of the number of tags within reception range and how often tags transmit their signal (Voegeli et al. 1998; Hightower et al. 2001; Simpfendorfer et al. 2002), which are both controlled by users. To note, collisions are less likely in coding schemes that use single-pulse methods with a small ratio of transmit times to interval times. Needless to say, researchers should also consider expected tag battery life, and either restrict inferences of survival, abundance, or site fidelity to the period before estimated tag failure, or account for the possibility that *DE* may abruptly or gradually decline after this period.

In mobile transect surveys, boat speed is probably the most important user-controlled factor affecting DE_{mobile}. At faster speeds, tags along a transect are within detection range for a shorter period of time, engine-related noise and flow-related noise of water moving past the hydrophone are greater (thereby reducing the signal:noise ratio) and, if the hydrophone is towed behind the boat, it travels at shallower depths where background noise levels are usually higher (Voegeli and Pincock 1996; Melnychuk and Christensen 2009). As surveys are generally conducted at the surface, the depth of tags also affects DE_{mobile}. This is particularly a concern for deep-water demersal species or for tags laying on the bottom either inside fish that have died or after having been defecated by a predator (Melnychuk and Christensen 2009).

Environmental and behavioral factors

Any factor that increases background noise near the same frequency as a tag signal will decrease the signal:noise ratio and therefore *DE*. Wind may be the dominant environmental factor that reduces detection ranges and efficiencies (Hobday and Pincock 2012). Hydrologic conditions that create noise include tidal currents, silt or other suspended matter, turbulent flow, and broken water surfaces due to waves or rain (Voegeli et al. 1998). Gradients within the water column like haloclines or thermoclines may decrease detection ranges due to refraction or reflection (Voegeli and Pincock 1996), or more rarely, might increase ranges. Seasonal variation in river discharge and related noise levels may lead to a systematic change in *DE* over this same period for receivers placed in rivers (Simpfendorfer et al. 2008; Melnychuk 2009a). Soniferous invertebrates or fishes, whether on or near receivers, can increase background noise levels (Simpfendorfer et al. 2002; Heupel et al. 2006; Heupel et al. 2008). Finally, human sources like boat traffic, dams, and fish farms can all generate noise and introduce air bubbles or suspended matter into the water column, thereby reducing *DE* (Voegeli et al. 1998; Hightower et al. 2001; Simpfendorfer et al. 2002; Heupel et al. 2006; Simpfendorfer et al. 2008). Variation over time in any of the above noise sources can result in diurnal or seasonal variation in *DE* and ranges (Hobday and Pincock 2012).

Detection probabilities can be reduced by physical obstructions between tag and receiver such as submerged aquatic vegetation (Hightower et al. 2001; Simpfendorfer et al. 2002), bottom topography like large rocks or coral reefs (Simpfendorfer et al. 2002; Farmer 2009), or aquatic life that grows on receivers (Heupel et al. 2008). Even without any direct sound barrier, the bottom type over which studies are conducted can affect *DE* through absorption of sound. Some substrates like soft sediments absorb sounds more than others (Heupel et al. 2006). Noise sources can be less problematic for signal detection when using narrow-band receivers than broad-band receivers due to the greater range of frequencies received on broad-band receivers, which can potentially decrease the signal to noise ratio.

Although it is usually assumed that *DE* should be homogeneous among individuals if they are tagged with the same type of transmitter, the behavior of tagged fish can affect *DE*. The depth of tagged fish may vary with respect to a deployed receiver. In some cases this has little effect on detecting tags and is even of biological interest (Block et al. 1997). In other cases, if tag depths are above or below the horizontal plane of greatest sensitivity for a receiver oriented vertically or existing environmental gradients (e.g., thermocline, halocline), detection ranges and *DE* may be reduced. Some tagged individuals may spend a greater proportion of time either outside the sampling area of a receiver array, in locations within it that are noisier (Hightower et al. 2001), or in locations near benthic structures or complex geomorphology that lead to obstructed transmissions (Simpfendorfer et al. 2008). Any of these possibilities could lead to reduced *DE* for this behavioral subset of the tagged population. Similarly, in migrating fish that cross a line of receivers, faster individuals will be within detection range for a shorter period of time, so on average may have reduced DE_{mig}.

EVALUATION METHODS

Although there are many factors that can reduce detection probabilities, a researcher may compensate for many of these during the design of a tracking study. When designing receiver grids for quantifying the presence of tagged fish within the array over time, receivers can be deployed densely to increase the overlap of their detection ranges and therefore minimize the probability of a tagged fish returning to the array but not being detected. Similarly, when designing a line array for detecting migrating fish, shorter distances between adjacent receivers will decrease the probability of a tagged fish crossing the line undetected. In mobile tracking studies, shorter distances between parallel transect lines and slower boat speeds will decrease the probability of not detecting a tag within the sampling area. There are trade-offs associated with these strategies for increasing *DE* within a particular area: for a fixed number of available receivers (or mobile transects) due to a fixed budget and/or time restraints, deploying them densely implies the total sampling area will necessarily be smaller. Similarly, slower boat speeds will probably reduce the size of the area searched.

Whichever measure of *DE* is of interest for study objectives, this probability should be quantified over a range of variables and conditions that may occur during the biological study. Users can evaluate background noise at specific locations using broad-spectrum hydrophones (Heupel et al. 2006), or can evaluate detection ranges at specific locations under a range of environmental conditions. Both of these measurement types are useful for designing tracking studies and both can be important indicators of *DE*, but neither are direct measures of *DE* that can be used to estimate a proportion of tagged fish that have died or are no longer in a study area.

Proportion of transmissions detected

Two factors that should always be taken into account when assessing DE_{single} are power output of the tag and tag-to-receiver distance. The test tag should have the same properties as the transmitters used to tag fish. The sensitivity of hydrophones may vary among receivers, or even decrease over time for a given receiver, so researchers may wish to evaluate DE_{single} for multiple receivers. Researchers should assess DE_{single} at several distances to determine the shape of the decay with increasing distance. A test tag can be deployed at fixed distances from a receiver, ensuring that a sufficient number of known signals are transmitted at each distance (the more transmissions the better; 50 transmitted signals will allow DE_{single} to be estimated at 2% increments). If multiple receivers are available, these can be simultaneously deployed at varying distances from a fixed-location test tag so this effect of distance is isolated from a possible confounding of testing different distances at different times. An alternative approach could be to deploy multiple tags at varying distances from a fixed-location receiver; in this case, tag transmission characteristics should be considered to avoid signal collisions.

As an example, tag tests were conducted in March 2005 over a period of 2.2 h in Howe Sound, a coastal fjord in British Columbia. Twelve VEMCO VR-2 receivers were deployed in a line, raised from the seabed at a depth of ~145 m each. From one end of the line (A), they were spaced at 0, 100, and every 50 m thereafter to a maximum distance of 600 m at the other end of the line (B). A boat was positioned above (A) and two test tags of differing acoustic power (VEMCO V7 and V9, 136 and 142 dB at 1 m, respectively) were lowered to a range of depths (2, 5, 10, 20, 40, 70, 100, and 130 m) for 15 min at each depth. This corresponded to 40 signals transmitted from each tag at each depth; the tags were offset at a fixed repeat rate so their signals would not collide. The proportion of signals detected on each receiver was calculated for each tag at each depth. The raw data are not shown (D. Welch, Kintama Research

Corp., unpublished data), but a Lowess bivariate fit was used to interpolate DE_{single} as a function of tag-to-receiver horizontal spacing and tag depth (Figure 1). The probability surfaces show that >90% of transmissions were detected at a wide range of distances and depths, but this declined sharply after a range of about 500 m for V9 tags and 400 m for V7 tags, and also declined sharply when test tags were in surface waters to a depth of 30 m (V9 tags) or 60 m (V7 tags). Both surfaces show irregular contour trends that are sensitive to DE_{single} at particular depths and distance combinations; these may simply result from chance, as relatively few transmissions were emitted at each tag placement. Different results are likely to be observed in other environments with different local conditions; researchers should not assume that published values for detection ranges or *DE* will hold for their particular study site.

When assessing DE_{single}, it is crucial that environmental conditions during test tag evaluations reflect those that occur during the biological study (Hobday and Pincock 2012). If favorable weather conditions occur during a test tag evaluation, detection ranges and DE_{single} could be considerably lower under less favorable conditions during the actual study. Three general approaches are possible:

(1) if tags whose coding schemes involve single, short pulses are used (or else tags with long pulse trains that have long delay intervals between trains), it is possible to deploy a test tag in the area with little interference to signal reception of fish tags (this is often called a sentinel tag). Decreases in DE_{single} for the test tag may reveal increased background noise or obstructions around the hydrophone; near-continuous on-site monitoring would allow such problems to be dealt with immediately, or accounted for in data analyses after the study. Conversely, if tags with coding schemes involving long pulse trains are used, then the presence of test tags in the area runs the risk of test tag signals colliding with fish tag signals. The following two options deal with this issue:

(2) test tag evaluations can be conducted under a range of specific, known environmental conditions at different times before and/or after the actual study. Then, specific environmental conditions observed during the study can be matched with the appropriate *DE* evaluations that occurred during the test period (Melnychuk and Hausch 2011). This approach has the advantage of explicitly recognizing temporally-varying detection ranges and efficiencies so that biological inferences can make use of this detailed information. Disadvantages include having to measure specific conditions during both test tag evaluations and at all times during the actual study, as well as the possibility that fish behavioral effects on *DE* may confound effects due to environmental conditions. Also, much effort is needed to characterize how DE_{single} changes under a range of conditions. Alternatively,

(3) DE_{single} can be "integrated" across all possible environmental conditions to provide an overall (i.e., average) measure. This involves deploying a test tag(s) for a long enough time to experience a range of different conditions, which is assumed to reflect the range of conditions during the actual study. Time-varying patterns in DE_{single} can still be quantified using this approach (Simpfendorfer et al. 2002; Heupel et al. 2006), although relating these to time-varying patterns during the study requires matching environmental conditions. If DE_{single} evaluations are conducted near the study area (or within the area, for test tags with single, short pulse coding schemes) at the same time as the actual study, then temporal variation during test tag assessments (due to weather, etc.) will likely be representative of that during the study.

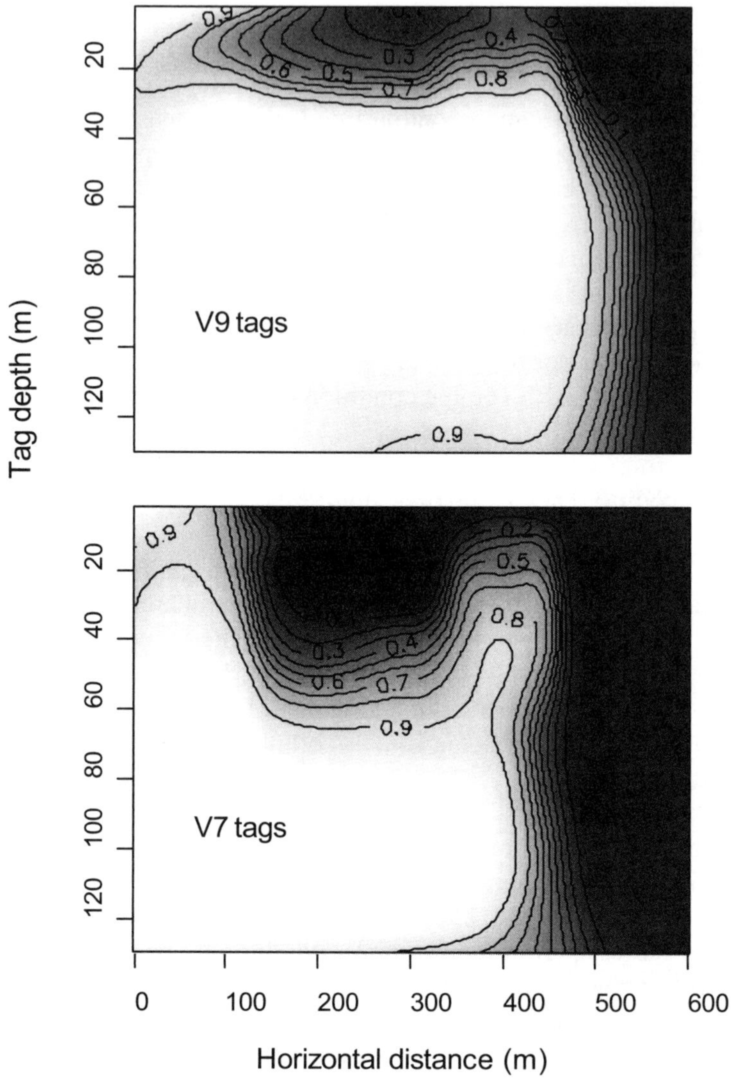

FIGURE 1. Detection probability of an individual transmission at different tag depths and tag-to-receiver distances. VEMCO VR-2 receivers were deployed at 145 m depth at various distances from two test tags (VEMCO V9 or V7). Detection probabilities were calculated as the proportion of 40 transmissions of each of two tags at each of eight depths that were successfully decoded. Probability surfaces are interpolated from a Lowess fit (smoothing parameter $\alpha = 0.2$), with shading ranging from high detection probability (white, 100% detection) to low detection probability (dark gray, 0% detection). Detection probability contours are overlaid onto shaded areas, ranging from <10% to >90%. Note the different scales of x and y-axes. Raw data were provided by D. Welch (Kintama Research Corp., Nanaimo, British Columbia).

In addition to temporal variation in environmental conditions, spatial variation in acoustic conditions can occur within the area. Variation in DE_{single} can occur among sub-areas as a result of variation in bottom topography, depth, sediment type, aquatic vegetation, and the influence of tides, river flow, or boat traffic. Detection efficiency assessments should be conducted across these spatially-varying conditions, either during the study if possible (so that test tag assessments are more reflective of current conditions during the study), or before and after the study if test tags with long pulse trains are used (to avoid interference from test tags). Test tag tows from a boat while recording GPS readings can be used to efficiently cover a large number of transmission locations in order to identify problem areas where tag signals are received less consistently (Hedger et al. 2008). This information can assist in strategies for re-deploying receivers to increase *DE* in these sub-areas, as well as to recognize lower *DE* in some sub-areas and account for the deficiency during analyses.

If the effect of multiple factors on DE_{single} can be evaluated, these should be analyzed in a common framework instead of one at a time. This permits an assessment of whether interactions between factors also affect DE_{single} (Farmer 2009). For example, the effects of horizontal spacing and depth on DE_{single} (Figure 1) were evaluated together. At horizontal distances <100 m, the effect of depth on DE_{single} was less than at distances >100 m (in this particular habitat at this particular time). If such an evaluation were extended over a wider range of environmental conditions (Farmer 2009), effects of several interactions could be revealed.

These methods of evaluating DE_{single} can be useful for designing appropriate spacing between receivers in tracking studies. In contrast, quantifying and accounting for DE_{single} is probably not as crucial when used for positioning algorithms of tagged fish within a receiver array (Simpfendorfer et al. 2008).

Movement around stationary arrays

If continuous contact with a tag is desired for accurately monitoring tag positions over time, receivers should be densely deployed with overlapping detection radii (Heupel et al. 2006) to ensure that $DE_{res} \approx 1$. If only occasional detections are required to address questions related to presence within an area, then receivers can be deployed further apart (Heupel et al. 2006) or as single units (for example, in the case of fish aggregating devices; Klimley and Holloway 1999) and in these cases it becomes especially important to estimate DE_{res}.

Test tag evaluations can be used to estimate DE_{res} over time. Test tags can be either deployed at fixed locations or towed by boat within a grid array. The depth and speed of towed tags should represent general movement patterns of the study species. Evaluations should occur over a range of weather or tidal conditions to quantify how these factors affect DE_{res}. Using GPS to record the position and speed of a towed test tag through the array, detection patterns on stationary receivers can be used to calibrate positional and speed estimates of tagged fish (Cash et al. 2005; Hedger et al. 2008). Similarly, mobile tracking within an array of stationary receivers can be used to follow a fish and evaluate how often it is detected on stationary receivers, based on known locations from manual tracks (Topping et al. 2006). This method eliminates biases from test tags that might not be towed at depths or speeds reflecting those of the study species.

Initially, a relatively long time interval can be considered to verify whether at least one test tag transmission is detected by at least one receiver. The test tag sequences can then be randomly sub-sampled, selecting a shorter time interval within the deployment or tow through the grid array and assessing whether tags are still detected at least once within this shorter interval. This can be repeated at shorter and shorter intervals, randomly sampling several sequences for each defined interval. At the extreme, evaluations at the shortest possible time interval become equivalent to assessing DE_{single} over the entire grid array.

Once DE_{res} is evaluated over a specified time interval, it can be interpreted as the probability that nondetection implies the fish is not present within the area (as opposed to present, but not detected over that interval). Longer time intervals will generally give higher DE_{res}, but shorter intervals are of interest to study fine-scale movements within and away from the array. An optimal interval can be selected that is as short as possible but still at or near 100% DE_{res}. This provides a high level of confidence in establishing presence, so that tags are consistently detected while they are within the area, and that absence of detection over some time period implies they have left the area. Such patterns of absence could be suggestive of forays outside of the area (Topping et al. 2006; Lindholm et al. 2007; Afonso et al. 2008) or of mortality (Hightower et al. 2001; Heupel and Simpfendorfer 2002). A high relocation or detection efficiency is required to make reliable estimates of the timing of these movement or mortality rates (Hightower et al. 2001) and to ensure high precision of parameter estimates (Pollock et al. 2004). Conversely, if detections of a tag within an array are sporadic, possible forays outside of the area would be confounded with low DE_{res}. If many tagged fish are present, signal collisions can lower DE_{res}, especially if tag repeat rates are frequent and pulse trains are relatively long. A longer sampling interval may be required in these cases to ensure that DE_{res} remains near 100%.

If users rely on a temporal sequence of detections to establish presence, sampling effort (the number or proportion of fully functioning receivers within the array) should be quantified over time (Klimley and Holloway 1999; Dewar et al. 2008). Variation in the number or position of receivers in a grid array over time could explain temporal variation in DE_{res}. Even if receivers are operational over some time period, the probability of detecting a given transmission may decrease over time due to biofouling (Heupel et al. 2008), so test tag evaluations should occur before, after, and (if possible) during a study period. The researcher could then identify possible limitations to detecting tagged fish at some time(s), and caveat their inferences of fish detection results accordingly. Inconsistent or time-varying estimates of *DE* from test tags (e.g., different measures before and after a study) are not necessarily problematic. The researcher can simply bracket a range of plausible biological results dependent on the range of observed *DE* estimates before, during, and after the study. For example, a low estimate of DE_{res} implies a greater number of tagged fish in the area at some time after expanding from the number detected, whereas a high estimate of DE_{res} implies fewer fish estimated, so a range of abundance could be reported dependent on the range of DE_{res} values estimated. Similarly, considering a wide range of possible values of DE_{mig} can be used to bracket reasonable estimates of survival in migration studies (Melnychuk 2009b; Welch et al. 2009, Welch et al. 2011).

Migrations past stationary receiver stations

Several measures can be taken to increase DE_{mig}. These include spacing receivers closer together on a line, using higher power tags, decreasing the intervals between tag pulses, and staggering receivers in a zig-zag pattern rather than in a straight line (Clements et al. 2005; zig-zag patterns can also provide information on direction of travel). All of these have associated trade-offs, so DE_{mig} will rarely be 100%. Estimating this probability can rely on a combination of field-based methods and mark–recapture modeling.

Multiple tows of a test tag past a line of receivers can be used to assess the detection frequency, as well as patterns in detection ranges. Tows should be done at a depth and speed reflective of the migrating fish. When multiple receivers are arranged in a line, it is especially important to conduct tows between adjacent receivers (Pecl et al. 2006). By knowing the position and speed of the test tag crossing the line, the temporal pattern in tag transmissions can be converted into a measure of approximate distance from tag to receiver line. Distances can be estimated as either perpendicular distance from tag to the line of receivers regardless of whether the tag crosses overtop a receiver or in a gap, or distance from tag to any particular receiver while crossing (accounting for the position crossed along the line).

As an example, tows with a VEMCO V9 test tag were conducted across two lines of acoustic receivers deployed on the seabed in Howe Sound in 2005 (Figure 2; unrelated to receiver lines discussed in previous examples and illustrated in Figure 1). VEMCO VR-2 and VR-3 receivers were paired at each position on a line to compare detection rates. The test tag had a fixed repeat rate, so the distance between successive transmissions was easily calculated from average boat speed. Most tows across the receiver lines resulted in several detections. One tow on line B had no detections on VR-2 receivers, and another on line B had no detections on either receiver type. There is some indication that tags were detected from further away while the boat approached the line than after it passed over (Figure 2), which could suggest that engine-related noise was greater aft of the boat than ahead. Al-

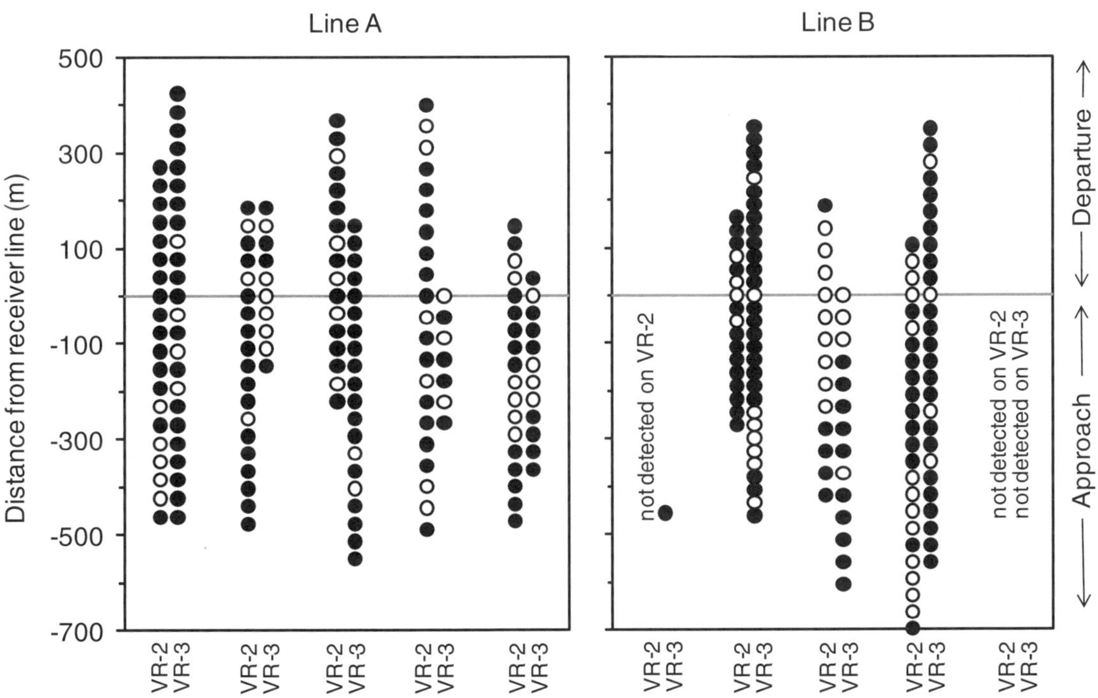

FIGURE 2. Detection sequences of 10 test tag tows past two stationary receiver lines. Approximate distance intervals were calculated from average boat speed (ranging from 7.4 to 12.7 km/h while crossing). VEMCO VR-2 and VR-3 receivers were paired at the same location. Empty circles show transmissions not detected, and black circles show transmissions detected. Test tags were generally deployed >500 m from the line while approaching and retrieved >500 m from it after passing; transmissions not detected either before the first detection or after the last detection are not shown.

though the number of tows was limited, the number of detections in a crossing event was inversely correlated with boat speed ($r = -0.39$; probably related to boat noise rather than tag speed) and inversely correlated with the distance to the nearest receiver when crossing the line ($r = -0.46$), as might be expected. There were insufficient tows conducted to provide a good estimate of DE_{mig} (and tow speeds were faster than juvenile salmon typically swim), but it is straightforward to conduct enough passes at a more representative speed to estimate this probability. Tows should be conducted across a range of environmental conditions and at different positions along the line to evaluate these effects on DE_{mig}. Tows by boat may underestimate DE_{mig} for a given speed and depth due to added background noise from the boat engine, and may overestimate DE_{mig} if tows are only conducted in good weather (which would generally result in high-*DE* situations).

In rivers, DE_{mig} can be estimated by snorkelers with test tags floating past receivers (Melnychuk and Hausch 2011). This has the advantage that DE_{mig} can still be estimated during periods when few fish migrate past receivers. As DE_{mig} often varies with river discharge, test tag passes should be conducted across a wide range of flow conditions, at least bracketing those occurring during fish migration periods.

Test tag passes provide an expectation of DE_{mig} for tagged fish, but this index could be biased due to depth, speed, and movement pattern differences. When the interest is in estimating survival, actual DE_{mig} (or p) can be estimated from fish tag detection data using CJS models. If survival rates during migrations are low such that few tagged fish cross a particular receiver line, or if DE_{mig} is relatively low at and after a detection line, large sample sizes may be required to achieve reasonable precision in estimates. If survival and DE_{mig} are both relatively high, reasonable precision can be achieved with a tagged sample size of 50–100 fish. Estimating DE_{mig} from actual fish detection data ensures that the depth and speed of tags during crossing events are representative of the study species.

If multiple groups of tagged fish are released and share at least some portion of their migration routes passing a site, DE_{mig} estimates can be constrained to be equal among groups at this location (Burnham et al. 1987; Lebreton et al. 1992; Melnychuk 2009a). This can increase sample sizes for estimating DE_{mig}, avoid over-fitting to sparse data, and provide estimates that are more precise than group-specific DE_{mig} estimates. To further increase the parsimony of CJS models, DE_{mig} estimates can be constrained as a function of an environmental covariate like river flow (Melnychuk 2009a), measured background noise, or geometrical properties of a receiver line. If receivers on a line are deployed closely to one another, we would expect fewer tagged fish to pass by undetected. The overlap of receiver coverage on a line can be assessed by determining the frequency of simultaneous detections on adjacent receivers (Voegeli et al. 1998).

Finally, DE_{mig} could be predicted based on the results of fine-scale range testing. If the relationship between DE_{single} and distance is quantified at several distance intervals with stationary test tag range testing or with test tag tows by boat, these probabilities can be used along with fish speed and tag interval time to predict DE_{single} for each transmission during a crossing event. The complement of the product of ($1\text{-}DE_{single}$) for all transmissions gives an overall DE_{mig}. Assessments of DE_{mig} from test tag passes, tows, or fine-scale range testing should not take the place of mark–recapture estimates if the latter are estimable, but can be used simultaneously to explore the effect of factors like current speed and receiver spacing on DE_{mig}. The researcher has more control over estimating the effect of these factors when using test tags than when using only detection data from tagged fish.

Area swept in mobile tracking studies

When the number and location of tags in some area at some time are unknown, mobile tracking surveys can be used to search for tags. Surveys usually consist of linear transects, typically arranged in parallel lines spaced some standardized distance apart. The probability of detecting a tag present in the area swept, DE_{mobile}, can be estimated at least three ways. First, similar to how DE_{single} may decline with increasing distance from tag to stationary receiver, we might expect that tags further away from the center line of a mobile transect are less likely to be detected (either at any given moment, or over the long term). When tags are detected during blind transects and their locations are triangulated, the perpendicular distance to the center line can be measured. When a sufficient number of tags are detected to produce a frequency distribution of perpendicular distances (≥20 or 30 tags), attenuation functions can be fit to these measurements to estimate how DE_{mobile} decreases with increasing distance from the center line (Anderson et al. 1979; Melnychuk and Christensen 2009).

Another method for estimating DE_{mobile} involves simultaneous use of a second sampling method to detect tagged fish. If sampling methods are independent, the probability of detecting a tag with both methods equals the product of the probabilities of detecting a tag with each method. For example, in mobile tracking laps around oil platforms on two days, Lowe et al. (2009) found more tags with a VEMCO VR-100 mobile receiver than were detected by an array of stationary VR-2 receivers over that same time period (the VR-100 typically has greater detection range). Different tags were heard on different systems, so neither system was perfect: of 100 fish originally tagged, 17 were detected only with mobile tracking, six were detected only on the VR-2 array during mobile tracking periods, and 15 were detected with both methods. The unknown number of tagged fish currently present in the area, X, can be estimated using the Lincoln-Petersen method (Seber 1982). We equate: $(15/X) = ((17 + 15)/X) \cdot ((6 + 15)/X)$, which requires that $X = 44.8$. Thus the best estimate of DE on the mobile tracking system is 32/44.8 = 0.714 (or, more directly, 15/21 = 0.714). The estimate of DE is biased high if the two methods are not independent but share positive covariation. This can occur if some behavioral subset of fish are relatively easy to detect on both systems (e.g., pelagic-oriented fish) while another subset is difficult to detect on both systems (e.g., benthic-oriented fish). Melnychuk and Christensen (2009) present a similar calculation for two mobile tracking systems operating simultaneously on the same tracking vessel.

Other studies have used multiple tagging technologies, which can be considered together for estimating DE_{mobile}. Enders et al. (2007) used a combination of hand-held radio telemetry, PIT tag telemetry, and underwater visual surveys to compare DE of sampling methods for studying the behavior of juvenile salmonids in small streams. Even if DE cannot be estimated explicitly for the second sampling method, the method can still be used to calibrate DE_{mobile} estimates. If acoustic-tagged fish are also given external markings like spaghetti tags, scuba diving can be used to visually verify whether tagged fish are present in an area swept by mobile tracking transects. Overhead and surrounding benthic structures may interfere with acoustic signal detection even though the fish is easily seen (or vice versa). Alternatively, telemetry can provide the calibration for other sampling programs. Edwards et al. (2007) used telemetry and time-depth data loggers for estimating the detection probability of aerial surveys for manatees. Korman et al. (2002) used radio telemetry to calibrate estimates of visual observer efficiency during snorkel surveys for steelhead trout spawners.

A third method for estimating DE_{mobile} returns to test tag evaluations. A tag can be deployed at a depth representative of the study species (especially important if fish are typically near the bottom surrounded by interfering structures), and transect passes can be repeatedly conducted past the tag to quantify the proportion of passes where the tag is detected. Since DE_{mobile} is expected to vary with boat speed (Melnychuk and Christensen 2009), tag depth, tag-to-receiver distance, and environmental conditions, test tag passes should be conducted across a range of these variables. If tagged fish are confined to some area (e.g., a lake or reservoir) and the entire area is sampled, then the probability of relocating tags on successive surveys can be calculated from the number missed but known to have been there because they were detected in later surveys (Hightower et al. 2001). This proportion might vary among sampling trips or seasonally. If mobile tracking is conducted in an area where acoustic conditions show little variation over time, then prior calibrations of test tag acoustic strength (i.e., measuring dB levels of test tag signals at specific distances) can be used to later estimate the approximate distance to a fish tag in the area (Block et al. 1997). Finally, since boat-related noise may dominate the total background noise in mobile tracking studies, maximizing the distance between boat and hydrophone or taking other measures to reduce boat noise reaching the hydrophone will aid in increasing DE_{mobile}.

CONCLUSIONS

Telemetry is a useful tool for studying the population dynamics and movement patterns of tagged fish, but if any inferences are to be made regarding survival or the presence of tagged fish in a study area, it is important to estimate the probability of detecting tagged fish. Assessment of detection efficiencies should correspond to study objectives, since different measures are necessary for addressing different kinds of biological questions. Evaluations based either on test tags or fish detection data should explicitly consider how this nuisance parameter changes under a range of technological, environmental, and behavioral factors. Increasing the precision and accounting for any temporal trends or spatial variation in detection probability estimates will lead to greater precision and reduced bias, respectively, in the estimates of demographic and movement parameters of fish populations.

ACKNOWLEDGMENTS

I thank Sam Johnston, Doug Pincock, Carl Walters, Colin Simpfendorfer, John Eiler, and two reviewers for comments on the Section. Detection data used for Figure 1 were provided by David Welch. Test tag tow data shown in Figure 2 were part of a mobile tracking program funded by the FishAmerica Foundation and the Pacific Salmon Foundation.

REFERENCES

Afonso, P., J. Fontes, K. N. Holland, and R. S. Santos. 2008. Social status determines behaviour and habitat usage in a temperate parrotfish: implications for marine reserve design. Marine Ecology-Progress Series 359:215–227.

Anderson, D. R., J. L. Laake, B. R. Crain, and K. P. Burnham. 1979. Guidelines for line transect sampling of biological populations. Journal of Wildlife Management 43:70–78.

Block, B. A., J. E. Keen, B. Castillo, H. Dewar, E. V. Freund, D. J. Marcinek, R. W. Brill, and C. Farwell. 1997.

Environmental preferences of yellowfin tuna (*Thunnus albacares*) at the northern extent of its range. Marine Biology 130:119–132.

Burnham, K. P., D. R. Anderson, G. C. White, C. Brownie, and K. H. Pollock. 1987. Design and analysis methods for fish survival experiments based on release-recapture. American Fisheries Society Monograph 5, Bethesda, Maryland.

Cash, K. M., T. W. Hatton, E. C. Jones, R. J. Magie, N. S. Adams, and D. W. Rondorf. 2005. Three-dimensional fish tracking to evaluate the removable spillway weir at Lower Granite Dam during 2003. Report by U.S. Geological Survey to the U.S. Army Corps of Engineers, Contract W68SBV00104592, Walla Walla, Washington.

Clements, S., D. Jepsen, M. Karnowski, and C. B. Schreck. 2005. Optimization of an acoustic telemetry array for detecting transmitter-implanted fish. North American Journal of Fisheries Management 25:429–436.

Comeau, L. A., S. E. Campana, and M. Castonguay. 2002. Automated monitoring of a large-scale cod (*Gadus morhua*) migration in the open sea. Canadian Journal of Fisheries and Aquatic Sciences 59:1845–1850.

Cormack, R. M. 1964. Estimates of survival from the sighting of marked animals. Biometrika 51:429–438.

Dewar, H., P. Mous, M. Domeier, A. Muljadi, J. Pet, and J. Whitty. 2008. Movements and site fidelity of the giant manta ray, Manta birostris, in the Komodo Marine Park, Indonesia. Marine Biology 155:121–133.

Edwards, H. H., K. H. Pollock, B. B. Ackerman, J. E. Reynolds, and J. A. Powell. 2007. Estimation of detection probability in manatee aerial surveys at a winter aggregation site. Journal of Wildlife Management 71:2052–2060.

Eiler, J. H. 1995. A remote satellite-linked tracking system for studing Pacific salmonids with radio telemetry. Transactions of the American Fisheries Society 124:184–193.

Enders, E. C., K. D. Clarke, C. J. Pennell, L. M. N. Ollerhead, and D. A. Scruton. 2007. Comparison between PIT and radio telemetry to evaluate winter habitat use and activity patterns of juvenile Atlantic salmon and brown trout. Hydrobiologia 582:231–242.

Farmer, N. A. 2009. Reef fish movements and marine reserve designs. Doctoral dissertation. University of Miami, Miami, Florida.

Grothues, T. M., K. W. Able, J. McDonnell, and M. M. Sisak. 2005. An estuarine observatory for real-time telemetry of migrant macrofauna: Design, performance, and constraints. Limnology and Oceanography: Methods 3:275–289.

Hedger, R. D., F. Martin, J. J. Dodson, D. Hatin, F. Caron, and F. G. Whoriskey. 2008. The optimized interpolation of fish positions and speeds in an array of fixed acoustic receivers. ICES Journal of Marine Science 65:1248–1259.

Heupel, M. R., K. L. Reiss, B. G. Yeiser, and C. A. Simpfendorfer. 2008. Effects of biofouling on performance of moored data logging acoustic receivers. Limnology and Oceanography: Methods 6:327–335.

Heupel, M. R., J. M. Semmens, and A. J. Hobday. 2006. Automated acoustic tracking of aquatic animals: Scales, design and deployment of listening station arrays. Marine and Freshwater Research 57:1–13.

Heupel, M. R., and C. A. Simpfendorfer. 2002. Estimation of mortality of juvenile blacktip sharks, *Carcharhinus limbatus*, within a nursery area using telemetry data Canadian Journal of Fisheries and Aquatic Sciences 59:624–632.

Hightower, J. E., J. R. Jackson, and K. H. Pollock. 2001. Use of telemetry methods to estimate natural and fishing mortality of striped bass in Lake Gaston, North Carolina. Transactions of the American Fisheries Society 130:557–567.

Hobday, A. J., and D. G. Pincock. 2012. Estimating detection probabilities for linear acoustic monitoring arrays. Pages 325–346 *in* J. R. McKenzie, B. Parsons, A. C. Seitz, R. Kopf, M. Mesa, and Q. Phelps, editors. Advances in fish tagging and marking technology. American Fisheries Society, Symposium 76, Bethesda, Maryland.

Holland, K. N., R. W. Brill, S. Ferguson, R. Chang, and R. Yost. 1985. A small vessel technique for tracking pelagic fish. Marine Fisheries Review 47:26–32.

Jolly, G. M. 1965. Explicit estimates from capture-recapture data with both death and immigration-stochastic model. Biometrika 52:225–247.

Kaplan, E. L., and P. Meier. 1958. Nonparametric estimation from incomplete observations. Journal of the American Statistical Association 53:457–481.

Klimley, A. P., and C. F. Holloway. 1999. School fidelity and homing synchronicity of yellowfin tuna, *Thunnus albacares*. Marine Biology 133:307–317.

Korman, J., R. N. M. Ahrens, P. S. Higgins, and C. J. Walters. 2002. Effects of observer efficiency, arrival timing, and survey life on estimates of escapement for steelhead trout (*Oncorhynchus mykiss*) derived from repeat mark-recapture experiments. Canadian Journal of Fisheries and Aquatic Sciences 59:1116–1131.

Lebreton, J. D., K. P. Burnham, J. Clobert, and D. R. Anderson. 1992. Modeling survival and testing biological hypotheses using marked animals: a unified approach with case studies. Ecological Monographs 62:67–118.

Lindholm, J., P. J. Auster, and A. Knight. 2007. Site fidelity and movement of adult Atlantic cod *Gadus morhua* at deep boulder reefs in the western Gulf of Maine, USA. Marine Ecology-Progress Series 342:239–247.

Lowe, C. G., K. M. Anthony, E. T. Jarvis, L. F. Bellquist, and M. S. Love. 2009. Site fidelity and movement patterns of groundfish associated with offshore petroleum platforms in the Santa Barbara Channel. Marine and Coastal Fisheries: Dynamics, Management, and Ecosystem Science 1:71–89.

Manly, B. F. G., and M. J. Parr. 1968. A new method of estimating population size, survivorship and birth rate from capture-recapture data. Transactions of the Society for British Entomology 18:81–89.

Melnychuk, M. C. 2009a. Estimation of survival and detection probabilities for multiple tagged salmon stocks with nested migration routes, using a large-scale telemetry array. Marine and Freshwater Research 60:1231–1243.

Melnychuk, M. C. 2009b. Mortality of migrating Pacific salmon smolts in southern British Columbia, Canada. Doctoral dissertation. University of British Columbia, Vancouver, B.C.

Melnychuk, M. C. 2012. Potential effects of acoustic tag strength variation on detection probabilities of migrating fish: recommended measurements prior to tagging. Pages 313–323 *in* J. R. McKenzie, B. Parsons, A. C. Seitz, R. Kopf, M. Mesa, and Q. Phelps, editors. Advances in fish tagging and marking technology. American Fisheries Society, Symposium 76, Bethesda, Maryland.

Melnychuk, M. C., and V. Christensen. 2009. Methods for estimating detection efficiency and tracking acoustic tags with mobile transect surveys. Journal of Fish Biology 75:1773–1794.

Melnychuk, M. C. and S. J. Hausch. 2011. Method for estimating detection probabilities of non-migrant tagged fish: applications for quantifying residualization rates. Transactions of the American Fisheries Society 140:1613–1628.

Melnychuk, M. C., and C. J. Walters. 2010. Estimating detection probabilities of tagged fish migrating past fixed receiver stations using only local information. Canadian Journal of Fisheries and Aquatic Sciences 67:641–658.

Meyer, C. G., Y. P. Papastamatiou, and K. N. Holland. 2007. Seasonal, diel, and tidal movements of green jobfish (Aprion virescens, Lutjanidae) at remote Hawaiian atolls: Implications for marine protected area design. Marine Biology 151:2133–2143.

Ng, C. L., K. W. Able, and T. M. Grothues. 2007. Habitat use, site fidelity, and movement of adult striped bass in a Southern New Jersey estuary based on mobile acoustic telemetry. Transactions of the American Fisheries Society 136:1344–1355.

Ohta, I., and S. Kakuma. 2005. Periodic behavior and residence time of yellowfin and bigeye tuna associated with fish aggregating devices around Okinawa Islands, as identified with automated listening stations. Marine Biology 146:581–594.

Pecl, G. T., S. R. Tracey, J. M. Semmens, and G. D. Jackson. 2006. Use of acoustic telemetry for spatial management of southern calamary Sepioteuthis australis, a highly mobile inshore squid species. Marine Ecology-Progress Series 328:1–15.

Perry, R. W., J. R. Skalski, P. L. Brandes, P. T. Sandstrom, A. P. Klimley, A. Ammann, and B. MacFarlane. 2010. Estimating survival and migration route probabilities of juvenile Chinook salmon in the Sacramento-San Joaquin River Delta. North American Journal of Fisheries Management 30:142–156.

Pine, W. E., K. H. Pollock, J. E. Hightower, T. J. Kwak, and J. A. Rice. 2003. A review of tagging methods for estimating fish population size and components of mortality. Fisheries 28:10–23.

Pollock, K. H., H. H. Jiang, and J. E. Hightower. 2004. Combining telemetry and fisheries tagging models to estimate fishing and natural mortality rates. Transactions of the American Fisheries Society 133:639–648.

Rechisky, E. L., D. W. Welch, A. D. Porter, M. C. Jacobs, and A. Ladouceur. 2009. Experimental measurement of hydrosystem-induced delayed mortality in juvenile Snake River spring Chinook salmon (*Oncorhynchus tshawytscha*) using a large-scale acoustic array. Canadian Journal of Fisheries and Aquatic Sciences 66:1019–1024.

Seber, G. A. F. 1965. A note on the multiple recapture census. Biometrika 52:249–259.

Seber, G. A. F. 1982. The estimation of animal abundance and related parameters, 2nd edition. Charles W. Griffin, London.

Simpfendorfer, C. A., M. R. Heupel, and A. B. Collins. 2008. Variation in the performance of acoustic receivers and its implication for positioning algorithms in a riverine setting. Canadian Journal of Fisheries and Aquatic Sciences 65:482–492.

Simpfendorfer, C. A., M. R. Heupel, and R. E. Hueter. 2002. Estimation of short-term centers of activity from an array of omnidirectional hydrophones and its use in studying animal movements. Canadian Journal of Fisheries and Aquatic Sciences 59:23–32.

Skalski, J. R., J. Lady, R. Townsend, A. E. Giorgi, J. R. Stevenson, C. M. Peven, and R. D. McDonald. 2001. Estimating in-river survival of migrating salmonid smolts using radiotelemetry. Canadian Journal of Fisheries and Aquatic Sciences 58:1987–1997.

Szedlmayer, S. T., and R. L. Schroepfer. 2005. Long-term residence of red snapper on artificial reefs in the northeastern Gulf of Mexico. Transactions of the American Fisheries Society 134:315–325.

Topping, D. T., C. G. Lowe, and J. E. Caselle. 2005. Home range and habitat utilization of adult California sheephead, *Semicossyphus pulcher* (Labridae), in a temperate no-take marine reserve Marine Biology 147:301–311.

Topping, D. T., C. G. Lowe, and J. E. Caselle. 2006. Site fidelity and seasonal movement patterns of adult California sheephead *Semicossyphus pulcher* (Labridae): An acoustic monitoring study. Marine Ecology-Progress Series 326:257–267.

Voegeli, F. A., G. L. Lacroix, and J. M. Anderson. 1998. Development of miniature pingers for tracking Atlantic salmon smolts at sea. Hydrobiologia 371/ 372:35–46.

Voegeli, F. A., and D. G. Pincock. 1996. Overview of underwater acoustics as it applies to telemetry. Pages 23–40 *in* E. Baras, and J. C. Phillipart, editors. Underwater Biotelemetry: Proceedings of the First Conference and Workshop on Fish Telemetry in Europe. University of Liège, Belgium.

Welch, D. W., G. W. Boehlert, and B. R. Ward. 2003. POST—the Pacific Ocean salmon tracking project. Oceanologica Acta 25:243–253.

Welch, D. W., M. C. Melnychuk, E. R. Rechisky, A. D. Porter, M. C. Jacobs, A. Ladouceur, R. S. McKinley, and G. D. Jackson. 2009. Freshwater and marine migration and survival of endangered Cultus Lake sockeye salmon smolts using POST, a large-scale acoustic telemetry array. Canadian Journal of Fisheries and Aquatic Sciences 66:736–750.

Welch, D. W., M. C. Melnychuk, J. Payne, E. L. Rechisky, A. Porter, G. Jackson, B. Ward, S. Vincent, and J. Semmens. 2011. In situ measurement of coastal ocean movements and survival of juvenile Pacific salmon. Proceedings of the National Academy of Sciences 108(21):8708–8713.

Acoustic Telemetry: Case Studies

Section 8.1

Use of Acoustic Telemetry to Evaluate Survival and Behavior of Juvenile Salmonids at Hydroelectric Dams: a Case Study from Rocky Reach Dam, Columbia River, USA

Tracey Steig and Christopher Holbrook

INTRODUCTION

Columbia and Snake river salmon and steelhead *Oncorhynchus* spp. populations have declined from historic levels due to several factors, including the operation of hydroelectric dams. While most downstream migrating juvenile salmon and steelhead pass safely through a single dam, cumulative mortality through several dams can be significant (Davidson 1965; Bell et al. 1967; Schweibert 1977). Mortality during dam passage is generally attributed to turbine entrainment or deep spillway passage, sometimes in combination with tailrace predation. Since the early 1980s, considerable effort has been devoted to the development of structures and project operations that safely bypass downstream migrants around dams. However, the success of a bypass structure depends on survival advantage through the bypass compared to other routes, and the fraction of the population that uses each route. Rigorous evaluations are needed to determine the effectiveness of fish passage improvements. This Section will review a case study describing the use of acoustic telemetry to evaluate fish passage at Rocky Reach Dam on the mid-Columbia River.

ESTIMATING SURVIVAL AND IDENTIFYING SOURCES OF MORTALITY

Simple methods can be used to estimate survival of a fish population as it passes a dam, however more complex approaches are required to show sources and mechanisms of mortality. At the simplest level, survival in river reaches containing dams can be compared to other river reaches lacking dams (Skalski et al. 1998; Holbrook et al., in press), or dam survival can be estimated directly by comparing survival of fish released above and below dams (Burnham et al. 1987; Skalski et al. 2001). While these methods quantify the effect of a dam on a population, they do not reveal sources of mortality.

Bypass structures and other passage improvements are designed to pass more fish through high survival routes leaving fewer fish to pass through low survival routes. Route-specific survival studies provide even greater information to identify sources of mortality and to quantify the fraction of the population that uses each route (e.g., turbines, spillways, bypass structures) (Skalski et al. 2002). Route-specific survival analyses have helped to identify passage problems and to prioritize passage improvements in the Columbia and Snake rivers. Altering route "choice," however, requires understanding of fish behavior as they approach and pass a dam.

Fish behaviors can be characterized during dam approach and passage from fine-scale fish position data. The spatial (and/or temporal) distribution and movement patterns of fish in a forebay can be used to evaluate the performance of fish passage structures (Cash et al. 2005; Walker et al. 2009; Walker et al. 2010). Such data can also be also used to identify optimal dam operations and to guide the development of new fish guidance and passage structures (Evans et al. 2008; Nestler et al. 2008).

ADVANTAGES OF ACOUSTIC TELEMETRY

Acoustic telemetry has become a popular tool for fish passage evaluations due to its ability to provide both highly accurate (e.g., sub-meter) two- or three-dimensional (2D or 3D) fish positions for behavioral evaluations and large-scale presence/absence data for survival evaluations. The long detection range (usually hundreds of meters) and ability to monitor continuously with autonomous receivers provides more precise estimates of survival and movement rates compared to traditional mark and recapture methods (e.g., coded wire tags, physical marks, passive integrated transponder (PIT) tags). This also improves calculation of behavioral metrics such as travel time between detection sites and forebay residence time. The long detection range also provides greater spatial coverage compared to PIT tags or radio tags. In addition, the position of the tagged fish can be determined to investigate the behavioral characteristics of individual fish as they approach and pass through the forebay of the dam. No other technology permits sub-meter positioning over large areas in water. Thus, acoustic telemetry is currently the only single tool that allows both large-scale fish survival and fine-scale behavior to be evaluated in a single study.

DESIGN CONSIDERATIONS FOR SURVIVAL STUDIES

The goal of survival studies is to determine whether an acoustically tagged fish survives through a given section of the river during downstream migration. Typically, downstream migrating salmonids are tagged with acoustic transmitters, released into the river, and then detected on receivers at downstream locations. The detection data are used to estimate survival between subsequent receiver pairs. Detections on downstream receivers also permit estimation of detection efficiency (i.e., detection probability) at each upstream receiver (Skalski et al. 1998).

Although estimating detection probabilities is necessary to obtain unbiased survival estimates, low detection probabilities (i.e., some fish pass a site undetected) result in less precise (i.e., more uncertain) survival estimates. Thus, greater detection probabilities are preferred and may even be required to meet certain study objectives.

There are four reasons that an acoustically tagged, actively migrating smolt may not be detected at a downstream detection site. First, the fish may not survive to the detection site. Second, a live fish may pass the site undetected with a functional tag. This may be due to many factors including insufficient tag detection range, the fish passes too quickly for its tag to be detected, or the fish passed in an area shadowed from any hydrophone detection. Third, the tag may fail prior to arrival at the detection array due to insufficient battery life or malfunction. Finally, the tag may be expelled from a live fish prior to passage. The first two cases will not affect survival estimates because mark–recapture models directly estimate both survival and detection probabilities. However, mark–recapture models assume that no tags are lost or fail prematurely. When these assumptions are violated, fish survival estimates will be lower than true fish survival because premature failure and tag loss are misidentified as mortality by the model.

A method to account for tag failure has been developed by Townsend et al. (2006), which adjusts survival estimates to minimize bias due to variable tag life. However, one of the most important aspects of survival studies is to provide the longest possible life from the acoustic tag to minimize the risk of premature failure. Tag life can be increased by reducing the signal transmit rate or reducing the signal pulse width. Slower transmit rates and shorter pulse widths will increase battery life because less energy is emitted per unit time. However, slower transmit rates and shorter pulse widths may reduce the probability that a tag is detected by a receiver.

High detection probabilities are preferred to minimize uncertainty in survival estimates. Detection probabilities can be increased by increasing the transmit rate and detection range of the tag. A minimum transmit rate can be set based on fish movement rates and detection ranges. However the range at which a tag is detected and identified in a given environment depends on the frequency, source level, and pulse width of the transmitted signal.

In quiet environments, lower frequency acoustic signals are generally preferred for acoustic telemetry systems because lower-frequency sound waves have lower signal attenuation in water. However, due to high levels of background mechanical noise below 300 kHz near hydroelectric dams (HTI 1997), signals in the 300–500 kHz range are better to achieve desired detection ranges for survival studies and for fine-scale fish position studies.

Longer detection ranges can also be achieved by using a greater source level and/or transmitting a longer pulse width signal. In general, doubling the transmit pulse width results in doubling the detection range. However, increasing pulse widths or source levels require more energy due to the tag transmitting a signal for longer time duration and/or transmitting at a greater intensity, which in turn reduces the life of the tag. Therefore there is an inverse relationship between detection range and tag life. Further discussions on this subject are presented in Section 7.1.

DESIGN CONSIDERATIONS FOR PASSAGE BEHAVIOR STUDIES

Fine-scale position data have been used to calculate a variety of behavioral metrics, including vertical and horizontal spatial distribution, dam passage location, travel time, residence time, passage zones, nearest approach passage, zones of influence for guidance or collection structures, and avoidance or attraction behavior. These metrics have been used to describe fish behavior at many Columbia and Snake river dams, including Rocky Reach and Rock Island dams (Steig et al. 2010), Wanapum and Priest Rapids dams (Sullivan et al. 2009), Lower Granite Dam (Cash et al. 2003) , and McNary Dam (Walker et al. 2009; Walker et al. 2010).

The ability to determine the 3D position of a tagged fish requires the detection of the same signal on at least four hydrophones (Ehrenberg and Steig 2003). Three-dimensional acoustic tag coordinates with sub-meter resolution require accurate knowledge of the individual hydrophone positions, time synchronization among hydrophones, and a measurement of sound velocity. In addition, the hydrophones detecting the tag signal must have a direct "line of sight" path to the tag. For 3D positioning, the difference in the arrival time of the transmission to each of four hydrophones is used to triangulate the exact location of the tag. Typically many sequential tag positions are derived, providing a time-series of locations. These positions are combined to represent the swimming path for each tagged fish, which is mapped and presented in a 3D display. Similarly, a 2D position (X, Y) can be obtained using triangulation across only three hydrophones. Another method available for determining 3D position is utilizing a tag with a pressure sensor to estimate depth in combination with the estimated 2D position.

Study objectives often require high position resolution to detect small changes in behavior. This can be achieved by utilizing relatively fast tag repetition rates which produce greater position resolution over time. For example, Johnston et al. (2004) tracked juvenile salmonids within a turbine unit at Wanapum Dam in 2004. With an acoustic tag that transmitted at 25 pulses/s, and water velocities of about 2 m/s, the system provided a 3D position approximately every 8 cm of travel. The estimated tag life for this tag operating at 25 pulses/s was 7–10 d. Clearly, such high repetition rates would not be suitable for a survival study that requires longer tag life. However, high repetition rates may be useful for tracking in high water velocities and confined spaces.

An exact measurement of the location of the tagged fish requires that signal arrival time differences, the receiving hydrophone locations, and the velocity of sound in water are known perfectly. Since these variables cannot be measured without error, the calculated 3D tag position will not be perfect. For this reason, a method has been developed to calculate this 3D position error (Ehrenberg and Steig 2002).

Accurate measurement of the time of signal arrival measurement is critical because error in signal arrival time will result in calculated tag position error. Acoustic receivers are designed to accurately measure arrival time of each acoustic tag pulse at each of the receiving hydrophones. Narrower pulse widths can provide more accurate measurements of signal arrival time compared to longer pulse widths (Ehrenberg and Steig 2002). One method of utilizing the detection range benefits of a longer pulse width and the increased precision of a narrower pulse width is by using an encoded signal. For example, a Barker coded wideband signal minimizes the adverse effects of multipath as compared to typical continuous wave signals. The Barker coded signal improves the accuracy of the pulse arrival time, which in turn improves the accuracy of the tag position estimate. Further discussion regarding the Barker coded signal are presented in Section 7.1.

The time of arrival must also be synchronized among all hydrophones in a positioning array. Some acoustic tag receivers, (e.g., model 290 Acoustic Tag Receiver (ATR); Hydroacoustic Technology Inc. (HTI), Seattle, USA) can interrogate multiple hydrophones simultaneously by attaching numerous hydrophones to a single receiver. The receiver clocks can also be synchronized across multiple receivers using Global Position Satellite (GPS) technology. For example, four ATRs monitoring 16 hydrophones each were synchronized at McNary Dam in 2009 (Walker et al. 2010). This allowed the 3D position to be calculated from any combination of four hydrophones among 64 synchronized hydrophones, providing continuous tracks across the large forebay area upstream of McNary Dam.

Deployment and operation of telemetry systems near dams is challenging due to greater depths and higher water velocities compared to typical river reaches. For this reason, hydrophones must be securely anchored or moored to minimize or eliminate movement during the study since unknown hydrophone movement can result in tag position measurement error. Optimal geometry for precise tag position estimate would be to space hydrophones equal distances apart in X, Y, and Z dimensions. Greater depths in front of dams can enhance 3D positioning by allowing hydrophone geometry to approach a 1:1:1 in the X, Y, and Z dimensions.

One of the parameters required for the calculation of the tag's 3D position, as previously stated, is the velocity of sound in water. A common method for estimating the velocity of sound in water is to use an equation by Mackenzie (1981) which uses water temperature, water salinity, and water depth in its calculation. In freshwater studies conducted in relatively shallow water depths, like this case study, water temperature is the dominant parameter. The ATRs used in this case study incorporated temperature sensors (accuracy to 0.1°C) within the hydrophones for accurately estimating the velocity of sound in water.

Since measurement errors are unavoidable, it is useful to consider some typical deployments to determine the expected accuracy of the 2D or 3D tracking system. The accuracy of the tag position estimate also depends on the relative location of the hydrophones receiving the tag signal and the location of the tag within the array. One approach for predicting location errors is the use of an analytical method developed by Ehrenberg and Steig (2002). The inputs to this analysis are the hydrophone locations, the tag location, the signal pulse width, the received signal to noise ratio at each hydrophone, and the velocity of sound in water. This method can be used to select hydrophone positions that minimize position errors prior to the actual deployment of the hydrophones.

CASE STUDY

Study Background

Chelan County Public Utility District (Chelan PUD) is the owner and operator of Rocky Reach Dam on the mid-Columbia River. In March of 2002, Chelan PUD signed a Habitat Conservation Plan (HCP) with federal, state and tribal fisheries agencies for Rocky Reach and Rock Island dams. The HCP is a 50 year agreement designed to address anadromous salmon survival issues at the dams. The goal of the HCP is to achieve 100% no net impact for each of the species covered by the agreement. The no net impact will be achieved through a combination of tributary habitat enhancement, hatchery compensation, and by meeting a juvenile project survival standard of ≥93% averaged over 3 years.

In this Section, we review a case study describing the use of acoustic telemetry to determine if fish passage goals were met at Rocky Reach Dam in 2010. The objectives of this study were to: 1) determine if survival of juvenile yearling Chinook salmon *Oncorhynchus tshawytscha* through the project met or exceeded the minimum survival standard of 93%; 2) compare survival through the various passage routes (i.e., powerhouse, surface collector, in-turbine screens, and spillway); 3) estimate the fraction of the migratory population that used the surface collector as a passage route; and 4) to provide behavioral information to ultimately increase survival of the salmon smolts through the project.

METHODS

Study Site

Rocky Reach Dam is located on the Columbia River, 11 km north of Wenatchee, Washington at river km 764. The portion of the river considered the Rocky Reach Project extends from 300 m below Rocky Reach Dam to 300 m below Wells Dam, which is the next upstream dam. The dam's spillway is perpendicular and its powerhouse parallel to river flow (Figure 1). The powerhouse is 332 m long and contains 11 vertical Kaplan turbines numbered from south to north. Each turbine has three rectangular intakes, 6.1 m wide by 15.2 m high at the headgate slot. The spillway is located upstream of the powerhouse, over 230 m long and has 12 automatic spill gates. Each gate is 15.2 m wide and approximately 18.3 m deep to the spill gate ogee. In 2003 a surface collector was installed at the south end of the powerhouse near Turbine Unit 1 to bypass downstream migrating salmon smolts around the dam (Figure 1). The surface collector consists of two entrances that are 6.1 m wide by 18.3 m tall with an entrance flow of approximately 170 m^3/s.

Figure 1. Aerial image shows the orientation of Rocky Reach Dam on the Columbia River, with the location and orientation of the surface collector and bypass screens at Units 1–2.

Acoustic Telemetry System

The acoustic telemetry system used for this study was the ATR tracking system. For this case study, 307 kHz tags and hydrophones with omni-directional beam widths were used. Each ATR system provides a precise time of arrival (accurate to within 1/12,000 of a sec) for each unique tag on each hydrophone. In addition, presence-absence data can be used to remotely track tagged fish in 3D with sub-meter resolution. The ATR system supports up to 16 individual hydrophones. All systems operated continuously from 22 April to 24 June 2010.

The acoustic tags used in this study were HTI model 795Lm Micro Acoustic Tags. The tags transmitted a 1 ms pulse (i.e., pulse width) every 3.8–6.1 s (i.e., pulse repetition rate or signal period). Tags weighed 0.65 g in air and had a verified average battery life of 32 d with these settings.

Hydrophone Installation

Rocky Reach Dam Detection and Positioning Arrays

A total of six ATR systems were installed at Rocky Reach Dam for both presence-absence detections and 3D positioning. The locations of all of the forebay hydrophones are shown in Figure 2. An underwater view of the south end of the Rocky Reach Dam powerhouse and the surface collector entrance is shown in Figure 3. Three ATR systems with 14–16 hydrophones each were used in the powerhouse forebay to monitor the west side of the river, the surface collector, and the bypass channel (Figure 2). Another two ATR systems with six hydrophones each were used in Rocky Reach gatewell slots for Units 1 and 2 (Figure 2). Lastly, another two ATR systems, with three to four hydrophones each, were used in the Rocky Reach spillway forebay to monitor the east side of the river (Figure 2). The 3D paths were calculated during transit through the powerhouse forebay. In addition, the arrival time of the last detected pulse for each fish across all hydrophones was used to determine the location of the tagged fish as they passed through the dam and to assign the passage route.

Hydrophones were either attached directly to the face of the dam or attached to 227 kg steel clumps and buoys to minimize potential movement during the study. The positions of the hydrophones attached to the dam were measured to within 2.5 cm in the X, Y, and Z. The positions of the hydrophones placed in the forebay were determined by transmitting a signal from the hydrophones fixed to the dam and receiving the signal on the hydrophones in the forebay. The measurement of the signal delays between hydrophones was used to calculate the forebay hydrophone positions. This method was virtually identical to the method used to calculate the 3D position of tagged fish.

In-River Detection Arrays

In-river detection sites were installed upstream of Rocky Reach Dam at Beebe Bridge (river km 813) and the Rocky Reach boat restriction zone (BRZ; river km 765). The Beebe Bridge detection site consisted of one ATR and four hydrophones installed across the river. The Rocky Reach BRZ detection site consisted of one ATR and nine hydrophones. ATRs were installed at two sites downstream of the dam, south of Wenatchee, Washington, near the Rock Island Hydro Park (river km 742), and upstream of Rock Island Dam at the BRZ (river km

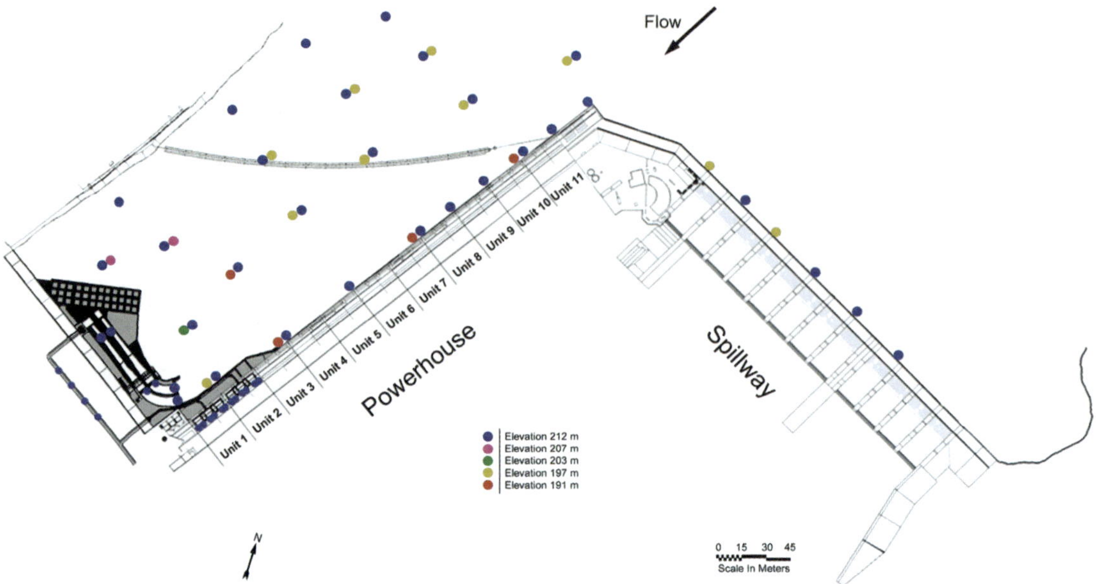

Figure 2. Plan view of Rocky Reach Dam forebay showing the location of hydrophones for monitoring acoustically tagged Chinook smolts for the 2010 study.

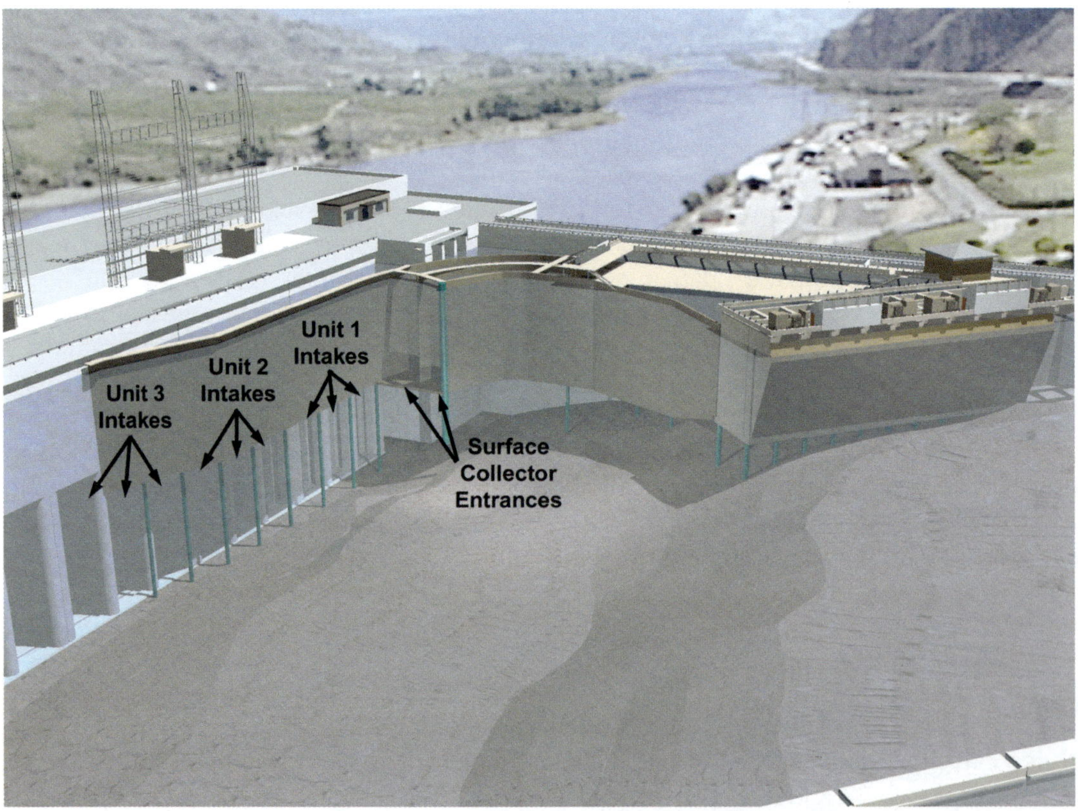

Figure 3. Underwater view of the south end of the powerhouse and the surface collector entrances at Rocky Reach Dam.

730). The Rock Island Hydro Park detection site consisted of one ATR and five hydrophones installed across the river. The Rock Island BRZ detection site consisted of one ATR and 10 hydrophones.

All in-river hydrophones were attached to 45 kg steel clumps and placed in the river at predetermined GPS coordinates. Accurate measurement of hydrophone locations was not necessary because these arrays were only used for presence-absence data, not fine-scale positioning. However, time-synchronization across hydrophones was necessary to provide cross-channel distribution of tagged fish.

Fish Release and Tagging

Actively migrating juvenile Chinook salmon were obtained from the Rocky Reach juvenile collection facility. Fish weight ranged from 22 to 125 g (mean 53 g) and length ranged from 135 to 219 mm (mean 168 mm). Acoustic tags were surgically implanted and fish were released at Wells Dam tailrace, Rocky Reach Dam surface collector, and Rocky Reach Dam tailrace in 15 release groups of 23–42 fish. Total Chinook smolts at each release location ranged from 376 to 503 fish. Steig et al. (2011) provide further details on fish handling and tagging procedures.

Data Processing

The primary data provided by the acoustic telemetry systems are the individual tag detection histories with precise times of detection on each hydrophone. Each of the signal detections are accurately time stamped to within 1/12,000 of a sec. Such precise time information provides the ability to calculate numerous parameters for each tagged fish, such as 1) time of first detection on each hydrophone, 2) time of last detection on each hydrophone, 3) hydrophone of nearest approach, and 4) nearest hydrophone where the fish was last detected. These values were used to calculate the: 1) horizontal distribution across an in-river array, 2) horizontal passage route distribution at the dam, 3) travel time between hydrophone arrays, and 4) correlations of upstream passage locations to dam passage routes.

Calculating 3D Positions

The method that is used to determine 3D position by an acoustic tag system follows the same basic principles employed by GPS technology. The acoustic tag transmits a signal which is received by at least four hydrophones. By knowing the positions of the four hydrophones and measuring the relative signal arrival times at the hydrophones, the locations of the tagged fish can be estimated. Specifically, if h_{ix}, h_{iy}, h_{iz} specify the x, y, z location of the ith hydrophone and F_x, F_y, F_z specify the unknown x, y, z location of the tagged fish, then the travel time from the tagged fish to the ith hydrophone, t_i is given by

$$t_i = \frac{1}{c}\sqrt{(h_{ix} - F_x)^2 + (h_{iy} - F_y)^2 + (h_{iz} - F_z)^2} \tag{1}$$

where c is the velocity of sound in water. Unfortunately the absolute travel time cannot be measured directly. However, the differences between the arrival times of the signal at the various hydrophones $(t_i - t_j)$ can be measured as shown by

$$t_i - t_j = \frac{1}{c}\left[\sqrt{(h_{ix} - F_x)^2 + (h_{iy} - F_y)^2 + (h_{iz} - F_z)^2} - \sqrt{(h_{jx} - F_x)^2 + (h_{jy} - F_y)^2 + (h_{jz} - F_z)^2}\right] \quad (2)$$

For four hydrophones, there are three such distinct signal arrival time difference equations. The system of nonlinear equations is determined by solving the tagged fish coordinates, F_x, F_y, F_z such that the mean squared difference between the measured $(t_i - t_j)$ and calculated time differences (right side of the equation above) are minimized. There are numerous methods for calculating this minimization. Although not calculated for this case study, a similar method could be used to calculate the 2D position, except ignoring the effect of depth on the tag position.

For this case study, the received tag signals were entered into a data file using the MarkTags© software. This program allows the user to automatically or manually save information from the received tag signals, which includes the tag ID, tag signal arrival time, signal strength, signal pulse width and system hydrophone threshold. Further analysis was conducted using the AcousticTag© software. This program was used to calculate the 3D tag positions from tag detections in the Rocky Reach Dam forebay. Numerous tracking parameters are available to increase 3D tracking accuracy (Bergé et al., in press).

Statistical Analyses

Sample Size Justification

According to the Chelan PUD's HCP agreement, the survival studies needed to meet certain scientific requirements to be considered a valid study. A survival study is considered valid if the standard error (SE) of the survival estimates is not more than 2.5% and invalid if the SE is greater than 3.5%. If the SE is between 2.5% and 3.5%, the HCP committee can include this study in the 3 years average at their discretion, given that other study protocols were met. Based on simulations, it was predicted that 400–500 tagged fish would be required per each release location to meet the SE precision requirements. The actual sample sizes of tagged fish per release location used in this case study ranged from 376 to 503 fish.

Estimating Project and Route Specific Survival

Survival through the project and route-specific survival were estimated using mark–recapture models based on the Cormack-Jolly–Seber model (Cormack 1964; Jolly 1965; Seber 1965) and the single-release model presented by Skalski et al. (1998). Model structures and fish-specific detection histories were tailored to the detection locations, dam passage routes, and release locations of this study. Likelihoods were constructed and models were fit to the data as described by Lady and Skalski (2009). Standard errors were calculated for the estimated parameters following the methods described by Skalski et al. (1998). Further discussions of the effects of reach length on survival are presented in Section 9.1.

Project passage survival was separately estimated for fish released during the daytime and nighttime using a paired release–recapture model introduced by Burnham et al. (1987) and adapted to telemetry studies by Skalski et al. (2001). The schematic of the paired release sites and sample sizes are shown in Figure 4. Capture histories for the release groups to Rock Island Hydropark and Rock Island BRZ were used to estimate the reach survivals in the paired

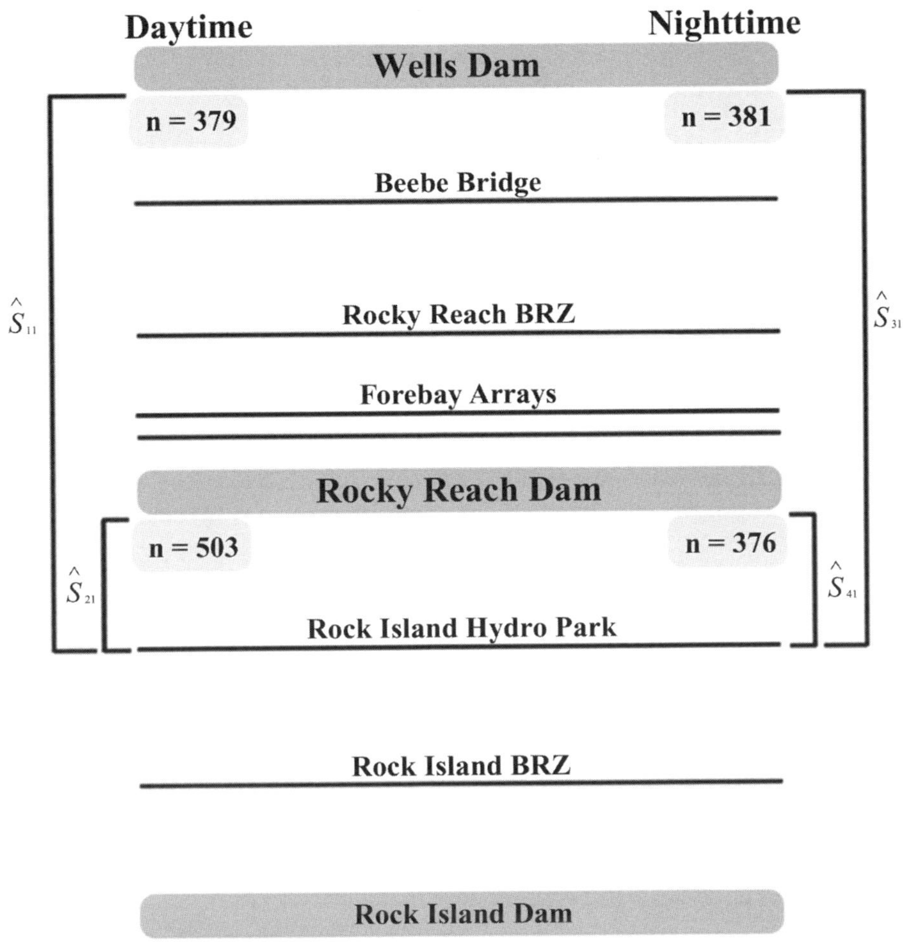

Rocky Reach Project Passage Survival: $\hat{S}_{RR-Day} = \frac{\hat{S}_{11}}{\hat{S}_{21}}$ and $\hat{S}_{RR-Night} = \frac{\hat{S}_{31}}{\hat{S}_{41}}$

FIGURE 4. Schematic of the paired-release design for daytime and nighttime releases used to estimate dam passage survival at Rocky Reach Dam.

release–recapture design. The reach survivals were corrected for a very small amount of premature tag failure using the method of Townsend et al. (2006). In all cases, the probability of a tag being active when the fish passed the detection arrays was ≥97%. The estimates of project passage survival were calculated using the ratio of the tag-life adjusted reach survivals from release to Rock Island Hydropark.

Route specific survival was determined by the multiple-release recapture model as described by Skalski et al. (2009). An additional release of tagged Chinook smolts was performed at the entrance to the surface collector at Rocky Reach Dam. The forebay of Rocky Reach Dam was instrumented with hydrophones and the dam passage routes were determined for each of the tagged Chinook smolts. The passage route survival along with the survival of the surface collector released Chinook were used to calculated route specific survival. The dam passage survival was calculated by combining each of the route specific survival estimates as described by Skalski et al. (2010a). Separate estimates of dam passage proportions and route-specific survivals were calculated for daytime and nighttime periods. The schematic for the triple release sites and sample sizes are presented in Figure 5.

Behavioral Study

Proportional Use of Passage Routes

To estimate the fraction of the population using each passage route, route-specific abundance was estimated from detection data using the Lincoln-Peterson single mark–recapture methods (Seber 1982). This method uses two detection arrays at a single location to provide

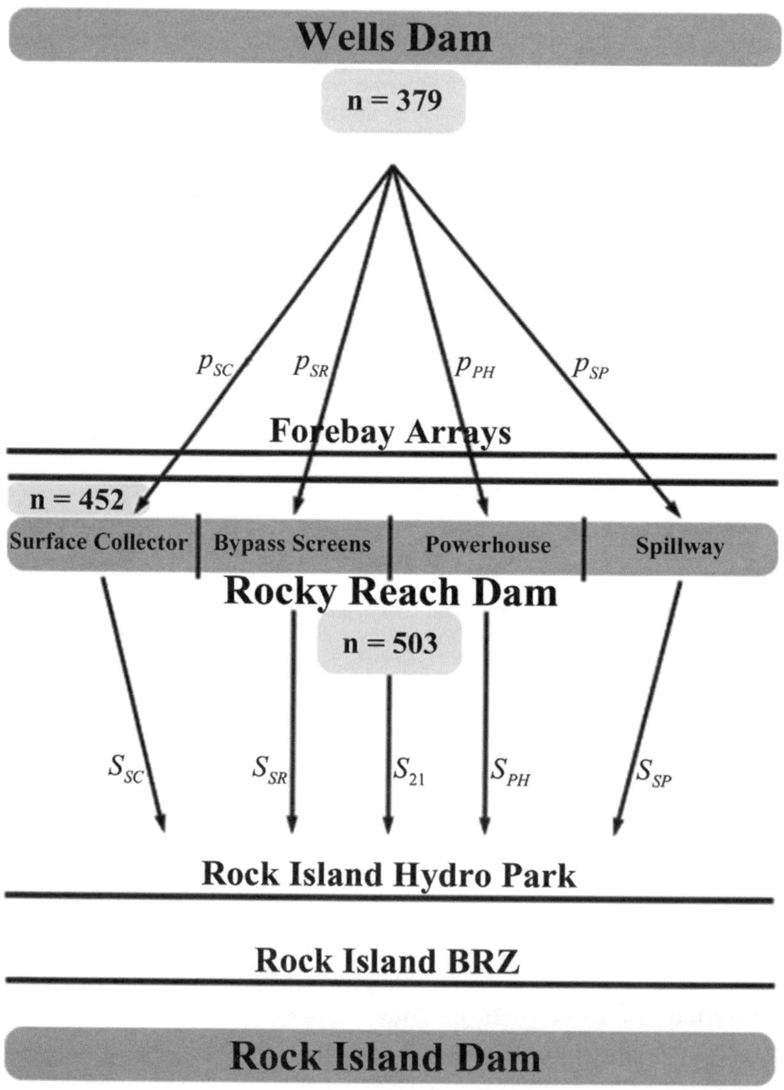

Figure 5. Schematic of the triple-release design used to estimate dam, project, and route-specific passage survivals and proportions.

two independent "surveys" of passing tagged fish. It was assumed that the population was closed and that detection probabilities were independent between the two detection arrays. To meet the latter, detection arrays were designed so that detection in the first array had no effect on whether a fish was or was not detected in the second array. As this method estimated arrival abundance at the dam, passage proportions were not influenced by upstream survival. Hence, passage percentiles estimated smolt passage, given their survival to the dam.

Passage route use was estimated separately for daytime and nighttime depending on the passage time. Daytime and nighttime periods were defined by nautical sunrise and sunset at Rocky Reach Dam during the study period. Route specific passage at Rocky Reach Dam was calculated for the surface collector, the bypass screens, Turbine Units 1–2 combined, Turbine Units 3–11 combined, and the spillway.

Spatial Distributions in the Forebay

To illustrate the distribution of Chinook smolts in the Rocky Reach Dam forebay, density plots of fish concentrations were produced. For this analysis, the forebay was divided into individual cells 15 m × 15 m × 3 m. The total number of tagged fish that entered each cell was summed once, regardless of how long each fish remained in a particular cell, and divided by the cell volume as described by Steig and Johnston (2010). This density, referred to as the singular detection density, provides a representative indication of the spatial distribution of tagged fish throughout an area. The equation for calculating the singular density is shown by

$$\sigma_{s\,m}^{\;k=1} = \frac{\sum_{l}^{j=1} d}{(X)(Y)(Z)} \tag{3}$$

where,

σ_S = time independent tagged fish density for cell k (singular density),
d = total number of tagged fish detected in a given cell, independent of time spent in the cell,
k = cell number,
m = total number of cells,
j = fish number,
l = total number of fish,
X = x dimension of each cell,
Y = y dimension of each cell,
Z = z dimension of each cell.

The density for each cell was calculated and interpolated by kriging (Davis 1986) between cells and overlaid on the plan view of Rocky Reach Dam. For illustrative purposes, the singular density was divided into 10 elevation slices. These singular density plots remove the masking effect of milling and present a more representative indication of the overall spatial distribution of smolts in the forebay.

RESULTS

Project Survival

Project passage survival for Chinook using daytime releases was estimated to be $\hat{S}_{RR\text{-}Day}$ = 0.952 ($\widehat{SE}$ = 0.017). For nighttime releases, project passage survival at Rocky Reach was estimated to be $\hat{S}_{RR\text{-}Night}$ = 0.898 ($\widehat{SE}$ = 0.020). The estimates of project passage survival for daytime and nighttime releases were significantly different (p = 0.038, two-tailed). A pooled estimate of project passage survival was calculated to be $\hat{S}_{RR}$ = 0.925 ($\widehat{SE}$= 0.014). The results from the 2010 Chinook survival study for Rocky Reach Dam are presented by Skalski et al. (2010b).

Route-specific Survival

Route specific survival results were based on passage during the daytime and nighttime periods as describe by Skalski et al. (2010b) and Steig et al. (2011). This is different than the project survival estimates, previously presented, which were based on releases during the daytime and nighttime periods (Skalski et al. 2010b; Steig et al. 2011). For this reason, project survival estimates using daytime and nighttime releases are not directly comparable to the daytime and nighttime route specific passage survival results.

Survival through the surface collector was estimated to be $\hat{S}_{SC}$ = 0.969 ($\widehat{SE}$ = 0.009). This estimate of absolute survival through the surface collector was used to convert the relative survival estimates to absolute route-specific survivals (Table 1). Survival estimates, for daytime and nighttime passage, through Units 3–11 were nearly significant (p = 0.062, two-tailed) and lower during the daytime (0.836 versus 0.919). Survival through the screens also appeared to be lower for the daytime passage smolts as compared to the nighttime (0.923 versus 0.989) but not significantly so (p = 0.143, two tailed). Survival through Units 1–2 were not significantly different between the daytime (0.919) and nighttime (0.890) (p = 0.692, two tailed). The results from the 2010 Chinook survival study for Rocky Reach Dam are presented by Skalski et al. (2010b).

Proportional Use of Passage Routes

Most juvenile Chinook passed through the surface collector (48%; Table 2; Figure 6) followed by the combined turbine Units 3–11 (30%). A greater fraction passed through the surface collector in 2010 than during previous studies (e.g., 27% in 2004 and 32% in 2005). Passage proportions were significantly different between daytime and nighttime at three of the four routes. There was also significantly greater passage through the surface collector during the nighttime (60%) than daytime (43%). Conversely, fewer fish used the bypass screens at nighttime as compared to daytime (2% versus 7%). Likewise, Chinook passage through Units 1–2 decreased during the nighttime (8%) as compared to the daytime (19%), contributing to most of the nighttime increase in surface collector passage. No Chinook passed through the spillways, since no spill occurred during the study period. The results from the 2010 Chinook behavioral study for Rocky Reach Dam are presented by Steig et al. (2011).

TABLE 1. Estimates of route-specific survival at Rocky Reach for yearling Chinook salmon during daytime and nighttime periods. Standard errors are in parentheses and two-tailed p-values calculated to test for a difference in survival.

Parameter	Absolute survival Daytime	Absolute survival Nighttime	*p*-value (2-tailed)
$S_{\text{Surface collector}}$	0.969 (0.009)		
$S_{\text{Bypass screens}}$	0.923 (0.042)	0.989 (0.015)	0.143
$S_{\text{Units 1-2}}$	0.919 (0.028)	0.890 (0.068)	0.692
$S_{\text{Units 3-11}}$	0.836 (0.030)	0.919 (0.033)	0.062

TABLE 2. Daytime versus nighttime passage total and percent route specific passage abundance estimates (standard error in parentheses) at Rocky Reach Dam in 2010.

Passage Location	Chinook			
	Daytime Passage		Nighttime Passage	
	Total (SE)	% (SE)	Total (SE)	% (SE)
Surface Collector	208.0 (0)	42.62% (2.24%)	144.0 (0)	60.00% (3.16%)
Bypass Screens	33.0 (0.180)	6.77% (1.14%)	5.0 (0)	2.08% (0.92%)
Units 1 and 2	93.0 (0)	19.06% (1.78%)	20.0 (0)	8.33% (1.78%)
Units 3-11	154.0 (0)	31.56% (2.10%)	71.0 (0)	29.58% (2.95%)
Spillway	0 (0)	0% (0%)	0 (0)	0% (0%)
Total		100.00%		100.00%

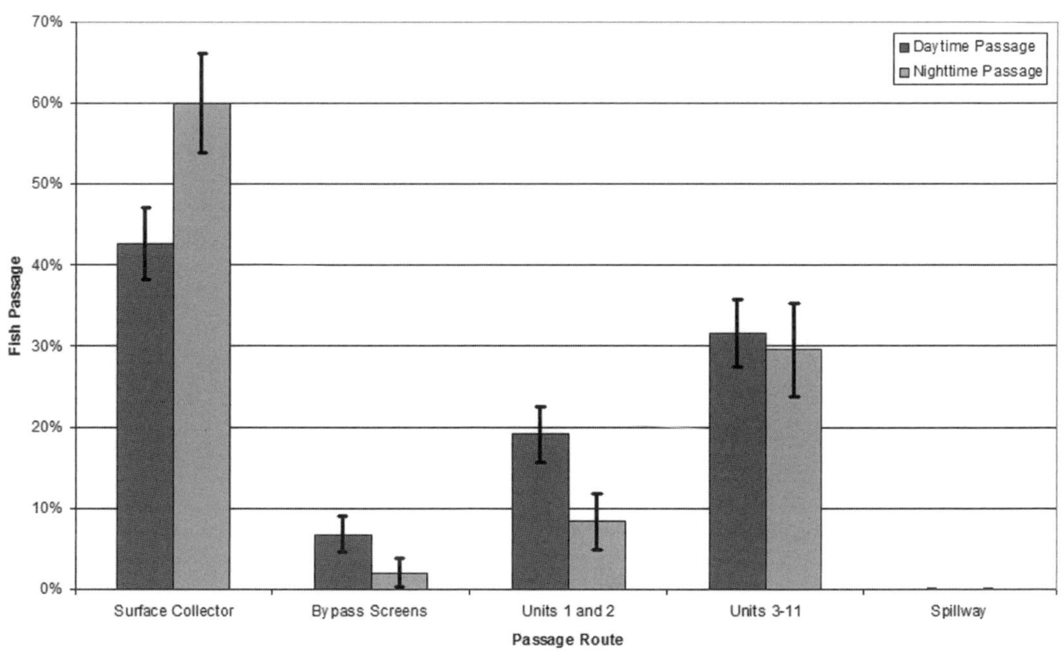

FIGURE 6. Chinook route specific passage estimates for daytime versus nighttime passage (with 95% CI) at Rocky Reach Dam in 2010.

Approach Distributions and Passage Routes

The horizontal distribution across the river was calculated for a detection array upstream of Rocky Reach Dam at the BRZ which is approximately 300 m upstream (Figure 1). Overall, 55% of the passage occurred along the west side of the river at the BRZ. Of fish that approached on the west side of the forebay, over half of them passed through the surface collector. Approximately 38% of the Chinook approached the dam in the center of the river at the BRZ and 47% of these passed through the surface collector.

The Chinook horizontal distribution at the BRZ was further evaluated for daytime and nighttime passage efficiency (Table 3 and Figure 7). Most fish remained on the west side and center of the river regardless of passage during the daytime or nighttime hours. In addition, a greater proportion of tagged fish that approached along the west side of the river passed through the surface collector during the nighttime hours (63%) as compared to daytime (46%).

Travel Times

The overall harmonic mean travel times from release in the Wells Dam tailrace to Beebe Bridge and to the Rocky Reach BRZ was found to be 0.39 d (SE = 0.02) and 3.41 d (SE = 0.12), respectively. Nighttime (0.62 d) release groups took twice as long to arrive at Beebe Bridge than daytime (0.29 d) release groups, and this difference was significantly different (Table 4; Figure 8). However, downstream of Beebe Bridge there were no significant differences in travel times between daytime and nighttime release groups at other detection sites.

Examples of 3D Fish Tracks

Three dimensional tracks provided detailed information about individual fish movements and behaviors. Examples of 3D paths for Chinook smolts traveling through the Rocky Reach powerhouse forebay are shown overlaid on a plan view of Rocky Reach Dam (Figures 9 and 10). The color of the fish tracks indicates the elevation in the water column and the color dots above each of the turbine units correspond to the flow through each unit and the time of fish

Table 3. Proportion of Chinook across the river at the Boat Restriction Zone with the associated route specific passage comparing daytime versus nighttime passage at Rocky Reach Dam, 2010.

	West Side of River	Center of River	East side of River
Daytime Passage	57.9%	34.9%	7.2%
SC	46%	39%	37%
Screen	7%	5%	6%
Units 1-2	18%	20%	20%
Units 3-11	28%	36%	37%
Nighttime Passage	47.9%	44.2%	7.9%
SC	63%	59%	47%
Screen	3%	0%	5%
Units 1-2	12%	6%	0%
Units 3-11	22%	35%	47%

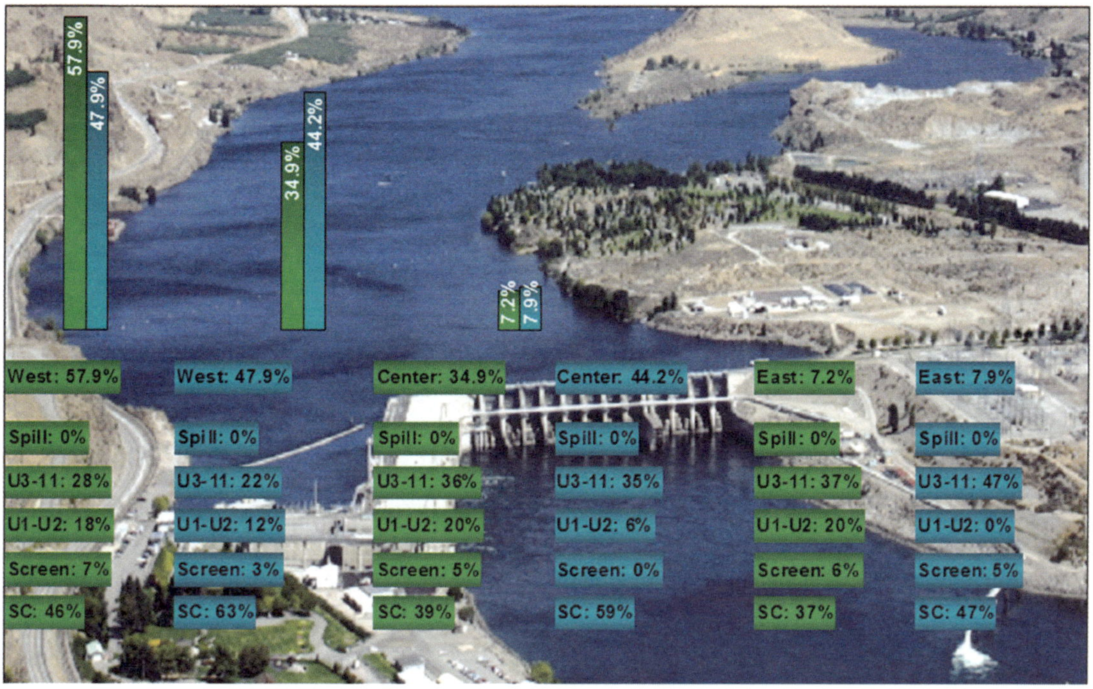

FIGURE 7. Proportion of Chinook across the river, directly upstream of Rocky Reach Dam at the Boat Restriction Zone, with the associated route specific passage for daytime passage (green) and nighttime passage (blue) at the dam in 2010. The proportions of detections have been grouped into three categories: west side, center, and east side of the river.

TABLE 4. Harmonic mean travel times by release group from Wells Dam tailrace releases in 2010 (measured in days, with standard error in parentheses).

	Travel Time d (SE)	
	Daytime Releases	Nighttime Releases
Beebe Bridge	0.29 (0.04)	0.62 (0.03)
Rocky Reach BRZ	3.27 (0.18)	3.57 (0.16)
Hydro Park	3.86 (0.21)	4.16 (0.18)
Rock Island BRZ	4.36 (0.21)	4.60 (0.19)

passage.

Chinook smolt ID 4467.30 was released on 5 May 2010 in the Wells Dam tailrace (Figure 9a). This fish passed through the surface collector during the daytime at 20:04 h on 7 May 2010 after spending 1 h in the powerhouse forebay. At time of tagging, this smolt was 205 mm long and weighed 103 g. This smolt exhibited a straight path behavior, traversing from near the west shore, across the forebay, directly to the surface collector entrance, which was its ultimate passage route. The entire path of the smolt was in the upper water column. Figure 9b shows the 3D path of Chinook smolt ID 4269.16, which was released in the Wells Dam tailrace on 7 May 2010. The time of surface collector passage was during the daytime at 10:33 h on 21 May 2010 after spending 45 min in the forebay. This smolt was 207 mm long and

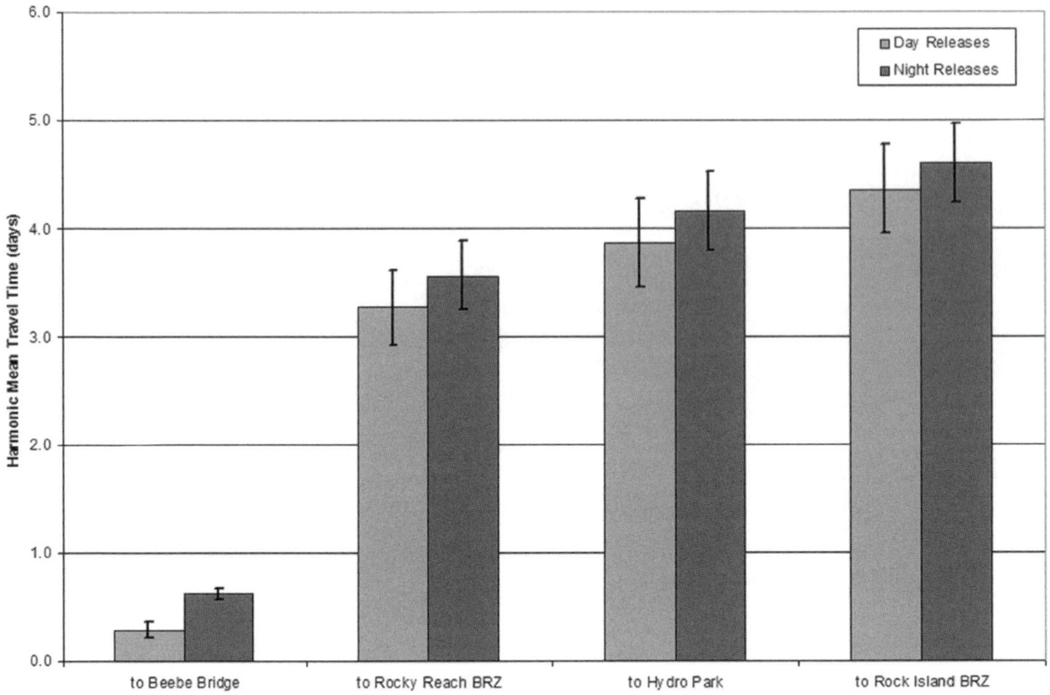

Figure 8. Harmonic mean travel time (with 95% CI) by release groups of Chinook from Wells Dam tailrace in 2010.

weighed 97 g and traveled in the upper water column throughout its entire transit across the forebay. There was milling behavior seen after the smolt passed under the forebay log boom. There was also milling seen just upstream of the surface collector entrance, suggesting a searching behavior.

Chinook smolt ID 6178.14 was released on 1 May 2010 in the Wells Dam tailrace (Figure 10a). This fish passed through the surface collector during the nighttime at 01:48 h on 8 May 2010 after spending 40 min in the powerhouse forebay. At time of tagging, this smolt was 170 mm long and weighed 45 g. This smolt exhibited a straight path behavior, traversing from the center of the river, across the forebay, directly to the surface collector entrance, where it ultimately passed. The path upstream of the log boom was in the lower half of the water column, and downstream of the log boom it traveled in the upper part of the water column. Figure 10b shows the 3D path of Chinook smolt ID 5319.24, which was released in the Wells Dam tailrace on 3 May 2010. The time of surface collector passage was during the nighttime at 20:48 h on 8 May 2010, after spending 34 min in the forebay. This smolt was 152 mm long and weighed 35 g and traveled in the upper water column throughout its entire transit across the forebay. The smolt showed a straight path along and relatively close to the powerhouse forebay wall.

On average, the Chinook that passed through the surface collector traveled higher in the water column as they approached the dam as compared to the Chinook that passed through the turbine units or the bypass screens (Steig et al. 2011).

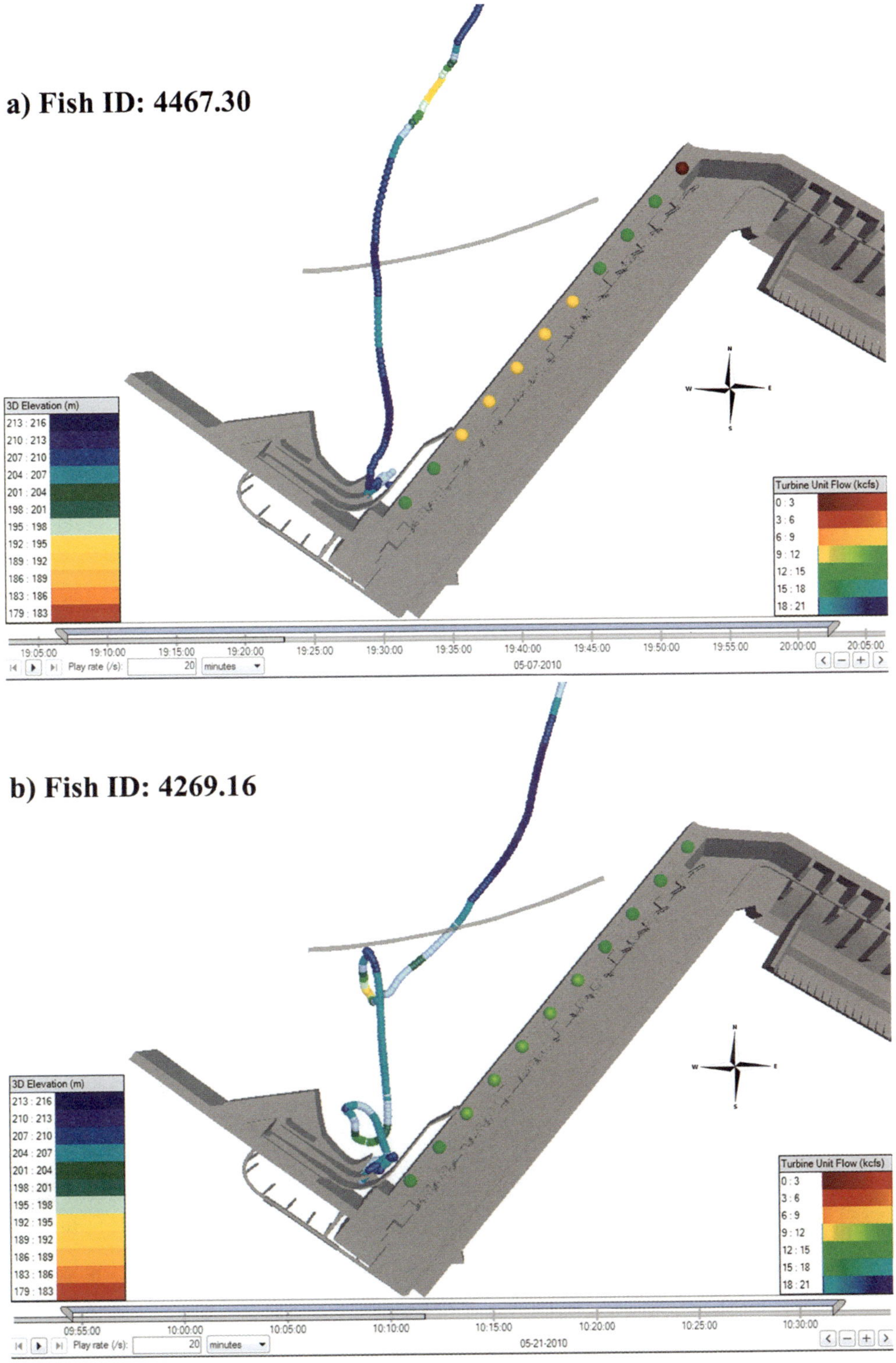

FIGURE 9. Examples of Chinook swimming pathways at Rocky Reach Dam in spring 2010. Each fish passed downstream through the surface collector during daytime hours; a) Fish ID: 4467.30 and b) Fish ID: 4269.16. Color corresponds to elevation in meters and flow through each turbine unit.

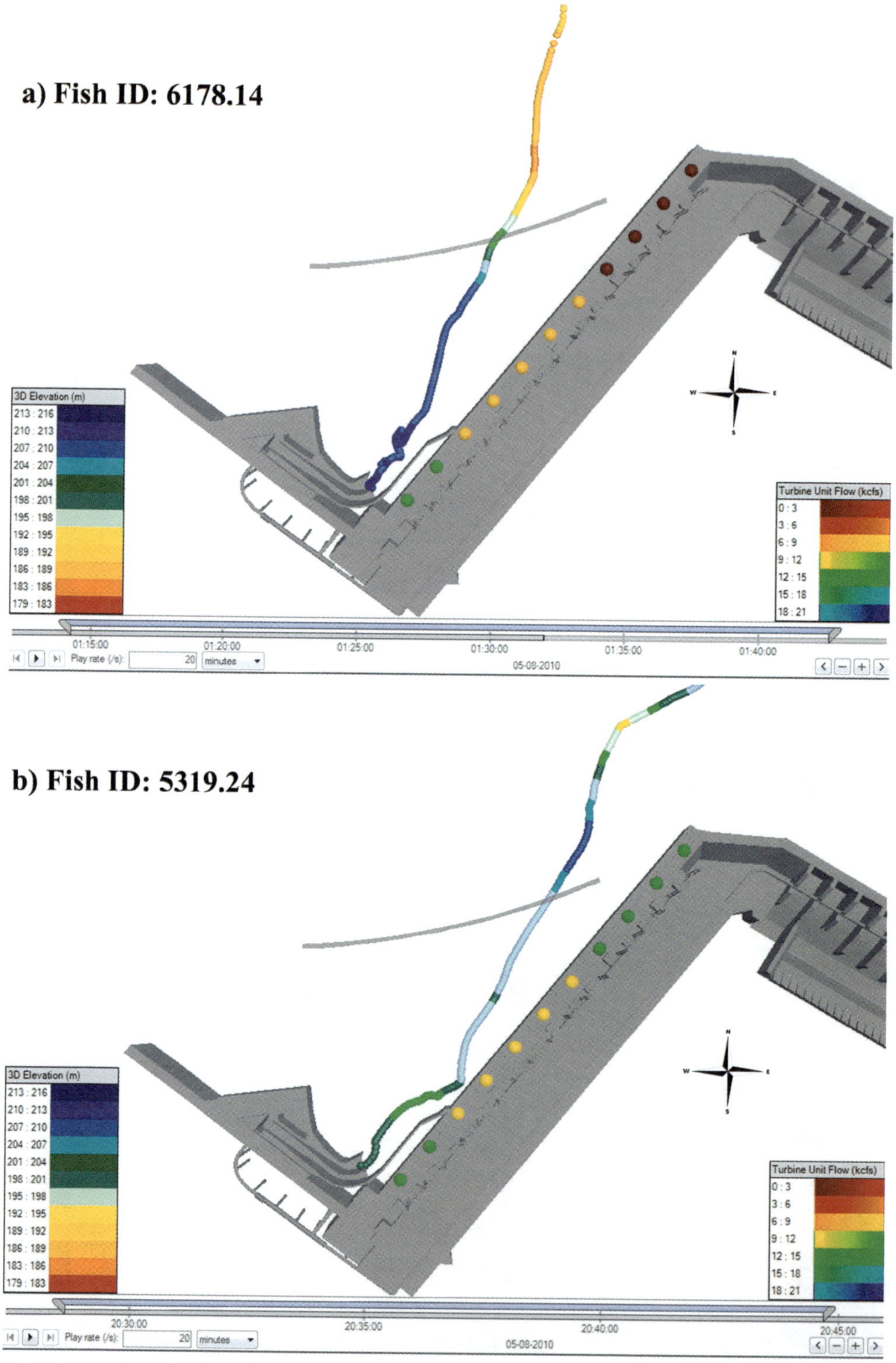

Figure 10. Examples of Chinook swimming pathways at Rocky Reach Dam in spring 2010. Each fish passed downstream through the surface collector during nighttime hours; a) Fish ID: 6178.14 and b) Fish ID: 5319.24. Color corresponds to elevation in meters and flow through each turbine unit.

Spatial Distributions in the Forebay

The singular density of tagged Chinook smolts passing through the forebay of Rocky Reach Dam were calculated separately for daytime and nighttime passage periods for each passage route (Figures 11 and 12). Figure 11 shows the singular density for the smolts passing through the surface collector during the daytime. In general, fish traveled primarily by two routes, along the powerhouse wall as well as along the side of the surface collector structure to the west of the entrance, and in the upper water column (i.e., elevation 204–213 m) (Figure 11). This characteristic was most likely the result of the smolts exhibiting a searching behavior by milling in the powerhouse forebay during the daytime. Most fish that passed at night approached the surface collector along the powerhouse wall, specifically in elevations 204–213 m (Figure 12). There were fewer Chinook in the lower water column (elevation 186–192 m), similar to the daytime passage densities (Figure 12).

DISCUSSION

The 2010 Chinook salmon survival study at Rocky Reach Dam was an investigative study in preparation for survival performance testing in 2011–2013. As stipulated in Chelan PUD's HCP agreement, survival studies need to meet certain scientific requirements to be considered valid. This case study met precision requirements (SE ≤ 0.025) with survival SE of 0.017, 0.020, and 0.014, for daytime, nighttime and pooled project survival estimates, respectively. Therefore this study met the scientific precision requirements to be considered a valid survival study. Likewise, since the SE is ultimately controlled by the sample sizes of tagged fish utilized in this study, the sample sizes were adequate to meet the study objectives.

Although the scientific precision requirements were met by this study, pooled project passage survival ($\hat{S}_{RR} = 0.925$) was slightly below the minimum required standard of ≥93% survival. In addition, this study would not have been considered valid in a performance survival test. Chelan PUD's HCP requires that the river flow during the study must be between the historical minimum 10% river flow rates and the maximum of 90% river flow rates. The 2010 spring river flow was below the minimum of 10% historical flow rates and thus, would not have been considered a valid study. It is generally considered that lower flow rates contribute to lower downstream smolt survival. Therefore Chelan PUD is confident that Chinook salmon survival at Rocky Reach Dam will meet the required minimum survival standard (≥93%) when flows are within the acceptable range.

Passage proportions and route-specific survival estimates suggest that project survival could further be increased if more fish could be passed through the surface collector bypass route. Indeed, the surface collector seems to be the safest passage route around the dam, since survival was higher through the surface collector than through turbine units or bypass screens (Table 1). However, about half of the fish still passed through turbines (Figure 6), where survival was lowest. Thus, project survival may be increased in the future if a greater fraction of the population can be passed through the surface collector.

The results of these behavioral studies might be helpful to improve smolt survival at Rocky Reach Dam in the future. For example, the upstream passage horizontal distribution at the Rocky Reach BRZ combined with the ultimate passage route at the dam (Figure 7) might be used to design upstream behavioral guidance devices. If there was a method to move Chi-

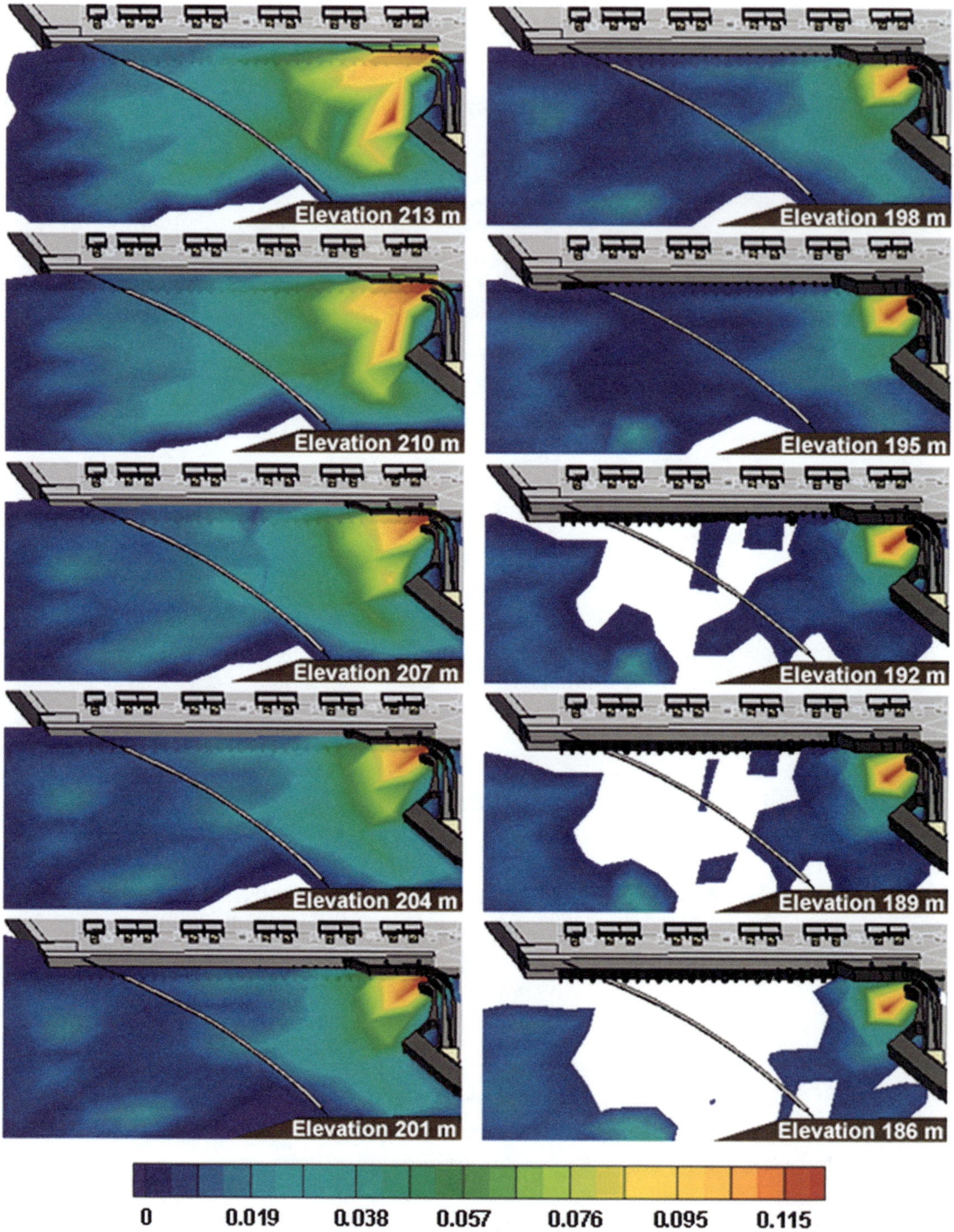

FIGURE 11. Singular densities (fish/m^3) of Chinook passage through the SC at Rocky Reach Dam during the spring of 2010. Density plan view figures are shown in 3 m slices from elevation 213 m (top left) to elevation 186 m (bottom right) and represent fish passage during daytime hours.

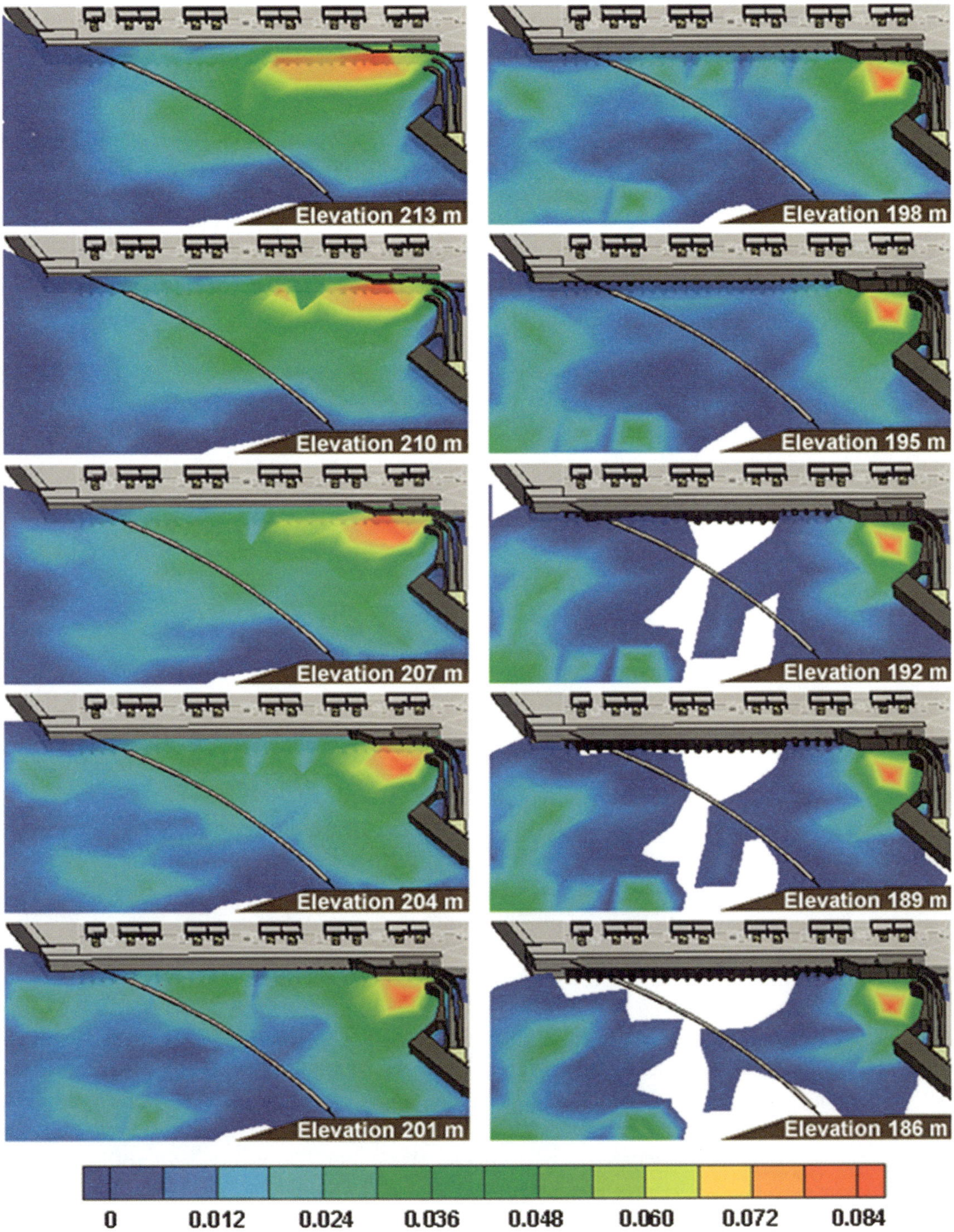

FIGURE 12. Singular densities (fish/m^3) of Chinook passage through the SC at Rocky Reach Dam during the spring of 2010. Density plan view figures are shown in 3 m slices from elevation 213 m (top left) to elevation 186 m (bottom right) and represent fish passage during nighttime hours.

nook smolts toward the west shoreline of the river at the BRZ, this may increase the surface collector passage and ultimately survival. The horizontal passage distribution at the dam provides information of where efforts might be concentrated to increase future bypass efficiency and survival. The 3D paths (Figures 9 and 10) and passage density results (Figures 11 and 12) provide better understanding of smolt behavior, which in turn could lead to improved methods to safely bypass smolts around the dam, ultimately increasing smolt survival.

To determine the efficacy of acoustic tag study results, it is common to compare these results with those obtained using other techniques. For this reason, Steig et al. (2004) compared the migration behavior and survival results between fish tagged with acoustic tags and fish tagged with PIT tags at Rocky Reach and Rock Island dams. Surface collector efficiency and harmonic mean travel times for acoustic and PIT tagged smolts, for three different species and two year classes of Chinook salmon smolts, were not significantly different ($p > 0.05$) for 13 of 16 comparisons. Rock Island Dam project survival for PIT tagged ($\hat{S}_{RI} = 0.956$, $\widehat{SE} = 0.025$) and acoustic tagged ($\hat{S}_{RI} = 0.952$, $\widehat{SE} = 0.026$) Chinook smolts was not significantly different ($p > 0.05$). The acoustic tag survival study required far fewer total tags and fish (798) than the PIT tagged survival study (90,000) for the similar precision of results. In this case, the results of the acoustically tagged smolts were nearly identical to the PIT tagged smolts, however the acoustic tag study required 89,202 fewer fish. Thus, acoustic telemetry is less labor-intensive and more cost-effective. Further, the smaller sample sizes required for acoustic telemetry studies means that fewer fish are removed from the migratory population.

Telemetry System Options and Flexibility

Despite the tradeoffs between detection range, position resolution, and tag life, the acoustic tag settings used in this study were suitable for both survival and behavior analyses. One advantage of the acoustic tags (model 795Lm) used in this study was that the pulse repetition rate and pulse width were user programmable. Pulse widths of 1–3 msec and pulse repetition rates of 3–8 s are commonly used to attain desired tag life and detection range in survival studies. Faster repetition rates can be utilized for finer-scale behavior studies. The values used for pulse width (1 msec) and repetition rate (3.8–6.1 s interval) in this study provided sufficient tag life (average 32 d) for minimal failure (<3%), yet detailed fish tracks were also obtained (Figures 11 and 12).

The tags in this study also used the period encoding technique to uniquely identify each tagged fish. It has been shown that period encoding is more efficient than other techniques (e.g., digital encoding) since all of the transmitted energy in the pulse is used for detection, tracking, and identification (Ehrenberg and Steig 2009). Another main advantage of the period encoded tag signal is that there are no destructive collisions between tag signals. Minimizing collisions is extremely important for successful survival studies with large sample sizes.

The success of an acoustic telemetry study can also depend on receiver reliability and accessibility for data access and troubleshooting. In this case study, ATR's were used for detecting tagged fish at the in-river sites. All of these sites were connected to a satellite internet connection and data were copied to a data server computer on an hourly basis. In addition, each of the ATR's could be remotely accessed and controlled. This connectedness provided real-time feedback regarding the operation of the remote sites and ensured high data quality. Similar studies have been conducted at other dams on the lower Columbia River utilizing autonomous single-hydrophone data loggers (HTI model 291) instead of ATRs. These data loggers operate with relatively low power draw utilizing DC power from batteries that are charged with solar panel charging systems. Recently, such data loggers have been networked to small net-book computers with cell phone connectivity for remote access and data transfer to an FTP site for data archiving (Walker et al. 2010). Thus, data loggers can allow autonomous deployments in remote locations where AC power is not available. Both types of installations have achieved a high degree of reliability for remote, unmanned data collection sites.

FUTURE DIRECTIONS

Acoustic telemetry has clearly become a valuable tool for fisheries research, particularly at hydroelectric dams in the Columbia and Snake rivers. However, research programs could still benefit from the development of smaller, more sophisticated tags. Even the smallest tags available today are too large for some migrating juvenile salmonids. With advancing technology, smaller, longer-life tags will allow survival and behavior studies to use smaller individuals of existing fish populations. In addition, smaller, more sophisticated, energy efficient, autonomous data loggers or receiver nodes are being developed. These nodes will provide longer time periods between replacing batteries and provide options to remotely connect with low cost cell phone technology.

ACKNOWLEDGMENTS

The authors would like to thank the following Chelan County PUD personnel for their assistance, coordination and advice in the completion of this study: Keith Truscott, Steve Hemstrom, Lance Keller and Barry Keesee. All contributed to the direction and successful completion of this study. In addition, numerous other Chelan County PUD Fish and Wildlife Department personnel, including Dave Beardsley and his tagging crew were instrumental in the success of this study. Special thanks are due to John Skalski and Rich Townsend of Columbia Basin Research for statistical expertise and survival analysis. Mention of trade names or commercial products does not imply endorsement by the U. S. Government. This article is Contribution 1651 of the U.S. Geological Survey Great Lakes Science Center.

REFERENCES

Bergé, J., H. Capra, H. Pella, T. Steig, M. Ovidio, E. Bultel and N. Lamouroux. In press. Accuracy of a hydro acoustic telemetry system in a large fast-flowing river: intrinsic and environmental determinants. Fisheries Research.

Bell, M. C., A. C. Delacy and G. J. Paulik. 1967. A compendium on the success of passage of fish through turbines. U.S. Army Corps of Engineers, Portland, Oregon

Burnham, K. P., D. R. Anderson, G. C. White, C. Brownie, and K. H. Pollock. 1987. Design and analysis methods for fish survival experiments based on release-recapture. American Fisheries Society Monograph 5.

Cash, K. M., T. W. Hatton, E. C. Jones, R. J. Magie, K. C. Mayer, N. M. Swyers, N. S. Adams, and D. W. Rondorf. 2003. Three-dimensional fish tracking to evaluate the removable spillway weir at Lower Granite Dam during 2002. U.S. Army Corps of Engineers, Walla Walla, Washington.

Cash, K. M, D. M. Faber, T. W. Hatton, E. C. Jones, R. J. Magie, N. M. Swyers, R. K. Burns, M. D. Sholtis, S. A. Zimmerman, J. S. Hughes, T. L. Gilbride, N. S. Adams, and D. W. Rondorf. 2005. Three-dimensional behavior and passage of juvenile salmonids at The Dalles Dam, 2004. U.S. Army Corps of Engineers, Portland, Oregon.

Cormack, R. M. 1964. Estimates of survival from the sighting of marked animals. Biometrika 51:429–438.

Davidson, F. A. 1965. The survival of the downstream migrant salmon at the power dams and in their reservoirs on the Columbia River. PUD of Grant County, Ephrata, Washington

Davis, J. C. 1986. Statistical and data analysis in geology. 2nd edition Wiley, New York

Ehrenberg, J. E., and T. W. Steig. 2002. A method for estimating the "position accuracy" of acoustic fish tags. ICES Journal of Marine Science 59:140–149.

Ehrenberg, J. E., and T. W. Steig. 2003. Improved techniques for studying the temporal and spatial behaviour of a fish in a fixed location. ICES Journal of Marine Science 60:700–706.

Ehrenberg, J. E., and T. W. Steig. 2009. A study of the relationship between tag-signal characteristics and achievable performances in acoustic fish-tag studies. ICES Journal of Marine Science 66:1278–1283.

Evans, S. D., N. S. Adams, D. W. Rondorf, J. M. Plumb, and B. D. Ebberts. 2008. Performance of a prototype surface collector for juvenile salmonids at Bonneville Dam's first powerhouse on the Columbia River, Oregon. River Research and Applications 24:960–974.

Holbrook, C. M., J. Zydlewski, and M. T. Kinnison. In press. Survival of migrating Atlantic salmon smolts through the Penobscot River, Maine, USA: a pre-restoration assessment. Transactions of the American Fisheries Society.

HTI. 1997. Phase 1 Report: Acoustic tag feasibility testing Rocky Reach Dam. Progress Report by Hydroacoustic Technology, Inc. to Chelan Company PUD, Wenatchee, Washington.

Johnston, S. V., M. A. Timko, G. W. Tritt, R. Heun, and P. G. Heisey. 2004. Pilot study for using acoustic tags to track juvenile salmonids in a turbine intake and the tailrace at Wanapum Dam in 2004. Report for Grant County Public Utility District No. 2, Ephrata, Washington.

Jolly, G. M. 1965. Explicit estimates from capture-recapture data with both death and immigration-stochastic model. Biometrika 52:225–247.

Lady, J. M., and J. R. Skalski. 2009. USER 4—User specified estimation routine. University of Washington, Seattle.

Mackenzie, K. V. 1981. Nine-term equation for the sound speed in the oceans. Journal of the Acoustical Society of America 70(3):807–812.

Nestler, J. M., R. A. Goodwin, D. L. Smith, J. J. Anderson, and S. Li. 2008. Optimum fish passage and guidance designs are based in the hydrogeomorphology of natural rivers. River Research and Applications, 24:148–168.

Schweibert, E. editor. 1977. Columbia River salmon and steelhead. American Fisheries Society, Special Publication 10, Bethesda, Maryland.

Seber, G. A. F. 1965. A note on the multiple recapture census. Biometrika 52:249–259.

Seber, G.A.F. 1982. The estimation of animal abundance. MacMillan, New York.

Skalski, J. R., S. G. Smith, R. N. Iwamoto, J. G. Williams, and A. Hoffmann. 1998. Use of PIT-tags to estimate survival of migrating juvenile salmonids in the Snake and Columbia Rivers. Canadian Journal of Fisheries and Aquatic Sciences 55:1484–1493.

Skalski, J. R., J. Lady, R. Townsend, A. E. Giorgi, J. R. Stevenson, C. M. Peven, R. D. McDonald. 2001. Estimating in-river survival of migrating salmonid smolts using radiotelemetry. Canadian Journal of Fisheries and Aquatic Sciences 58:1987–1997.

Skalski, J. R., R. Townsend, J. Lady, A. E. Giorgi, and J. R. Stevenson. 2002. Estimating route-specific passage and survival probabilities at a hydroelectric project from smolt radiotelemetry studies. Canadian Journal of Fisheries and Aquatic Sciences 59:1385–1393.

Skalski, J. R., R. A. Buchanan, R.L. Townsend, T.W. Steig, and S. Hemstrom. 2009. A multiple-release model to estimate route-specific and dam passage survival at a hydroelectric project. North American Journal of Fisheries Management 29:670–679.

Skalski, J. R., R. L. Townsend, T.W. Steig, and S. Hemstrom. 2010a. Comparison of two alternative approaches for estimating dam passage survival of salmon smolts. North American Journal of Fisheries Management 30:831–839.

Skalski, J. R., R. L. Townsend, and T. W. Steig. 2010b. Survival, diel passage, and migration dynamics of yearling Chinook salmon smolts at Rocky Reach Dam in 2010. Final report for Chelan County Public Utility District No. 1, Wenatchee, Washington., by Columbia Basin Research, School of Aquatic and Fishery Sciences, University of Washington, Seattle.

Steig, T. W., J. R. Skalski, and B. H. Ransom. 2004. Comparison of acoustic and PIT tagged juvenile Chinook, steelhead, and sockeye salmon (*Oncorhynchus* spp.) passing dams on the Columbia River, USA. Pages 275–286 in M. T. Spedicato, G. Lembo, and G. Marmulla, editors. Aquatic telemetry: advances and applications. Food and Agriculture Organization of the United Nations, FAO, Rome.

Steig, T. W. and S. V. Johnston. 2010. Innovative techniques for analyzing three-dimensional data. IEEE Catalog Number: CFP10OCE-CDR, ISBN: 978–1-4244–4333-8, Library of Congress: 2009902396.

Steig, T. W., P. A. Nealson, K. K. Kumagai, B. J. Rowdon, J. R. Selleck and C. Tunnicliffe. 2010. Route specific passage of juvenile Chinook, sockeye and steelhead salmon using acoustic tag methodologies at Rocky Reach and Rock Island Dams in 2009. Final report for Chelan County Public Utility District No. 1, Wenatchee, Washington., by Hydroacoustic Technology, Inc, Seattle.

Steig, T. W., K. K. Kumagai, B. J. Rowdon, C. P. Mott and C. Tunnicliffe. 2011. Route specific passage of juvenile Chinook and steelhead salmon using acoustic tag methodologies at Rocky Reach and Rock Island Dams in 2010. Final report for Chelan County Public Utility District No. 1, Wenatchee, Washington.

Sullivan, L. S., C. D. Wright, S. E. Rizor, M. A. Timko, C. A. Fitzgerald, M. L. Meagher, J. R. Skalski, and R. L. Townsend. 2009. Analysis of juvenile Chinook, steelhead and sockeye salmon behavior using acoustic tags at Wanapum and Priest Rapids dams, 2008. Draft report for Grant County Public Utility District No. 2, Ephrata, Washington.

Townsend, R. L., J. R. Skalski, P. Dillingham, and T. W. Steig. 2006. Correcting bias in survival estimation resulting from tag failure in acoustic and radiotelemetry studies. Journal of Agricultural, Biological, and Environmental Statistics, Vol. 11 2:183–196.

Walker, C. E., R. E. Reagan, J. M. Plumb, T. W. Hatton, N. M. Swyers, M. D. Sholtis, E. C. Jones, and S. J. Brewer. 2009. Juvenile salmonid survival and passage at McNary Dam during tests of temporary spillway weirs, 2008. Pages 1–91 *in* N. S. Adams and T. L. Liedtke, editors. Juvenile salmonid survival, passage, and egress at McNary Dam during tests of temporary spillway weirs, 2008. Anadromous Fish Evaluation Program Report 2008-W68SBV80448890. U.S. Army Corps of Engineers, Walla Walla, Washington.

Walker, C. E., R. E. Reagan, T. W. Hatton, S. D. Evans, N. M. Swyers, M. D. Sholtis, J. C. Charrier, and S. J. Brewer. 2010. Juvenile salmonid survival and passage at McNary Dam during tests of temporary spillway weirs, 2009. Pages 1–79 *in* N.S. Adams and T. L. Liedtke, editors. Juvenile salmonid survival, passage, and egress at McNary Dam during tests of temporary spillway weirs, 2009. U. S. Geological Survey Report to the U. S. Army Corps of Engineers, Walla Walla, Washington.

Section 8.2

Considerations for Tagging and Tracking Fish in Tropical Coastal Habitats: Lessons from Bonefish, Barracuda, and Sharks Tagged with Acoustic Transmitters

Karen J. Murchie, Andy J. Danylchuk, Steven J. Cooke, Amanda C. O'Toole, Aaron Shultz, Chris Haak, Edward J. Brooks, and Cory D. Suski

INTRODUCTION

Acoustic telemetry is proving to be a very useful tool for understanding the spatial ecology of fish and invertebrates that use tropical coastal marine environments (e.g., Simpfendorfer et al. 2002; Stark et al. 2005; Lindholm et al. 2006; Gordon and Seymour 2009). Given that radio telemetry is not effective in salt water systems due to signal attenuation (Pincock and Voegeli 1992), the development of acoustic telemetry has increased the capacity to examine the movement patterns and habitat use of highly mobile marine organisms, such as bony fishes and sharks (e.g., Meyer et al. 2007, 2009). The advent of remote logging receivers has also enabled data to be collected continuously, further increasing the capacity to examine factors such as diurnal movement patterns (e.g., Murchie et al. 2010) as well as the influence of stochastic events (e.g., tropical storms; Heupel et al. 2003) on the spatial ecology of fish in tropical coastal environments. Although telemetry has yielded a better understanding of the spatial ecology of fishes, there can be many hurdles to overcome when using acoustic telemetry in tropical coastal environments. For example, high water temperatures, shallow intertidal environments, hurricanes, and predominantly open systems can all influence the ability to address specific questions related to the movement patterns of fishes.

Our team has been studying the spatial ecology of a variety of fish species in the coastal waters of the Bahamian Archipelago since 2003. During this time we have encountered and overcome a number of challenges ranging from the collection and tagging of animals through to the deployment and maintenance of telemetry equipment, with most challenges related to the environment (Table 1). Here we summarize our experience in the form of a case study focused on bonefish *Albula vulpes*, great barracuda *Sphyraena barracuda*, and sharks (lemon sharks *Negaprion brevirostris*, Caribbean reef sharks *Carcharhinus perezi*, tiger sharks *Galeocerdo*

Table 1. Unique challenges when using acoustic telemetry in shallow tropical coastal environments.

Challenges	Summary of Challenge
Air temperature	Thermal stress for staff; Desiccation of fish while handling or performing surgery
Water temperature	Thermal stress for fish during capture, holding and tagging
Salinity	Corrosion of equipment
Depth (shallow)	Small reception ranges for tracking fish
Depth (deep)	Deploying and recovering receivers at great depths
Tides and currents	Shifting receivers that can result in equipment loss; unsafe conditions for divers
Storms and hurricanes	Noise which reduces reception range of arrays; Loss of equipment
Biofouling	Growth of algae and encrusting invertebrates on remote receivers
Predators	Predation on tagged fish; Threat to humans
Humans	Harvest of tagged fish; Theft and vandalism of telemetry equipment
Openness and connectivity	Fish can go anywhere—difficult to define array boundaries
Remoteness	Availability of anesthetics; Time required to ship and/or receive equipment; risk management
Lack of standard techniques	Fish capture techniques unique
Lack of natural history data	Environmental tolerances unknown

cuvier). Where possible, we direct the reader's attention to other studies that have considered the solution to environmental challenges. We also discuss the many applications, potential and realized, of acoustic telemetry for the study of marine life in tropical coastal habitats.

PROJECT BACKGROUND

Initially our telemetry studies were based around elucidating the spatial ecology of bonefish; a highly prized sport fish throughout their worldwide distribution (Pfeiler et al. 2000). Despite their known economic value (a billion dollar per year industry in the Florida Keys (Humston 2001) and 141 million dollars per year in The Bahamas (BFFA 2010)), large gaps exist in the scientific literature surrounding basic bonefish biology, making conservation and management strategies challenging (Ault 2008). As our questions regarding the movement patterns of bonefish evolved, so did the scale and structure of our studies. Specifically, we began using a multispecies approach to examine predator–prey relationships between bonefish, barracuda, and lemon sharks, as well as basic movement patterns of the predators alone. Much like bonefish, little is known about the biology of barracuda. Because of their abundance throughout their circumtropical distribution within the western Atlantic Ocean, Caribbean Sea, and Indo-Pacific regions, barracuda likely play a role as important apex predators in tropical waters (de Sylva 1963). Understanding the movement patterns of barracuda may also lead to an understanding of why some barracuda accumulate ciguatoxins while others do not; a human health issue of great concern in tropical communities (Bottein Dechraoui et al. 2005). Knowledge of shark movement patterns is increasingly important as populations are threatened worldwide due to overfishing and other human disturbances (Stevens et al. 2000; Baum et al. 2003). Because of their position in the food chain, sharks likely play an important role in structuring and maintaining diverse marine communities (Baum and Myers 2004;

Myers et al. 2007). As such, data on the movement patterns of lemon sharks, Caribbean reef sharks, and tiger sharks can help identify essential habitats needed to buffer shark populations against disturbance and declines in abundance.

CAPTURE AND TAGGING METHODOLOGY

A primary assumption of telemetry studies is that the postrelease behavior of tagged individuals is representative of the population of inference. As the process of surgical implantation of electronic tags has the potential to introduce bias to the sample and alter aspects of fish growth (Martin et al. 1995), physiology (Jepsen et al. 1998), swimming ability (Wagner and Stevens 2000), and survival (Jepsen et al. 1998), it is important to follow best practices at all stages of handling (i.e., capture, pre- and postoperative care, and tagging; see Brown et al. 2010). Here we describe the capture and tagging techniques that we have employed for use on three different groups of fish species. Given the different capture and surgical/attachment techniques associated with each group (see Table 2) we discuss them individually and then summarize the commonalities at the end. It should be noted that for all intracoelomic implantations, surgical equipment and transmitters were disinfected in a 5% povidone-iodine solution prior to surgery to minimize the risk of infection. We do not report information on other surgical details such as incision size etc., as incision size is dependent on the size of the transmitter, the number of sutures is dependent on the incision size, and the suture needle size and type is dependent on the size of the animal. All suture material used was absorbable monofilament (PDS II, Ethicon, Johnson and Johnson, New Jersey).

TABLE 2. Summary of species studied, acoustic transmitters used, and method of transmitter attachment.

Species	Transmitter type*	Method of transmitter attachment
Bonefish (*Albula vulpes*)	V13 coded	Intracoelomic implantation
	V9AP-2L coded acceleration & pressure sensors	Intracoelomic implantation
	V9 continuous pingers	Gastric implantation
Great barracuda (*Sphyraena barracuda*)	V13 coded	Intracoelomic implantation
	V9A-2L coded acceleration & pressure sensors	Intracoelomic implantation
Lemon sharks (*Negaprion brevirostris*)	V16-4L coded	Intracoelomic implantation
Caribbean reef sharks (*Carcharhinus perezi*)	V16 coded	Intracoelomic implantation
	V16 coded with temperature & pressure sensors	Intracoelomic implantation
	V9AP-2L coded acceleration & pressure sensors	External attachment
Tiger sharks (*Galeocerdo cuvier*)	V16 coded	Intracoelomic implantation

*note that all acoustic transmitters were manufactured by Vemco Inc., Shad Bay, NS

Bonefish

Although bonefish are the pinnacle species for many recreational anglers, catch-per-unit-effort can be low making catching bonefish via hook and line an unsuitable technique for collecting sufficient numbers of individuals for tagging purposes. In fact, Larkin et al. (2008) emphasized this by stating "owing to the difficulty of obtaining bonefish for our study, we employed a professional bonefish captain to help catch bonefish and then bring them to holding pens with recirculating seawater at the University of Miami's Rosenstiel School…" To overcome the problem of capturing large numbers of bonefish slated for transmitter implantation, we have successfully refined the use of a seine net deployed at the mouth of tidal creeks to intercept schools of bonefish on the incoming and outgoing tides. Used as a block net, a large (>30 m in length) soft, small mesh (3.2 cm or smaller) seine net can be stretched across the entire width of a creek mouth to impede the movement of bonefish. The use of small mesh nets helps to avoid gilling or entanglement of bonefish (see Murchie et al. 2009). When a school of bonefish approaches, the two terminal ends of the net are quickly pulled together to encircle the fish. Field assistants positioned along the shoreline or in the shallows can also help herd the bonefish into the net as it is closed. Dip nets, cradles and bare (wet) hands can then be used to transfer individual fish from the net to flow-through holding pens (1.3 m × 0.8 m × 1.25 m tall, 3.1 cm extruded plastic mesh) submerged in a minimum of 0.6 m of water, where the fish remain until sorting and transmitter implantation. Dip nets and bare hands can also be used to quickly remove nontarget species, thus minimizing the stress and potential mortality of bycatch. Collecting a large number of bonefish in one capture event allows for the selection of fish of a certain size and sex based on the specific criteria or purpose of the study. This technique also allows for all transmitters to be deployed within the same short period of time, avoiding potential complications related to small numbers of tagged fish being released at temporally discontinuous intervals. Moreover, deploying transmitters using temporally concentrated effort helps to increase efficiencies associated with field logistics, especially continuity with field staff (in particular, trained surgeons; see Wagner and Cooke 2005).

When working with bonefish, pre- and postoperative care is as important as the intracoelomic implantation itself. Holding fish in flow-through pens provides fish time to recover in ambient, well-oxygenated seawater prior to surgery, as well as following the surgical procedure itself. Because of the tidal nature of many tropical shallow coastal environments, it is often necessary to move the holding pens to deeper areas where the water is cooler and more oxygenated. This is especially the case during slack low tide in the summer months where the water draining from shallow tidal creeks can be greater that 35°C (Murchie et al. 2011). It is also important to have multiple pens available to distribute the fish and not hold fish at high densities for fear of local depletion of dissolved oxygen as well as stress associated with confinement with a large number of conspecifics (Murchie et al. 2009). Because great barracuda and sharks coexist with bonefish in shallow tropical waters, it is important to remain vigilant when holding bonefish in pens and chase away any potential predators (Cooke and Philipp 2004).

Surgeries to implant transmitters can be conducted on a stable boat with ample room for the surgeon, assistants, anesthetic bath, and field surgery table. Conducting surgery on a boat rather than on shore provides easy access to the holding pens for retrieving fish slated for transmitter implantation as well as returning fish to the pens to recover from the anesthetic. Pumps and any other equipment requiring power can be run via a power inverter connected to the boat battery. A canvas top for the boat can provide shade from the hot tropical sun for field staff, as well as provide shade to the fish during the surgical procedure.

Because of its ease of transport and use, we chose tricaine methanesulfonate (MS-222) for anesthetizing bonefish prior to surgery. Creating a situation where the fish is immobile can allow the surgeon to work expeditiously, reduce the duration of anesthesia, and ultimately expedite recovery. Studies that do not anesthetize fish when implanting transmitters (e.g., Humston et al. 2005) may confound their results if stresses associated with the surgical procedure influence postrelease behavior and survival. Regardless of the type of anesthetic, we suggest that an experienced field staff member be specifically assigned to the task of anesthetizing the fish so that they can provide focused care and promptly determine when the fish is ready for surgery (i.e., stage 4 of anesthesia; Summerfelt and Smith 1990).

For bonefish, we used a portable surgery table constructed out of a plastic tote (approx. 100 L, Figure 1a). The lid of the tote was inverted to act as the surgery platform and fitted over the opening of the bin below. Small holes were drilled in the lid to allow water to drain back into the bin, with the bin acting as a sump. A small bilge pump (approx. 175 L/h) was placed in the bottom of the bin and the outflow tube was fitted through a small hole in the vertical wall of the bin near the top. The bilge pump was used to provide continual irrigation of the fish's gills and to keep the skin of the fish damp during surgery. Fish were placed on a rectangular piece of high-density foam during the surgical procedure, providing a soft

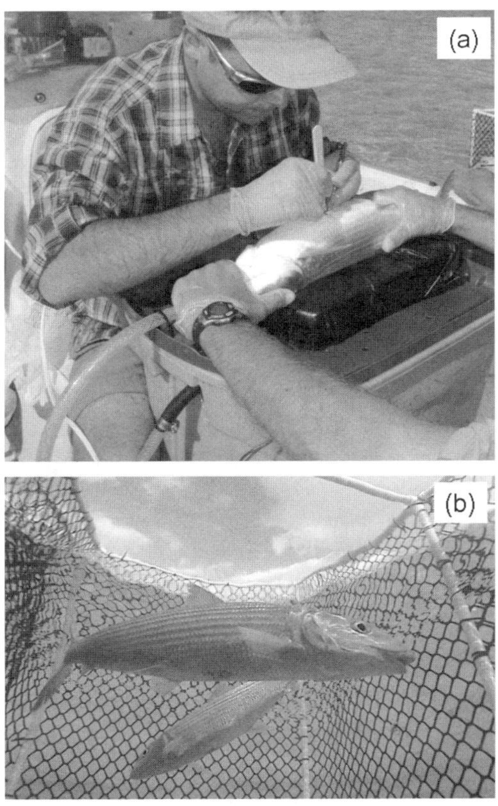

Figure 1. (a) Intracoelomic transmitter implantation on bonefish *Albula vulpes*. (b) Bonefish recover from surgery in a flow through net pen.

surface that minimized mucus loss. Once surgery was completed, fish were returned to a holding pen to recover from the anesthetic (Figure 1b). Because bonefish typically reside in schools, individuals surgically implanted with transmitters were released with a large number of conspecifics that were also captured using the block net. Releasing fish in a large school could potentially reduce the likelihood of short-term postrelease mortality via predation by barracuda and sharks.

One alternative to the intracoelomic implantation of transmitters in bonefish is gastric tagging. Gastric tagging employs the use of a small plunger to insert the acoustic tag through the esophagus and into the stomach of the bonefish (see Cooke and Philipp 2004; Danylchuk et al. 2007a). Benefits to this technique are that the fish do not need to be anesthetized, which greatly reduces the duration of the procedure as well as recovery time for the tagged fish. Potential challenges to the use of gastric tagging include the possible regurgitation of the transmitter, the need to use smaller tags (both in diameter and length), as well as reduced duration for tracking since the tag will eventually be defecated.

Although bonefish naturally reside in shallow tropical flats that can become very warm in the summer months, water temperature should still be considered a potential stressor especially when combined with the capture and surgical implantation of transmitters (see Beyers and Rice 2002; Murchie et al. 2011). At higher water temperatures metabolic rates increase (Brett 1995), and stress responses of fish are intensified (Wilkie et al. 1997). Indeed, at the warmest water temperatures during tagging (29°C; end of August), the survival rate of tagged bonefish diminished to less than 43% within two weeks following surgery, compared to an average survival rate of 80% for fish tagged at cooler water temperatures (20–25°C; November–March) (Figure 2). Unless the purpose of the study is to examine variation in movement patterns and postrelease mortality as it relates to water temperature, we would strongly discourage the capture and deployment of transmitters in bonefish during the hot, tropical summer months (i.e., July–September).

Great Barracuda

Unlike bonefish, using rod and reel to capture barracuda for telemetry studies can be quite effective. Angling for barracuda typically does not require a highly skilled angler (especially when spinning gear is used), is relatively inexpensive, and enables standardized sampling from a variety of habitats including mangrove creeks and tidal flats, to deeper offshore areas. Although there is risk of bycatch, anglers can target specific species by regulating the type of lure, bait, lure retrieval, and trolling speed. Success, however, is often dictated by environmental conditions such as water temperature, season, habitat, and tides how these factors influence the behavior of the targeted species (Wall et al. 2009). Angling may also present a biased sample of the population by selecting for more aggressive, healthier, or genetically vulnerable individuals (Cooke et al. 2007). A size bias may also result, depending upon the size and type of hook, lure and bait used. Fish that are captured via angling may be exposed to physiological stress, injury, barotrauma, and predation due to an angling event (Skomal and Chase 2002; Bartholomew and Bohnsack 2005; Danylchuk et al. 2007b), which could hinder the chance of survival when combined with the stress associated with the surgical procedure (including anesthesia) to implant acoustic transmitters. Nevertheless, the effects of capture can be easily reduced by landing the fish quickly, using barbless hooks, employing correct handling methods, and fishing at lower water temperatures, all of which are practices that will

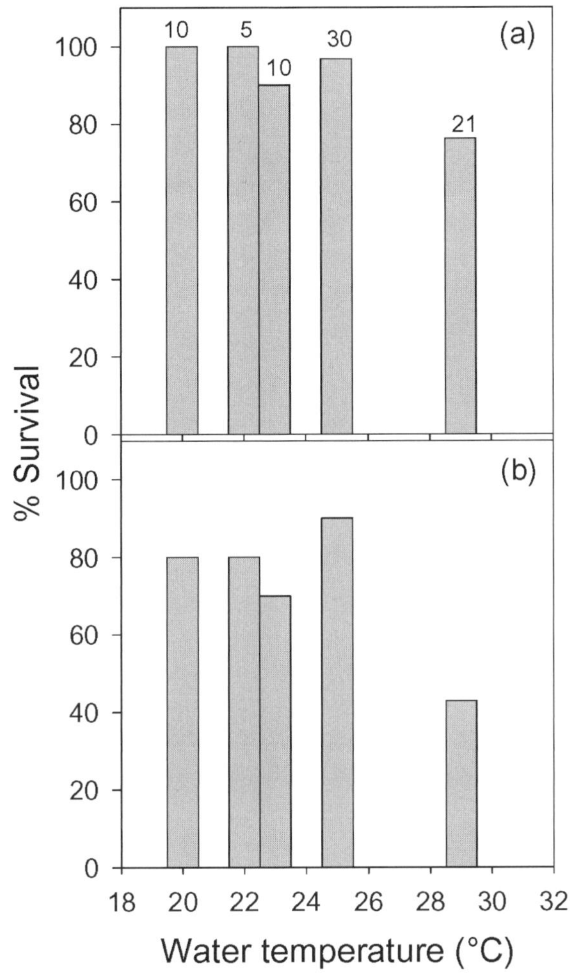

FIGURE 2. Survival of bonefish *Albula vulpes* tagged at a range of water temperatures after (a) 48 h and (b) 2 weeks. Numbers above the bars represent the total sample size of bonefish at each tagging temperature.

increase the chance of the postrelease survival (Schaeffer and Hoffman 2002; Bartholomew and Bohnsack 2005; Cooke and Suski 2005).

When capturing barracuda for intracoelomic implantation of acoustic tags, we have found that trolling with heavy action recreational gear and artificial lures at speeds of 6–9 knots to be most effective. Because great barracuda are excellent burst swimmers, trolling at higher speeds reduces the number of bycatch species encountered such as yellowtail snapper *Ocyurus chrysurus*. Barracuda have proven to be relatively tolerant of physiological stress and injury associated with capture using rod and reel (O'Toole et al. 2010a), however it is not recommended that barracuda be angled at high water temperatures during the summer months prior to the performance of a surgery. It is important to restrict the amount of time the fish is played and landed; a large mesh cradle can be used to handle great barracuda in a safe and appropriate manner. Further, barracuda seldom exhibit excessive bleeding, rarely are deeply hooked in critical areas, and tend to recover well after surgery (O'Toole et al. 2010a, 2010b).

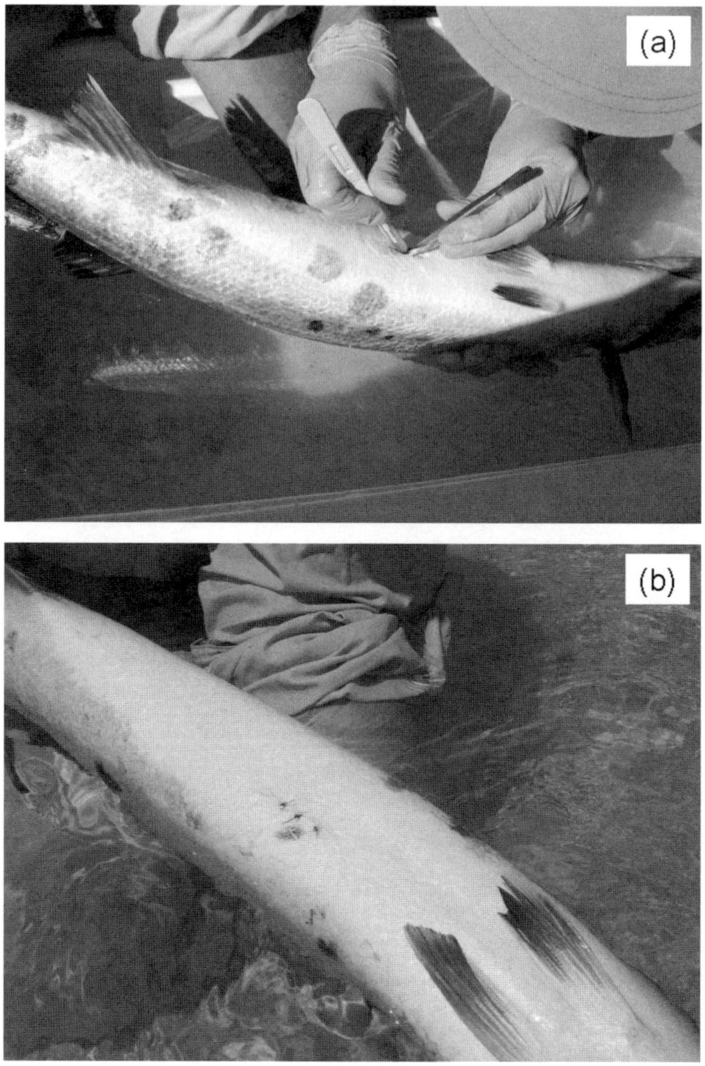

Figure 3. Intracoelomic transmitter implantation on barracuda *Sphyraena barracuda* (a) in a cooler on the boat and (b) kneeling in the tidal flats.

As stated above, we recommend anesthetization prior to surgery. For barracuda, surgeries can take place either on the boat with an assistant holding the fish in a cooler of water, or surgeries can occur in shallow tidal flats where the assistant kneels and holds the fish directly in the ocean (Figure 3a and b, respectively). While barracuda are predators, we recommend holding the fish either in a cooler of fresh seawater or in the ocean directly, until they are fully recovered before release.

Sharks

The intra- and inter-specific size range of sharks in tropical coastal waters varies considerably, as does the specific region of the coastal zone in which the sharks inhabit. As such, the most effective capture method for sharks intended to be deployed with acoustic transmitters will be dependent on a number of experimental, site, and species specific variables. The two most important variables include the ability of the target species to withstand the physiological and physical stress associated with capture, and the abundance of the target species. The short and long term consequences of the intense anaerobic exercise associated with capture events is poorly understood for most species of sharks (Skomal 2007). Where research exists it has focused of the effects of recreational catch and release angling in predominantly temperate species of sharks (e.g., Brill et al. 2008). All research suggests a linear relationship between hooking duration and the magnitude of the homeostatic disruption and as such hooking and handling durations for transmitter attachment or surgery should be minimized to ensure postrelease viability.

For sharks inhabiting shallow (<1 m) coastal flats (e.g., juvenile lemon sharks) conventional recreational angling equipment with rod and reel can be an effective and efficient capture method (Murchie et al. 2010). Clear, shallow coastal waters allows for active sight-fishing of individual sharks, which greatly reduces the amount of field time needed for capture sharks for tagging. Although longlines (see below) can also be used in shallow water, there is the potential for multiple sharks to be caught, increasing the hooking duration and likely the physiological stress associated with capture for some individuals (if sharks are to be processed sequentially). Using heavy rods and reels can permit the rapid retrieval of sharks as a means to reduce landing times and physiological impacts prior to tagging with acoustic transmitters.

In deeper coastal waters, longlines offer the most efficient and adaptable way of capturing sharks (e.g., Caribbean reef and tiger sharks), and can also yield additional diversity, relative abundance and demographic data at the same time. Longlines use a large number of baited hooks that catch sharks efficiently, especially when abundances are relatively low. Mainline lengths of longlines can vary with the number of hooks to be deployed as well as the length of the gangions. To avoid tangling of two sharks hooked side by side, the spacing between gangions must be at least 2.5 times the gangion length. Longer gangions provide a larger range of movement for the captured animal but can reduce the number of hooks that can be placed on any given length of line. Fewer hooks will allow for shorter hooking durations as lines can be checked or hauled quickly and regularly; however more hooks will increase the catch rate especially if the target species is less abundant. Hook timers (e.g., Lindgren-Pitman, Pompano Beach, Florida) can easily be incorporated into the gangion to provide a quantitative way of monitoring hooking duration that can, in turn, provide guidance for suitable transmitter candidates. It should be ensured that mainlines have sufficient floatation or anchor points to reduce dragging and tangling among natural structures such as coral reefs. Tangled lines lead to reduced movement capacity for captured sharks which in turn increase the magnitude of the physiological stress and reduces the viability of candidate animals for tagging.

Only jaw-hooked sharks should be candidates for transmitters given the physical trauma usually associated with gut hooking (Borucinska et al. 2002). Circle hooks are thought to reduce the instances of gut hooking (Cooke and Suski 2004; Kerstetter and Graves 2006). For example, 82% of all sharks captured on 16/0 Mustad circle hooks during three years of seasonal longline surveys in The Bahamas were hooked in the jaw thus providing a high proportion of candidate animals (Brooks et al., unpublished data). In general, hook removal tools, although effective, should not be used as they can inflict additional trauma on the jaw. A less traumatic option is to rotate the circle hook through the jaw until the barb is visible from the exit wound. The hook can then be cut behind the barb and the shank rotated out of the jaw. Hooks should be large enough to encircle the jaw bone of the shark and allow the point to protrude from the exit wound as far as the barb. Hooks that are too small will embed in the cartilage of the jaw and will not be removed easily.

Acoustic transmitters can be surgically implanted in the peritoneal cavity or attached to sharks externally via an anchor system. Acoustic transmitters can also be fed to sharks and subsequently retained in the gastric tract, however the retention times using this method are very variable and range from 24 h to 34 d (Brunnschweiler 2009). The method of attachment is primarily determined by the required retention time of the transmitter, the likelihood that the shark will attempt to dislodge an externally attached tag, and the overall size of the shark since surgeries may prove logistically unfeasible for large specimens (>4 m in length). To date, acoustic telemetry studies suggest that the majority of sharks surgically implanted with transmitters recover with no discernable long-term effects (see Chapman et al. 2005; Meyer et al. 2009).

To implant sharks, the candidate animal should be kept in the water. For juvenile lemon sharks captured in shallow flats areas or mangroves creeks on rod and reel, the animal can be restrained by hand (Figure 4a). For larger specimens (i.e., Caribbean reef and tiger sharks) captured by longline, individuals should be restrained next to the boat by securing the gangion to the bow and a tail rope to the stern (Figure 4b). Where possible, the shark should be oriented head towards the current to allow passive irrigation of the animal's gills. Once secured, the shark should be inverted (ventral side up), and held until the onset of tonic immobility. Tonic immobility is a reversible, coma like stasis displayed by a number of elasmobranchs (Henningsen 1994), and allows the surgery to proceed without the use of anesthetic. In the rare case that the candidate species does not exhibit tonic immobility, anesthetic compounds can be used (e.g., MS-222, Heupel and Heuter 2001).

In some cases where only short term deployments or specific experimental parameters require it, tags can be attached externally, usually with a dart tag, or in some cases, securely wired to either the first or second dorsal fin. External attachment removes the possibility of mortality due to surgery, is generally easier to perform and can be conducted, in some cases, without actually capturing the shark. This technique does however pose significant challenges in terms of transmitter retention. Acoustic transmitters are generally attached to a shark using a dart tag attachment however the retention times for most types of dart tags are relatively short, and tag shedding is a common problem even for tags that do not have an electronic transmitter attached (Dicken et al. 2006). Standard SSD dart tags (Hallprint, Australia) with no electronic transmitters, applied to Caribbean reef sharks in the Bahamas were shed in less than six months, however, the retention time for nylon headed PDL dart tags (Hallprint, Australia), has been much longer (Brooks, unpublished data). Tag shedding is likely due to a number of different factors which include bio-fouling which increases the hydrostatic drag of the tag and tissue necrosis which weakens the anchor point. It has also been hypothesized that the physical and often violent nature of mating in most shark species could cause a tag to be shed.

To externally attach acoustic acceleration transmitters (model V9AP; Vemco Inc., Shad Bay, NS) to Caribbean reef sharks, two holes were punched through the lower third of the first dorsal fin using 12 gauge piercing needles and one polyethylene backing plates as a guide. Leaving the needles in place, a second backing plate was fitted over the opposite side of the dorsal fin over the 'sharp' end of the piercing needles. The distance between the holes was measured to be exactly that of the distance between the two attachment eyes of the transmitter. Two 30 cm lengths of 0.5 mm stainless steel locking wire (Loos & Co, Naples, Florida) were passed through the eye of the transmitter at either end. The wire was wrapped twice

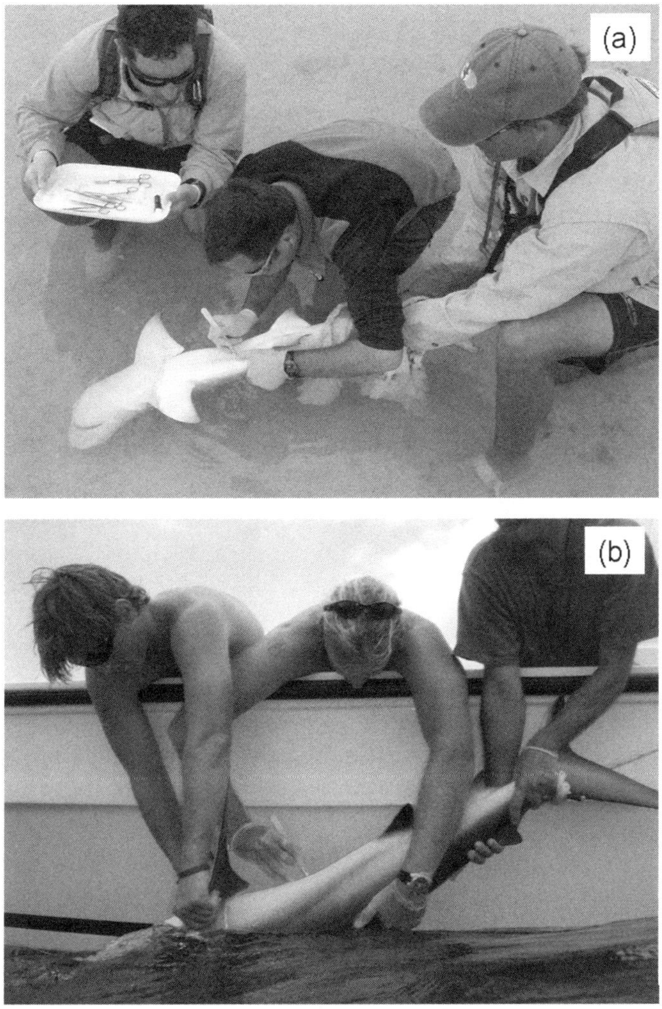

Figure 4. Intracoelomic transmitter implantation on sharks occurring (a) in the tidal flats, and (b) from the side of a boat in deeper waters.

through the eye at the mid point so the two loose ends of approximately 15 cm length lay at 90 degrees to the long axis of the tag. Each pair of wire ends were passed through the piercing needles in the dorsal of the shark and the needles removed. The wire ends were tightened until the transmitter and backing plates rested snuggly against the dorsal fin, and excess wire was trimmed (Figure 5).

General summary of capture and tagging

All fish surgically implanted or affixed with acoustic telemetry devices should be provided the best possible care throughout the capture, tagging, and recovery process to ensure the highest chance of survival and the least amount of influence on postrelease behavior. This includes using adequately trained personnel for surgeries and best current surgical practices (e.g., aseptic techniques, choice of tools, anesthetization, surgical incision, and wound clo-

Figure 5. External attachment of an acceleration transmitter.

sure) within the limits of remote field conditions. While weather and environmental conditions can present challenges to field-based surgical procedures, the alternative of relocating specimens back to the laboratory can increase stress effects associated with handling and holding times. Because all species discussed above have the potential to be preyed upon, they require adequate recovery time prior to release. As such, the use of flow through net pens, or dedicating additional staff and time to the postoperative care of tagged individuals is required. We also advocate the timing of tagging programs occurs during periods of cooler water temperatures to minimize the additional effects of thermal stress to the capture, handling, and tagging procedures. The methods described above have resulted in excellent wound healing in recaptured specimens, and the continued monitoring (3+ years) of a number of tagged individuals.

TRACKING TELEMETERED FISH

Manual tracking

Unlike radio telemetry, where signals from transmitters are detected by an aerial receiving antenna, manual tracking of regular acoustic transmitters (i.e., not including pop-up satellite tags, archival tags) requires a hydrophone to be placed in the water. When transmitter-implanted fish are able to reside in very shallow water (e.g., intertidal areas of tropical flats and mangrove creeks), the detection range of the manual hydrophone can be impaired due to a lost line of sight with the transmitter (Heupel et al. 2008). In situations such as these, we have used manual tracking to determining presence-absence only at predetermined discrete stations and during tidal periods that will maximize the likelihood of a detection if a tagged bonefish is present (i.e., so as to not wrongly conclude tagged fish are absent because of limited capacity for their detection). However, a major limitation of this technique is that determining the presence or absence of tagged individuals can be confounded by the fact that personnel and gear cannot be in multiple places at once.

Where water depth is sufficient, manual tracking of individuals can be very informative in answering questions regarding habitat selection and home range (Morrissey and Gruber 1993a; b, respectively). In our experience, effective continual tracking of bonefish in depths of 1–30 m for well over 14 h can be achieved by using a small boat equipped with both directional and omnidirectional hydrophones. Continuous pinger-style transmitters are typically easier to follow than coded transmitters (i.e., with delayed transmission), but have the disadvantage of a considerably shorter battery life than that of coded tags (i.e., 21 d versus 700 d). Ultimately, the type of transmitters used as well as method of manual tracking should be dependent on the questions being asked, the nature of the environment, and a preliminary understanding of the movement patterns of the fish species being studied.

Passive arrays

The open nature of tropical marine coastal environments presents a challenge for configuring fixed remote acoustic receiver arrays. Great consideration should first be given to whether a fixed receiver array or even acoustic telemetry is the best technology to employ when attempting to examine the spatial ecology of highly mobile marine fishes. Some preliminary understanding of the spatial movement patterns of the focal species can be advantageous when making decisions regarding the methods for telemetry. When limited or no previous data are available for a focal species, conducting a small pilot study can help determine the scale and scope of the array needed to address specific questions related to a fish's spatial ecology.

For our work in The Bahamas, the initial goal was to use a fixed remote acoustic telemetry array to examine different elements of the spatial ecology of bonefish and juvenile lemon sharks associated with shallow tidal flats and mangrove creeks. Conducting a pilot study that involved deploying a small number of remote receivers and releasing a limited number of study fish surgically implanted with coded transmitters allowed us to determine whether questions related to site fidelity and movement patterns among habitat types could be adequately addressed with this technology. The pilot study also allowed us to test different mooring systems for the remote receivers especially because it was advantageous to ensure that the transducer of the receiver was submerged for the longest possible duration through the entire tidal cycle. The final design for mooring the receivers employed a segment of steel rebar cemented firmly into the opening of a concrete block; this allowed the receiver to be securely positioned in a number of different orientations depending on water depth and coastline morphology (Figure 6). In areas where water depths always exceeded 1 m, the mooring system could be deployed so that the receiver was vertical in the water column. In cases where water depths became shallow because of tides, the substrate was excavated to ensure the mooring system and receiver remained submerged, or in some instances, such as narrow rocky channels, the mooring system was laid horizontally on the substrate.

Our initial array for studying bonefish and lemon sharks was configured in a series of curtains of receivers extending perpendicular from shore, allowing us to determine along-shore movements among shallow flats and tidal mangrove creeks. To examine finer-scale movement patters within mangrove creek systems, we also deployed receivers at choke points,

FIGURE 6. (a) Deployment of an acoustic receiver using scuba. (b) Underwater positioning of an acoustic receiver.

such as in narrow creek mouths, to measure the movement of tagged fish in and out of these systems. Where possible, receivers at choke points were integrated into curtains to maximize the scope of detection while using the least number of receivers possible. In all cases, the range of detection was determined for each receiver to ensure that curtains would not allow a tagged individual through without being detected. Because biofouling can have an impact on the performance ability of the receivers (Heupel et al. 2008), regular visits to the mooring units were made to determine if anti-fouling measures were necessary. Since the accumulation of algae and other growth occurred mainly on the concrete blocks, with limited fouling on the receivers themselves, no anti-fouling products were applied. The minimal level of fouling observed was likely due to the relatively nutrient poor waters of tropical regions as well as the smooth surface of the receivers.

As we expanded our research to include great barracuda and larger, more coral reef orientated sharks (i.e., Caribbean reef sharks, tiger sharks), we continued to use the same mooring technique even though the receivers were deployed in considerably greater water depths (up to 40 m) and further offshore (>5 km). By firmly affixing the receivers to rebar and concrete blocks, the entire unit could be place on the substrate with limited chance that it would move with tidal currents and wave action. Although this technique required scuba (self contained underwater breathing apparatus) diving to deploy and retrieve receivers, the benefit was that we did not need to add surface or subsurface floats because of the clear tropical waters and the ability to get within a few meters of the deployment site using GPS. Not using floats to indicate position reduced the likelihood that the receivers would be tampered with (see Domeier 2005). Over the course of more than three years, only three out of 54 receivers were lost, likely because they were removed by local fishers as evident by the presence of cut cable ties that were used to attach the receivers to the rebar.

The design of an array is a function of the specific ecological or biological questions the project is trying to answer, as well as available resources (Heupel et al. 2006). We have used a combination of curtains and nonover-lapping grids. Forming curtains of overlapping receivers at strategic points along a hypothesized movement route will allow the testing of a hypothesis with fewest resources. By placing curtains at natural choke points, the number of receivers required to form a curtain can be reduced. The distance between curtains can be fine tuned to the expected spatial scale of the movements. However, this is not the best array design for sensor tags given the likely low number of detections. Instead, a nonoverlapping grid is more appropriate for sensor tags, yielding a higher number of expected detections as an area of seabed is continuously monitored. The spacing of the grid will depend of the scale of movements a species is likely to make, and the available number of acoustic receivers (see Heupel et al. 2006).

APPLICATIONS OF BIOTELEMETRY IN TROPICAL COASTAL SYSTEMS

Protected areas

Understanding the spatial ecology of fishes and invertebrates inhabiting tropical coastal habitats is vitally important for the design and management of marine protected areas (MPAs). Marine protected areas have been advocated as effective relatively ‘low cost’ fisheries management tool that can provide ecological and fishery benefits (Murray et al. 1999; Roberts et al. 2001; Polunin 2002); especially in areas such as the tropics were biodiversity can be high and the ability to care for coastal habitats often limited. From a fisheries perspective, MPAs, and in particular marine reserves, can help conserve essential habitat and promote the buildup of biomass, which, in turn, may result in the spillover of adults into adjacent fishing grounds and the downstream export of larvae to more distant fished areas (Murray et al. 1999).

Although acoustic telemetry has become a standard tool for studying the spatial ecology of fishes, only in the past decade has this technology been used to study the function of MPAs (e.g., Heupel and Simpfendorfer 2005; Schmiing et al. 2009), and rarely has it been used to assist in boundary demarcation based on the identification of critical habitat and site fidelity of species to be protected (e.g., Chapman et al. 2005; Kerwath et al. 2009). Using acoustic telemetry for this particular application can provide greater resolution when compared to conventional passive external tagging, and can operate at a scale that enables sufficiently precise positioning to determine how much time a fish spends inside and outside specific habitats or moving across MPA boundaries (Davis 2004). Matching acoustic telemetry with the mapping of the coastline and substrate characteristics can help to identify critical habitat for tropical marine fishes. In turn, this knowledge of movement patterns, home range sizes and habitat use can then be used to determine whether proposed and existent boundaries of MPAs will be able to promote the maintenance and possibly the increase of stock biomass, and ultimately meet the management objectives for conservation.

One of the objectives of our research in the waters surrounding South Eleuthera, The Bahamas, was to determine whether a proposed no-take marine reserve (Dahlgren 2002; Danylchuk 2003) would offer protection for bonefish. As such, bonefish surgically implanted with coded acoustic transmitters were collected from two focal mangrove creeks within the proposed boundary of the reserve at various intervals spanning more than three years. Examining the spatial ecology of bonefish across several years has allowed us to examine seasonal variation in movement patters, and helped to identify critical reproductive habitats that would fall outside the current iteration of the boundaries for the proposed MPA (Danylchuk et al., unpublished data)

Locating spawning sites

Some groups of marine fishes in the tropics have evolved a reproductive strategy that involves migrating to relatively discrete locations where they form large aggregations for spawning (Domeier and Colin 1997; Claydon 2004). Such migrations to and from spawning aggregations sites can greatly broaden the spatial extent of a fishes home range; however, their time at aggregations sites can be relatively short compared to time spent in other habitats not associated with reproduction. Because movements to spawning aggregation sites occurs in such discrete periods and often a fair distance away from nonreproductive habitats, identifying the timing and location of spawning aggregations for tropical coastal marine fishes can be difficult without the use of acoustic telemetry. Understanding the dynamics of spawning aggregations has important implications for the conservation and management of fish stocks because fish that predictably aggregate at specific times and locations can be vulnerable to intense fishing pressure and overharvesting (Coleman et al. 1996; Domeier and Colin 1997; Roberts and Hawkins 1999; Sala et al. 2003). A well-studied group of fishes that form transient, site-specific spawning aggregations are the Serranidae, such as the red hind *Epinephelus guttatus* (Sadovy et al. 1994; Beats and Friedlander 1998) and Nassau grouper *Epinephelus striatus* (Bolden 2000; Whaylen et al. 2004). Documenting the location and spatial and temporal dynamics of grouper spawning aggregations has helped to characterize the biology and ecology of this important group of coral reef fishes (Whaylen et al. 2004), as well as highlighted the need to learn more about other marine fishes that aggregate to spawn (Sadovy and Domeier 2005).

A major component of our research on bonefish in The Bahamas was using acoustic telemetry to document the timing and location of spawning (Danylchuk et al., in press). Anecdotal evidence from the Caribbean and The Bahamas, as well as initial data from our telemetry work, suggested that bonefish may spawn in large aggregations in deep offshore waters, however the spatial ecology of bonefish related to spawning had never been formally quantified. Data from our initial array showed individuals moving towards the end of Cape Eleuthera in the winter and early spring (Murchie 2010), consistent with reports from the literature on their movements associated with spawning. When extending our array, we took this into account, along with accounts from local fisherman who witnessed large schools of bonefish near deeper water. The extended array included a combination of receivers at choke points, large curtains radiating out from a peninsula of land at the junction between shallow and deeper offshore waters, and overlapping 'nets' of receivers that allowed us to determine the direction of movement once bonefish moved away from the shoreline into open water. As data were collected from the array throughout the course of the suspected spawning season (see Figure 7), we made slight adjustments to the positioning of the receivers to maximize the likelihood of locating both prespawning and spawning aggregations for bonefish. This, in combination with manual tracking, proved to be instrumental for determining the timing and location of aggregations sites, especially because bonefish move offshore to spawn at night (Danylchuk et al., in press).

Biotoxins and ocean health

Acoustic telemetry can be a valuable tool in answering unique questions such as linking the spatial ecology of reef fish to the presence and of biotoxins. Ciguatera fish poisoning is caused by a naturally occurring toxin that bioaccumulates in apex predators that has the potential to make humans very ill if reef fish with high concentrations of ciguatoxin are consumed (Lehane and Lewis 2000; Dickey and Plakas 2010). Telemetry can be combined with tissue biopsies (tissue is analyzed for harmful concentrations of ciguatoxin) to help understand relationships between the spatial ecology of fish (e.g., barracuda) and the potentially harmful toxin levels.

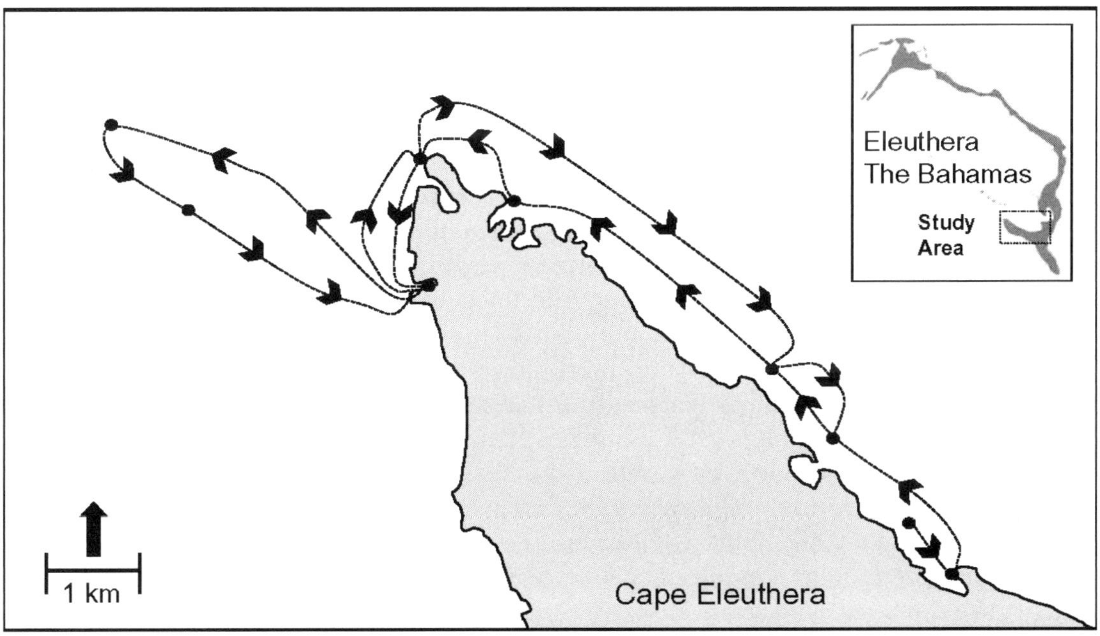

FIGURE 7. Generalized movements of bonefish 8899 in one week (December 2–9, 2008) in the study area along the coast of Cape Eleuthera, The Bahamas (N 24°50'05" and W 76°20'32"). Bonefish 8899 displays representative movements during winter months from within shallow tidal creeks to deeper offshore areas, presumably for spawning. The dashed lines with chevrons indicate the direction of the movement from one receiver (solid black circles) to the next. The inset map displays the entire island of Eleuthera with the study area highlighted.

Energy flow and connectivity

Energy is the currency of life, and having an appreciation of how energy flows through an ecosystem is one of the most basic ecological principles (Smith 1992). The active transport of nutrients via foraging migrations made by fish promotes energy flow across distinct habitat boundaries and effectively connects communities (Valentine and Heck 2005; Gaines et al. 2007). Acoustic telemetry can be used to elucidate the distribution of free-swimming fish among various habitats in their natural environment (Cooke et al. 2004), and provide information that is paramount to understanding how tropical coastal ecosystems function. Acoustic telemetry can also be used to examine energy flow at the organismal level by providing a method to approximate the cost of activity in a fish's energy budget. The use of animal-borne acceleration data loggers is gaining popularity for studying activity costs in free-swimming fish (Wilson et al. 2007). While these data loggers require retrieval to access the data (Ropert-Coudert and Wilson 2005), recent advances in onboard processing have allowed for the production transmitters capable of encoding and transmitting tri-axial accelerometer data efficiently (see Murchie et al. 2011; O'Toole et al. 2010b). Contribution to the production of a bioenergetics model for any tropical fish species not only provides fisheries managers with a useful tool for fish production, but would further assist scientists in understanding coastal ecosystem dynamics.

Fishing mortality

Fish captured by commercial and recreational fisheries are often released in accordance with harvest regulations or voluntarily due to conservation ethic or other reasons. The assumption with releasing such fish is that the majority survive and do so with few sublethal consequences. However, that assumption is rarely tested. Acoustic telemetry has great potential for monitoring post release survival and behavioural alterations in both the recreational and commercial sectors. A recent review by Donaldson et al. (2008) revealed that acoustic telemetry is increasingly being used to study postrelease mortality and can have many advantages over other methods such as holding fish in pens, tanks or cages. However, there are relatively few examples exist in the marine environment. Cooke and Philipp (2004) and Danylchuk et al. (2007a) used pinger acoustic transmitters to track bonefish after release by recreational anglers and demonstrated post release predation and short term behavioral alterations. Not only did such studies generate mortality data, but they revealed factors that were associated with mortality which provided anglers and managers with potential strategies for improving postrelease survival. Commercial bycatch studies that use acoustic telemetry have been less common; however, our group has used acoustic accelerometer tags to evaluate the fine-scale post release behavior of Caribbean reef sharks captured by long line. Given the many conservation and management concerns associated with postrelease mortality and sublethal alterations in the commercial and recreational sector, particularly in tropical marine systems that have received relatively little research attention, there is a great need for additional research and acoustic telemetry is one of the best ways to generate meaningful data.

CONCLUSIONS AND FUTURE OPPORTUNITIES

Acoustic telemetry is a powerful tool for gaining insight into the habitat use, life histories, intra- and inter-specific interactions of tropical marine organisms, as well as the potential impacts of natural and anthropogenic disturbances. Through our work on diverse taxa (e.g., bonefish, barracuda, and multiple species of sharks) we faced many challenges including dealing with different deployment dates of transmitters and receivers, as well as optimizing data collection rates and working with an array that aims to cover study questions focused on many species. We also encountered environmental challenges unique to tropical coastal ecosystems. Collectively, our experiences revealed that an adaptive approach to the use of acoustic telemetry is necessary as a means to compensate for certain limitations of this technology for understanding the movements of fish in these environments.

Future use of acoustic telemetry in tropical coastal ecosystems can include pairing the fine-scale movements of organisms with detailed environmental data such as water temperature and salinity to provide critical information for climate change modeling. There is also the opportunity to gain information on smaller species as transmitter technology and battery miniaturization develops. Because of the cost of acoustic telemetry gear, and strains on resources especially when working in remote coastal areas, we advocate communication and cooperation with other researchers in the area, and other larger groups such as the Ocean Tracking Network to look at broader scale movements. We also support the recommendation of Grothues (2009) that scientists and equipment manufacturers engage in a dialogue encouraging the use of a common code scheme and cross-equipment communication to maximize the opportunity to track fish in coastal marine systems.

ACKNOWLEDGMENTS

The authors gratefully acknowledge C. Maxey and the staff, students, and volunteers of the Cape Eleuthera Institute and The Island School for logistical support and assistance with field work. The various projects described here were supported by grants from Bonefish and Tarpon Trust, the Charles A. and Anne Morrow Lindbergh Foundation, the Patagonia World Trout Initiative, the Canadian Foundation for Innovation, the Ontario Research Fund, the Save Our Seas Foundation, Carleton University, University of Massachusetts Amherst, the Cape Eleuthera Foundation, and the University of Illinois. K. Murchie was supported by a Natural Sciences and Engineering Research Council CGSD fellowship. S. Cooke was supported by the Canada Research Chairs program. The authors also thank two anonymous reviewers for valuable feedback on an earlier version of the manuscript.

REFERENCES

Ault, J. S., editor. 2008. Biology and management of the world tarpon and bonefish fisheries. CRC Press, Boca Raton, Florida.

Bahamian Flats Fishing Alliance (BFFA). 2010. The economic impact of flats fishing in The Bahamas. Gainesville, Florida.

Bartholomew, A., and J. A. Bohnsack. 2005. A review of catch-and-release angling mortality with implications for no-take reserves. Reviews in Fish Biology and Fisheries 15:129–154.

Baum, J. K., and R. A. Myers. 2004. Shifting baselines and the decline of pelagic sharks in the Gulf of Mexico. Ecology Letters 7:135–145.

Baum, J. K., R. A. Myers, D. G. Kehler, B. Worm, S .J. Harley, and P. A. Doherty. 2003. Collapse and conservation of shark populations in the Northwest Atlantic. Science 299:389–392.

Beats, J., and A. Friedlander. 1998. Evaluation of a conservation strategy: a spawning aggregation closure for red hind, Epinephelus guttatus, in the U.S. Virgin Islands. Environmental Biology of Fishes 55:91–98.

Beyers, D. W., and J. A. Rice. 2002. Evaluating stress in fish using bioenergetics-based stressor-response models. Pages 289–320 *in* S. M. Adams, editor. Biological indicators of aquatic ecosystem stress. American Fisheries Society, Bethesda, Maryland.

Bolden, S. K. 2000. Long-distance movement of Nassau grouper (*Epinephelus striatus*) to a spawning aggregation in the central Bahamas. Fishery Bulletin 98:642–645.

Bottein Dechraoui, M. Y., J. A. Tiedeken, R. Persad, Z. Wang, H. R. Granade, R.W. Dickey, J. S. Ramsdell. 2005. Use of two detection methods to discriminate ciguatoxins from brevetoxins: Application to great barracuda from Florida Keys. Toxicon 46:261–270.

Borucinska, J., N. Kohler, L. Natanson, and G. Skomal. 2002. Pathology associated with retained fishing hooks in blue sharks, *Prionace glauca*, with implications for their conservation. Journal of Fish Diseases 25:515–521.

Brett, J. R. 1995. Energetics. Pages 1–68 *in* C. Groot, L. Margolis, and W. C. Clarke, editors. Physiological ecology of Pacific salmon. UBC Press, Vancouver.

Brill, R., P. Bushnell, S. Schroff, R. Seifert, and M. Galvin. 2008. Effects of anaerobic exercise accompanying catch-and-release fishing on blood-oxygen affinity of the sandbar shark (*Carcharhinus plumbeus*, Nardo). Journal of Experimental Marine Biology and Ecology 354:132–143.

Brown, R. S., S. J. Cooke, G. N. Wagner, and M. B. Eppard. 2010. Methods for surgical implantation of acoustic transmitters in juvenile salmonids: a review of literature and guidelines for techniques. U.S. Army Corps of Engineers, Portland, Oregon.

Brunnschweiler, J. M. 2009. Tracking free-ranging sharks with hand-fed intra-gastric acoustic transmitters. Marine and Freshwater Behaviour and Physiology 42:201–209.

Chapman, D. D., E. K. Pikitch, E. Babcock, and M. S. Shivji. 2005. Marine reserve design and evaluation using automated acoustic telemetry: a case study involving coral reef-associated sharks in the Mesoamerican Caribbean. Marine Technology Society Journal 39:42–55.

Claydon, J. 2004. Spawning aggregations of coral reef fishes: characteristics, hypotheses, threats, and management. Oceanography and Marine Biology: An Annual Review 42:265–302.

Coleman, F. C., C. C. Koenig, and L. A. Collins. 1996. Reproductive styles of shallow-water grouper (Pisces:Serranidae) in the eastern Gulf of Mexico and the consequences of fishing spawning aggregations. Environmental Biology of Fishes 47:129–141.

Cooke, S. J., and C. D. Suski. 2004. Are circle hooks effective tools for conserving freshwater and marine recreational catch-and-release fisheries? Aquatic Conservation: Marine and Freshwater Ecosystems 14:299–326.

Cooke, S. J., and D. P. Philipp. 2004. Behavior and mortality of caught-and-released bonefish (*Albula* spp.) in Bahamian waters with implications for a sustainable recreational fishery. Biological Conservation 118:599–607.

Cooke, S. J., S. G. Hinch, M. Wikelski, R. D. Andrews, T. G. Wolcott, and P. J. Butler. 2004. Biotelemetry: a mechanistic approach to ecology. Trends in Ecology and Evolution 19:334–343.

Cooke, S. J., and C. D. Suski. 2005. Do we need species-specific guidelines for catch-and-release recreational angling to effectively conserve diverse fishery resources? Biodiversity and Conservation 14:1195–1209.

Cooke, S. J., C. D. Suski, K. G. Ostrand, D. H. Wahl, and D. P. Philipp. 2007. Physiological and behavioral consequences of long-term artificial selection for vulnerability to recreational angling in a teleost fish. Physiological and Biochemical Zoology 80:480–490.

Dahlgren, C. 2002. Marine protected areas in The Bahamas. Bahamas Journal of Science 9:41–49.

Danylchuk, A. J. 2003. Fisheries management in South Eleuthera: can a marine reserve help save the "holy trinity?" Proceedings of the Gulf and Caribbean Fisheries Institute 56:169–178.

Danylchuk, A. J., S. E. Danylchuk, S. J. Cooke, T. L. Goldberg, J. Koppelman, and D. P. Philipp. 2007a. Post-release mortality of bonefish (*Albula* spp.) exposed to different handling practices in South Eleuthera, Bahamas. Fisheries Management and Ecology 14:149–154.

Danylchuk, S. E., A. J. Danylchuk, S. J. Cooke, T. L. Goldberg, J. Koppelman, and D. P. Philipp. 2007b. Effects of recreational angling on the post-release behaviour and predation of bonefish (*Albula vulpes*): the role of equilibrium status at the time of release. Journal of Experimental Marine Biology and Ecology 346:127–133.

Danylchuk, A. J., S. J. Cooke, T. L. Goldberg, C. D. Suski, K. J. Murchie, S. E. Danylchuk, A. Shultz, C. R. Haak, E. Brooks, A. Oronti, J. B. Koppelman, and D. P. Philipp. In Press. Aggregations and offshore movements as indicators of spawning activity of bonefish (*Albula vulpes*) in The Bahamas. Marine Biology.

Davis, J. B. 2004. Acoustic tracking of fish: how continuous data on fish movement could change the planning of MPAs. MPA News 5:1–3.

de Sylva, D. P. 1963. Systematics and life history of great barracuda *Sphyraena barracuda* (Walbaum). University of Miami Press, Coral Gables, Florida.

Dicken, M. L., A. J. Booth, and M. J. Smale. 2006. Preliminary observations of tag shedding, tag reporting, tag wounds, and tag bioufouling for raggedtooth sharks (*Carcharias taurus*) tagged off the east coast of South Africa. ICES Journal of Marine Science 63:1640–1648.

Dickey, R. W., and S. M. Plakas. 2010. Ciguatera: a public health perspective. Toxicon 56:123–136.

Domeier, M. L. 2005. Methods for the deployment and maintenance of an acoustic tag tracking array: an example from California's Channel Islands. Marine Technology Society Journal 39:74–80.

Domeier, M. L., and R. L. Colin. 1997. Tropical reef fish spawning aggregations: defined and reviewed. Bulletin of Marine Science 60:698–726.

Donaldson, M. R., R. Arlinghaus, K. C. Hanson, and S. J. Cooke. 2008. Enhancing catch-and-release science with biotelemetry. Fish and Fisheries 9:79–105.

Gaines, S. D., B. Gaylord, L. Gerber, A. Hastings, and B. Kinlan. 2007. Connecting places: The ecological consequences of dispersal in the sea. Oceanography 20:90–99.

Gordon, M. R., and J. E. Seymour. 2009. Quantifying movement of the tropical Australian cubozoan *Chironex fleckeri* using acoustic telemetry. Hydrobiologia 616:87–97.

Grothues, T. M. 2009. A review of acoustic telemetry technology and a perspective on diversification in coastal tracking arrays. Pages 77–90 *in* J. L. Nielsen, H. Arrizabalaga, N. Fragoso, A. Hobday, M. Lutcavage, and J. Sibert, editors. Tagging and Tracking of Marine Animals with Electronic Devices. Reviews: Methods and Technologies in Fish Biology and Fisheries, Volume 9. Springer, London.

Henningsen, A. D. 1994. Tonic immobility in 12 elasmobranchs: use as an aid in captive husbandry. Zoo Biology 13:325–332.

Heupel, M. R., and R. E. Heuter. 2001. Use of an automated acoustic telemetry system to passively track juvenile blacktip shark movements. Pages 217–236 *in* J. R. Sibert and J. L. Nielsen, editors. Electronic Tagging and Tracking in Marine Fisheries. Springer, New York.

Heupel, M. R., C. A. Simpfendorfer, and R. E. Hueter. 2003. Running before the storm: blacktip sharks respond to falling barometric pressure associated with Tropical Storm Gabrielle. Journal of Fish Biology 63:1357–1363.

Heupel, M. R., and C. A. Simpfendorfer. 2005. Using acoustic monitoring to evaluate MPAs for shark nursery areas: the importance of long-term data. Marine Technology Society Journal 39:10–18.

Heupel, M. R., J. M. Semmens, and A. J. Hobday. 2006. Automated acoustic tracking of aquatic animals: scales, design and deployment of listening station arrays. Marine and Freshwater Research 57:1–13.

Heupel, M. R., K. L. Reiss, B. G. Yeiser, and C. A. Simpfendorfer. 2008. Effects of biofouling on performance of moored data logging acoustic receivers. Limnology and Oceanography: Methods 6:327–335.

Humston, R. 2001. Development of movement models to assess the spatial dynamics of fish populations. Doctoral thesis. Rosentiel School of Marine and Atmospheric Science, University of Miami, Miami, Florida.

Humston, R., J. J. Ault, M. M. Larkin, and J. Luo. 2005. Movements and site fidelity of the bonefish *Albula vulpes* in the northern Florida Keys determined by acoustic telemetry. Marine Ecology-Progress Series 291:237–248.

Jepsen, N., K. Aarestrup, F. Økland, and G. Rasmussen. 1998. Survival of radio-tagged Atlantic salmon (*Salmo salar* L.) and trout (*Salmo trutta* L.) smolts passing a reservoir during seaward migration. Hydrobiologia 371/372:347–353.

Kerstetter, D. W., and J. E. Graves. 2006. Effects of circle versus J-style hooks on target and non-target species in a pelagic longline fishery. Fisheries Research 80:239–250.

Kerwath, S. E., E. B. Thorstad, T. F. Næsje, P. D. Cowley, F. Økland, C. Wilkie, and C. G. Attwood. 2009. Crossing invisible boundaries: the effectiveness of the Langebaan Lagoon marine protect area as a harvest refuge for a migratory fish species in South Africa. Conservation Biology 23:653–661.

Larkin, M. F., J. S. Ault, R. Humston, J. Luo, and N. Zurcher. 2008. Tagging of bonefish in south Florida to study population movements and stock dynamics. Pages 301–320 *in* J. Ault, G. Kelley, and R. Humston, editors. The world biology of tarpon and bonefish, CRC Press, Boca Raton, Florida.

Lehane, L., and R. J. Lewis. 2000. Ciguatera: recent advances but the risk remains. International Journal of Food Microbiology 61:91–125.

Lindholm, J., A. Knight, L. Kaufman, and S. Miller. 2006. Site fidelity and movement of the parrotfishes *Scarus coeruleus* and *Scarus taeniopterus* at Conch Reef (northern Florida Keys). Caribbean Journal of Science 42:138–144.

Martin, S. W., J. A. Long, and T. N. Pearsons. 1995. Comparison of survival, gonad development, and growth between rainbow trout with and without surgically implanted dummy radio transmitters. North American Journal of Fisheries Management 15:494–498.

Meyer, C. G., T. B. Clark, Y. P. Papastamatiou, N. M. Whitney, and K. N. Holland. 2009. Long-term movement patterns of tiger sharks *Galeocerdo cuvier* in Hawaii. Marine Ecology-Progress Series 381:223–235.

Meyer, C. G., K. N. Holland, and Y. P. Papastamatiou. 2007. Seasonal and diel movements of giant trevally *Caranx ignobilis* at remote Hawaiian atolls: implications for the design of Marine Protected Areas. Marine Ecology-Progress Series 333:13–25.

Morrissey, J. F., and S. H. Gruber. 1993a. Habitat selection by juvenile lemon sharks, *Negaprion brevirostris*. Environmental Biology of Fishes 38:311–319.

Morrissey, J. F., and S. H. Gruber. 1993b. Home range of juvenile lemon sharks, *Negaprion brevirostris*. Copeia 1993:425–434.

Murchie, K. J., S. E. Danylchuk, C. E. Pullen, E. Brooks, A. D. Shultz, C. D. Suski, A. J. Danylchuk, and S. J. Cooke. 2009. Strategies for the capture and transport of bonefish, *Albula vulpes*, from tidal creeks to a marine research laboratory for long-term holding. Aquaculture Research 40:1538–1550.

Murchie, K. J. 2010. Physiological ecology and behavior of bonefish (*Albula vulpes*) in tropical tidal flats ecosystems. Doctoral thesis. Carleton University, Ottawa.

Murchie, K. J., E. Schwager, S. J. Cooke, A. J. Danylchuk, S. E. Danylchuk, T. L. Goldberg, C. D. Suski, and D. P. Philipp. 2010. Spatial ecology of juvenile lemon sharks (*Negaprion brevirostris*) in tidal creeks and coastal waters of Eleuthera, The Bahamas. Environmental Biology of Fishes 89:95–104.

Murchie, K. J., S. J. Cooke, A. J. Danylchuk, and C. D. Suski. 2011. Estimates of field activity and metabolic rates of bonefish (*Albula vulpes*) in coastal marine habitats using acoustic tri-axial accelerometer transmitters and intermittent-flow respirometry. Journal of Experimental Marine Biology and Ecology 396:147–155.

Murchie, K. J., S. J. Cooke, A. J. Danylchuk, S. E. Danylchuk, T. L. Goldberg, C. D. Suski, and D. P. Philipp. 2011. Thermal biology of bonefish (*Albula vulpes*) in Bahamian coastal waters and tidal creeks: an integrated laboratory and field study. Journal of Thermal Biology 36:38–48.

Murray, S. N., R. F. Ambrose, J. A. Bohnsack, L. W. Botsford, M. H. Carr, G. E. Davis, P. K. Dayton, D. Gotshall, D. R. Gunderson, M. A. Hixon, J. Lubchenco, M. Mangel, A. MacCall, D. A. McArdle, J. C. Ogden, J. Roughgarden, R. M. Starr, M. J. Tegner, and M. M. Yoklavich. 1999. No-take reserve networks: sustaining fishery populations and marine ecosystems. Fisheries 24:11–25.

Myers, R. A., J. K. Baum, T. D. Shepherd, S. P. Powers, and C. H. Peterson. 2007. Cascading effects of the loss of apex predatory sharks from a coastal ocean. Science 315:1846–1850.

O'Toole, A. C., A. J. Danylchuk, C. D. Suski, and S. J. Cooke. 2010a. Consequences of catch-and-release angling on the physiological status, injury, and immediate mortality of great barracuda (*Sphyraena barracuda*) in The Bahamas. ICES Journal of Marine Science 67:1667–1675.

O'Toole, A. C., K. J. Murchie, C. Pullen, K. C. Hanson, C. D. Suski, A. J. Danylchuk, and S. J. Cooke. 2010b. Locomotory activity and depth distribution of adult great barracuda (*Sphyraena barracuda*) in Bahamian coastal habitats determined using acceleration and pressure biotelemetry transmitters. Marine and Freshwater Research 61:1446–1456.

Pfeiler, E., D. Pardon, and R. E. Crabtree. 2000. Growth rate, age and size of bonefish from the Gulf of California. Journal of Fish Biology 56:448–453.

Pincock, D. G., and F. A. Voegeli. 1992. Quick course in underwater telemetry systems. VEMCO Limited, Shad Bay, Nova Scotia.

Polunin, N. V. C. 2002. Marine protected areas, fish and fisheries. Pages 293–318 *in* P. J. B. Hart and J. C. Reynolds, editors. Handbook of fish and fisheries, Volume II. Blackwell Scientific Publications, Oxford, U.K.

Roberts, C. M., J. A. Bohnsack, F. Gell, J. P. Hawkins, and R. Goodridge. 2001. Effects of marine reserves on adjacent fisheries. Science 294:1920–1923.

Roberts, C. M., and J. P. Hawkins. 1999. Extinction risk in the sea. Trends in Ecology and Evolution 14:241–246.

Ropert-Coudert, Y., and R. P. Wilson. 2005. Trends and perspectives in animal-attached remote sensing. Frontiers in Ecology and the Environment 3:437–444.

Sadovy, Y., and M. Domeier. 2005. Are aggregation-fisheries sustainable? Reef fish fisheries as a case study. Coral Reefs 24:254–262.

Sadovy, Y. A. Rosario, and A. Roman. 1994. Reproduction in an aggregating grouper, the red hind, *Epinephelus guttatus*. Environmental Biology of Fishes 41:269–286.

Sala, E., O. Aburto-Oropeza, G. Paredes, and G. Thompson. 2003. Spawning aggregations and reproductive behavior of reef fishes in the Gulf of California. Bulletin of Marine Science 72:103–121.

Schaeffer, J. S., and E. M. Hoffman. 2002. Performance of barbed and barbless hooks in a marine recreational fishery. North American Journal of Fisheries Management 22:229–235.

Schmiing, M., P. Afonso, F. Tempera, and R. Santos. 2009. Integrating recent and future marine technology in the design of marine protected areas—the Azores as a case study. Oceans 2009-Europe, Horta, Portugal.

Simpfendorfer, C. A., M. R. Heupel, and R. E. Hueter. 2002. Estimation of short-term centers of activity from an array of omnidirectional hydrophones and its use in studying animal movements. Canadian Journal of Fisheries and Aquatic Sciences 59:23–32.

Skomal, G. B., and B. C. Chase. 2002. The physiological effects of angling on post-release survivorship in tunas, sharks and marlin. Pages 135–138 *in* J. A. Lucy and L. Studholme, editors. Catch and release in marine recreational fisheries. American Fisheries Society, Symposium 30, Bethesda, Maryland.

Skomal, G. B. 2007. Evaluating the physiological and physical consequences of capture on post-release survivorship in large pelagic fishes. Fisheries Management and Ecology 14:81–89.

Smith, R. L. 1992. Elements of ecology, 3rd Edition. Harper Collins Publishers Inc., New York.

Stark, K., G. Jackson, and J. Lyle. 2005. Tracking arrow squid movements with an automated acoustic telemetry system. Marine Ecology-Progress Series 299:167–177.

Stevens, J. D., R. Bonfil, N. K. Dulvy, and P. A. Walker. 2000. The effects of fishing on sharks, rays, and chimaeras (chondricthyans), and the implications for marine ecosystems. ICES Journal of Marine Science 57:476–494.

Summerfelt R. C., and L. S. Smith. 1990. Anesthesia, surgery and related techniques. Pages 213–272 *in* C. B. Schreck and P. B. Moyle, editors. Methods for fish biology. American Fisheries Society, Bethesda, Maryland.

Valentine, J. F., and K. L. Heck, Jr. 2005. Perspective review of the impacts of overfishing on coral reef food web linkages. Coral Reefs 24:209–213.

Wagner, G. N., and E. D. Stevens. 2000. Effects of different surgical techniques, suture material and location of incision site on the behavior of rainbow trout (*Oncorhynchus mykiss*). Marine and Freshwater Behaviour and Physiology 33:103–114.

Wagner, G. N., and S. J. Cooke. 2005. Methodological approaches and opinions of researchers involved in the surgical implantation of telemetry transmitters in fish. Journal of Aquatic Animal Health 17:160–169.

Wall, C. C., F. E. Muller-Karger, and M. A. Roffer. 2009. Linkages between environmental conditions and recreational king mackerel (*Scomberomorus cavalla*) catch off west-central Florida. Fisheries Oceanography 18:185–199.

Whaylen, L., C. V. Pattengill-Semmens, B. X Semmens, P. G. Bush, and M. R. Boardman. 2004. Observation of the Nassau grouper, Epinephelus striatus, spawning aggregation site in Little Cayman, Cayman Islands, including multi-species spawning information. Environmental Biology of Fishes 70:305–313.

Wilkie, M. P., M. A. Brobbel, K. Davidson, L. Forsyth, and B. L. Tufts. 1997. Influences of temperature upon the postexercise physiology of Atlantic salmon (*Salmo salar*). Canadian Journal of Fisheries and Aquatic Sciences 54:503–511.

Wilson, R. P., E. L. C. Shepard, and N. Liebsh. 2007. Prying into the intimate details of animal lives: use of a daily diary on animals. Endangered Species Research 4:123–137.

Section 8.3

Ontogeny of a Sonic Telemetry Program for Atlantic Salmon *Salmo salar*

FREDERICK G. WHORISKEY

INTRODUCTION

The Atlantic salmon *Salmo salar* is an iconic anadromous species of great social and cultural importance (Dunfield 1985; Beland and Bielak 2002). The potential value of the annual recreational fishery for Atlantic salmon in North America probably exceeds CAN $175 million (Whoriskey and Glebe 2002). In addition, salmon are important for subsistence in some communities and for food and ceremonial purposes for First Nations and Tribes. Atlantic salmon spawn and rear as juveniles in rivers draining to the North Atlantic Ocean in both Europe and North America. After 1–5 years in freshwater, the juveniles become "smolts" and undertake feeding migrations to sea that can last for 1–5 years before returning to home rivers to spawn. Atlantic salmon from North America can return from the ocean to home rivers for their first spawning after one (termed "grilse," or 1 Sea Winter (SW) fish) or two or more (Multi Sea Winter fish (MSW)) years. In southern portions of North America, the MSW fish are predominantly female, whereas the 1 SW fish are mostly males (O'Connell et al. 2006). Atlantic salmon do not necessarily die after spawning, and postspawn adults can migrate back to the sea and return to home rivers for repeat spawnings in subsequent years. Conservation-oriented management efforts for Atlantic salmon are targeted at ensuring that enough females survive to spawn to seed freshwater habitats with target levels of egg depositions (Chaput et al. 1998).

Atlantic salmon populations are at historic low levels, especially in North America (WWF 2001; Amiro 2003; ICES 2009). There were multiple contributors to the declines, including lost productive capacity in freshwater habitat, blocked freshwater migration routes, and overfishing. More recently, the survival rates of salmon during their ocean feeding migration have dropped precipitously and have stayed low, for unknown reasons (O'Neill et al. 2000; Cairns 2001). This recent increase in at-sea mortality rates has severely impacted Atlantic salmon across the species' range, but especially in southern North America where a number populations have been listed or are being considered for listing as endangered by national authorities (Federal Register 2009; DFO 2009).

North American Atlantic salmon that spend two or more years at sea before maturing for the first time are believed primarily to migrate to feeding areas off of Greenland (Ritter 1989; Reddin et al. 1988; Reddin and Short 1991; but see Dadswell et al. 2010). Mortality can occur at any point in the ocean, although present evidence points to periods of higher mortality in the early stages of migration (Friedland et al. 1998, 2000; Kocik et al. 2009). The younger, smaller salmon are likely more susceptible to predation and diseases as they transition from freshwater to the sea (e.g., Thorstad et al. 2007; Kocik et al. 2009). To migrate to sea the juvenile salmon abandon their territorial, benthic life style for that of a pelagic fish. This transition may make them more susceptible to predators encountered in freshwater, and to physical hazards on the downstream journey (e.g., rapids, dams and waterfalls). Once in the estuary, the animals enter salt water for the first time and could die if they have poor osmoregulatory ability. In addition, they encounter a novel predator suite for which they may not yet have developed appropriate antipredator behavior. Thus, elevated mortalities are predicted in the earliest stages of the migration.

Telemetry has become a powerful tool for documenting animal movements and survival (Priede and Swift 1992). Radio-telemetry has been successfully used with both terrestrial and freshwater animals. Unfortunately, radio signals are blocked in seawater, so marine animals are tracked using sonic telemetry, which also works in freshwater. However, the detection range of sonic signals in freshwater is typically less than that of radio signals. Sonic telemetry at this time is the best solution for researchers attempting to follow the movements between fresh and salt water of anadromous fishes.

A number of studies have documented Atlantic salmon movements in freshwater using both radio-telemetry and sonic telemetry methods (e.g., Thorstad et al. 2011). As interest grew in the marine phase of the Atlantic salmon's migration, investigators shifted to sonic telemetry. In Europe, active tracking studies were conduced. In these, a vessel with a mobile receiver remained in contact with the tagged fish as they migrated, and followed smolts through estuaries and for limited distances into surrounding coastal areas (Moore et al. 1995, 2000, Thorstad et al. 2004). The development of passive receivers that could be moored in fixed places for extended periods opened up new potential for large-scale sonic telemetry work. Judicious placement of receivers permitted quantitative estimates to be obtained of the rates of loss of salmon across different portions of the migration route. Thus, it offered the potential to localize places and times of highest salmon mortality, which are unknown at present. From this information, it may be possible to infer the mortality cause. At the very least, it could identify geographic areas of heavy mortality, permitting investigators to efficiently employ scarce resources during the search for causes. Acquiring adequate sonic telemetry data requires a research program that spans a broad geographic range, and is replicated over as many years as possible to pick up natural environmental variation. This variability can then be related to documented changes in migration and survival patterns.

Starting in the early 1990s, the Atlantic Salmon Federation (ASF), a NonGovernmental Organization (NGO) based in Canada, began a sustained program of sonic telemetry work with Atlantic salmon. Early studies were focused on limited questions, and helped foster the development of more affordable and robust technologies (e.g., Carr et al. 1997; Lacroix and McCurdy 1996). Subsequently, the ASF collaborated with Canada's Department of Fisheries and Oceans to establish the first large-scale coastal array of passive sonic receivers, which was used to track endangered inner Bay of Fundy region salmon (Lacroix et al. 2004a; Lacroix and Knox 2005; Lacroix 2008). Besides providing valuable information on this unique lineage of salmon, this study validated the concept of large scale coastal receiver arrays and refined the logistics for deployment. The work also helped stimulate the development of other major telemetry initiatives (Pacific Ocean Shelf Tracking Program; Ocean Tracking Network).

The early successes led to a more ambitious ASF tracking effort that began in 2003 and focused on Atlantic salmon watersheds draining to the Gulf of St. Lawrence. The scientific objectives of the ASF program were to:

- Document ocean migration routes of smolts from different rivers.
- Establish quantitative, replicated estimates of smolt/postsmolt mortality rates, by migration stage (freshwater, estuary, across the Gulf of St. Lawrence).
- Compare the characteristics and behavior of survivors to fish believed to have died during the monitored migration period, for inferences about potential mortality causes.
- Explore evidence for social dynamics (synchronized movements of fish from different rivers) as determinants of salmon survival.
- Sustain the effort over an extended period to document changes in survival and migration patterns that resulted from naturally occurring environmental variations ("natural experiments").

This case study describes the development of the ASF sonic telemetry research program.

METHODS

Study sites

The work focused on four rivers draining into the Gulf of St. Lawrence (Figure 1). A number of practical and logistical considerations played into this choice. These included:

- The four rivers (Saint John (North Shore Québec), Cascapedia, Restigouche, and Miramichi Rivers) together may annually account for about 20% of all the MSW Atlantic salmon returns to North America (ICES 2009).
- All have major recreational fisheries; hence the work had local interest.
- There were stock assessment programs of varying scales on each of these rivers. The assessments were undertaken by government agencies, local watershed groups, First Nations, and/or the ASF and provided information on smolt outputs for these rivers, and in some cases data on adult returns. The rotary screw fish traps used in the smolt evaluations provided animals in excellent shape for sonic tagging (e.g., Music et al. 2010). In anticipation of the smolt tracking work, the ASF developed a program that provided some of the smolt wheels to these monitoring programs. This built collaborations with science teams on the rivers in advance of the telemetry program.
- The most northern and southern of the study rivers were separated by a distance of approximately 600 km. It was hoped that this would be sufficient to pick up systematic differences in migration patterns or survival due to global warming trends.

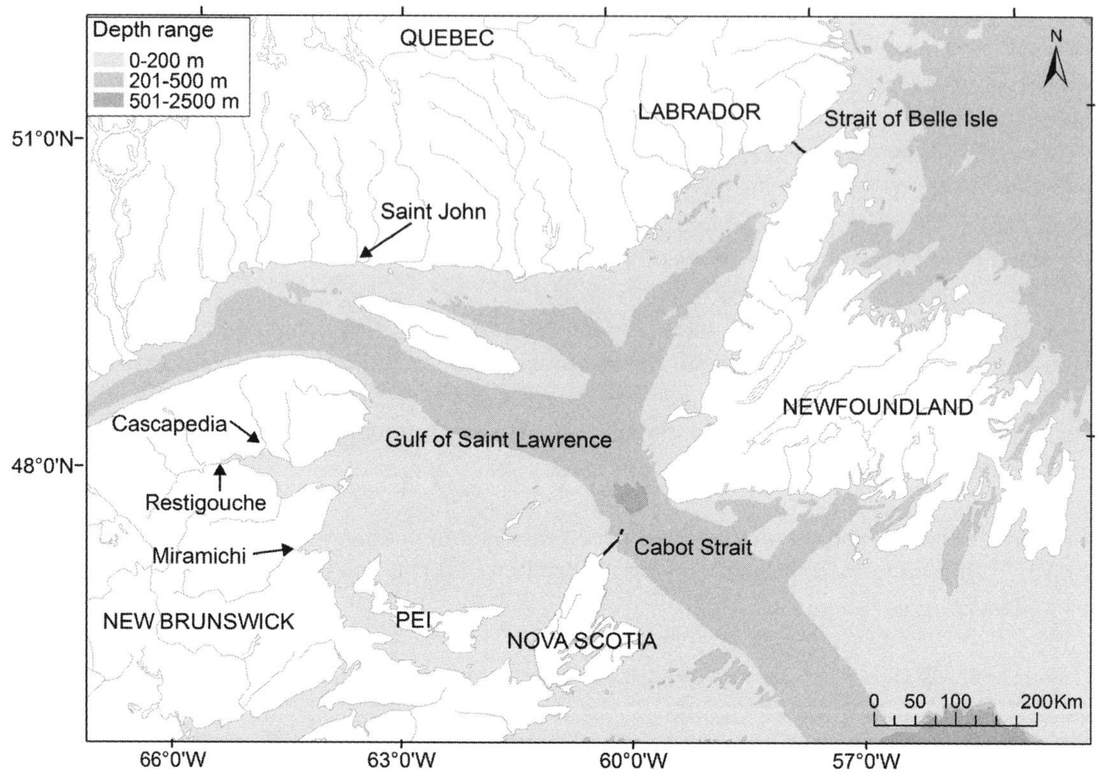

Figure 1. Map of the study area, showing the locations of the four rivers where salmon smolts were tagged, and the acoustic receiver arrays in the Strait of Belle Isle and the Cabot Strait.

- Some data are available on the movements of salmon from these rivers from historic external marking programs (Ritter 1989; Caron 1983).

- Measures of the genetic relatedness of the fish among some of these rivers are available (Dionne et al. 2008), providing an opportunity to derive inferences about the role of genetics in determining migration route.

- Logistically, it was relatively easy and cost-effective to work within the region. All sites were easily accessible by boat or road. Also, because smolt migrations are cued by water temperatures (reviewed in Thorstad et al. 2011), runs began and finished at some southern sites before they commenced in the north. This made it possible to move some of the telemetry equipment from the south to the north in the same year, keeping capital costs down.

- Significant portions of the recreational anglers on these rivers sponsored the program by contributing to the purchase and installation of sonic tags in individual fish. In return for this, they received a tracking history of their sponsored fish, and regular updates on progress of the program as a whole. The cost of up to 50% of the tags deployed each year came from this annual subscription campaign. Its existence facilitated approaches to Foundations for matching funds and provided direct communication to user groups.

Fish tagging

Smolts to be tagged were captured in freshwater in rotary screw fish traps operated on the target rivers. Smolts from the Miramichi River were obtained from the two major branches of the river (in the main stem of the Little Southwest Branch of the Northwest Miramichi River near Catamaran Brook, and Rocky Brook of the Southwest Branch). Restigouche River smolts were captured in the Kedgwick tributary. For the other two rivers, the smolts used were captured at mid to lower main stem sites.

Smolts >13.1 cm fork length (FL) were chosen as they were able to accommodate the tags (Lacroix et al. 2004b; Jepsen et al. 2002). VEMCO (Halifax, Nova Scotia) V9–6 L sonic pingers were used (21 mm length, 1.6 g weight in water, approximately 74 d battery life, 20–60 s randomized transmission rate, with 40 s nominal rate). Smolts were anesthetized (clove oil, 40 mg/l), and the tags surgically implanted in the body cavity by making a small ventral surface incision. The incision was closed by two, occasionally three, Ethilon sutures, and the fish were allowed to recover for a few hours in a holding pen placed in the river before being released.

Receiver placements

The program used VEMCO VR2 and/or VR2W equipment. The choice of equipment was dictated by previous experience with the company, the reliability, ease of operation, relatively low cost of the products, and the size of the receivers that made them easy to transport and position using various deployment strategies.

For each river system, as a minimum an array of was positioned at: 1) the point where salt water ended and the estuary began (the Head of Tide [HoT]), 2) at the estuary exit, and 3) at points in the Gulf of St. Lawrence. This permitted estimating mortality rates for the smolts during the freshwater, estuary, and initial ocean migration. Targeted spacing distances among receivers were 800 m, which experience had shown generally provided overlapping coverage and ensured high detection probability for tagged fish that passed by the array. For the rivers in which we were working, it took a maximum of three receivers to provide full coverage of the channel at the Head of Tide. Depths at the Head of Tide at all sites were less than 10 m. Numbers of receivers required to cover the estuary exits ranged from one (Saint John (North Shore Québec) River) to 27 (Restigouche River estuary exit in the Baie des Chaleurs). Depths of deployment sites in the estuary exits ranged from about 5 m (Saint John (North Shore Québec) and Miramichi Rivers) to about 40 m (Restigouche River).

Methods used to deploy receivers depended on the site. In river and Head of Tide areas of heavy recreational fishing, in advance of the fishing season a diver would take the receiver fixed inside a concrete block to the bottom, and cover the block with rocks leaving just the nub of the hydrophone exposed. This approach was also used in the estuary exit of the Saint John (North Shore Québec) river where an extensive fishery for brook trout *Salvelinus fontinalis* was pursued at the time of smolt migration. Such receivers would be retrieved by a diver after all tagged fish had passed by, at a time which did not conflict with fishing activity. At Head of Tide and estuary exit areas, receivers were suspended at a depth of about 3 m on a rope between a hazard-yellow surface float and an anchor on the bottom. To avoid currents twisting and severing the ropes, swivels were woven into the rope, and the rope was shackled to 1.5 m of chain fixed to the anchor. The surface floats made it easy to verify that the receivers were on station and served as a focal point of discussion with local people to explain the project and its objectives.

On the Restigouche and Cascapedia Rivers, the Head of Tide receiver placements fell within traditional fishing areas of the Listugujj First Nation and the Gesgepegiac First Nation, respectively. We worked with agents of the Listugujj Rangers and the Gesgepegiac First Nation Fisheries Department to deploy the units within their fishing areas. These groups in turn provided surveillance for the receivers, and explained to interested members of the First Nations the purpose of the equipment. On the Miramichi River, ASF field teams deployed the equipment in association with the well organized and influential Miramichi Salmon Association (MSA). The MSA conducted an annual public relations campaign on the science work underway in the watershed including the sonic telemetry program.

In the Gulf of St. Lawrence in 2007, after a pilot effort in 2006, an array of 22 receivers was placed to completely cover the northern exit of the Gulf (the Strait of Belle Isle, Figure 1) between Green Island Cove Newfoundland, and L'Anse a Loup, Labrador. This has been deployed from mid-June to October of each year since 2007. Pack-ice and icebergs are common in this Strait in winter and spring, and tidal currents can exceed 7.4 km/h (4 knots). Initial deployments occurred each year when suitable ice conditions prevailed; typically this line was installed around 18 June. Local fishermen were engaged to assist with the deployments. They provided vessels, mooring suggestions, and helped explain the project in local communities. Receivers were attached at a depth of about 7 m on a rope suspended from twin hazard-yellow surface float to a 20 or 40 kg commercial trawl anchor on the bottom. A 1.5 m length of chain from the anchor was attached via a shackle to the line leading to the surface buoys. Both anchor types performed successfully. Twin floats were provided to ensure that there would be a marker in case one of the buoys broke off, and to ensure that the buoys would surface in slack water. The strong currents in the Strait would submerge the buoys during peak tidal flows. Commercial fishermen in the area recommended that each mooring use a rope length that was twice that of the depth at the deployment site. The upper portion attached to the surface buoy was a sinking line, whereas the lower portion attached to the anchor was a floating line. The two lines were attached to each other with a swivel.

Time of data retrieval from the lines varied over the years. In the initial years of the program, the receivers in the rivers and the estuaries were retrieved by divers or by hauling the line and anchor system, and their data downloaded within about six weeks after their initial deployment. However, over time, additional concurrent Atlantic salmon studies using sonic telemetry were initiated in some of the rivers to study other life stages of Atlantic salmon (returning adults migrating upriver; spent adult salmon ("kelts") migrating back to sea in the spring of the year after their spawning period), and we began to work collaboratively with teams studying other fishes in some of our study sites (e.g., the Department of Fisheries and Oceans working with striped bass *Morone saxatilis* on the Miramichi River; the Ocean Tracking Network with results from the Strait of Belle Isle line covering multiple species released by various groups). As this occurred, receiver arrays remained in the water for longer than needed to document smolt/postsmolt movements. To minimize the risk of lost data in the case of missing receivers, these arrays were typically downloaded about six weeks after their initial deployment, returned to the water, and downloaded again when they were finally retrieved usually in late autumn.

In 2009, the Ocean Tracking Network (http://oceantrackingnetwork.org/) at Dalhousie University began install an array of sonic receivers in the Cabot Strait, the southern exit of the Gulf of St. Lawrence. A partial array covering the area from Cape North on Cape Breton, Nova Scotia, out to and slightly beyond St. Paul Island (Figure 1) was deployed in 2009. It was comprised of 37 stations, of which 31 are equipped with Benthos 875t acoustic receivers, VR2W receivers and trawl floats on a mooring designed originally by Canada's Department of Fisheries and Oceans—Bedford Institute of Oceanography. Five stations are equipped with ORE Offshore Port MFE acoustic releases, VR2W receivers, and Romor CROM (syntactic foam collar) floats. A final station was equipped with an ORE Offshore Port MFE release, a VR4W receiver and a Romor CROM float. The VR4W receiver is equipped with an acoustic modem that is designed to permit a remote upload of data without having to retrieve the receiver. The equipment was deliberately mixed to test the performance of different combinations under the conditions found here. This line was operational during the smolt migration period in 2010. Eventually, the OTN plans to place a curtain of VEMCO VR4 acoustic receivers across the entire distance of the Cabot Strait. However, the great depths (>1000 m) in parts of this Strait will probably require that the receiver units be placed on risers of significant lengths; balancing the length of these risers with local currents will require much experimental work. In addition, density stratification occurs in the water column during certain portions of the year here, and has the potential to block tag signals from moving from one water mass to another. It may also interfere with communications via acoustic modems with the receivers.

RESULTS AND DISCUSSION

Work began on the Miramichi River in 2003, with an extension of activity to the Restigouche River in 2004, and the Cascapedia and Saint John (North Shore Québec) Rivers in 2005 (Table 1). Gulf of St. Lawrence arrays first became fully operational in 2007. The first year on the Miramichi River was used to develop methods for deploying the telemetry equipment in a large river system and stimulate interest within the membership and supporters of Atlantic Salmon Federation. The other rivers were added as our experience and resources permitted, and as the organization's confidence grew that the program could be successful and afforded.

The phasing in over time of new rivers and of additional receiver arrays at increasing distances from the tagging sites was also planned to stimulate long-term support for the work. Each year brought new results about Atlantic salmon migration, maintaining donor interest. This interest was the key to generating an extended (eight years and continuing) data series on movements and survival of fish across portions of the migration route. This time series will permit the correlation of natural but unpredictable environmental variations with smolt movements and survival. Had all elements of the tracking program been implemented at once, it is doubtful that a nongovernmental organization such as the ASF whose annual budgets vary according to contributions and consequently focuses on short-term activity, could have sustained the effort.

By 2006 when all rivers systems were fully on line, more than 200 salmon smolts were tagged each year (2006–2010 average of 241 per year). A total of 1,644 have been tagged from the program's inception to 2010 (Table 1). Our target was to tag between 80 and 100 per year in the larger Restigouche and Miramichi Rivers systems, and 40–50 per year in the smaller Cascapedia and Saint John (North Shore Québec) Rivers. However, low fish availability in some rivers in some years reduced the numbers we could tag, and the tags originally targeted for these sites were distributed to other rivers.

Table 1. A. Numbers of Atlantic salmon smolts tagged in the study rivers, by year. B. shortest distances (km) from the tagging sites in each river to the Head of Tide (HoT), estuary exit, and to the Strait of Belle Isle (SoBI). Two tagging sites were used on the Miramichi River (Rocky Brook, main stem of Little Southwest branch near Catamaran Brook).

A.

River	2003	2004	2005	2006	2007	2008	2009	2010
Miramichi	100	100	50	52	80	80	80	80
Restigouche	0	100	73	70	100	46	80	80
Cascapedia	0	0	8	81	47	40	54	40
Saint John	0	0	8	30	25	47	44	49
All rivers	100	200	139	233	252	213	258	249

B.

River	Tag site to HoT	HoT to estuary exit	Estuary exit to SoBI
Miramichi			
Rocky Brook	128	56	800
Catamaran Brook	36	63	800
Restigouche	115	125	800
Cascapedia	8	6	800
Saint John	20	3	665

Relatively consistent patterns of detection of tagged smolts were evident among years for each of the river systems. For example, for the Miramichi River system, about 84% of the smolts annually tagged successfully passed the head of tide receiver array after traversing freshwater (range 63–91%; Figure 2). Similarly, in this system an average of 55% (range 43–69%) and 23% (range 15–29%) respectively, of the fish originally released made it to the estuary exit and Strait of Belle Isle array in the Gulf of St. Lawrence. Similar data have been obtained from the other river systems.

One intriguing finding was that the Strait of Belle Isle is an important migration route for Atlantic salmon en route to feeding grounds off of Greenland. Significant fractions of sonically tagged fish from the Miramichi (Figure 2), Restigouche, and Cascapedia Rivers have been detected crossing this array. By contrast, to date only a single fish from the Saint John (North Shore Québec) River has been detected exiting the Strait of Belle Isle, and no tagged smolts from these four rivers were detected in 2010 by the Cabot Strait Ocean Tracking Network partial array.

The detection data provides minimum estimates of survival. To conserve battery life, the sonic tags transmit on an intermittent basis. Depending on fish rates of travel, a tag may or may not signal when a fish is in range of a receiver array. In this work there were a number of instances where fish were missed on one array then subsequently detected in the next array along the migration route, which clearly shows that the fish survived to cross the first array. It is also possible that other individuals that successfully crossed a first array without being detected died before reaching the next array. Thus the data on detections are generally an un-

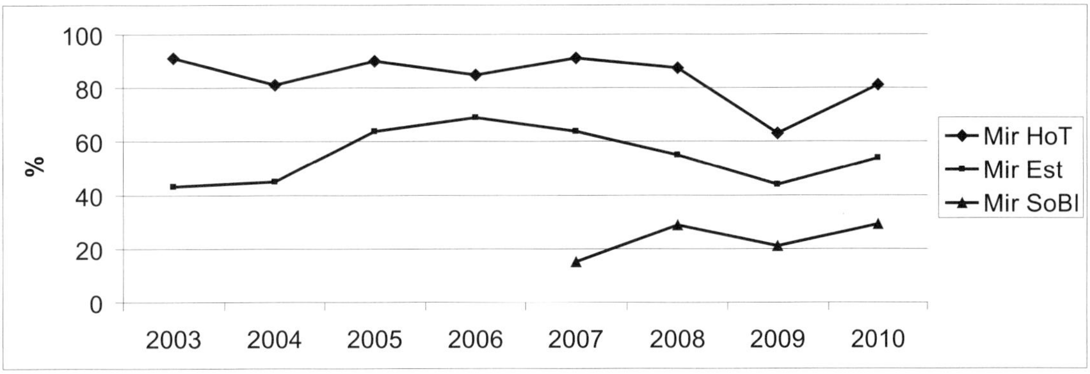

FIGURE 2. Percentages of Miramichi (Mir) River smolts tagged and released in fresh water that were detected at the head of tide (HoT), the estuary exit (Est), and at the Strait of Belle Isle (SoBI) between 2003 and 2010.

derestimate of the numbers of fish surviving to a particular migration route point. There is also a possibility that changes in environmental conditions (e.g., fast current flows, storms) could vary the distances at which a receiver detects a signaling tag. This also could result in the raw detection data providing underestimates in the numbers of fish surviving to a given migration point. At present, only a quarter of the Cabot Strait is fitted with receiver equipment, and no tagged smolts were detected in the single year of operation. The successful completion of the array and more years of work will be required to assess its use as an Atlantic salmon migration pathway and to refine survival estimates accordingly. Finally, we cannot know at the time we tag the smolts what their life history patterns will be. Some of the tagged fish will probably mature after a single year at sea. The ocean feeding areas of the 1 SW fish are believed to be close to their river of origin. Thus a failure to detect such individuals on the arrays at the exit of the Gulf of St. Lawrence may not indicate that these fish had died, but rather never left the Gulf.

The challenge now is to translate the detection information into estimates of survival rates. We are presently working with mark–recapture models to establish estimates and confidence limits for survivals of smolt from all rivers to particular points in their migration. We are also beginning to routinely employ "sentinel tags." These are tags that fire at predictable fixed intervals and which are moored at constant distances from selected receivers within an array. They provide information about changing detection efficiencies of receivers for given arrays. A comprehensive analysis of this information may assist in developing more robust estimates of mortality rates along the migration routes.

Results of this work will generate survival estimates that can inform international efforts to model salmon abundance at sea and to successfully manage salmon from particular rivers that are harvested in multiple interceptory fisheries along their migration routes (e.g., ICES 2009). The telemetry work is also documenting the ocean habitats used by Atlantic salmon, which will inform recovery plans for endangered populations.

The receiver arrays that have been deployed so far cover approximately half of the distance between the study rivers and Greenland feeding areas. Additional telemetry lines will be required to generate survival estimates to Greenland, and strategies may need to be developed to extend the battery life of tags (e.g., custom programming on-off times of tags so that they are active when approaching telemetry arrays, but off to conserve battery life otherwise). At present, for smolts using the Strait of Belle Isle route with no on-off tag programming, the tags will have about a month of life left after they cross this Strait's array.

Given the long distances separating the present receiver arrays, the knowledge gained of the movements for the salmon smolts is necessarily on a coarse scale. Deploying large numbers of receiver arrays to document movements on a finer scale may not be practicable or affordable. Innovative receiver deployments may permit finer-scale descriptions of these movements. Deploying receivers on autonomous vehicles could provide episodic coverage in areas that do not have arrays of receivers. Similarly, large marine animals such as seals or sharks that occasionally surface could be fitted satellite positioning systems and mobile receivers with satellite linkages. They could provide information on tagged animals that they have been near during the course of their movements. Research into this possibility is presently underway (S. Iverson, Dalhousie University, personal communication).

ACKNOWLEDGMENTS

Many individuals have assisted with this research program. They were associated with a number of organizations, including the Cascapedia Society, Club Hill Camp, Inc., the Department of Fisheries and Oceans, the Gesgepegiac First Nation, the Listuguuj First Nation, the Ministère des Ressources naturelles et de la faune, Québec, the Miramichi Salmon Association, and the Restigouche River Watershed Management Committee. The ASF field teams included, M. Best, P. Brooking, J. Carr, G. Doucette, and M. Lefevre, plus variable numbers of summer students. I thank them all for their assistance. I also thank Joan Trial, an anonymous reviewer, and the editors for comments on the manuscript, and the editors for their assistance and patience.

REFERENCES

Amiro, P. G. 2003. Population status of inner Bay of Fundy Atlantic salmon (*Salmo salar*), to 1999. Canadian Technical Report of Fisheries and Aquatic Sciences No. 2488.

Beland K. F., A. T. Bielak. 2002. Atlantic salmon fisheries in eastern North America: The prince and the pauper. Pages 61–76 *in* K. D. Lynch, M. L. Jones and W. W. Taylor, editors. Sustaining North American salmon: perspectives across regions and disciplines. American Fisheries Society, Bethesda, Maryland.

Caron, F. 1983. Migration vers l'Atlantique des post-saumoneaux (*Salmo salar*) du Golfe du Saint-Laurent. Le Naturaliste canadien 110:223–227.

Cairns, D.K. 2001. An evaluation of possible causes of the decline in pre-fishery abundance of North American Atlantic salmon. Canadian Technical Report of Fisheries and Aquatic Sciences No. 2358.

Carr, J. W., G. L. Lacroix, J. M. Anderson, and T. Dilworth. 1997. Movements of non-maturing cultured Atlantic salmon (*Salmo salar*) in a Canadian river. ICES Journal of Marine Science 54:1082–1085.

Chaput, G., J. Allard, F. Caron, J. B. Dempson, C.C. Mullins, and M. F. O'Connell. 1998. River-specific target spawning requirements for Atlantic salmon (*Salmo salar*) based on a generalized smolt production model. Canadian Journal of Fisheries and Aquatic Sciences 55:246–261.

Dadswell, M. J., A. D. Spares, J. M. Reader, and M. J. W. Stokesbury. 2010. The North Atlantic subpolar gyre and the marine migration of Atlantic salmon *Salmo salar*: the 'Merry-Go-Round' hypothesis. Journal of Fish Biology 77:435–467.

DFO (Department of Fisheries and Oceans) 2009. List of 97 aquatic species currently protected under SARA. Available: http://www.dfo-mpo.gc.ca/species-especes/species-especes/salmon-atl-saumon-eng.htm (May 2011).

Dionne, M., F. Caron, J. J. Dodson, and L. Bernatchez. 2008. Landscape genetics and hierarchical genetic structure in Atlantic salmon: the interaction of gene flow and local adaptation. Molecular Ecology 17:2382–2396.

Dunfield, R. W. 1985. The Atlantic salmon in the history of North America. Canadian Special Publication of Fisheries and Aquatic Sciences No. 80.

Federal Register. 2009. Endangered and Threatened Species; Determination of Endangered Status for the Gulf of Maine Distinct Population Segment of Atlantic Salmon. Federal Register Volume 74 (No. 117, Friday, June 19, 2009): 29344–29387. http://www.nero.noaa.gov/nero/regs/frdoc/09/09SalmonES.pd.

Friedland, K., L. P. Hansen, and D. A. Dunkley. 1998. Marine temperatures experienced by postsmolts and the survival of Atlantic salmon, *Salmo salar*, in the North Sea area. Fisheries Oceanography 7:22–24.

Friedland, K. D., L. P. Hansen, D. A. Dunkley, and J. C. MacLean. 2000. Linkage between ocean climate, post-smolt growth, and survival of Atlantic salmon (*Salmo salar* L.) in the North Sea area. ICES Journal of Marine Science 57:419–429.

ICES (International Council for the Exploration of the Sea) 2009. Report of the Working Group on North Atlantic Salmon (WGNAS). ICES CM 2009/ACOM:06 REF. DFC.

Jepsen, N., A. Koed, E. B. Thorstad, and E. Baras. 2002. Surgical implantation of telemetry transmitters in fish: how much have we learned? Hydrobiologia 483:239–248.

Kocik, J., J. P. Hawkes, T. F. Sheehan, P. A. Music and K. F. Beland. 2009. Assessing estuarine and coastal migration and survival of wild Atlantic salmon smolts from the Narraguagus River, Maine using ultrasonic telemetry. Pages 293–310 *in* A. J. Haro, K. L. Smith, R. A. Rulifson, C. M. Moffitt, R. J. Klauda, M. J. Dadswell, R. A. Cunjak, J. E. Cooper, K. L. Beal and T. S. Avery, editors. Challenges for diadromous fishes in a dynamic global environment. American Fisheries Society, Symposium 69, Bethesda, Maryland.

Lacroix, G. L. 2008. Influence of origin on migration and survival of Atlantic salmon (*Salmo salar*) in the Bay of Fundy, Canada. Canadian Journal of Fisheries and Aquatic Sciences 65:2063–2079.

Lacroix, G., D. and P. F. McCurdy. 1996. Migratory behaviour of post-smolt Atlantic salmon during initial stages of seaward migration. Journal of Fish Biology 49: 1086–1101.

Lacroix, G., P. F. McCurdy, and D. Knox. 2004a. Migration of Atlantic salmon postsmolts in relation to habitat use in a coastal system. Transactions of the American Fisheries Society 133:1455–1471.

Lacroix, G., D. Knox, and P. F. McCurdy. 2004b. Effects of implanted dummy acoustic transmitters on juvenile Atlantic salmon. Transactions of the American Fisheries Society 133:211–220.

Lacroix, G. L., and D. Knox. 2005. Distribution of Atlantic salmon (*Salmo salar*) postsmolts of different origins in the Bay of Fundy and Gulf of Maine and evaluation of factors affecting migration, growth and survival. Canadian Journal of Fisheries and Aquatic Sciences 65:2063–2079.

Moore, A., E. C. E. Potter, H. J. Milner, and S. Bamber. 1995. The migratory behaviour of wild Atlantic salmon smolts in the estuary of the River Conwy, North Wales. Canadian Journal of Fisheries and Aquatic Sciences 52:1923–1935.

Moore, A., G. L. Lacroix and J. Sturlaugsson. 2000. Tracking Atlantic salmon post-smolts in the sea. Pages 49–64 *in* D. H. Mills, editor. The ocean life of Atlantic salmon: environmental and biological factors influencing survival. Fishing News Books, Oxford.

Music, P. A., J. P. Hawkes, and M. S. Cooperman. 2010. Magnitude and causes of smolt mortality in rotary screw fish traps: an Atlantic salmon case study. North American Journal of Fisheries Management 30:713–722.

O'Connell, M. F., J. B. Dempson, and G. Chaput. 2006. Aspects of the life history, biology, and population dynamics of Atlantic salmon (*Salmo salar* L.) in eastern Canada. Canadian Science Advisory Secretariat Research Document 2006/014.

O'Neill, S., J. Ritter and K. Robichaud-LeBlanc. 2000. Proceedings of a workshop on research strategies into the causes of declining Atlantic salmon returns to North American Rivers. Canadian Stock Assessment Secretariat Proceedings Series 2000/18.

Priede, I. G., and S. M. Swift, editors. 1992. Wildlife telemetry: Remote monitoring and tracking of animals. Ellis Horwood, New York.

Reddin, D. J., D. E. Stansbury, and P. B. Short. 1988. Continent of origin of Atlantic salmon (*Salmo salar* L.) caught at west Greenland. Journal du conseil International pour l'Exploration de la Mer 44:180–188.

Reddin, D. G., and P. B. Short. 1991. Postsmolt Atlantic salmon (*Salmo salar*) in the Labrador Sea. Canadian Journal of Fisheries and Aquatic Sciences 48:2–6.

Ritter, J. M. 1989. Marine migration and natural mortality of North American Atlantic salmon (*Salmo salar* L.). Canadian Manuscript Report of Fisheries and Aquatic Sciences No. 2041.

Thorstad, E. B., F. Økland, B. Finstad, R. Sivertsga, P. A. Bjørn, and R. S. McKinley. 2004. Migration speeds and orientation of Atlantic salmon and sea trout post-smolts in a Norwegian fjord system. Environmental Biology of Fishes 71:305–311.

Thorstad, E. B., F. Økland, B. Finstad, R. Sivertsgård, N. Plantalech, P. A. Bjørn, and R. S. McKinley. 2007. Fjord migration and survival of wild and hatchery-reared Atlantic salmon and wild brown trout post-smolts. Hydrobiologia 582:99–107.

Thorstad, E. B., F. Whoriskey, A. Rikardsen, and K. Aarestrup. 2011. Aquatic nomads: the life and migrations of the Atlantic salmon. Pages 1–32 *in* Ø. Aas, S. Einum, A. Klemetsen and J. Skurdal, editors. Atlantic salmon ecology. Wiley-Blackwell, Oxford, U.K.

Whoriskey, F. G., and J. Glebe. 2002. The Atlantic salmon recreational angling industry:economic benefits. Pages 77–91 *in* K. D. Lynch, M. L. Jones, and W. W. Taylor, editors. Sustaining North American salmon: perspectives across regions and disciplines. American Fisheries Society, Bethesda, Maryland.

WWF (World Wildlife Fund). 2001. The status of wild Atlantic salmon: a river by river assessment. World Wildlife Fund, Washington, D.C.

Data Management and Analysis

Section 9.1

Time-to-Event Analysis as a Framework for Quantifying Fish Passage Performance

Theodore Castro-Santos and Russell W. Perry

INTRODUCTION

The ability to monitor movements of animals using radio- and acoustic-telemetry has made it possible to observe continuous behavioral processes that determine key life history events. In this way telemetry offers similar improvements over mark-and-recapture methods as cinematography does over still photography. Imagine trying to reconstruct an entire movie from a series of 10 frames, or even 100; and compare that with watching a feature-length film. Clearly, the film provides a far more complete representation of the script. In the same way, collecting continuous records of free-ranging animals presents researchers with an opportunity to improve understanding of the causes and contexts of the movements of tagged individuals.

Researchers are also presented with a challenge: how to quantify continuous observations in ways that have biological relevance? Specifically, there is a need for a theoretical framework from which models and hypotheses can be developed that identify causal mechanisms underlying key behaviors and life history events. The purpose of this Section and the next (Section 9.2) is to demonstrate how telemetry studies can be designed and their data analyzed to optimize understanding of those causal mechanisms by taking full advantage of the temporal and spatial information that telemetry provides.

One of the most important behavioral applications of telemetry has been to study the passage of diadromous fishes past hydroelectric dams. In the United States, government agencies and utilities spend >$80,000,000 annually building and evaluating structures designed to improve survival and expedite passage of migratory fish. Various forms of acoustic-, and radio-telemetry (including passive integrated transponder (PIT) telemetry), are the primary tools used to evaluate the performance of these structures, and the combination of pressing concerns with significant funding has produced important advances in telemetry technology. Because of the importance of fish passage questions to telemetry technology, the methods described in this paper will be cast in this context. Readers should be aware, however, that the same concepts and methods have a much broader relevance, and analogous situations exist throughout the behavioral sciences (Allison 1995; Anderson et al. 2005; Williams et al. 2001).

COMMON METHODS USED TO QUANTIFY FISH PASSAGE

One problem that has long vexed fishway engineers is that there are no broadly accepted standards for what constitutes effective fish passage. Success is sometimes defined by evidence of reproduction by species that were previously denied access to spawning habitat. For populations that are limited by available habitat, this fact alone might constitute a dramatic improvement and hence be deemed sufficient. A slightly more rigorous method is to actually count the number of individuals passing a structure. This can be done with human or automatic counters, and provides an index that managers can use to identify trends.

Both of the above methods have some serious limitations, however, and neither is very effective for quantifying passage performance of downstream migrants. In this case, the focus of managers and engineers is to divert fish away from dangerous passage routes like turbines, and into safer routes. This is accomplished by reducing passage rates through the dangerous route (e.g., using racks or screens) and/or increasing passage rates through the desired route (e.g., through placement and design of fish passage structures). Here 'rate' refers to the proportion of the population passing per unit time. For each individual, this equates to the instantaneous probability of passage, which varies over time as the fish is exposed to different parts of the forebay and to different environmental or operational conditions. Thus, the probability of an individual passing a given route is determined by the duration of its exposure to that route relative to other available routes and to passage rates through each route.

Although the importance of passage route selection has been recognized since the 19th century, rigorous studies of the problem were nonexistent until the latter half of the 20th century. Importantly, these initial efforts to assess route selection and survival predate effective telemetry technology. Mark–recapture techniques, however, were already well-developed and their use was widespread in the fisheries profession. The availability of these methods, coupled with a prevalent perception that the greatest danger associated with dam passage was turbine-induced mortality, led to the development and broad application of multinomial models to quantify passage and survival (Burnham et al. 1987; Skalski et al. 2002). Thus passage performance came to be defined by the proportion of fish using a given route, along with the associated survival, and the varying rates of passage through the available routes were largely ignored.

An unintended consequence of focusing on passage proportions instead of rates is that the continuous time series collected during telemetry studies has not been used to its fullest extent. Moreover, multinomial models do not provide an efficient framework for understanding mechanisms driving passage route selection. Fish may be exposed to multiple conditions before passage, and focusing on overall proportion passing a given route can preclude explicit exploration of the roles of fluctuating environmental variables (diel period, generation, or deliberate experimental manipulations of flow fields) on route selection and other aspects of passage performance. This is particularly troubling because it is at this spatial and temporal scale at which behavior and hence passage performance are determined. This means that quantifying route selection rather than time-based passage rates can be misleading, making it difficult for managers and engineers to develop effective solutions.

Another problem with simple enumeration of passage proportions is that this can make it difficult to account for covariates that change over time, such as the spill condition, generation, temperature, etc. These factors can complicate analyses—how should we account for the fact that the different treatment conditions occur at different and variable times as the fish attempt to pass the obstacle? Do fish pass at the same rate during their first hour of exposure as compared with their second hour? If we can't control arrival times, how can we differentiate between treatment effect and changing passage rates? Because passage at obstacles is a time-based process, almost any study of fish passage is likely to encounter time-varying covariates, and analytical methods that do not explicitly include these processes will always contain systematic error. The event time methods we describe below do account for these processes, which is a key strength to this approach.

In this Section and the next, we describe the use of two complementary analytical frameworks for measuring passage at a hydroelectric facility. Time-to-event analysis (commonly called 'survival analysis') quantifies covariate effects (including time-varying covariates) on passage rates and route selection (this Section), and mark–recapture methods (Section 9.2) quantify the effects of route selection and migratory delay on survival. Taken together, these two metrics provide the information needed to optimize passage performance and survival. Telemetry data on individual movements provide the requisite level of detail needed to apply both analytical frameworks to the same data set.

A SIMULATED FISH PASSAGE STUDY

Study design

To illustrate both the rationale for and value of time-based performance evaluations we developed an individual-based model to simulate downstream migrating salmonids at a hydropower dam (Figure 1). In this scenario, the dam has only two passage routes: turbines (expected to confer greater mortality risk) and spillway (presumably a safer and therefore preferred route). In our hypothetical scenario, managers suspected that increased spill might reduce the proportion of fish passing through turbines, and thereby simultaneously improve survival and reduce migratory delays. To assess this, an experiment was designed to evaluate passage rates through the two routes under high- and low-spill conditions.

The simulated experiment was designed such that spill condition alternated between 'Low' and 'High' with a new treatment applied every 48 h. To control for seasonal effects, the study period was stratified into three 4-d blocks with each block consisting of a 'High' and a 'Low' treatment. The resulting pattern of treatments was 'Low-High, High-Low, Low-High', with the final condition persisting until all fish had passed or died. Simulated smolts were marked with transmitters and released in lots of 50 fish at the beginning of each treatment and at a nominal distance upstream (Figure 2).

Fish Movements

In this simulation we wanted the fish to approach the forebay of the dam volitionally, presumably following similar paths to those selected by free-ranging salmon smolts—this is a common feature of telemetry studies (Zabel 2002; Zabel and Anderson 1997). Importantly, however, this approach means that arrival times varied: we assigned arrival times based on a lognormal distribution of travel time from the upstream release location.

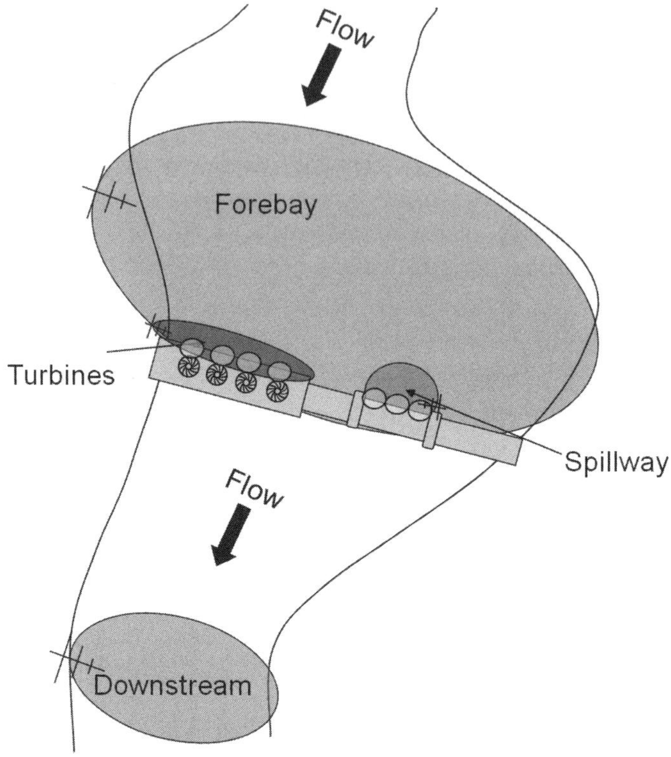

FIGURE 1. Schematic of a simple telemetry study to describe passage at a hydroelectric dam. Approach was measured by the forebay antenna (or antenna array) and the two entry zone antennas (turbines and spillway). Passage was documented by short-range (PIT) antennas, and verified by the downstream antenna. An additional antenna array was included to detect fish entering the marine environment (not shown here).

Once a smolt enters a forebay, it must first locate and then enter or reject a passage route. There are three time-based components to this process (Figure 3). Each component is associated with a particular zone. The 'approach zone' comprises the entire forebay, the region from which the fish must approach either passage route. Throughout this Section we use 'forebay' and 'approach zone' interchangeably. Each passage route has an associated 'entry zone', within which fish are able to detect a potential route of passage. Detection may be governed by hydraulic or other physical cues such as light, vision, etc. The final zone is the passage route itself. Fish that enter fishways do not necessarily pass—in the case of upstream passage this is a major component of passage failure. For most downstream passage routes at dams, however, passage is assured once a fish has entered a passage route and is entrained into a velocity field that it cannot escape. For simplicity, therefore, we will limit our discussion to the approach and entry zones, with the assumption that entry and passage are synonymous in this case.

Having once entered a zone, a fish may advance to a subsequent zone, or it may depart that zone without advancing. When a fish departs a zone it is once again available to approach and enter any of the subsequent zones. For example, a fish that arrives at the Spillway entry zone may return to the Forebay approach zone after some interval, at which point it is once

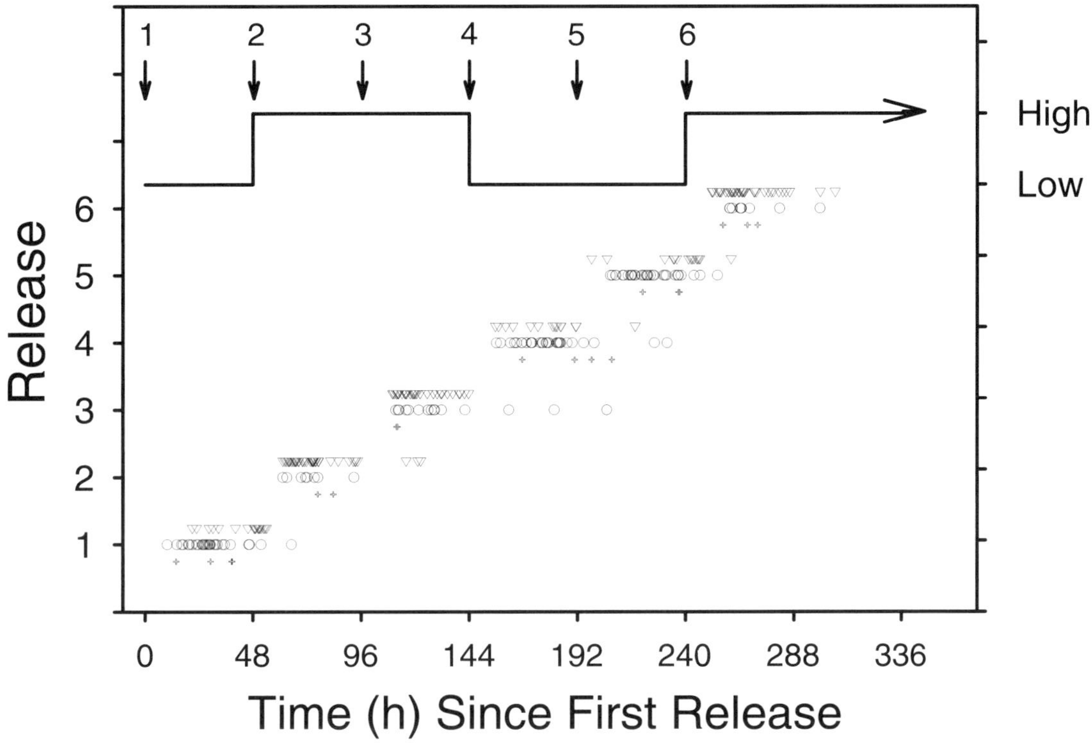

FIGURE 2. Modeled experimental design and release schedule (solid line and arrows). Releases occurred every 48 h and exposure time to each condition was a function of release time, arrival time, and time to pass. Scatters show modeled passage times through turbines (triangles) and spill (circles). Crosses indicate fish that died in the forebay. Note that some fish from a given release were exposed to both spill conditions.

again able to approach either zone. The time spent occupying each zone is important because it determines the amount of time available to advance to the next zone. We will refer to the time spent in a given zone as 'retention time' and its inverse, the rate at which fish depart the zone as 'rejection rate.' Passage through safe routes is expedited when approach and entry rates are maximized and rejection rate is minimized for the safe route, and when approach and entry rates are minimized and rejection rate is maximized for the unsafe route.

We represent these rates and their associated event times as follows (Table 1 and Figure 3): 'approach time' (T_{A_e}) refers to the time required to arrive at a given entry zone once a fish has entered the forebay, and 'entry time' (T_{E_e}) refers to the time required to enter (and in this case pass) a passage route once that route has been detected. Note that here the subscript '*e*' refers to an entry zone: in this example, this assumed values of '*T*' for the turbines and '*S*' for the spillway route, respectively. Each of these zones also has an associated 'retention time' (T_{R_A} for the approach zone (Forebay), T_{R_T} for the Turbine entry zone, and T_{R_S} for the Spillway entry zone; Table 1). Note that the Forebay has a single retention time; this is because both entry zones fall within the Forebay, and fish that are present within the Forebay are candidates for approach to either entry zone. Finally, note that each of the event times described here is the product of an associated rate which follows an inverse relationship (% h^{-1})—thus low

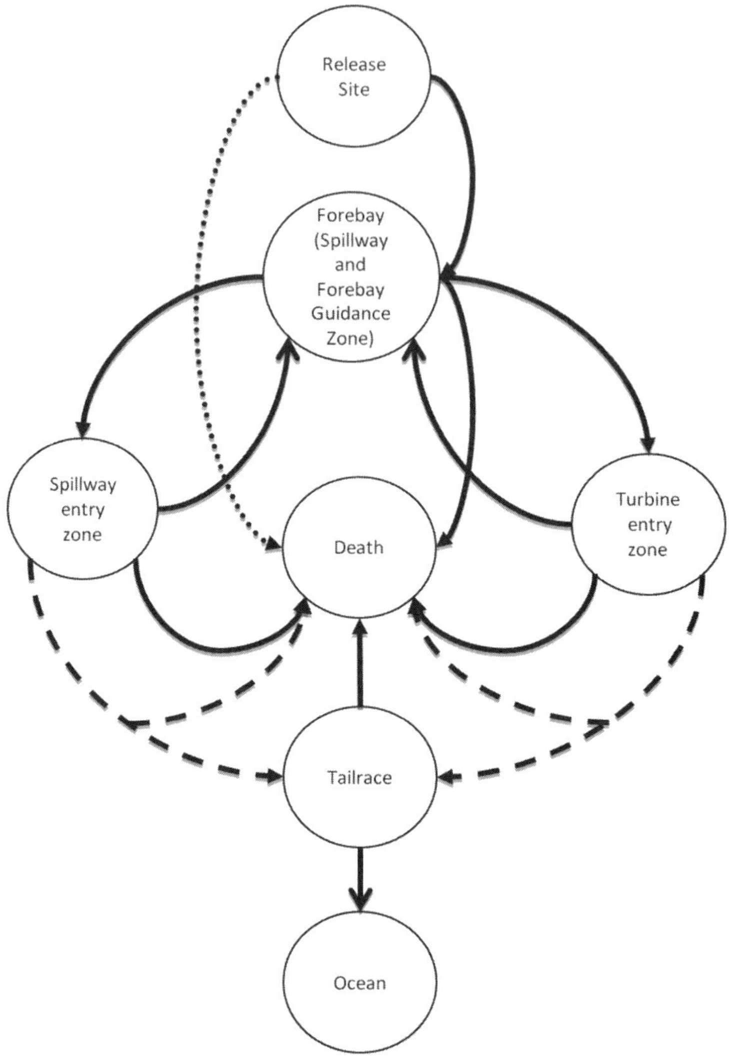

Figure 3. Schematic of simulation model showing possible movement and passage behaviors. Note that mortality can occur in any location (this is covered in detail by Perry et al. Section 9.2), but is excluded from this schematic for simplicity. Dashed connectors indicate passage events. The time-to-event analysis methods described here can be used to quantify any and all of these rates; however in this hypothetical scenario we have no way of definitively identifying time of death. Because of this, all death events within the forebay and entry zones are treated as censored observations when calculating rates of approach, entry, and retention.

event rates produce long event times and vice versa.

Movement of individuals between approach and entry zones was simulated by comparing randomly drawn approach, entry, and retention times between competing possible movements (Table 1). For example, whether a fish moved to the Turbine or Spillway entry zone from the forebay approach zone was determined by first randomly drawing T_{A_S} and T_{A_T} from characteristic time distributions and then moving the fish to the zone with the lesser approach time. That is, if $T_{A_S} < T_{A_T}$ then the fish moved from the forebay approach zone to the spillway entry zone in T_{A_S} hours, and if $T_{A_T} < T_{A_S}$ then the fish moved to the Turbine entry

Table 1. Symbols and parameters used to describe various time-based functions in the fish passage simulation model. Units are in hours.

Symbol	Behavior	Distribution	Parameters
T_T	Travel time from release point to Forebay approach zone	Lognormal: exp(Normal (μ_T, σ_T)	$\mu_T = 3.00$; $\sigma_T = 0.50$
T_{A_T}	Approach time to turbine entry zone	Weibull$(\theta_{A_T}, \sigma_{A_T})$	$\theta_{A_T} = 0.85$; $\sigma_{A_T} = 4.00$
T_{A_S}	Approach time to Spillway entry zone	Weibull$(\theta_{A_S}, \sigma_{A_S})$	
	Low spill condition		$\theta_{A_S} = 0.85$; $\sigma_{A_S} = 8.00$
	High spill condition		$\theta_{A_S} = 0.85$; $\sigma_{A_S} = 2.50$
T_{R_A}	Retention time in Approach Zone	Fixed, unlimited	∞
T_M	Time of death in forebay	Weibull(θ_M, σ_M)	$\theta_M = 1.00$; $\sigma_M = 100$
T_{E_T}	Entry time into turbine route	Weibull$(\theta_{E_T}, \sigma_{E_T})$	$\theta_{E_T} = 1.25$; $\sigma_{E_T} = 5.00$
T_{R_T}	Retention time in Turbine entry zone	Weibull$(\theta_{R_T}, \sigma_{R_T})$	$\theta_{R_T} = 0.50$; $\sigma_{R_T} = 1.50$
T_{E_S}	Entry time into Spillway route	Weibull$(\theta_{E_S}, \sigma_{E_S})$	
	Low spill condition		$\theta_{E_S} = 1.25$; $\sigma_{E_S} = 5.00$
	High spill condition		$\theta_{E_S} = 1.25$; $\sigma_{E_S} = 2.00$
T_{R_S}	Retention time in Spillway entry zone	Weibull$(\theta_{R_S}, \sigma_{R_S})$	
	Low spill condition		$\theta_{R_S} = 0.50$; $\sigma_{R_S} = 0.75$
	High spill condition		$\theta_{R_S} = 1.00$; $\sigma_{R_S} = 5.00$

zone in T_{A_T} hours. Movement of fish from an entry zone to either the forebay approach zone or the passage structure was simulated in a similar fashion. For example, if $T_{E_e} < T_{R_e}$ the fish entered and passed the structure. If $T_{E_e} > T_{R_e}$, however, the fish moved from the entrance zone back to the forebay approach zone, where it was once again subject to moving into either entrance zone. A fish's movement history upstream of the dam was terminated when it either entered a passage structure or died in the forebay. The time of death in the forebay (T_M) was drawn from an exponential distribution (Table 1) when the fish first entered the forebay, and movements upstream of the dam continued until the fish either passed the dam or died in the forebay, whichever occurred first. To simplify the simulation, fish were not allowed to move upstream out of the forebay, which was accomplished by setting T_{R_A} to infinity.

In this way the model simulated actual movements of migrating fish as they pass through a forebay, with multiple exposures of varying duration possible to either or both passage routes. Furthermore, as is discussed more fully below, some fish were exposed to more than one treatment condition. This occurred because of variability in times of arrival, approach, entry, and retention, as well as the relationship between release schedule and experimental design.

Transmitters and Receivers

In addition to the above simulation, we simulated the location and efficiency of the telemetry receivers (Figure 1). Simulated radio (or acoustic) receivers were placed such that detection zones coincided with the approach zone and each entry zone. In this case, the approach zone encompassed the entire forebay of the dam, and no overlap occurred between the two entry zones. Tags were set to transmit every 5 s, and detection at each receiver was simulated as a random binomial. Probability of detection was equal between the two entry zone receivers (P_{detect} = 0.20 transmission^{-1}), but was lower for the forebay receiver (P_{detect} = 0.10 transmission^{-1}). Note that although these detection rates seem low, they correspond to detection probabilities of 93% per minute spent in the entry zones and 72% per minute in the forebay (see Section 7.5 and Section 9.2 for details on detection probability). This is important, because below we will describe the methods used to estimate entry and exit times from each entry zone as well as the forebay. Imperfect and variable detection probabilities are common in practice, and we wanted our simulation to capture the error this phenomenon introduces.

We also included monitors at each passage route. This could be done using PIT telemetry or short-range radio-telemetry (acoustic telemetry methods that allow for fine-scale 2- and 3-dimensional tracking can also work for this application). Here we assume PIT telemetry; because PIT tags fire almost instantaneously, observed passage times reflect the true simulated passage time. In actual studies, however, passage events are sometimes missed. As with the radio receivers, the passage route receivers were also assigned random binomial probabilities of detection (P_{detect} = 0.95 through the Turbines and P_{detect} = 0.75 through the Spillway). Thus it was possible for simulated fish to pass a given route and not be detected, just as happens in practice. Two additional receivers were deployed downstream (one in the tailrace (P_{detect} = 0.95) and one in the 'Marine Environment' (P_{detect} = 0.70), Section 9.2). These made it possible to identify fish that passed the dam undetected, provided they survived to the next receiver.

ANALYSIS

There are many possible ways to analyze the data produced by the simulation described above. As suggested in the Introduction, one might simply look to see if any smolts are arriving to the marine environment. If so, then passage is possible. Clearly, this is not a satisfactory measure. More typically, evaluations will quantify proportion passing through each route. This may be acceptable if the only issue of concern is survival through a given route. It is insufficient, however, if the intent is to understand effects of operations, or experimental, environmental, or biological conditions on passage rate. This is because passage route selection is not an instantaneous event, but instead is the result of the competing, continuous processes.

We structured our model to produce a more biologically realistic description of passage behavior and performance. Under this model, route selection results from a random process that is contingent on exposure to the different routes. By properly designing and analyzing telemetry experiments and data, it is possible to understand not only where the fish are passing, but also a) to what extent route selection is driven by the discrete components of aproach, entry, and retention; and b) the effects of treatments on each of these components, and how this ultimately affects route selection. Simply enumerating passage route selection overlooks these essential underlying components and so fails to accurately quantify the causal processes. The analyses in this Section and Section 9.2 allow for comparison of the two methods and highlight the differences between these two approaches.

Data Management and Parameters

Because we adopted an approach that simulates actual telemetry data, the first step in preparing the analysis was to organize the data. Typically, database software and applications serve this purpose (Section 9.4). In addition to the telemetry data, the database also contained a table with information for each individual fish (in this case only Tag Code and release time). For the purposes of this paper we assumed that no other variables of interest existed, however it is common to also include such variables as temperature, length and weight of each fish, along with any comments specific to that individual that might affect performance (Section 9.3). The simulated telemetry data were imported into the same database, with identifying fields for receiver and download (see Section 9.4). At this point the data were sorted on ID and time to produce a time-series of each fish.

The next task was to determine appropriate indicators to identify when fish entered and left a particular zone. For this simulation, the only zones of interest were the forebay ('Approach Zone') and the entry zones for the turbine and spillway passage routes. If every transmission had been detected when a fish was present in one of these zones, then we could have set a threshold interval between detections equal to the transmission rate of the tags: any missed reads would have indicated that the fish had left the zone of interest. Typically, though, many transmissions go undetected, and this approach is not meaningful. A better approach is to use the data to identify gaps in detections. This is done by calculating the interval between successive detections at a given location. Sequential transmissions within a single exposure period will typically be some small multiple of the transmission time. Longer intervals will be increasingly rare, and a threshold can be set at an arbitrary interval duration, at a particular quantile (say, the 99th percentile of interval duration), or at a visually or statistically distinguishable inflection point or bimodality in the data.

Interval distributions from the simulated receivers are presented as a histogram in Figure 4. The percent of observations is presented on a log scale—this helps to clarify the threshold interval for distinguishing missed reads from new exposures. Among fish in the forebay (Approach Zone) there was no clear threshold, so 400 s was used. There does appear to be some multimodality at the entry zones (indeed, we know this exists)—given the data, a reasonable threshold to define new exposures should be at about 280 s. In this hypothetical scenario, we feel that these thresholds represent a reasonable if somewhat arbitrary balance between the empirical data and the apparent behavior of the fish. Although more rigorous methods are possible (such as using moving-point regression to identify an inflection point) we feel that the graphical approach described here is adequate and more convenient for most real-world applications.

We refer to the retention period within a given zone as an 'exposure,' which is analogous to a 'state' in multi-state and Markov chain analyses (Lindsey 2004). Having established the thresholds that delineate discrete exposure events, the next step was to group observations together so that first and last observations could be established for each exposure. To do this, new fields were added to the data table that identified distinct exposures in each zone. We used database software (Microsoft Access and a Visual Basic routine) to step through the data and identify intervals in each detection zone greater than the threshold. When such an interval was

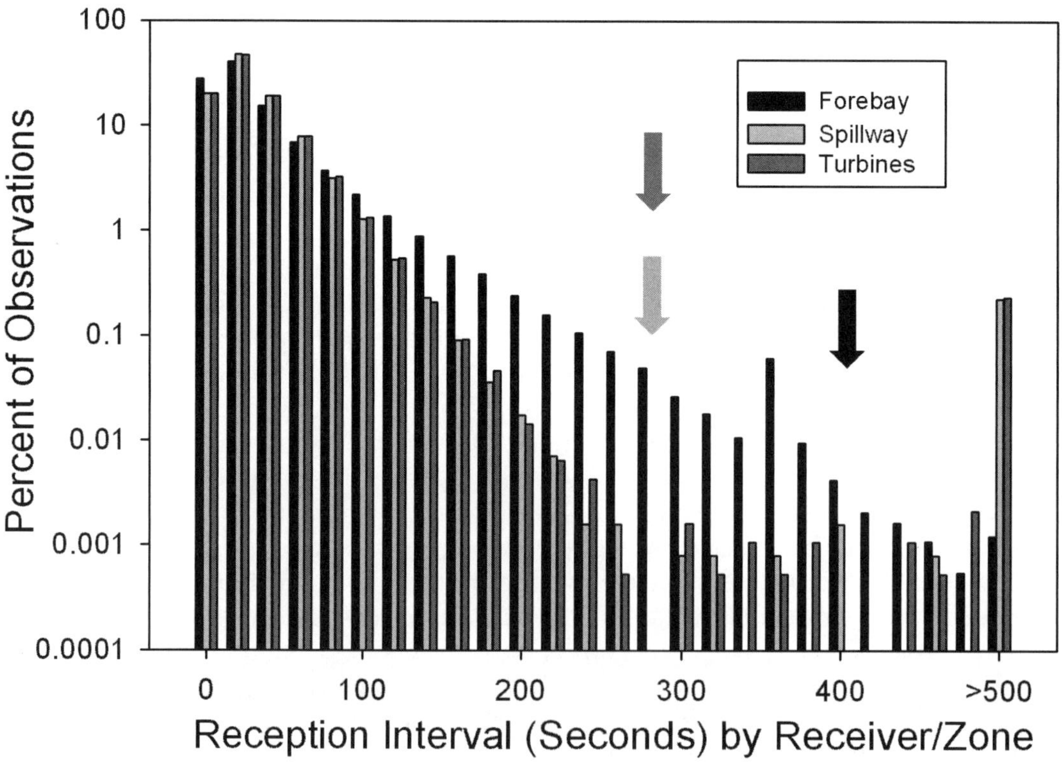

FIGURE 4. Reception intervals (s) of tags transmitting in forebay and each entry zone. Arrows indicate threshold levels used to delineate new exposures to each zone.

encountered, the exposure variable was incremented by 1, with the event time being reset to zero each time a new exposure was initiated. Note that by using this method it was not necessary for a fish to be detected at another location to identify a new exposure—we were able to infer exposure objectively using only the intervals between successive detections. Because the Spillway and Turbine entry zones were nested within the forebay, those receivers were included in calculating the intervals for approach zone exposures. Only the Spillway and Turbine radio and PIT antennas were used to identify Entry Zone exposures.

Delineating Exposure States with Competing Risks and Time Varying Covariates

This situation whereby fish can exit a given zone through more than one route is called a 'competing risks' process in the Survival Analysis literature (Allison 1995). Fish that enter the approach zone can experience one of three fates: they can approach one of the two entry zones, or they can depart the forebay without encountering either entry zone (Figure 3—note that forebay departure in this case includes mortality, failed tags, etc.). Thus three distinct rates can be estimated in the forebay: approach to the turbines, approach to the spillway, and abandonment of the forebay. When a fish experiences one of these three fates, it constitutes a complete observation of that event time, and an 'incomplete' observation for each of the other two event times (incomplete observations are termed 'censored' in the survival analysis literature—we will maintain this convention). In other words, when calculating the fraction of a population passing on a given time interval, an individual remains in the denominator (called the 'risk set') so long as it is present in the approach zone, but only complete observations get assigned to the numerator for estimating rates. Censored observations remove the individual from the risk set at their last known exposure time. In this way all available fish contribute to the observed rate of passage through each zone. This is one aspect of time-to-event analyses that separates it from other analytical methods and highlights its particular strength: all individuals contribute to rate calculation, regardless of fate or passage route.

The rate at which fish abandon a given zone is of fundamental interest, because it determines the duration of exposure and hence opportunity for approach or passage. Retention time within each zone defines the risk set for that zone. Later, we will present results as cumulative distribution functions ($F(T)$) for approach and attraction, as well as their complement (so-called 'Survivorship functions': $S(T) = 1\text{-}F(T)$) for retention. This provides an intuitive graphical description of the process we are quantifying.

Arrival at an entry zone defines an obvious change of state and so creates a logical censoring mechanism with respect to retention in the approach zone. Fish that passed without being detected by entry zone antennas, but that were subsequently observed at downstream antennas were also censored at their last observation within the approach zone. There is no single interpretation of why detections stop at a given location, however. Indeed, it is important to recognize that in this simulation absence of observations cannot be conclusively assigned to any particular fate. The consequences of this, as well as techniques for minimizing bias in survival estimates are detailed in Section 9.2. In our simulation, forebay mortality and tag failure did occur, and although the simulation did not allow the fish to move upstream, the efficiency of the forebay antenna was such that we could not differentiate between missed reads and forebay abandonment. Thus we considered any gap of 400 s or more that was not followed by observations at downstream antennas to constitute forebay abandonment. This was reasonable, because either mechanism could have removed fish from the risk set: they were no longer available to be guided to the entrance zones, and so this realistically described the amount of time available to locate one or both entry zones.

Note that this approach captures the limitations of telemetry technology. Also, it is important to recognize that bias can be introduced if the interval threshold is set too low, causing new exposures to be assigned to fish that did not actually leave the forebay. This will tend to cause an underestimate of forebay retention time. We deliberately included this error here with respect to forebay abandonment to demonstrate this potential consequence of insufficient receiver coverage.

As with the approach zone, fish that entered an entry zone were also subject to competing risks. In this case, fish either passed or returned to the forebay. For entry rates, passage time constituted a complete observation; final detections during presences that ended without an observed passage event were censored (incomplete) observations. Again, retention (i.e., rejection rates) had the opposite relationship, with passage times constituting censored observations and the last entry zone observation before returning to the approach zone constituting complete observations (Tables 1 and 2, Figure 3).

Time of passage was identified by the short-range (PIT) loggers at each entrance. As with any telemetry system, these receivers were not 100% efficient. This meant that fish could pass undetected. In part to control for this, as well as providing a mechanism for estimating detection efficiency (Section 9.2), an additional simulated receiver was placed in the tailrace. Detections on the tailrace receiver that occurred within the lag threshold for an entry zone presence (plus a minimum travel time from the passage route receiver and the Tailrace receiver) were included in the final entry zone presence. This made it possible to identify passage events for exposures that did not end in passage route detection. Final observations for fish that passed undetected (but that were subsequently detected in the tailrace) were censored with respect to both passage and retention times, because actual passage time was unknown. Fish that passed undetected and that were not detected downstream were assumed to have died in the forebay and so did not contribute to passage rate estimates (but see Section 9.2 for a more rigorous treatment of survival estimates).

For the purposes of this example we assumed that fish that rejected an entry zone and returned to the forebay must begin the process of searching for an entry zone once again. To account for this, exposures within the approach zone incremented when the fish departed an entry zone. Approach and entry times were measured from the beginning of each exposure, and end times occurred whenever the fish transitioned between zones.

Data from several of the possible fates are described in Table 2. The first individual entered the forebay 26 h after release, entered the Turbine Entry Zone 8.96 h later (35 h after release) and passed through the turbines after another 0.36 h. Thus we have a complete approach event to the turbine zone, but an incomplete (censored) observation of approach to the Spillway zone and of forebay retention (Approach Censor = 1). We then have a complete observation for Turbine passage (Censor = 1), but an incomplete (censored) observation for retention near the turbine zone.

The second individual arrived at the forebay after 23.61 h, 3.32 h later appeared to leave the forebay without arriving at either entry zone (Approach Censor = 0 for both zones). This fish re-entered the forebay 27.04 h into the experiment and after 2.19 h arrived at the Spillway entry zone (Approach Censor = 2), and passed after a further 1.05 h (Spillway Censor = 1).

Table 2. Data from 5 simulated fish showing exposure grouping variables for approach (Forebay), and Turbine and Spillway entry zones. Begin and End times are shown for each exposure and indicate time since release. Associated Event Times (T1 and T2) were calculated from the beginning of each exposure. Censoring variables indicate whether a fish successfully arrived at an entry zone (Approach) or passed (Entry). In the approach zone, Censor = 0 indicates forebay death or departure; Censor = 1 indicates arrival at the Turbines entry zone; Censor = 2 indicates arrival at the Spillway entry zone, and Censor = 3 indicates change of covariate value (Treatment). For entry zones, Censor = 0 indicates failure to pass during that exposure; Censor = 1 indicates passage terminates the exposure, and Censor = 2 indicates change of a covariate value.

	Exposure			Time					Censor	
ID	Forebay	Turbines	Spillway	Begin	End	T1	T2	Treatment	Approach	Entry
1	1			26.22	35.18	0.00	8.96	Low	1	
1	1	1		35.18	35.55	0.00	0.36	Low		1
2	1			23.61	26.92	0.00	3.32	Low	0	
2	2			27.04	29.23	0.00	2.19	Low	2	
2	2		1	29.23	30.28	0.00	1.05	Low		1
3	1			19.03	24.66	0.00	5.63	Low	1	
3	1	1		24.66	25.90	0.00	1.24	Low		0
3	2			25.91	26.02	0.00	0.12	Low	2	
3	2		1	26.02	26.07	0.00	0.04	Low		0
3	3			26.07	31.69	0.00	5.62	Low	1	
3	3	2		31.69	31.83	0.00	0.14	Low		0
3	4			31.84	32.68	0.00	0.84	Low	1	
3	4	3		32.68	36.32	0.00	3.64	Low		0
3	5			36.33	38.75	0.00	2.42	Low	0	
4	1			31.23	31.87	0.00	0.65	Low	2	
4	1		1	31.87	32.21	0.00	0.34	Low		0
4	2			32.21	36.73	0.00	4.52	Low	1	
4	2	1		36.73	38.66	0.00	1.93	Low		0
4	3			38.66	39.83	0.00	1.17	Low	2	
4	3		2	39.83	39.99	0.00	0.15	Low		0
4	4			40.01	40.57	0.00	0.56	Low	1	
4	4	2		40.57	42.73	0.00	2.16	Low		0
4	5			42.73	45.39	0.00	2.65	Low	1	
4	5	3		45.39	48.00	0.00	2.61	Low		2

Table 2. Continued.

ID	Exposure			Time				Treatment	Censor	
	Forebay	Turbines	Spillway	Begin	End	T1	T2		Approach	Entry
4	5	3		48.00	48.19	2.61	2.80	High		0
4	6			48.19	48.32	0.00	0.13	High	1	
4	6	4		48.32	48.35	0.00	0.03	High		0
4	7			48.35	48.50	0.00	0.15	High	2	
4	7		3	48.50	54.12	0.00	5.62	High		1
5	1			11.87	19.59	0.00	7.72	Low	2	
5	1		1	19.46	19.59	0.00	0.13	Low		0
5	2			19.60	23.11	0.00	3.51	Low	2	
5	2		2	23.11	25.31	0.00	2.20	Low		0
5	3			25.32	27.99	0.00	2.67	Low	1	
5	3	1		27.99	28.14	0.00	0.15	Low		0
5	4			28.18	36.25	0.00	8.07	Low	2	
5	4		3	36.25	37.06	0.00	0.81	Low		0
5	5			37.07	42.78	0.00	5.72	Low	1	
5	5	2		42.78	42.90	0.00	0.12	Low		0
5	6			42.90	48.00	0.00	5.10	Low	3	
5	6			48.00	48.02	5.10	5.12	High	2	
5	6		4	48.02	49.39	0.00	1.37	High		1

The third fish entered the forebay, approached the Turbines, but then rejected the Turbine entry zone and returned to the forebay. Subsequently it approached the Spillway entry zone, but again failed to pass. Finally, after two further failed approaches to the Turbine entry zone, and having spent almost 20 h in the forebay, this fish disappeared—presumably having died or moved upstream; the distinction is immaterial to this analysis.

The fourth example shows how data are handled in the presence of a time-varying covariate. Here, the fish entered the forebay and approached each entry zone multiple times. As part of the study design (Figure 2), the spillway level switched from Low to High after 48 h. At that time the fish was present in the Turbine entry zone. Because the spill level might influence passage and rejection rates at either entry zone these data were censored at this time (Entry Zone Censor = 2), and a new record added to the dataset to reflect the changed condition. The clock continued to run, however, and so the event time for this second observation began at the time during the exposure when the condition changed (T1). This next observation was also censored because the fish did not pass until after two more entry zone exposures. Thus we have two censoring events during a single entry zone exposure: the first occurring with the changed spill condition (a time-varying covariate), and the second occurring later in the exposure when the fish returned to the forebay. Importantly, the first part of the entry zone exposure (0–2.6 h) contributed to the estimate of passage rate (termed 'hazard' in the survival analysis literature) under the Low Spill condition, and the second part (2.6–2.8 h) under the High Spill condition. The effect of the treatment was quantified as the proportional change in passage rate under each condition. The final example is similar to the fourth example, except that in this case the spill condition changed while the fish was in the approach zone. This constituted a fourth censoring level (Censor = 3) or fate for this zone, whereas the entry zones had only 3 possible censoring levels.

The arrangement of data shown in Table 2, with separate columns for start and end times, is the appropriate configuration for performing proportional hazards regression using the so-called 'counting process' style of input. Other methods exist, but we find this most convenient for performing analyses in the presence of time-varying covariates. We performed all modeling and analysis using SAS software (SAS 2003), and used Cox's proportional hazards regression (the PHReg procedure) to quantify treatment effect on each of the rates described above. Allison (1995) provides a thorough review of how to structure code and censoring for competing risks scenarios using the counting process style of input shown in Table 2.

Regression Model

The regression model estimated from the above procedure is a proportional hazards model, i.e. it describes the relative effect of a unit change in variables X_i on the log-hazard function:

$$\ln h_i(t) = \ln(\lambda_0\,(t)) + \beta_1 X_{i1} + \ldots + \beta_k X_{ik} \tag{9.1}$$

where $\lambda_0\,(t)$ is the baseline hazard function, or the instantaneous rate when all covariates are set to zero; the hazard (approach rate, entry rate, or rejection rate) of the ith observation is affected by a set of covariates ($X_1, \ldots, X_k$); and the magnitude of the effect is described by the coefficients ($\beta_1, \ldots, \beta_k$). Thus, each event type is associated with a unique baseline hazard function that can be affected by a distinct set of covariates. This is called a proportional hazards model because the hazard is proportional to covariate values on any given time-step. For example, for individuals i and j equation (1).1 can be rewritten as

$$\frac{h_i}{h_j} = \exp(\beta_1(X_{i1} - X_{j1}) + ... + \beta_k(X_{ik} - X_{jk})). \tag{9.2}$$

In our example with only one binary covariate, this means that an individual exposed to High Spill (Treatment = 1) will have an expected passage rate at time t equal to $\exp(\beta_i)$ times the passage rate of an individual exposed to Low Spill (Treatment = 0) at the same time. Note that the $\lambda_0\ (t)$ terms cancel out (9.1), because the proportional effect of the covariate is assumed to be the same across all time intervals (Allison 1995). This assumption of proportionality across time is one of the few key assumptions underlying this method. Excellent texts are available describing details on how to test for this, and what to do when the assumption cannot be met (Allison 1995; Hosmer and Lemeshow 1999).

Note that in this case we stratified by forebay exposure. This is because approach is a recurring event. We assume that each time a fish enters or returns to the forebay that the clock is reset to zero. It is very possible, though, that approach time distribution will change with each exposure to the forebay, and stratifying controls for both this effect and for any possible dependence among event times by a single individual. Another way to control for previous experience is to include previous exposure time as a covariate in each time step. In this way any 'learning' or other cumulative behavioral effects can be quantified.

RESULTS

Passage times and routes of each simulated fish are presented in Figures 2 and 5. Note that in several cases fish were exposed to more than one spill condition. Also the total opportunity for exposure varied among fish and treatments. Although an equal number were released under high- and low-spill conditions, some fish were released at the beginning of a treatment and some in the middle of each treatment period. Finally, the time to arrive at the forebay varied, which potentially further complicated interpretation of results, but which is explicitly handled using the methods described here.

Overall passage results are summarized in Table 3. The upper part of the table summarizes proportions passing each route and associated residence time by release; the lower part summarizes overall proportions and mean forebay residence time by spill condition at time of passage. We provide these data here in part to illustrate the misleading inference that can result from considering only proportions passing different routes, and to highlight the advantages of using proportional hazards regression (Table 4). Note that Table 3 presents the actual simulated data: fish with unobserved fates are assigned their known fate even though this would not be possible in an actual field study. Careful scrutiny of these data reveals some interesting patterns. First, note the change in proportion passing the two routes between release 2 and 3. Both releases occurred under the High Spill condition (Figure 2), but a higher proportion from the third release pass through the turbines. This occurred because several fish from the third release did not pass until after the condition had changed to Low Spill.

Changes in residence time are also evident, with lower overall residence time for the 2nd, 3rd, and 6th releases, except for turbines on the 3rd release, which had relatively large residence

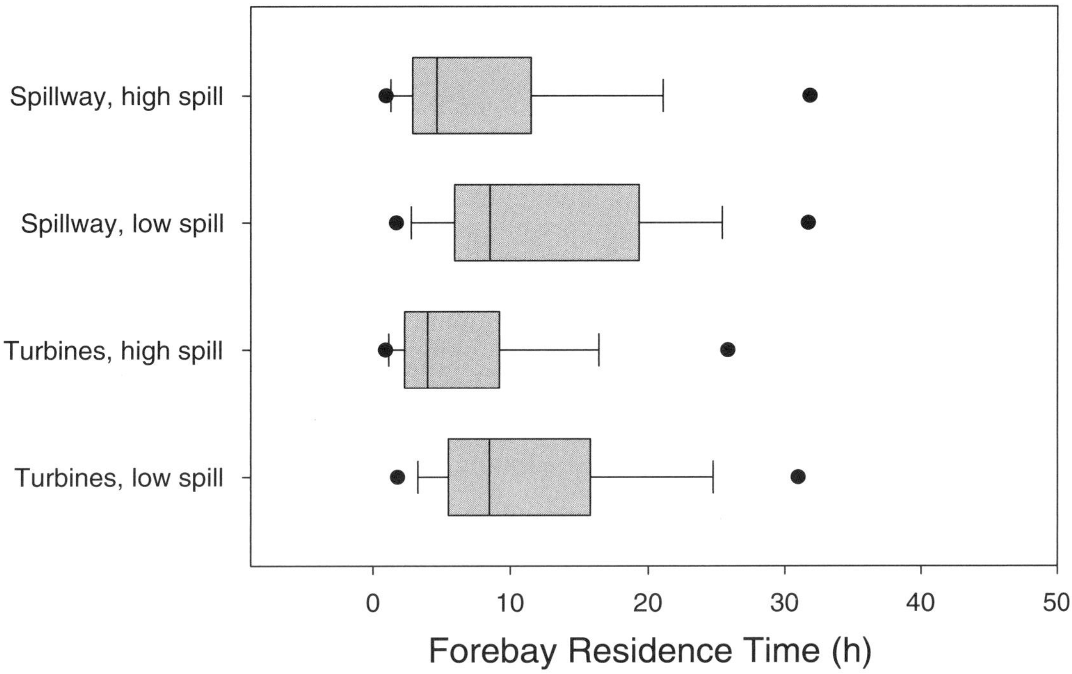

FIGURE 5. Forebay residence time by passage route and spill condition. Data are medians (vertical lines), first and third quartile (boxes) 10th and 90th percentiles (whiskers) and 5th and 95th percentiles (points). Note that residence times for both routes are reduced under High Spill condition. This would reasonably cause one to conclude that turbine passage rates increased under high spill. However this was not the case: instead, this pattern is a result of improved approach rates and greater retention and passage through the spillway—the actual rate of approach, retention, and passage through the turbines was uniform under all conditions (Figures 6 and 7), and the apparent reduction is the result of reduced opportunity for passage through the turbines under High Spill conditions. From this example, one can easily see how failure to appropriately quantify passage rates can lead to an inaccurate interpretation of treatment effects.

time. This is informative, because in the simulation, the only rates affected by spill condition were approach to, passage through, and retention within the Spillway zone. Thus the increased residence times in the turbine entry zones occurred because of reduced passage through the spillway, and for no other cause (Figure 5). A study that limited its analysis to proportional passage and overall residence time would have identified greater proportion passing through the turbines under Low Spill condition. This could reasonably (but wrongly) be interpreted as the Low Spill condition causing greater retention near the turbines and increased rates of approach and passage through the turbines, when in fact, these rates were independent of spill. In this model elevated turbine passage occurred only because there was more time available for turbine passage; actual rates of turbine passage remained unchanged between the two spill levels (Figures 6 and 7).

Similar patterns were evident for overall passage route selection. Although release designs were balanced, overall passage under the High Spill condition was 14% greater than under the Low Spill condition.

As described above, we see greater passage times expected through each route and also greater forebay mortality under Low Spill than under High Spill (Table 3; Figure 5). One

TABLE 3. Sample size and passage route by release. "FB" denotes fish that died in the forebay. The upper part of the table groups data by release group, the lower part groups data by passage route. Note that these are the true data produced by our simulation, and therefore all fish are accounted for. The data are presented here to allow for comparisons with the results produced by the 'imperfect knowledge' associated with imperfect reader efficiencies.

		Proportion			Mean Cumulative Residence Time (h)		
Release	Total	FB	Turbines	Spillway	FB	Turbines	Spillway
1	50	0.08	0.58	0.34	12.77	10.40	16.64
2	50	0.04	0.16	0.80	8.78	5.57	4.37
3	50	0.04	0.32	0.64	6.31	10.08	6.61
4	50	0.08	0.64	0.28	19.08	13.43	11.58
5	50	0.06	0.62	0.32	13.93	10.25	17.84
6	50	0.06	0.20	0.74	4.76	5.42	5.19
Spill Condition			Proportion				
Low	129	0.09	0.70	0.21	15.38	11.45	10.33
High	171	0.04	0.21	0.75	6.35	7.74	7.98

TABLE 4. Output of Cox's Proportional Hazards Regression models of approach, retention, and entry. The hazard ratio is a measure of the relative change in rate among treatments (Spill = 0 for Low Spill and Spill = 1 for High Spill): a significant value greater than 1 indicates greater rates associated with the High Spill condition (equation 1.1). Likewise note that greater approach and rejection rates imply shorter passage and residence times, respectively. Δ AIC values indicate the reduction in AIC for the fitted model compared with the null model (i.e., no spill effect).

	N			Coefficient			
Model	Total	Censored	Complete	Spill	Hazard Ratio	$P >$ ChiSq	Δ AIC
Approach zone							
Spillway approach	816	349	467	1.162 ± 0.117	3.198	<0.0001	96
Turbines approach	816	410	406	0.166 ± 0.114	1.181	0.146	0
Forebay rejection	816	57	759	0.244 ± 0.427	1.276	0.514	–1
Entry zone							
Spillway entry	353	147	206	1.077 ± 0.228	2.936	<0.0001	28
Spillway rejection	353	205	148	–1.241 ± 0.155	0.289	<0.0001	68
Turbines entry	407	120	287	–0.010 ± 0.214	0.990	0.963	–2
Turbines rejection	407	283	124	0.121 ± 0.132	1.129	0.362	–1

might understandably interpret these results as indicating that turbine passage rates increased and forebay mortality rates decreased in the presence of high spill. This is an example of the misleading inferences that can be drawn when competing risks are not explicitly taken into account: in fact these effects were absent from the simulation. Instead, this error is the result of equating passage times with rates that don't take into account the fact that under High Spill fish passed the spillway sooner, and so were not available to pass through the turbines or to die in the forebay.

The output from the Proportional Hazards models is summarized in Table 4 and Figures 6 and 7. As stated above, the only effect of spill in the simulation was to increase approach and entry rates to the Spillway passage route, and to increase retention time (reduce rejection rate) within the spillway entry zone; with approach, entry, and retention unaffected with respect to the turbine zone. The proportional hazards model realistically and accurately captured the time-varying treatment effect, and this is supported both by hypothesis testing and model selection criteria (Table 4; Figures 6 and 7).

Although not shown here, the confidence intervals of each of the estimated curves encompass the true underlying functions. This raises an important point: the confidence interval for each curve is a function of the size of the risk set at any given moment. As the population decreases (i.e., toward the right-hand tail), the confidence intervals inevitably grow. This means that researchers must give careful consideration to the power of the study design to identify effects and estimate overall delay for the entire population: estimates will be most precise for fish that pass soonest, but the last fish to pass will usually have the greatest error around their passage time estimate.

The one case in which systematic error did occur in our analysis was with respect to Forebay Retention. Here, the underlying function shown in Figure 6 includes only expected forebay mortality rates. It excludes the possibility that fish will leave the forebay, because no such behavior was included in the simulation. The simulation did include receiver efficiency, however, which was set to 0.10 for the forebay receiver. This led to an erroneous estimated forebay departure rate of approximately 13% per hour, i.e., a fish present in the forebay for only one hour had a 13% chance of having no successful detections for a period of 400 s or longer. Recall that the method used to estimate departures from the forebay was based on empirical receiver data (Figure 4)—the low frequency of observations with lags >400 s created the appearance of a reasonable point to assume that fish had left the forebay. In fact, this did not occur, and so this method caused some error in the Forebay Residence Time estimate. We included this error deliberately here, because it realistically reflects the lack of knowledge that researchers are often faced with, and we wanted to demonstrate the consequence of this error. In this case, the error is minor: no difference in retention occurred by treatment, and the error incurred by ignoring detection efficiency was <10%.

DISCUSSION

This Section presents an example and rationale for time-to-event analysis of behavioral data from the perspective of a fish passage problem. Although issues of fish passage have been recognized for centuries, there has until recently been no unifying framework of theory with which to standardize studies and compare passage performance across conditions, regions, or species. It is our hope that this portion, and this Section in particular, will serve to help estab-

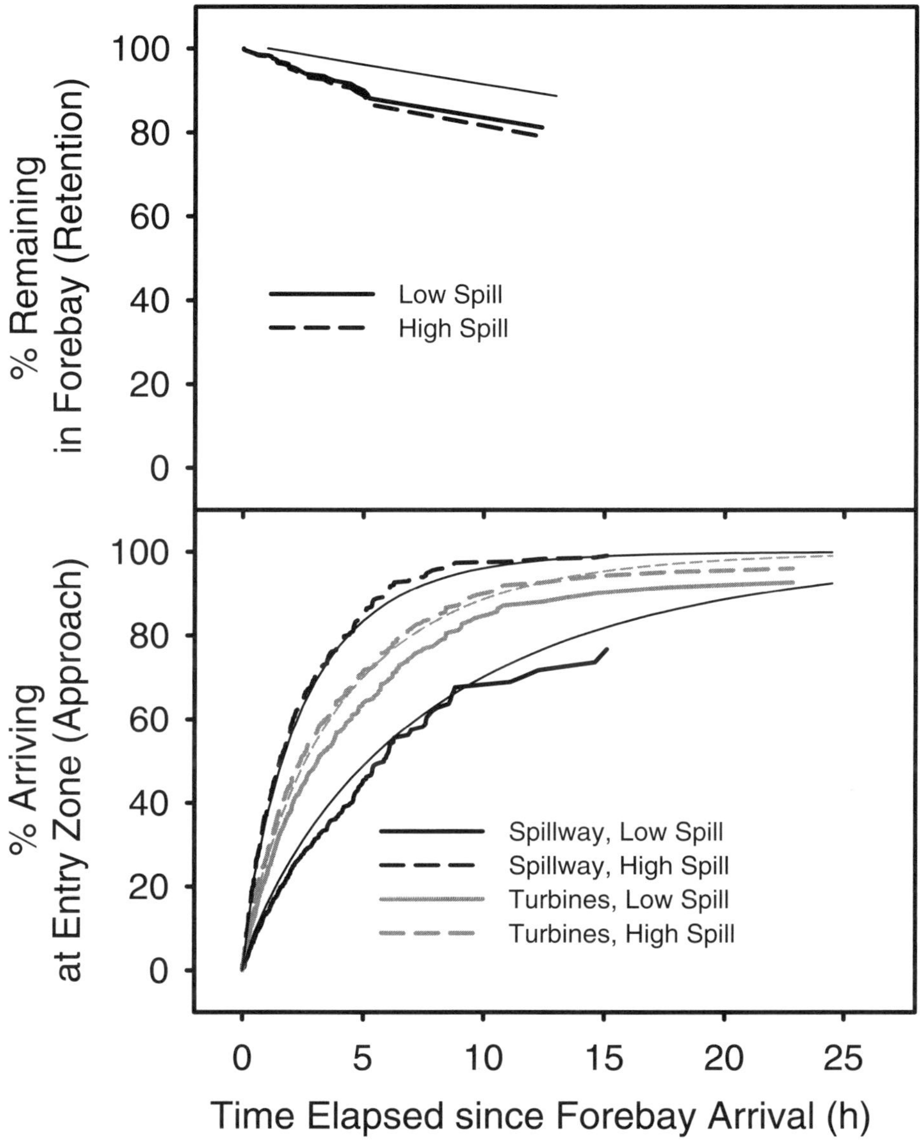

FIGURE 6. Forebay retention (upper panel) and approach rate (lower panel) estimates from regression models (thick curves) and known true functions (thin curves). Under the simulation, spill level had no effect on forebay retention or approach time to the turbines, but approach to the Spillway Entry Zone was faster under High Spill. This figure shows that the regression model was able to accurately capture the underlying pattern, despite the presence of a time-varying covariate. Note the contrast with the multinomial approach of Table 3, which seems to suggest that spill affects rates of turbine passage.

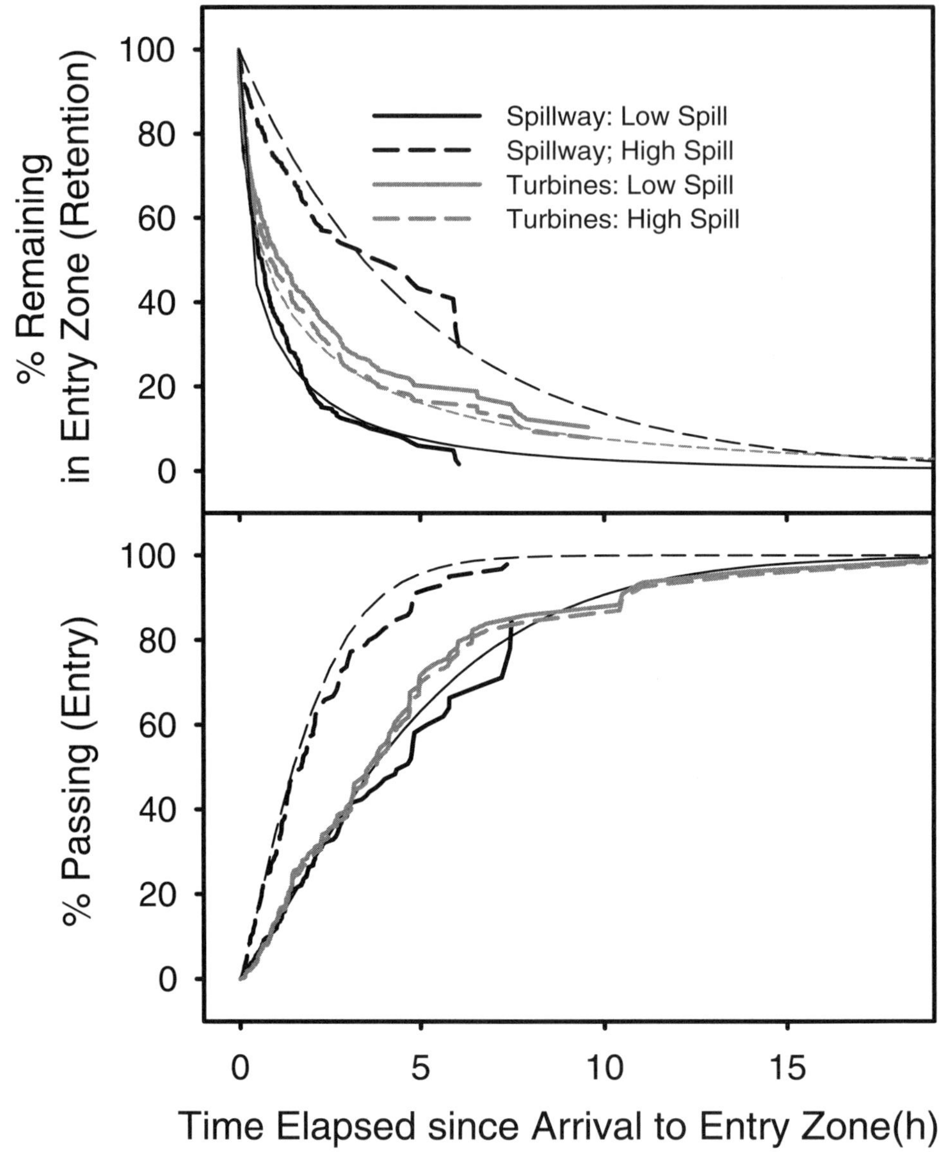

FIGURE 7. Entry zone retention (upper panel) and entry rate (lower panel) estimates from regression models (thick curves) and known true functions (thin curves). Under the simulation, spill level had no effect on retention near or entry through the turbines . However, both retention near and entry through the Spillway Entry Zone were greater under High Spill. Also note that entry through the turbines under both spill conditions was identical to entry through the spillway under the low spill condition. As with Figure 6, this Figure shows that the proportional hazards regression model was able to accurately capture the underlying pattern, despite the presence of a time-varying covariate.

lish such standards, and provide users with the necessary information to begin to assemble the data needed to improve fish passage bioengineering.

Here we have demonstrated that these methods can be particularly useful for telemetry data, where continuous observations are available on individuals. The applicability of these methods is much broader, however. Although not yet well-known in the behavioral ecology literature, the methods described here commonly form the basis for engineering (materials testing), medical (clinical trials), and actuarial applications (life tables). Indeed, most of the development of these methods has served the latter two applications. Their value for ecological applications has been gaining recognition, however, including observations of foraging behaviors (Muenchow 1986), social interactions (Moya-Larano and Wise 2000), and species invasions (Reino et al. 2009).

Importantly, application of these methods is not restricted to quantifying time-based rates. They have also been used successfully to measure swimming performance (Castro-Santos 2004, 2005; Haro et al. 2004) and ascent of fishways (Castro-Santos et al. 2009), where failure occurs in units of distance rather than time. Time- and distanced-based rates can even be coupled to control for variability in both motivation and swimming ability among organisms attempting to traverse barriers (Castro-Santos 2004).

Adequacy of the Model

Because this Section was largely intended as a demonstration and guide, the underlying model includes only the complexity needed to demonstrate the value of the approach. In field situations, many other covariates may exist that could affect passage rate: diurnal changes in behavior, changes in generation, river flow, temperature, and migratory motivation to name just a few. Studies should be designed, and power analyses performed, to establish appropriate release schedules and sample sizes to ensure adequate power to detect and quantify covariate effects.

Implications for Study Design

Perhaps because of the lack of a unifying theory of fish passage, evaluations and study designs vary widely (Bunt et al. 2011; Roscoe and Hinch 2010). Very often much more information is collected than is necessary to identify and describe the relevant behavioral responses. This has led to significant advances in telemetry technology, but the scale and quantity of information, as well as the cost of the studies, are often out of proportion to the questions being asked. By carefully identifying the temporal and spatial scale of behaviors leading up to passage it is possible to develop study designs that quantify the relevant behaviors without accumulating unnecessary data and expense.

With this in mind, researchers must also be careful not to accidentally discard important levels of resolution. Receivers should be set up to document detections on the finest temporal scale possible. Sometimes receiver memory limitations require researchers to impose a record-filtering protocol such that continuous exposures within a zone are grouped onto a single line of data, and some receivers offer this filtering as a standard feature. This type of processing is best done after data collection using a database protocol (see Section 9.4, and exposure identification protocol described above). In the current example, we used time intervals between each transmission to identify important behaviors (departure times for each zone). If we had grouped and filtered our data in the field we would have lost this level of resolution and thus lost the opportunity to identify crucial behaviors. Filtering and discarding raw data can render experiments useless—the researcher typically doesn't realize this problem until after the fact, and much effort and expense have been wasted through indiscriminate data filtering.

Value of the Approach

The simulation exercise presented here highlights the rationale for using event-time analysis as an appropriate and sufficient method for quantifying behaviors associated with fish passage. As mentioned above, the total exposure time varied among fish and treatments. Some fish were released at the beginning of a treatment and some in the middle of each treatment period; some were released under high- and some under low-spill conditions. Finally, the time to arrive at the forebay varied, which potentially further confounded interpretation of results. Multinomial (mark–recapture) models excel at quantifying the outcomes of survival and passage processes (Skalski et al. 2002), but have limited ability to quantify mechanisms driving the passage process (Wilson et al. 1991). For example, even if exposure time is included in a multinomial model it may not be a sufficient method to quantify underlying mechanisms driving the passage process because of the underlying competing risks: route selection is not simply binary, it is the result of a process whereby retention and passage events compete, and one event precludes the potential of the other. Multinomial models can quantify differences in the proportion of fish passing a route by including covariate effects, but without the mechanistic insights provided by event-time analysis. Given the high cost of telemetry tags and the detailed behavioral information typically ignored by multinomial models, using event time models allows researchers to extract the maximum information from costly studies.

Implications for Fish Passage Technology

The central benefit of this approach is that it allows researchers to identify underlying causes for passage rates. In this example, passage over the spillway was increased because rates of approach to and entry into the spillway route were improved under High Spill, and also because retention within the spillway zone was increased. Often, however, it is possible for opposing patterns to occur, say approach and retention improve, but entry rate decreases. This could give the impression that no benefit occurs (and indeed this is what a multinomial approach would indicate), where in fact two of the three components did improve. Bioengineers that are aware of this improvement can then shift their focus to solving the final issue (entry)—a problem with very different solutions from approach or retention.

The benefits of this kind of information have direct field relevance. It has long been known that the lateralis and visual systems allow fish to detect accelerations in the surrounding flow field, and Haro et al. (1998) applied this concept to develop a downstream passage structure that accelerated rates of attraction. This concept was verified by Kemp et al. (2005, 2006) and Enders et al. (2009), and has recently been applied successfully to expedite passage on rivers on both coasts of North America. Others are working to generalize this concept to hydraulic conditions appropriate for guiding fish through approach zones (Coutant 2001a, 2001b; Goodwin et al. 2006; Nestler et al. 2008). Verification and calibration of these approaches can only be achieved if measures of performance and behavior match the scale of these theoretical and physical approaches.

Neuro- and behavioral biologists characterize effective stimuli as being appropriate and sufficient. A stimulus that is either inappropriate or insufficient will not elicit a response. This is analogous to statistical methods as well. Any statistical model is of necessity only an approximation, but if the model is to reliably predict behavior it must be appropriate (quantifies the relevant underlying behaviors) and sufficient (captures that aspect of the behavior that is necessary to describe the outcome). More to the point, engineered solutions for fish passage should consider the behaviors that a given device is expected to affect or take advantage of. Do we expect the device to improve approach, retention, or entry? Once an expectation is defined, it becomes possible to quantify performance at the appropriate scale, and to watch for unexpected side-effects. Similarly, evaluations can more accurately identify sources of problems when structures fail to pass fish adequately.

Because the processes that lead to passage are time-based, and contain multiple competing outcomes, any statistical method that fails to describe these behaviors will necessarily be either inappropriate or insufficient or both. The event-time analysis methods described here are both appropriate and sufficient to describe fish passage performance. Because they are intrinsically appropriate, they are readily adapted to any fish passage situation or scale, and for this reason form a logical basis of a unifying framework for evaluations. If these methods are adopted as a standard, it will become possible to make objective comparisons and advance (or perhaps introduce) the scientific basis of fishway design.

REFERENCES

Allison, P. D. 1995. Survival analysis using the SAS system: a practical guide. SAS Institute, Cary, North Carolina.

Anderson, J. J., E. Gurarie, and R. W. Zabel. 2005. Mean free-path length theory of predator–prey interactions: Application to juvenile salmon migration. Ecological Modelling 186:196–211.

Bunt, C.M., T. Castro-Santos, and A. Haro. 2011. Performance of fish passage structures at upstream barriers to migration. River Research and Applications 28:457–478.

Burnham, K. P., D. R. Anderson, G. C. White, C. Brownie, and K. H. Pollock. 1987. Design and analysis methods for fish survival experiments based on release–recapture. American Fisheries Society, Monograph 5, Bethesda, Maryland.

Castro-Santos, T. 2004. Quantifying the combined effects of attempt rate and swimming capacity on passage through velocity barriers. Canadian Journal of Fisheries and Aquatic Sciences 61:1602–1615.

Castro-Santos,T. 2005. Optimal swim speeds for traversing velocity barriers: an analysis of volitional high-speed swimming behavior of migratory fishes. Journal of Experimental Biology 208:421–432.

Castro-Santos, T., A. Cotel, and P. W. Webb. 2009. Fishway evaluations for better bioengineering—an integrative approach. Pages 557–575 *in* A. J. Haro, K. L. Smith, R. A. Rulifson, C. M. Moffit, R. J. Klauda, M. J. Dadswell, R. A. Cunjak, J. E. Cooper, K. L. Beal, and T. S. Avery. Challenges for diadromous fishes in a dynamic global environment. American Fisheries Society, Symposium 69, Bethesda, Maryland.

Coutant, C. C. editor. 2001a. Behavioral technologies for fish guidance. American Fisheries Society, Symposium 26, Bethesda, Maryland.

Coutant, C. C. 2001b. Turbulent attraction flows for guiding juvenile salmonids at dams. Pages 57–77 *in* C. C. Coutant, editor. Behavioral technologies for fish guidance. American Fisheries Society, Symposium 26, Bethesda, Maryland.

Enders, E. C., M. H. Gessel, and J. G. Williams. 2009. Development of successful fish passage structures for downstream migrants requires knowledge of their behavioural response to accelerating flow. Canadian Journal of Fisheries and Aquatic Sciences. 66:2109–2117.

Goodwin, R. A., J. M. Nestler, J. J. Anderson, L. J. Weber, and D. P. Loucks. 2006. Forecasting 3-D fish movement behavior using a Eulerian-Lagrangian-agent method (ELAM). Ecological Modelling 192:197–223.

Haro, A., T. Castro-Santos, J. Noreika, and M. Odeh. 2004. Swimming performance of upstream migrant fishes in open-channel flow: a new approach to predicting passage through velocity barriers. Canadian Journal of Fisheries and Aquatic Sciences 61:1590–1601.

Haro, A., M. Odeh, J. Noreika, and T. Castro-Santos. 1998. Effect of water acceleration on downstream migratory behavior and passage of Atlantic salmon smolts and juvenile American shad at surface bypasses. Transactions of the American Fisheries Society 127:118–127.

Hosmer, D. W. and S. Lemeshow. 1999. Applied survival analysis: regression modeling of time to event data. Wiley, New York.

Kemp, P. S., M. H. Gessel, B. P. Sandford, and J. G. Williams. 2006. The behaviour of Pacific salmonid smolts during passage over two experimental weirs under light and dark conditions. River Research and Applications 22:429–440.

Kemp, P. S., M. H. Gessel, and J. G. Williams. 2005. Fine-scale behavioral responses of Pacific salmonid smolts as they encounter divergence and acceleration of flow. Transactions of the American Fisheries Society 134:390–398.

Lindsey, J. K. 2004. Statistical analysis of stochastic processes in time. Cambridge University Press, Cambridge, U.K.

Moya-Larano, J. and D. H. Wise. 2000. Survival regression analysis: a powerful tool for evaluating fighting and assessment. Animal Behaviour 60:307–313.

Muenchow, G. 1986. Ecological use of failure time analysis. Ecology 67:246–250.

Nestler, J. M., R. A. Goodwin, D. L. Smith, J. J. Anderson, and S. Li. 2008. Optimum fish passage and guidance designs are based in the hydrogeomorphology of natural rivers. River Research and Applications 24:148–168.

Reino, L., J. Moya-Larano, and A. C. Heitor. 2009. Using survival regression to study patterns of expansion of invasive species: will the common waxbill expand with global warming? Ecography 32:237–246.

Roscoe, D. W. and S. G. Hinch. 2010. Effectiveness monitoring of fish passage facilities: historical trends, geographic patterns and future directions. Fish and Fisheries 11:12–33.

SAS. 2003. SAS Version 9.13. SAS Institute, Inc., Carey, North Carolina.

Skalski, J. R., R. Townsend, J. Lady, A. E. Giorgi, J. R. Stevenson, and R. D. McDonald. 2002. Estimating route-specific passage and survival probabilities at a hydroelectric project from smolt radiotelemetry studies. Canadian Journal of Fisheries and Aquatic Sciences 59:1385–1393.

Williams, B. K., J. D. Nichols, and M. J. Conroy. 2001. Analysis and management of animal populations. Academic Press, San Diego.

Wilson, J. W., A. E. Giorgi, and L. C. Stuehrenberg. 1991. A method for estimating spill effectiveness for passing juvenile salmon and its application at Lower Granite Dam on the Snake River. Canadian Journal of Fisheries and Aquatic Sciences 48:1872–1876.

Zabel, R. W. 2002. Using "travel time" data to characterize the behavior of migrating animals. American Naturalist 159:372–387.

Zabel, R. W. and J. J. Anderson. 1997. A model of the travel time of migrating juvenile salmon, with an application to Snake River spring Chinook salmon. North American Journal of Fisheries Management 17: 93–100.

Section 9.2

Using Mark-Recapture Models to Estimate Survival from Telemetry Data

Russell W. Perry, Theodore Castro-Santos, Christopher M. Holbrook, and Benjamin P. Sandford

INTRODUCTION

Analyzing telemetry data within a mark–recapture framework is a powerful approach for estimating demographic parameters (e.g., survival and movement probabilities) that might otherwise be difficult to measure. Yet many studies using telemetry techniques focus on fish behavior and fail to recognize the potential of telemetry data to provide information about fish survival. The sophistication of both mark–recapture modeling and telemetry has dramatically improved since the 1980s, largely due to technological advancements in computing power (for mark–recapture models) and electronic components (for telemetry). Such advances now allow mark–recapture models to take advantage of the detailed information that telemetry techniques can provide.

The key feature of mark–recapture models is simultaneous estimation of detection and survival probabilities. With telemetry, a "capture" event consists of detecting a given tag code one or more times at a specific location or time. By contrast, in some studies interest may focus on the probability of detecting a single tag transmission (see Sections 7.2 and 9.1). Compared to conventional mark and recapture methods, telemetry methods often have greater detection probabilities due to large detection ranges, increased "effort" (i.e., continuous monitoring with autonomous receivers), and ability to simultaneously monitor multiple locations. Nonetheless, perfect detectability is rare in telemetry studies because both random (e.g., from electronic noise) and nonrandom processes (e.g., receiver loses power temporarily) can allow a fish to pass a receiver undetected. Failure to account for imperfect detection can lead to serious bias in survival estimates. When using telemetry to estimate survival, it is therefore critical to explicitly estimate detection probabilities to ensure unbiased estimates of survival (see Section 7.2). Fortunately, using telemetry techniques and mark–recapture models together yields the best of both worlds: Well-designed telemetry systems deliver high detection probabilities that result in precise estimates from small sample sizes. Mark–recapture models ensure estimates of the demographic parameters are unbiased with respect to the detection process.

In this Section, we illustrate the flexibility with which mark–recapture models and telemetry systems can be tailored to unique situations in aquatic environments. We describe simple to complex mark–recapture models and the associated telemetry system design needed to support each model. We simulate a tagged population of fish migrating downstream past a dam and apply different models to this simulated data set. Although we present a specific application, the general framework and flexibility of using mark–recapture models with telemetry data applies to many situations. We then discuss the assumptions associated with each model, and give examples showing how assumptions can be violated. Since our goal is to introduce the application of mark–recapture models to telemetry data, we do not present a complete treatment of the statistical theory or software used to implement mark–recapture models. This Section will be useful to readers familiar with either telemetry or mark–recapture models to design telemetry studies within a mark–recapture framework.

Telemetry system design (e.g., the spatial arrangement of antennas or hydrophones) dictates the parameters that can be estimated using a mark–recapture model. More complex mark–recapture models require more detailed information from a telemetry system. Therefore, study design is crucial to successfully apply mark–recapture models to telemetry data and involves 1) identifying the demographic parameters of interest, 2) designing a mark–recapture model to estimate these parameters, and 3) implementing a telemetry system that provides detection data required by the mark–recapture model. In our experience, many telemetry studies fail to follow these important steps. Therefore, the organization of this Section emphasizes the critical linkage between telemetry system design and mark–recapture model design.

A SIMULATED FISH PASSAGE STUDY

As described in Section 9.1, we simulated a fish passage study and then applied both time-to-event and mark–recapture analyses to the same simulated data set. Section 9.1 describes the simulation from the perspective of time-to-event analysis; here we describe details of the simulation relating to mark–recapture modeling. In summary, simulated tagged fish were released upstream of a dam, arrived at the dam over time during experimental treatments consisting of either a low spill or high spill dam operation, and then migrated downstream to the ocean. The proportion of tagged fish passing through the dam's turbines and spillway arises as a function of 1) treatment-specific rates of passage, and 2) arrival timing in the forebay (see Section 9.1).

Survival was simulated as a function of three different mortality processes occurring at different locations in the study area: 1) mortality upstream of the dam, which was a function of residence time in the forebay, 2) mortality as a result of passing through either turbines or spillway, and 3) mortality between the dam and river mouth at ocean entry. Specifically, mortality upstream of the dam is determined by time-dependent mortality in the forebay relative to the amount of time required to pass the dam (see Section 9.1). The probability of surviving dam passage was set at 0.70 for the turbines and 0.90 for the spillway. Survival from the dam to the river mouth declined exponentially over time as $S = e^{-rt}$ where $r = 0.005$ and t is the date of dam passage. This pattern was selected to simulate the effects of migratory delay on physiological and ecological preparedness for the marine environment (McCormick et al. 1998). Fates of individual fish were simulated by using random number generators to draw Bernoulli outcomes (i.e., we "flipped a coin") with associated survival probabilities (i.e., a probability of "heads").

As described in Section 9.1, 300 fish were released upstream of the dam, but we also simulated a release of 300 tagged fish in the tailrace corresponding to release times upstream of the dam. Fish released in the tailrace were considered a "control" group because their survival between the dam and river mouth was influenced only by in-river mortality processes (i.e., day of year). In contrast, for fish passing the dam, survival between the dam and river mouth was influenced by both mortality due to dam passage and mortality due to in-river processes.

The telemetry system was simulated to detect fish in the forebay, within each passage route (turbines and spillway), in the tailrace just downstream of the dam, and at the river mouth to detect fish that survived to the ocean (see Section 9.1 for details). Detection probabilities often vary among monitoring sites due to variable effort (e.g., number and detection range of antennas) and site configuration (e.g., channel width). Therefore, we assigned different detection probabilities to each monitoring site. The probability of detecting a tagged fish was set at 0.95 for the turbines, 0.75 for the spillway, 0.80 for the tailrace, and 0.70 at the river mouth. As with survival, detections of individual fish were simulated as random draws from a Bernoulli process with associated detection probabilities. Detection probabilities in actual field studies may be higher or lower than those simulated here.

By simulating the data, we know exactly the survival, passage, and detection fates of each individual. Therefore, by comparing estimates from the mark–recapture model to the calculations based on the known fates, we can verify that the estimated parameters are unbiased.

APPLYING MARK-RECAPTURE MODELS TO THE SIMULATED DATA SET

Given the telemetry system deployed to detect fish in our simulated study, a number of different mark–recapture models could be fit to these data. It is important to recognize that the complexity of any mark–recapture model is limited by the complexity of the telemetry system used to monitor the tagged population. Thus, models may be only as complex as the telemetry system allows, but may be simplified as desired by pooling or ignoring detection data appropriately. We illustrate this concept by first fitting simple mark–recapture models to telemetry data and showing which parameters can be estimated. We then show how adding information from the telemetry system allows for more complex models with additional parameters.

THE CORMACK–JOLLY–SEBER MODEL

The Cormack–Jolly–Seber (CJS) model has been the foundation of mark–recapture modeling for over forty years (Cormack 1964; Jolly 1965; Seber 1965). Two types of parameters are estimated by the CJS model: S_i, the probability of surviving from telemetry station i to i + 1, and p_i the probability of being detected at station i conditional on surviving to station i. For our simulated telemetry study, a CJS model can be constructed using detections at two sites: one at the dam (combining detections within the spillway and turbines) and one at the river mouth (Figure 1). This is the minimum telemetry system design that allows us to obtain any estimate of survival, and for the moment, we ignore information from other telemetry stations in our study. As will be seen, however, the CJS model for this telemetry design reveals little about our simulated population.

Relating the CJS parameters to our study, S_1 estimates the probability of fish surviving from the release point to a passage route entrance (spillway or turbines), and S_2 represents

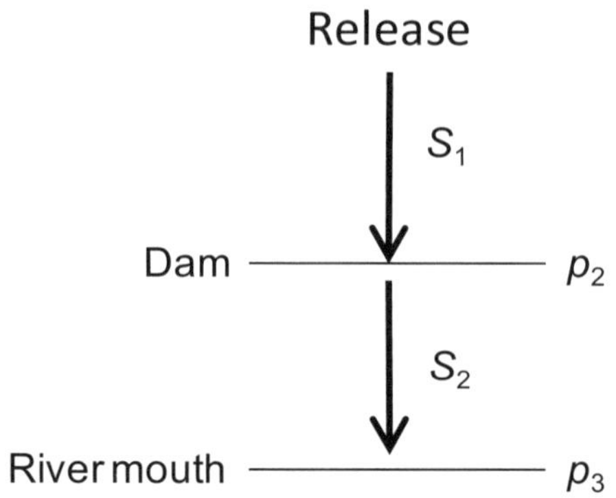

FIGURE 1. Schematic of CJS model showing survival (S_1 and S_2) and detection probabilities (p_2 and p_3) that can be estimated from the telemetry arrays installed at the dam and river mouth. In a CJS model, only the joint probability, $\lambda = S_2 * p_3$ can be estimated in the last reach. For this example, p_3 is estimated from a separate model for replicate arrays (described in detail below). Given an estimate for p_3, S_2 can be obtained as λ/p_3.

survival of fish from a passage route entrance to the river mouth. For detection, p_2 represents the average detection probability of both passage routes (spillway and turbines), and p_3 represents detection probability at the river mouth (Figure 1). By convention, the release site is considered the first detection location and p_1 is set to 1.

With the telemetry design for this model, we cannot estimate all of the parameters just described. Detection probability at telemetry station i is estimated from detections at locations downstream of station i. Therefore, the detection probability at the last station cannot be estimated. Since detection probability is needed to estimate survival, survival alone cannot be estimated through the final reach. The best we can do is to estimate $\lambda = S_2 p_3$, the joint probability of surviving and being detected at the river mouth. This confounding of the parameters limits inferences about survival in the final reach.

Maximum Likelihood Estimation

Here, we briefly present the statistical theory used to estimate the parameters of all mark–recapture models, using our simple CJS model as an example. Most mark–recapture models are based on the multinomial distribution, a probability distribution describing the probability of occurrence of events falling into discrete categories. In our case, the categories are detection histories of each fish's migration through the study area. Detection histories are composed of an alpha-numeric string describing whether a fish was detected at each location in the study area. For the CJS model, a "1" means a fish was detected at a given location and a "0" means it was not detected. For the simple CJS model above, a "101" history means the fish was released upstream of the dam ("1"), not detected in either the spillway or turbines ("0"), but subsequently detected at the river mouth ("1").

The set of all possible detection histories forms the categories of a multinomial distribution. In turn, each detection history has a probability of occurrence that depends on the underlying survival and detection probabilities (Table 1). For example, for the detection history "101" to have occurred, a fish must have survived to a passage route (S_1), not been detected in

Table 1. All possible detection histories, their probability of occurrence (π_i), and the number of fish with each capture history in the simulated data set (n_i) for our simple CJS model.

Capture history	Probability of occurrence (π_i)	n_i
111	$S_1{*}p_2{*}\lambda$	54
101	$S_1{*}(1-p_2){*}\lambda$	11
110	$S_1{*}p_2{*}(1-\lambda)$	183
100	$(1-S_1)+S_1{*}(1-p_2){*}(1-\lambda)$	52

either passage route ($1-p_2$), survived to the river mouth (S_2), and detected at the river mouth (p_2). Thus the probability of observing detection history "101" is the product of these event probabilities: $S_1 * (1-p_2) * S_2 * p_3 = S_1 * (1-p_2) * \lambda$.

For any given data set of detection histories, there is a unique set of parameter values that is most likely to have produced the observed detection history frequencies. To find the most likely parameter values, the next step in mark–recapture analysis is to define the likelihood function. For multinomial models, the likelihood takes the form

$$L(\boldsymbol{\theta} \mid n_i, R) \propto \prod_i \pi_i^{n_i}$$

where $L(\boldsymbol{\theta} \mid n_i, R)$ is the likelihood of the parameters (θ) given the data (n_i, R): π_i is the probability of observing each detection history, n_i is the number of fish with each detection history, and R is the total number of fish released. Given the probabilities of each detection history and the frequencies of occurrence in the simulated data set (Table 1), the full likelihood function for this example is:

$$L(\boldsymbol{\theta} \mid n_i, R) \propto (S_1 p_2 \lambda)^{54} (S_1(1-p_2)\lambda)^{11} (S_1 p_2(1-\lambda))^{183} ((1-S_1)+S_1(1-p_2)(1-\lambda))^{52}$$

The likelihood function yields the relative likelihood of one set of parameter values over another for a given data set of detection histories. By calculating and comparing likelihoods among different values for S_1, p_2, and λ, we can identify the most likely parameter set for the data. These "maximum likelihood estimates" will be closest to the true values of the parameters.

In practice, two approaches are used to find the maximum likelihood estimates: 1) deriving equations that provide the maximum likelihood estimates ("analytical" approach), and 2) using an optimization routine (e.g., Newton-Raphson method) that will systematically calculate the likelihood of different parameter values to find the set of parameter values that maximizes likelihood ("numerical" approach). The two methods are akin to estimating the mean using $\Sigma x_i/n$ (the "analytical" approach) or by using least-squares methods (i.e., the value of the mean that minimizes the total sum of squares—the "numerical" approach). The numerical approach is by far the most common means of obtaining maximum likelihood estimates because analytical expressions for maximum likelihood estimates do not exist for many mark–recapture models. However, both approaches can be used for the CJS model. We adapted the maximum likelihood estimators of the CJS model from Seber (1982) and Burnham et al. (1987) and present those equations here to illustrate both approaches.

The maximum likelihood estimators for detection probability (p_i), survival probability (S_i), and joint survival-detection probability (λ) of the CJS model are:

$$\hat{p}_i = \frac{r_i}{r_i + z_i}$$

$$\hat{S}_{i-1} = \frac{\hat{M}_i}{\hat{M}_{i-1}}$$

$$\hat{M}_i = \frac{m_i(r_i + z_i)}{r_i} = \frac{m_i}{\hat{p}_i}$$

$$\hat{\lambda} = \frac{r_{k-1}}{m_{k-1}}$$

Where variables are defined as follows:

i = detection site, $i = 1, 2, \ldots, k$, numbered from upstream to downstream ($i = 1$ is the release site, and $k = 3$ in our simple CJS model).

m_i = number of fish detected at site i.

r_i = number of fish detected downstream of site i (i.e., at sites $i + 1 \ldots k$), of those detected at site i.

z_i = number of fish *not* detected at site i, but detected downstream of site i.

Note that $M_1 = R$ (the number of fish released), $p_1 = 1$, and $r_i + z_i$ = total number of fish detected downstream of site i and therefore, known to be alive at site i.

The maximum likelihood estimators for the CJS model make intuitive sense. The detection probability is the fraction of fish detected at site i of those known to have migrated past site i. The parameter $\hat{M}_i$ estimates the number of fish alive at site i, which is the number fish detected at site i divided by the detection probability at site i. The survival probability from site i–1 to site i is the estimated number of fish alive at site i divided by the estimated number of fish alive at the previous site, i–1. These maximum likelihood estimators show the direct connection between the survival probabilities and the numbers of fish detected at each location. Standard errors for these estimators can be found in Seber (1982) and Burnham et al. (1987).

Applying the Model to the Simulated Data Set

The frequencies of the capture histories in Table 1 provide all the information needed to apply the CJS maximum likelihood estimators:

$$m_2 = n_{111} + n_{110} = 237,$$

$$r_2 = n_{111} = 54,$$

$$z_2 = n_{101} = 11,$$

$$\hat{M}_2 = 237/0.8307 = 285.3,$$

$$M_1 = R = 300$$

In Table 2, we compare the analytical estimates from the simulated data set to the known fates of fish surviving to each detection site over the entire study period. We also fit the CJS model with the software program USER (User-Specified Estimation Routine, Lady and Skalski 2009). This program maximizes the likelihood via an optimization routine and returns maximum likelihood estimates and their standard errors.

The analytical and numerical estimators produced the same parameter estimates (Table 2), and both are within 1.1 percentage points of the known-fate estimates. Note that $\hat{\lambda}$ accurately estimates $S_2 * p_3$ from the known fate data. An important insight here is that our interpretation about S_2 from λ depends completely on assumptions about the value of p_2. Given that p_2 cannot be estimated with the current design, if we had assumed perfect dectability ($p_3 = 1$), then our estimate of S_2 is 0.230 when the true value is really 0.365. Similarly, our estimate of S_1 would have been negatively biased if we had not used the mark–recapture model but instead assumed that $p_2 = 1$. In this case, substantial negative bias is the consequence of assuming perfect detectability. This example highlights the importance of explicitly estimating detection probabilities using a mark–recapture model in order to obtain unbiased estimates of survival.

Extensions to the Simple CJS Example

In the simple example above, we did not use telemetry arrays located in the forebay or in the immediate tailrace as shown in Figure 1 of Section 9.1. However, adding data from those receivers allows us to estimate new parameters and improve precision of existing parameters. For example, by including the forebay receiver detections as an additional occasion, we can separately estimate survival 1) between release and the forebay, and 2) between the forebay and the entrances to the spillway or turbines. Spatial partitioning of survival is the most common reason to add new receiver locations to a telemetry system.

Table 2. Known-fate survival of simulated fish, analytical estimates of parameters, and numerical estimates obtained from maximizing the likelihood via an optimization routine.

Parameter	Known-fate estimates	Analytical estimates	Numerical estimates (SE)
S_1	282 of 300 = 0.940	285.3/300 = 0.951	0.951 (0.048)
p_2	238 of 282 = 0.844	54/(54 + 11) = 0.831	0.831 (0.047)
S_2	103 of 282 = 0.365	NA	NA
p_3	65 of 103 = 0.631	NA	NA
λ	65 of 282 = 0.230	54/237 = 0.228	0.228 (0.027)

Extending the model with an additional sampling occasion can also improve precision of survival estimates. For example, antennas in the tailrace would likely detect transmitters in fish that had died during dam passage, making it impossible to estimate survival over such a short distance. Nonetheless, including this detection location in the model could substantially improve the precision of p_2 and S_1 because precision of the parameters is influenced by the number of surviving fish detected downstream of the dam. In the current model, only the 65 fish detected at the river mouth were used to estimate p_2 (Table 1). In contrast, in the simulated data set, 217 fish were detected in the tailrace. It makes sense to use these detections in the tailrace to improve precision of S_1.

To use the tailrace array, a detection location can be added to the CJS model as the next sampling location after the dam (see Figure 1). Then the survival probability for the reach between the dam and tailrace array is set to 1, assuming that all fish passing the dam will also pass the tailrace array, whether alive or dead. With this approach, detections on both the tailrace and river-mouth telemetry stations are used to estimate p_2 and S_1, and all other parameters are interpreted as above. When this model is applied to the data set, $\hat{S}_1 = 0.940$ with a standard error of 0.018. This estimate is exactly equal to the known-fate estimate for S_1, but the standard error is less than half that obtained from our original model (Table 2)—this represents a substantial improvement in precision. Another alternative approach is to lump the dam and tailrace detections together into one occasion. However, maintaining separate occasions provides spatially-explicit detection probabilities that may be insightful for evaluating the performance of the telemetry system. Further, the approach used here can be applied to more complex models that estimate route-specific survival (see Multistate models). These extensions to our simple CJS model illustrate the flexibility of using telemetry in conjunction with mark–recapture models.

Assumptions

CJS models are subject to seven assumptions. These assumptions relate to inferences to the population of interest, error in interpreting transmitter signals, and statistical fit of the data to the model's structure:

1) Tagged individuals are representative of the population of interest. For example, if the target population is wild-origin Chinook salmon *Oncorhynchus tsawytscha* then the sample of tagged fish should be drawn from that population. If hatchery-origin Chinook salmon must be tagged then inferences about survival of wild fish from hatchery fish must be based on subject matter arguments. Other areas that require representation include size distribution, health profiles, and migration timing. In many telemetry studies this may be difficult to perfectly achieve, requiring caveats or discussion for proper inference. For example, the size of an acoustic- or radio-tag may preclude tagging the smallest individuals of a particular population due to tag-burden concerns, or a study may need to be implemented during a three-week window of a five-week migration distribution. As we've noted, careful study design and candid discussion of limitations maximizes the value of inference.

2) Survival of tagged fish are the same as that of untagged fish. For example, the tagging procedures or detection of fish at telemetry arrays should not influence survival or

detection probabilities. If the transmitter negatively affects survival, then estimates of survival rates will be negatively biased relative to untagged fish. Also, using our simulated example, if the tagging process or transmitter presence affect swimming performance, tagged fish may not pass through spillway or turbine routes in the same proportions as untagged fish.

3) All sampling events are instantaneous. When the spatial or temporal scale of sampling occasions approaches the scale of time or distance between occasions, then it becomes difficult to correctly attribute mortality to a specific sampling period. For example, for spatial mark–recapture models, if the detection range of a telemetry station is 200 m, but the distance between telemetry stations is only 100 m, it becomes difficult to determine where mortality occurs. Thus, when estimating survival through space, sampling should take place over a short distance (e.g., hundreds of meters) relative to the distance between telemetry stations (e.g., kilometers). Likewise, when estimating survival over time, sampling occasions should be short (e.g., one day) relative to the time between sampling occasions (e.g., weeks). One way to address this assumption is to place telemetry stations in locations, if possible, where tagged fish move relatively quickly past an array and do not reside with the detection range for an extended period.

4) The fate of each tagged fish is independent of the fate of other tagged fish. If fish exhibit some sort of group behavior (i.e., schooling), sample size is effectively reduced and precision estimates become biased downward.

5) The prior detection history of a tagged fish has no effect on its subsequent survival. For example, in mark–recapture studies requiring physical recapture of animals, the capture process may influence subsequent survival. For telemetry, this assumption is usually satisfied by the passive nature of detecting tags.

6) All tagged fish alive at a sampling location have the same detection probability.

7) All tags are correctly identified and the status of tagged fish (i.e., alive or dead) is known without error. This assumes fish do not lose their tags and that the tag is functioning while the fish is in the study area. Additionally, this assumption implies that all detections are of live fish, that dead fish are not detected and interpreted as live, and that spurious detections (e.g., false positive detections) are not interpreted as live fish. This assumption is violated in cases where a detection location occurs close enough below a point-source mortality (e.g., dam passage) such that dead fish are detected at the downstream telemetry array. The consequence of violating this assumption is that mortality in one reach is attributed to a downstream reach.

Assumptions 5 and 6 can be formally tested using χ^2 Goodness of Fit tests known as Test 2 and Test 3 (Burnham et al. 1987). Both Test 2 and 3 are implemented as a series of contingency tables. Test 2 is informally known as the "recapture test" because it assesses whether detection at an upstream array affects detections at subsequent downstream arrays (assumption 6). Test 3 is known as the "survival test" because it assesses assumption 5 that fish alive at array i have the same probability of surviving to array $i + 1$. With telemetry data, these tests are sometimes uninformative because detection probabilities are so high that cell frequencies are too small for valid contingency table statistics (i.e., many zero frequencies in the "nondetected" category).

Tag failure can be evaluated to formally test Assumption 7 that tags do not fail prior to fish exiting the study area. A controlled tag life study can be conducted to estimate the probability of tag failure at any point in time after tags were turned on. The methods of Townsend et al. (2006) can then be used to estimate the average probability that a tag was alive while fish were in the study area. If tags fail prior to exiting the study area, then information from the tag life study can be used to adjust survival estimates for the probability of tag failure (Townsend et al. 2006).

It should be noted that nearly all telemetry studies violate most or all of these assumptions to some degree. It is logical to assume that the handling, tagging, and release protocols of a study, though carefully implemented, still have biological and behavioral effects on tagged individuals. This does not mean that inference cannot be made from the results of the study, only that researchers must report, and take care in how they present, the "challenges" of their data and results and put proper caveats on the potential limitations on their inference.

USING REPLICATED TELEMETRY ARRAYS TO ESTIMATE DETECTION PROBABILITY

In our simple CJS model, we were unable to estimate survival from the dam to the river mouth (S_2) without making assumptions about detection probability at the last telemetry station (p_3). To estimate p_3 and S_2, we need an additional telemetry array downstream of our final array. In some cases, another receiver could simply be installed further downstream, and the model extended accordingly to estimate S_2, p_3, and a new $\lambda = S_3 * p_4$. However, since the final station is already at the mouth of the river, our best option is to implement two independent telemetry stations at the same location. Let us assume that this detection station always consisted of two sets of independent telemetry stations, but that we did not explicitly use this information in our CJS model. The receivers at this location are so close together, that we can safely assume that no mortality occurs between them. As with the tailrace example above, one approach is to modify the CJS model by adding an additional reach and setting the survival probability to one. Another option is to use a separate model based on the Lincoln-Peterson method (Seber 1982) that estimates detection probability from the two monitoring stations.

The detections at each array yield a two-digit capture history defining whether fish were detected on both arrays ("ab"), or only one or the other array ("a0" or "0b"). The detection history probabilities are given in Table 3. Because we have an independent likelihood for the final monitoring station, the total likelihood for combined CJS and dual-array model is the product of the two likelihoods (i.e., $L_1 * L_2$). The idea is to use the dual-array model to estimate p_3, and then feed p_3 into the likelihood for the CJS model, allowing us to estimate S_2. The advantage of combining the models is that the resulting estimates incorporate uncertainty arising from both models. This technique is analogous to the robust design of Pollock (1982). Applying the combined CJS with dual array model to the simulated data, we find that $\hat{p}_{3a}$ = 0.40 (SE = 0.077), $\hat{p}_{3b}$ = 0.39 (SE = 0.076), $\hat{p}_3$ = 0.634 (SE = 0.083), and $\hat{S}$ = 0.359 (SE = 0.064). Our estimates for p_3 and S_2 agree well with the known fates (Table 2). By adding the dual array to the telemetry system and extending the model with the second likelihood, we are now able to estimate survival of tagged fish through the entire system. We know that 34.3% of

Table 3. Probability of occurrence for each detection history of the dual-array model, where p_a and p_b are the detection probability on the first array and second array. The probability of being detected by at least one of the arrays is $p_3 = 1 - (1 - p_{3a})(1 - p_{3b})$. Also given is the number of fish with each capture history in the simulated data set.

Detection history	Probability of occurrence (π_i)	n_i
ab	$p_{3a}p_{3b}/p_3$	16
a0	$p_{3a}(1-p_{3b})/p_3$	25
0b	$(1-p_{3a})p_{3b}/p_3$	24

simulated fish (103 of 300) survived to the ocean, whereas our estimate using a mark–recapture model is $\hat{S}_1\hat{S}_2 = 0.951*0.359*100 = 34.1\%$. This example shows how the design of the telemetry system is directly linked to the structure of the mark–recapture model.

Assumptions

The primary additional assumption of the replicate array model is that the two telemetry arrays are statistically independent. That is, fish detected by one array must constitute a random sample of fish passing through the detection field of the second array. If the replicate arrays are not independent, then estimates of detection probability may be biased, which may also induce bias in estimates of survival. In the example above, this assumption can be violated if the detection fields of the dual arrays do not encompass the entire channel cross section, or if they were both using the same power source and service was interrupted simultaneously. Perry and Skalski (2008) assess the consequences of failing the assumption of independent arrays and provide guidelines for configuration of telemetry arrays.

PARTITIONING COMPONENTS OF SURVIVAL

Survival between the dam and the ocean arises as a function of multiple mortality processes. In our simulation, fish experienced mortality due to passage through the dam and additional mortality during migration in the river below the dam. The estimate of S_2 above includes both of these processes, but what if we were interested in separating mortality due to dam passage from other sources of mortality? As previously mentioned, estimating survival to the tailrace using telemetry arrays is difficult because all tagged fish, alive and dead, may be detected at this location. The solution is to add a second release site immediately below the dam. The idea is that the tailrace release group experiences only mortality unrelated to dam passage in the river segment between the dam and the river mouth. Survival estimates from both release sites can then be used to partition S_2 for the upstream release into dam and in-river components (Figure 2). This is the "paired-release" model first defined by Burnham et al. (1987) and tailored to radio telemetry by Skalski et al. (2001). Although we present a specific example of the paired release model, the model has general application to any two groups of fish that experience some common source of mortality, with one of the groups experiencing some additional source of mortality that we wish to estimate.

To implement the paired release model for this study, we re-write S_2 as a function of its underlying components and add subscripts for clarity (Figure 2):

$$S_{2,1(\text{dam \& river})} = S_{2,1(\text{dam})} S_{2,1(\text{river})} \tag{1}$$

where $S_{2,1(\text{dam \& river})}$ is the survival probability through reach 2 for release site 1 (see Figure 2) and "dam & river" indicates that this survival probability is comprised of mortality due to passing the dam and also due to factors unrelated to dam passage in the stretch of river below the dam. The right hand side of Eqn. 1 separates the mortality sources into survival probabilities due to each source. Survival from the tailrace to the ocean is estimated from the second release group ($S_{2,2(\text{river})}$). If we assume $S_{2,1(\text{river})} = S_{2,2\ (\text{river})}$ then Eqn. 1 can be rearranged to yield:

$$S_{2,1(\text{dam})} = \frac{S_{2,1(\text{dam \& river})}}{S_{2,1(\text{river})}} = \frac{S_{2,1(\text{dam \& river})}}{S_{2,2(\text{river})}} \tag{2}$$

This model now has four likelihoods and eight unique parameters: a two-reach CJS model for the upstream release (3 parameters), a one-reach CJS model for the tailrace release (1 parameter), and two dual-array models (2 parameters each) to estimate $p_{3,i}$ for each release location. Note that the parameter $S_{2,1(\text{dam})}$ is not part of the likelihood, but rather is derived as a function of model parameters. Derived parameters are often of main interest, as illustrated in this example. However, since these parameters are not explicitly estimated, the standard error of derived parameters is usually estimated using the Delta method (Seber 1982).

The paired release model estimates the known-fate proportions closely (Table 4) and yields insights about mortality processes affecting our simulated population. The estimates suggest that mortality in the river below the dam is greater than mortality due to dam passage.

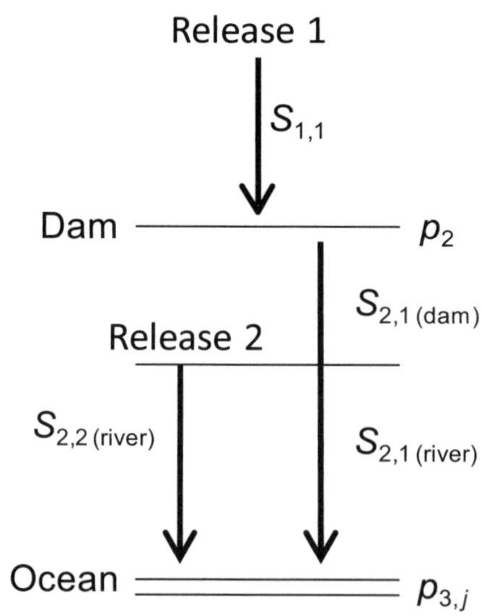

Figure 2. Schematic of the paired release model showing $S_{i,j}$ for reach i and release location j. The dashed line indicates the release location in the tailrace where survival is partitioned into dam-related and in-river components. Note that $S_{2,1(\text{dam})}$ and $S_{2,1(\text{river})}$ can be estimated only by assuming that $S_{2,1(\text{river})} = S_{2,2(\text{river})}$.

TABLE 4. Estimates of selected parameters from the simulated data set compared against the known fates of fish for the paired-release model shown in Figure 2.

Parameter	Known-fate estimate	Maximum likelihood estimate (SE)
$S_{2,1(dam)}$	224 of 282 = 0.794	0.777 (0.179)
$S_{2,1\ (river)}$	103 of 224 = 0.460	0.463 (0.068)

Assumptions

The paired release model is subject to the same assumptions of the CJS model plus the following additional assumption:

1) Survival is equal for upstream and tailrace release groups between the tailrace release point and the first downstream telemetry array. Inequality between these groups can result in either positive or negative bias, depending on the direction of the inequality (see Equation 2). This assumption implies that any handling and tagging mortality has been expressed prior to release.

MULTISTATE MODELS

In our example of fish passing through a dam, we have yet to address the fact that fish passed through different routes and under different treatments of dam operations. What is the survival of fish passing through each route? What fraction of the population passes through each route? Do the treatments influence passage and survival? Multistate models offer a flexible framework in which to answer these questions (Nichols and Kendall 1995; Lebreton and Pradel 2002). We first present the general framework of multistate models, and then show how our example can be cast as a multistate model. Last, we extend the model to include different kinds of state variables and illustrate how the multistate model can be used as a basis for evaluating hypotheses.

Like CJS models, multistate models estimate survival and detection probabilities. However, multistate models also allow individuals to be stratified into groups (states) to estimate survival and detection parameters separately for each group. Individuals are allowed to move (transition) from one state to another between sampling occasions. Multistate models estimate the probability that an individual transitions from one state to another, and also the state-specific survival probabilities. For example, these models were first designed to examine movement probabilities of animals among different geographic locations and survival probabilities for each geographic location (Brownie et al. 1993; Schwarz et al. 1993).

To describe the model, we use the notation of Brownie et al. (1993), but this notation is later adapted for models specific to our simulated study. The fundamental parameters estimated by the multistate model are:

ϕ_i^{rs} = joint probability of surviving from sampling occasion i to $i + 1$ and moving from state r at sampling occasion i to state s at occasion $i + 1$.

p_i^s = probability of detection in state s at occasion i.

Given multiple states, it is convenient to express these parameters in matrix form, here using three states for simplicity:

$$\phi_i = \begin{bmatrix} \phi_i^{11} & \phi_i^{12} & \phi_i^{13} \\ \phi_i^{21} & \phi_i^{22} & \phi_i^{23} \\ \phi_i^{31} & \phi_i^{32} & \phi_i^{33} \end{bmatrix} \quad \text{and} \quad \boldsymbol{p}_i = \begin{bmatrix} p_i^1 & 0 & 0 \\ 0 & p_i^2 & 0 \\ 0 & 0 & p_i^3 \end{bmatrix}.$$

When all states are sampled at every occasion and animals move among all states, all parameters are estimable except for the final interval, similar to the CJS model. In the parameterization above, ϕ_i^{rs} includes the underlying probabilities of both surviving and moving between states. To estimate the underlying parameters, the model can be reparameterized as a function of S_i^r, the probability of surviving from occasion i to $i + 1$ conditional on being in state r at occasion i; and Ψ_i^{rs}, the probability of transitioning from state r at occasion i to state s at $i + 1$, conditional on surviving to $i + 1$. Using the three-state example, the reparameterization is

$$S_i^r = \phi_i^{r1} + \phi_i^{r2} + \phi_i^{r3} \quad \text{and}$$

$$\Psi_i^{rs} = \frac{\phi_i^{rs}}{S_i^r}.$$

Note that Ψ_i^{rs} for a particular state s must be specified as one minus the sum of the others because the transition probabilities are constrained to sum to one.

The Route-Specific Survival Model

The route-specific survival model introduced by Skalski et al. (2002) is a multistate model designed to estimate the proportion of fish passing through each route at a dam, and then survival conditional on the route of passage. In our study, we have two passage routes, yielding the model shown in Figure 3.

The route-specific survival model requires replicate telemetry arrays with each passage route in order to estimate route-specific detection probabilities. With only a single detection array in each route, some fish will pass through a route undetected and although detected downstream, the route of passage of these fish is unknown. Without replicate telemetry arrays, we can estimate the average detection probability over both routes, but if detection probabilities differ between routes, then estimates of the proportion of fish passing each route will be biased. By installing two sets of antennas in each passage route, the detection probability of each route can be estimated, allowing the remaining parameters to be estimated without bias.

Assumptions

Multistate models are subject to the same assumptions as the CJS model plus two additional assumptions:

1) In order to separate state-specific survival from state transition probabilities, we must assume that all mortality occurs first and then fish transition “instantaneously” at

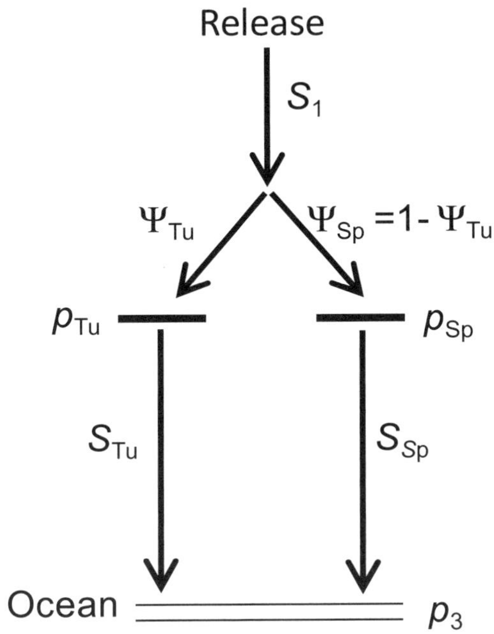

FIGURE 3. Schematic of the route-specific survival model (Skalski et al. 2002) showing route passage probabilities (Ψ) for the turbines (Tu) and spillway (Sp), route-specific detection probabilities (p_{Tu}, p_{Sp}), and survival from the dam to the ocean conditional on having passed through the turbines (S_{Tu}) or spillway (S_{Sp}). Route-specific detection probabilities are estimated by the dual array model presented earlier.

the end of the sampling period. For the route-specific model described above, this assumption is fulfilled because mortality occurs upstream of the dam prior to fish passing (transitioning) through either the spillway or turbines. An example where this assumption may be violated is when a river splits into multiple channels and the goal is to estimate the probability of entering a given river channel. In this case, if monitoring stations are located considerably downstream of the river junction, transition probabilities may be biased if survival probabilities differ in the two channels downstream of the river junction (see Perry 2010).

2) There is no error in assignment of states to individuals. For example, for the route-specific model described above, we assume that no fish are wrongly assigned to a passage route.

TAILORING MARK–RECAPTURE MODELS TO STUDY GOALS

Armed with the suite of models discussed thus far, we have the ability to tailor the telemetry system and mark–recapture model to answer specific questions in a study. In our simulated study, the dam was operated at high and low spill discharge implemented over six 48-h treatment periods. The primary objective is to determine dam operations that maximize rates of passage through the safest possible route. A priori, it is assumed that survival through turbines is lower than the spillway. Therefore, dam operations may affect survival at the population level by influencing the proportions of fish passing through each route. We also suspect a temporal component to survival, with fish passing the dam later in the study having lower survival to the ocean. We would like to quantify this relationship to link migration delay above the dam to survival below the dam.

How should we structure the model to assess these hypotheses? One can envision a number of *ad hoc* approaches to address the questions at hand. The first that comes to mind is constructing separate models for each of the six release groups above the dam. Most of the fish pass within a given treatment period, but some do not. This approach would likely capture major differences among treatments, but the estimates are not unique to a given treatment and represent a mixture of fish passing during multiple treatments. Another approach is to construct a separate model for each treatment, assigning fish to a treatment based on when they pass the dam. This approach is undesirable because we must exclude fish that are not detected in passage routes and estimating S_1 is impossible (by definition, only fish surviving to a passage route are included).

The answer is to generalize the multistate model to represent different kinds of state variables. State variables need not represent geographical locations. States can be defined as any classification variable that can be identified at the time of recapture. For example, multistate models have been used to estimate transitions among weight classes (Letcher and Horton 2009) and between breeding and nonbreeding status (Nichols et al. 1994). In our case, we define an additional state variable representing each 48-h time period. Thus, fish detected in a passage route were assigned to one of the six 48-h time periods; the low-spill treatment occurred during time periods 1, 4, and 5 and the high-spill treatment during the other time periods (see Figure 2 in Section 9.1). Although fish not detected in passage routes cannot be assigned to one of the six time periods, the probability function for this detection history incorporates the possibility that fish could have passed during any of the six time periods.

This design yields 12 unique combinations of the state variables (2 routes × 6 time-periods), and thus 12 unique transition and survival probabilities. We must also construct 12 dual array likelihoods to estimate the detection probabilities for each route in each time period. Next, we want to estimate the fraction of the population that passes through each route during each time period. The branching process in Figure 4 shows how this model can be structured. The parameter τ_t estimates the probability of fish passing the dam during time period t, $\Psi_{\mathrm{Tu},t}$ estimates the probability of passing through the turbines given the fish passed the dam during time period t, and $S_{\mathrm{Tu},t}$ is the probability of surviving from the dam to the ocean given passage through the turbines during time period t.

At this point in designing the model, $S_{\mathrm{Tu},t}$ and $S_{\mathrm{Sp},t}$ includes both mortality due to dam passage and in-river mortality below the dam. Therefore, we included the second release in the tailrace, so that we could estimate $S_{\mathrm{Tu},t\,(\mathrm{dam})}$ and $S_{\mathrm{Tu},t\,(\mathrm{river})}$ using a paired release design. Of primary concern here is whether mortality due to dam passage differs between routes.

This model now has 68 parameters, many more than previous models: The multistate model for the upstream release has 24 parameters, the CJS models for fish released below the dam has 6 parameters, and the replicate array models estimate 38 parameters. Since parameters are estimated for many subgroups, precision of parameters under this full model will be poor. Furthermore, we would like to evaluate hypotheses about the effect of dam operations on survival and use of passage routes. We will use model selection approaches based on Akaike's Information Criteria (AIC) to both improve parameter precision and evaluate biological hypotheses (Burnham and Anderson 2002). Among candidate models, the model with the smallest AIC value represents the model with the most favorable tradeoff between precision and accuracy of the estimates (i.e., over-fitting versus under-fitting). The difference

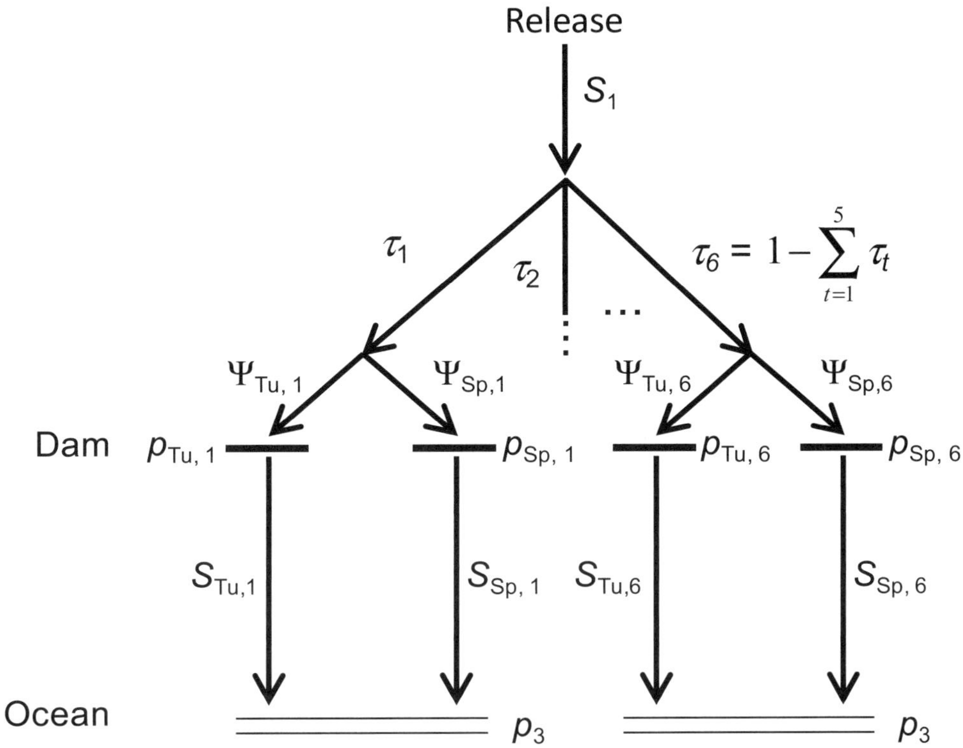

FIGURE 4. Schematic of multistate model for estimating passage (Ψ) and route-specific survival (S_{Tu} and S_{Sp}) probabilities conditional on the time period of dam passage.

in AIC values (ΔAIC) between two models represents the degree of support for one model over another. As a general rule of thumb, ΔAIC < 2 indicates little or no evidence that either model is more appropriate than the other (Burnham and Anderson 2002).

To compare models using AIC, we constructed a number of reduced models that represent different hypotheses about the parameters. For example, detection probabilities are estimated for each route and time period, but if detection probabilities do not vary over time then estimating a single detection probability for all time periods will improve precision by pooling information over all time periods. Comparing AIC of the full model against a model with detection probability constant over time provides evidence of support for one hypothesis over another.

We used a step-down approach to form and compare models (Lebreton et al. 1992). First, we evaluated whether detection probabilities varied by time period or release location. We compared the full model where all detection probabilities were estimated separately for each time period and release location to a model where detection probabilities for each route were constant over time and among release locations (denoted model "$p(.)$," see Table 5). The best-fit model for detection probability was then used as a basis for evaluating remaining hypotheses. Next, we compared three models for the tailrace release group to evaluate whether survival in the reach below the dam was constant over time, unique for each time-period, or declined exponentially over time (i.e., $S = e^{-rt}$). Using the best fit model based on detection probabilities and downriver survival, we then evaluated whether survival for each passage

TABLE 5. Model selection statistics for evaluating different hypotheses about detection (p), survival (S), and routing probabilities (Ψ). Statistics are as follows: NLL = negative log likelihood, AIC = Akaike's Information Criterion, and ΔAIC = difference in AIC between model i and the lowest-AIC model.

Model	Number of parameters	NLL	AIC	ΔAIC
Full	63	107.4	340.9	52.7
$p(.)$	35	120.1	310.1	21.9
$p(.), S_{2,2\ (\text{river})}(.)$	30	137.4	334.9	46.7
$p(.), S_{2,2\ (\text{river})}(\exp(t)),$	30	120.6	301.2	13.0
$p(.), S_{2,2\ (\text{river})}(\exp(t)), S_{\text{route, (dam)}}(\text{treat})$	23	124.4	294.8	6.6
$p(.), S_{2,2\ (\text{river})}(\exp(t)), S_{\text{route, (dam)}}(.)$	21	124.4	290.8	2.6
$p(.), S_{2,2\ (\text{river})}(\exp(t)), S_{\text{route,(dam)}}(.), \Psi_{\text{Tu}}(\text{treat})$	17	127.1	288.2	0.0
$p(.), S_{2,2\ (\text{river})}(\exp(t)), S_{\text{route,(dam)}}(.), \Psi_{\text{Tu}}(.)$	16	165.0	362.0	73.8

route was 1) constant over time, 2) constant within each treatment but different between treatments, and 3) unique for each time period. Finally, using the best-fit model at this stage, we evaluated the same three hypotheses for passage probabilities (i.e., constant for each route, treatment-specific, or time-specific). We then present the estimates for the best fit overall model.

The AIC-best model (Table 5) included constant detection probabilities at each location, survival below the dam declining exponentially with time (Figure 5), no effect of treatments on route-specific survival, but differences between treatments in the proportion of fish passing through the turbines. The probability of passing the turbines (Ψ_{Tu}) was much higher during the low-spill treatment compared to the high-spill treatment, and the estimates match the known-fates well (Table 6). Survival was lower through the turbines than through the spillway. Although some of the parameter estimates differ considerably from the known fates, the standard errors are large and the estimates fall well within the 95% confidence intervals (i.e., the estimate ± 1.96 standard errors). The exponential rate of decline in downstream survival also fit the known-fate data quite well (Figure 5). Last, the best-fit model selected by AIC followed the structure of the true model from which the data were generated.

Dam-passage survival can be estimated as a function of the route-specific survival and passage probabilities as

$$S_{2,1\ (\text{dam})} = \Psi_{\text{Tu}} S_{\text{Tu,(dam)}} + (1 - \Psi_{\text{Tu}}) S_{\text{Sp,(dam)}}.$$

In our simulated population, the estimate of $S_{2,1}$ (dam) was 0.671 (SE = 0.105) during the low-spill treatment, but 0.906 (SE = 0.123) during the high spill treatment. Since route-specific survival did not differ between treatments, the lower overall survival during the low-spill treatment can be attributed to a higher fraction of tagged fish passing through the turbines during this treatment.

OTHER COMPONENTS OF MARK-RECAPTURE MODELING

In this Section, we have illustrated the flexibility of using mark–recapture models to analyze telemetry data. The examples provided here provide a launching pad for developing

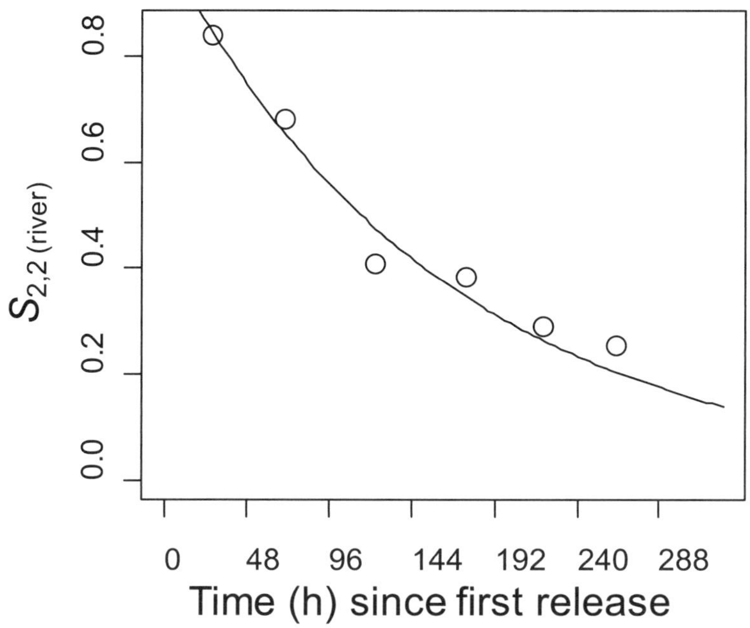

FIGURE 5. Known-fate proportions surviving to the ocean for each tailrace release group (circles) compared to the fitted survival function from the best-fit model (line).

TABLE 6. Maximum likelihood estimates (MLE) of selected parameters from the lowest-AIC model from the simulated data set compared against the known fates of fish.

Parameter	Known-fate estimate	MLE (SE)
$\Psi_{\text{Tu, low spill}}$	90 of 117 = 0.769	0.773 (0.044)
$\Psi_{\text{Tu, high spill}}$	36 of 165 = 0.218	0.217 (0.034)
$S_{\text{Tu,(dam)}}$	84 of 126 = 0.667	0.575 (0.121)
$S_{\text{Sp,(dam)}}$	140 of 156 = 0.897	0.998 (0.146)
r	0.005	0.006 (0.0007)

mark–recapture models to apply to your own telemetry studies. However, given the introductory nature of this Section, we have yet to discuss a number of topics that every practitioner of mark–recapture modeling must address. In closing, we briefly identify some of these topics so integral to mark–recapture modeling.

Evaluating Goodness of Fit

All models are capable of producing estimates, but not all estimates are good estimates in the sense of optimizing the balance between precision and bias. In our model selection example, we used AIC to identify the "best" model, but we still need to determine if the best model is a good one. Evaluating goodness of fit is necessary to assess how well the model fits the data. Lack of fit may be induced by overdispersion or by failure of model assumptions. Overdispersion is defined as more variability in the data than is expected given the model structure and is common when the parameter values are dependent upon some state or trait that is not incorporated in the model. Typically, when data are overdisperse, estimates from a mark–recapture model will be unbiased, but variance estimates will tend to be negatively biased. Therefore, it is important to test for overdispersion and adjust (inflate) the variance when overdispersion has been detected. There are a number of methods for estimating overdispersion (Lebreton et al. 1992; White et al. 2001), but most involve estimation of $\hat{C}$, the variance inflation factor. When the data fit the model's structure perfectly, then $\hat{C} = 1.0$. Lack of fit is indicated by $\hat{C} > 1$, in which case the common practice is to inflate the variance, as well as model selection statistics, by a factor of $\hat{C}$ (Burnham and Anderson 2002).

Mark–Recapture Software

Most biologists lack the mathematical and computer programming experience that are required to construct, fit, and diagnose mark–recapture models of practical scale without the help of automated software routines. Fortunately, computer software packages are available for constructing models and fitting them to the data. A good software program should: 1) be flexible (allow custom models to be built), and 2) provide standard errors and profile likelihood confidence intervals of parameter estimates (see Lebreton et al. 1992), including derived parameters. Integrated diagnostics and model selection criteria are also preferred. In our examples we used the software program USER (Lady and Skalski 2009) to fit the models to the data (i.e., maximize the likelihood function) and obtain estimates for derived parameters (e.g., $S_{2,1(dam)}$). USER is a flexible environment for model development because the model structure and parameters are completely user-specified. Thus, any custom model can be tailored to a specific telemetry study and then fit to the data. However, complete specification of model likelihoods can be a cumbersome task and requires a minimum level of quantitative savvy, especially for large, complex models.

Perhaps, the most common mark–recapture software program is MARK (White and Burnham 1999). MARK has many well-developed model structures, including diagnostic tools, goodness of fit tests, and model selection criteria (White et al. 2001). However, defining the structure of complex or unique models can be cumbersome or impossible with MARK. Recent development of the R package RMark (Laake and Rexstad 2011) has improved the ability to build models and perform model selection with MARK, but requires some knowledge of the R language.

MARK and USER are simply two of several available software packages that have enhanced the field of mark–recapture modeling by simplifying the cumbersome process of model building and fitting. Behind every graphical user interface, however, is a host of complex processes that affect the output. Practitioners should be familiar with likelihood estimation (including variances and confidence intervals), model selection, and linear model terminology. We discourage use of software without an understanding of the underlying statistical theory, such as optimization routines, link functions, model selection criteria, and goodness of fit tests. When used appropriately, these tools can free biologists from the time-consuming task of model construction and allow them to focus on the biological hypotheses behind the models.

Sample Size and Precision

Determining sample size required to achieve some desired level of precision is an important aspect of study design, particularly for telemetry studies where scare resources must be allocated between transmitters and telemetry monitoring stations. So then, how many fish need to be tagged to obtain a given level of precision? The answer depends on the model structure, the number of sampling occasions, the true values of the parameters, and biological variation in the parameters. For example, lower detection probabilities require higher sample sizes to obtain a given level of precision.

To determine sample sizes, simulation studies can be used to estimate precision of estimated parameters (Devineau et al. 2006). The first step is to assume a set of detection and survival probabilities that might be obtained in a given study. These values can be obtained from a pilot study, literature search, or expert opinion. Next, the expected value of capture history frequencies is calculated based on a given sample size and the assumed detection and survival probabilities. Finally, the mark–recapture model is applied to this simulated data set to yield the expected standard errors of survival and detection probabilities.

Fortunately, a number software programs exist to facilitate sample size and precision analysis. The software program SampleSize (Lady et al. 2003) is a simple-to-use program for determining precision and sample size with CJS and paired-release models. Program MARK also contains a simulation module that can be used for conducting sample size analysis for many of the models implemented in MARK. These software packages allow users to quickly run a number of different scenarios that vary parameter values, model structures, and sample sizes to develop a robust study design likely to achieve desired levels of precision.

Modeling Biological Hypotheses

In our application of the route-specific survival model to the simulated fish passage study, we introduce the idea of fitting mark–recapture models that represent different biological hypotheses. For example, model selection results showed strong evidence that riverine survival was not constant, but declined exponentially over time (Table 4; Figure 5). This conclusion may lead biologists to further evaluate time-dependent mortality processes, such as physiological condition, water temperature, or changes in predator densities.

Hypothesis-driven model selection begins with a general model (i.e., full model) as a representation of the underlying biological system. We then develop a set of reduced models that each represents a specific set of constraints. Each constraint represents an alternative view (i.e., hypothesis) of the biological system. Identifying candidate models requires knowledge of the biological system, and is therefore not a statistical exercise (Lebreton et al. 1992).

Reduced models may estimate parameters as functions of categorical or continuous covariates, such as rearing history or size at release. The simplest constraints involve categorical covariates that are fixed throughout the study. This would be used, for example, to determine if rearing history (e.g., hatchery, wild) has a "significant" effect on survival. Multistate models offer the ability to evaluate similar hypotheses when the categories (i.e., states) can change for a given fish during the study. For example, in our route-specific survival model, the route state for an individual fish could not be known at the time of release. Further, the route assignment was never known for fish that passed the dam undetected. Still, model selection in the multi-state framework allowed us to test the hypothesis that survival differed among passage routes (Table 4). In this way, multi-state models are a powerful tool for examining the effects of categorical state variables on demographic parameters.

Categorical predictor variables can also represent characteristics of the sampling occasion, rather than of the tagged fish. In a classic example, Lebreton et al. (1992) showed that survival of European dippers *Cinclus cinclus* was lower during flood years versus nonflood years. Analogous examples for migrating salmon might be to compare survival through river reaches that are freshwater versus estuarine, or through industrialized versus natural habitats. In the dipper example, sampling took place at relatively regular intervals (annually). However, telemetry receivers are never uniformly spaced throughout a study system. Therefore, to appropriately compare survival among individual river reaches, reach-specific survival estimates must be standardized (by reach length) within the likelihood. An intuitive approach is to estimate instantaneous survival instead of reach-specific survival, allowing direct comparison and hypothesis testing. Further, any other reach-specific survival rate (e.g., reach-specific survival or survival rate per river km) can also be derived from the instantaneous rates.

Survival is a key demographic parameter that describes the dynamics of a population. Combined with other data sources and analyses (e.g., time-to-event analysis in Section 9.1), telemetry and mark–recapture models can be used to understand the influence of external processes on populations (e.g., hydro-system regulation, habitat alternation, artificial rearing, water quality improvements). These analytical frameworks provide powerful tools for analyzing telemetry data, and we hope that our introduction to these tools encourages more biologists to consider them as an integral component of telemetry studies.

REFERENCES

Brownie, C., J. E. Hines, J. D. Nichols, K. H. Pollock, and J. B. Hestbeck. 1993. Capture-recapture studies for multiple strata including non-Markovian transitions. Biometrics 49:1173–1187.

Burnham, K. P., D. R. Anderson, G. C. White, C. Brownie, and K. H. Pollock. 1987. Design and analysis methods for fish survival experiments based on release-recapture. American Fisheries Society, Monograph 5, Bethesda, Maryland.

Burnham, K. P. and D. R. Anderson. 2002. Model selection and multimodel inference: a practical information-theoretic approach. Springer, New York.

Cormack, R. M. 1964. Estimates of survival from the sighting of marked animals. Biometrika 51:429–438.

Devineau, O., R. Choquet, J. D. Lebreton. 2006. Planning capture-recapture studies: straightforward precision, bias, and power calculations. Wildlife Society Bulletin 34: 1028–1035.

Jolly, G. M. 1965. Explicit estimates from capture-recapture data with both death and immigration-stochastic model. Biometrika 52:225–247.

Laake, J., and E. Rexstad. 2011. RMark—an alternative approach to building linear models inMARK. (Appendix C). *In* E. Cooch and G. White editors, Program MARK: a gentle introduction, 9th editor. [Online]. Available: http://www.phidot.org/software/mark/docs/book/ (September 2011).

Lady, J. M. and J. R. Skalski. 2009. USER 4: User-specified estimation routine. Available: http://www.cbr.washington.edu/paramest/docs/user/UserManual/UserManual.pdf (August 2011).

Lady, J. M., P. Westhagen, J. R. Skalski. 2003. SampleSize1.1: Sample size calculations for fish and wildlife survival studies. Available: http://www.cbr.washington.edu/paramest/docs/samplesize/manual/user.pdf (April 2012).

Lebreton, J. D., K. P. Burnham, J. Clobert, and D. R. Anderson. 1992. Modeling survival and testing biological hypotheses using marked animals: A unified approach with case studies. Ecological Monographs 62:67–118.

Lebreton, J. D., and R. Pradel. 2002. Multistate recapture models: modeling incomplete individual histories. Journal of Applied Statistics 29:353–369.

Letcher, B. H., and G. E. Horton. 2009. Seasonal variation in size-dependent survival of juvenile Atlantic salmon (*Salmo salar*): performance of multistate capture-mark-recapture models. Canadian Journal of Fisheries and Aquatic Sciences 65:1649–1666.

McCormick, S. D., L. P. Hansen, T. P. Quinn, and R. L. Saunders. 1998. Movement, migration, and smolting of Atlantic salmon (*Salmo salar*). Canadian Journal of Fisheries and Aquatic Sciences 55:77–92.

Nichols, J. D., and W. L. Kendall. 1995. The use of multi-state capture–recapture models to address questions in evolutionary ecology. Journal of Applied Statistics 22:835–846.

Nichols, J. D., J. E. Hines, K. H. Pollock, R. L. Hinz, and W. A. Link. 1994. Estimating breeding proportions and testing hypotheses about costs of reproduction with capture-recapture data. Ecology 75:2052–2065.

Perry, R. W. 2010. Survival and migration dynamics of juvenile Chinook salmon (*Oncorhynchus tshawytscha*) in the Sacramento-San Joaquin River Delta. Doctoral dissertation. University of Washington, Seattle, Washington.

Perry, R. W. and J. R. Skalski. 2008. The design and analysis of salmonid tagging studies in the Columbia Basin. Volume XXIII: Effects of array configuration on statistical independence of replicate telemetry arrays used in smolt survival studies. Technical report to BPA, Project No. 198910700. U.S. Department of Energy, Division of Fish and Wildlife. Portland, Oregon.

Pollock, K. H. 1982. A capture-recapture design robust to unequal probability of capture. Journal of Wildlife Management 46:757–760.

Schwarz, C. J., J. F. Schweigert, and A. N. Arnason. 1993. Post-release stratification in band-recovery models. Biometrics 44:765–785.

Seber, G. A. F. 1965. A note on the multiple recapture census. Biometrika 52:249–259.

Seber, G. A. F. 1982. The estimation of animal abundance. MacMillan, New York, New York.

Skalski, J. R., S. G. Smith, R. N. Iwamoto, J. G. Williams, and A. Hoffmann. 1998. Use of PIT-tags to estimate survival of migrating juvenile salmonids in the Snake and Columbia Rivers. Canadian Journal of Fisheries and Aquatic Sciences 55:1484–1493.

Skalski, J. R., J. Lady, R. Townsend, A. Giorgi, J. R. Stevenson, C. M. Peven, and R. D. McDonald. 2001. Estimating in-river survival of migrating salmonid smolts using radiotelemetry. Canadian Journal of Fisheries and Aquatic Sciences 58:1987–1997.

Skalski, J. R., R. Townsend, J. Lady, A. Giorgi, J. R. Stevenson, C. M. Peven, and R. D. McDonald. 2002. Estimating route-specific passage and survival probabilities at a hydroelectric project from smolt radiotelemetry studies. Canadian Journal of Fisheries and Aquatic Sciences 59:1385–1393.

Townsend, R. L., J. R. Skalski, P. Dillingham, and T. W. Steig. 2006. Correcting bias in survival estimation resulting from tag failure in acoustic and radiotelemetry studies. Journal of Agricultural, Biological, and Environmental Statistics 11(2):183–196.

White, G. C., and K. P. Burnham. 1999. Program MARK: survival estimation from populations of marked animals. Bird Study 46(Supplement):120–138.

White, G. C., K. P. Burnham, and D. R. Anderson. 2001. Advanced features of Program Mark. Pages 368–377 *in* R. Field, R. J. Warren, H. Okarma, and P. R. Sievert, editors. Wildlife, land, and people: priorities for the 21st century. Proceedings of the Second International Wildlife Management Congress. The Wildlife Society, Bethesda, Maryland, USA.

Section 9.3

Developing a Quality Assurance Plan for Telemetry Studies: a Necessary Management Tool for an Effective Study

Jill M. Hardiman, Chris E. Walker, and Tim D. Counihan

INTRODUCTION

Telemetry has been used to answer various questions associated with research, management, and monitoring programs and to monitor animal behavior and population dynamics throughout the world. Many telemetry projects have been developed to study the passage, behavior, and survival of migrating adult and juvenile salmonids at hydroelectric projects on the mainstem Columbia and Snake rivers (Skalski et al. 2001a, 2001b; Skalski et al. 2002; Keefer et al. 2004; Goniea et al. 2006; Plumb et al. 2006). Telemetry based field evaluations of the survival of salmon through hydroelectric projects are costly because of the technology (tags, telemetry systems, infrastructure, etc.) and personnel required to conduct the evaluations. Given the cost of implementing these projects, and the financial and conservation implications of the decisions made from the research results (e.g., forgone electricity production and conservation of threatened and endangered animals), ensuring quality data are collected by documenting all procedures, training, data checks, and that sound protocols and quality assurance and control procedures are in place is paramount.

Telemetry studies can pose unique data collection, processing, and analysis challenges. For instance, inferences about entire populations of animals are made from study animals that are captured, held, and tagged at disparate locations. Consequently great care must be taken to ensure that any potential biases that could arise from field procedures must be minimized (Peven et al. 2005). Interrogations of released study animals are remotely conducted by telemetry systems throughout the study area. The continuous recording of telemetry systems can result in large numbers of detections over a short time frame and the potential for false positive detections from records that are weak or erroneous. Thus, there is the potential to generate large data sets (many thousands of lines) that require significant postprocessing. Data reduction can be done using software or programming code within a software package or manually to discern noise from valid data and pull out the pertinent information for analysis. In either case, consistent well-documented procedures need to be in place to ensure quality results and allow for repeatability of study methods.

The development of a Quality Assurance Plan (QAP) will help document the procedures necessary to minimize potential errors or bias and provide quality data and ensure their implementation. Sound project management encompasses many facets that include identifying appropriate study designs, sample sizes and precision targets, equipment, personnel, protocols, and the level of training required to achieve the study objectives. These considerations will help to ensure data integrity, transparency of data collection and processing procedures, repeatability, data sharing, and proper interpretation of results for managers and decision makers. We will present examples of procedures we have used for telemetry studies on the behavior and survival of juvenile salmonids in the Columbia River Basin to illustrate how quality assurance measures can be incorporated into a QAP.

WHAT IS A QUALITY ASSURANCE PLAN?

A QAP is a formal document that describes processes that ensure reliability of work or aspects of a project or scientific study to satisfy stated objectives and expectations. The QAP should be a "stand alone" document which should be included or stored with the final report, products, and data for the research or monitoring project and made available to all personnel and stakeholders involved. The QAP differs from a study plan or proposal by focusing on the processes for ensuring data quality throughout the development, management, and implementation of the project. It will contain appendices documenting proposals (or study plans), budgets, Standard Operating Procedures (SOPs), training logs, equipment records, data sheets, data processing programs, and field site visit documentation for QC checks associated with the research project.

The QAP can be considered a guidance document to ensure that proper planning and procedures are in place for obtaining quality data. The QAP describes how quality assurance (QA) and quality control (QC) activities are applied to ensure quality data collection, management, and analysis to assure that final results are of the type and quality needed. A QAP should clearly describe the project objectives and all activities involved with data acquisition, whether the data are from direct measurement activities in the field, collected from other sources, or compiled from computerized databases, programs, and information systems. This requires early planning to eliminate approaches that are not appropriate or efficient, thereby saving time and potentially reducing poor data collection or data loss and decreasing the overall project costs. Implementing a plan as stated increases efficiency and provides for early detection of problems. The development of a QAP allows key personnel and stakeholders (i.e., data users, data producers, decision makers, etc.) to be involved in the planning process to ensure that their needs are clearly defined and addressed prior to project implementation. While the time spent for planning and development of a QAP may seem extensive and costly, the penalty for ineffective planning and poor project implementation has the potential for much greater costs. A QAP is an effective tool to be used by researchers and managers to communicate and evaluate project procedures and methodology to ensure all processes are in place for proper project implementation, thus further ensuring stated project objectives and goals will likely be achieved.

Quality assurance, quality control, and quality assessment are all components of a QA plan. Schroder and Shampine (1992) described these terms as follows:

Quality assurance (QA)—The systematic management of data-collection systems by using prescribed guidelines and criteria for implementing technically approved methods and policies. Quality assurance incorporates a comprehensive plan that outlines the overall process for providing a product or service that will satisfy the given requirements for quality.

Quality control (QC)—The specific operational techniques and activities used to ensure the required quality of data. Quality control consists of the application of technical procedures to achieve prescribed standards of performance and to document the quality of collected data. Quality-control data that do not meet required standards are used to evaluate and implement corrective actions necessary to improve performance to acceptable levels.

Quality assessment—The overall process of assessing the quality of environmental data by (1) reviewing appropriate implementation of QA policies and procedures and (2) analyzing the QC data. Quality assessment encompasses measurable and un-measurable factors that affect the quality of environmental data. Assessment of these factors may indicate limitations that require modifications to protocols or standard operating procedures for sample collection and analysis, or that affect the desired interpretation and use of the environmental data. Quality-assurance, quality-control, and quality-assessment systems complement each other to provide a comprehensive QA program that ensures that quality objectives are identified and integrated into all levels of water quality activities.

We present a general outline for developing a QAP (Table 1), using examples from telemetry studies for survival analysis of salmonids in the Columbia River Basin as a guideline. There are many additional resources available for development of a QAP (Jones 1999; Schroder and Shampine 1992; U.S. EPA 2001, 2002). This example, based on components of Columbia River field studies, demonstrates the importance of using a QAP to avoid potential data collection and analysis problems and to efficiently manage and analyze large data sets in a transparent and repeatable manner.

Development of a Quality Assurance Plan

A QAP is developed prior to data collection, thus requiring early project planning and a clear understanding of the project objectives and activities needed. It is recommended that a team be assembled with the necessary expertise to identify and plan the necessary tasks to address the management or research objectives. The team will write the plan, submit it to an appropriate audience for peer review, and make revisions as needed prior to project implementation. In-season changes can be made as needed and should be well documented within the QAP and then approved by the writing team with a final version updated as needed. The examples presented here elaborate on the outline presented in Table 1. They are provided as case scenarios and are not intended to encompass the level of detail that may be necessary for all projects. Each project and individual QAP should be formulated at a level appropriate to the study objectives, desired results, and precision necessary for that study.

SECTION DETAIL

Project Management

The project management section provides a general description of the overall organization of the project, including who will use these data, who will collect it, and what it will be used for. The first page of the QAP is typically the Title and Approval sheet. This includes

Table 1. Outline of Quality Assurance Plan sections and specific components.

Section	Components	Description
A – Project Management	A1 – Title and approval sheet	Date with signatures
	A2 – Table of contents	List of QAP elements
	A3 – Key personnel	Identify key individuals, agencies, their roles, distribution list for QAP
	A4 – Problem definition	State specific problem or decision to be addressed
	A5 – Introduction	Background, purpose, and scope of study
	A6 – Project description	Proposal or statement of work, clear objectives
	A7 – Data quality objectives	Metrics and expected precision
	A8 – Training	SOPs, special training, certification
	A9 - Documentation	Description of management of project documents and records
B – Data Collection	B1 – Sampling design and methods	Experimental design, sampling frequencies, sample size, replication
	B2 – Data flow	Overall data flow chart
	B3 – QAQC	Field and office activities
	B4 – Instrument/Equipment	Inspection, testing, calibration, maintenance
C – Data Processing, Review, and Storage	C1 – Processing steps	Programs, database design, organization
	C2 – Proofing steps	Auto- and manual-proofing
	C3 – QAQC data checks	Missing data, errors, etc.
	C4 – Performance checks	For manual proofers and auto proofer programs
	C5 – Final analysis	Statistics, model development
	C6 – Final products	Reports, presentations, manuscripts
	C7 – Data management	Data tracking and metadata
D – Assessment and Oversight	D1 – Periodic assessment	Evaluate QAP implementation
	D2 – Modifications	Update all changes made
	D3 – Documentation	Log changes, approvals
E – Data Verification and Validation	E1 – Adherence to assumptions	Clearly state assumptions and checks
	E2 – Reconciliation	Resolution summary of changes
F – Reporting	F1– Reporting and archiving	Final report documents, storage locations

the project title and the authorizing signatures and dates ensuring acceptance of the plan and the date on which it becomes effective. In this section the staff and agencies involved in the research, key personnel, and their specific roles will be listed as well as a distribution list for the QAP. A clear description of the problem statement or research objectives, why the project is important, the background information leading to the research objectives, and the project's geographic extent are all included. Often this section will reference the project proposal or scope of work which includes the specifics of statistical analyses and experimental design.

The data quality objectives should address questions about the planned statistical analyses, power of statistical tests, bias, and precision of the estimates. The study question(s) should be clearly defined *a priori* as this will drive the study design, sample size, and data processing time. The data quality objectives should clearly state details of the study plan, including the required sample size, precision of estimates, and analytical assumptions needed to meet the defined objectives. Many technical journal articles and texts are available on study design development that emphasize tradeoffs between management, monitoring, and research designs specific to telemetry studies, as well as basic study design principles for fishery science (Burnham et al. 1987; Skalski et al. 2002; Peven et al. 2005; Devineau et al. 2006; Guy and Brown 2007).

Example: Columbia and Snake rivers salmonid survival studies

One aspect of the study design should be an elucidation of both the experimental or management question and any potential confounding factors or sources of bias. Some examples of these for telemetry studies of juvenile salmon behavior and survival in the Columbia and Snake rivers include: randomization of treatments (e.g., dam operations, time of release, etc), randomization of personnel to maintain consistent treatment of tagged fish, random assignment of tags from various manufactured batches of tags, and randomized release times and locations. Additionally, fish collection procedures should be scrutinized and evaluated as to how the tagged individuals are representative of the population of interest (e.g., collection occurring during the bulk of the out-migration, tagged fish similar size to run-of-river fish). Some of the issues listed above are specific to survival studies, for instance the equal treatment of all study organisms to minimize any potential bias from handling, release time, and release conditions. Other issues are ubiquitous design concepts in scientific investigations, such as accounting for, minimizing, and understanding the potential for bias in study metrics.

Another important component of the QAP is the identification of training needed to perform all tasks involved in the project. Identifying training needs ahead of time allows for proper development of SOPs for all tasks associated with data collection, equipment setup and maintenance, and data processing, as well as providing adequate time for training personnel. Development of SOPs, proper training, and documentation of completed training necessary for all data collection procedures (e.g., tagging and transport of fish, proper netting techniques, equipment calibration, etc.) minimizes the potential for data collection biases. A necessary follow-up to training procedures are site visits to ensure that personnel are properly implementing all procedures in the field. Site visits will help recognize and correct potential "protocol drift." Implementing site visits earlier rather than later is recommended so that any deviations from the proper procedures (e.g., protocol drift) are discovered and remedial action is taken. This also allows re-training of employees early in the study so that errors are not carried through the entire study period. Further, site visits can identify systematic or logistical problems so they can be fixed early in the study.

The creation of formal SOPs provides documentation of project procedures that can then be repeated by others. When used in conjunction with a training log, formal SOPs provide documentation of training provided to ensure consistent quality data. This also creates accountability of personnel for understanding and performing procedures properly and establishes a guideline with which to perform QAQC checks in the field or office.

Example: SOPs from Columbia River survival studies

We present a list of SOPs that have been established and used for telemetry and survival studies of fish in the Columbia River Basin (Table 2) as an example of some of the procedures that can be developed.

Data Collection

The data collection section should cover all aspects of the study design implementation, ensuring all methodology is in place for proper data collection, handling, and analysis and that all QC procedures are employed and properly documented. The sample design and general methods should be covered in the proposal or scope of work and can be referenced with any necessary additional details in this section. This can be in the form of a methods section which refers to how the individual SOPs are used, and when and how data are collected and transferred. This section will address sample design and implementation issues such as sample sizes needed to address stated objectives and goals, as well as corrective actions to be taken if problems arise. Some key elements to include are how often data are collected (e.g., frequency of tagged individual releases, frequency of telemetry data collection), how sample organisms are collected and handled, the geographic scope and time frame that the study represents, and how the QAQC procedures ensure proper data collection. A data flow chart can be used to summarize data collection and management identifying the processing and proofing steps along the way. An example of a data flow chart and data flow description from field studies conducted in the Columbia River Basin is presented below.

Example: Data Flow Description from a Columbia River Study

An example of how telemetry data can be transferred and processed will provide a general understanding of how each step in the flow of data from the field to the office is supported and evaluated by QAQC activities (Figure 1). In the field, telemetry arrays are set up for collecting data from tagged individuals. Each of the arrays is comprised of telemetry receivers to record detection data. The study individuals are tagged and released for interrogation at telemetry arrays. Designated staff transfer or download data from the telemetry receivers regularly (daily or weekly) or data can be downloaded remotely via satellite connection. The data collected are transferred to computers housed in a laboratory or office and uploaded onto a network drive or file directory. Each data transfer procedure has a documentation system to ensure data sets are complete, i.e., all lines of data are transferred and time frames are completely covered without data gaps. These data checks at the time of transfer are also a check on the field equipment, ensuring all receivers are collecting data and enabling any equipment failure to be investigated promptly. All interim data files are merged into one unprocessed detection data file for the entire study period.

Table 2. List of some standard operating procedures from a telemetry field study for fish.

Title
Data Check-in
Uploading Data from Telemetry Receivers
Tag Testing Protocol
Tag Allocation
Fish Netting
Transporting Fish
MS-222 Preparation
Gastric Tag Implantation
Surgical Tag Implantation
Pit Tag Implantation
Sacrificed Fish Procedures
QAQC Procedures for Pre-release Tag Check
Release Procedures
Total Dissolved Gas Monitoring
Total Dissolved Gas Monitor Calibration
Dissolved Oxygen Meter
Use of Digital Scale
Sanitization of buckets
Tag Life Study

Tag implantation and release crews also collect data during tagging and release procedures, such as length, weight, date, location of release, and release times. All tagging and release data are transferred to a laboratory or office setting and are logged as received, proofed, and entered electronically on a computer network drive or file directory. Environmental data (e.g., water quality) are recorded for each animal holding area and release site to facilitate evaluation of physiological stress due to holding or release site conditions. All environmental data records also transferred to a laboratory or office setting and are logged as received and proofed, and are entered electronically on a computer network drive or file directory. All interim release data files are merged into one data file for the entire study period.

Once all the field data has been transferred and compiled into release and detection databases, proofing of the individual records can begin. The release data and the unprocessed detection data are merged into a single database for the entire study period. The merged database is then subjected to a proofing process. This process can start with a proofing program ("auto-proofing program") written in a database or data programming language (e.g., SAS, R, Matlab) or manual proofing. Establishing criteria for what constitutes a valid data entry is a key element of efficiently processing telemetry data and allows for repeatability by others using the data. The goal of a proofing program is to eliminate erroneous detections while maintaining valid detections. Detection data from tagged individuals are identified and flagged for further examination by trained employees if they do not satisfy established criteria (e.g., power levels, frequency of detections within a geographic region, geographic sequence). Alternatively, depending on the accuracy and precision desired and time allotted for data processing, these records can be removed from the final data set as not fitting the established

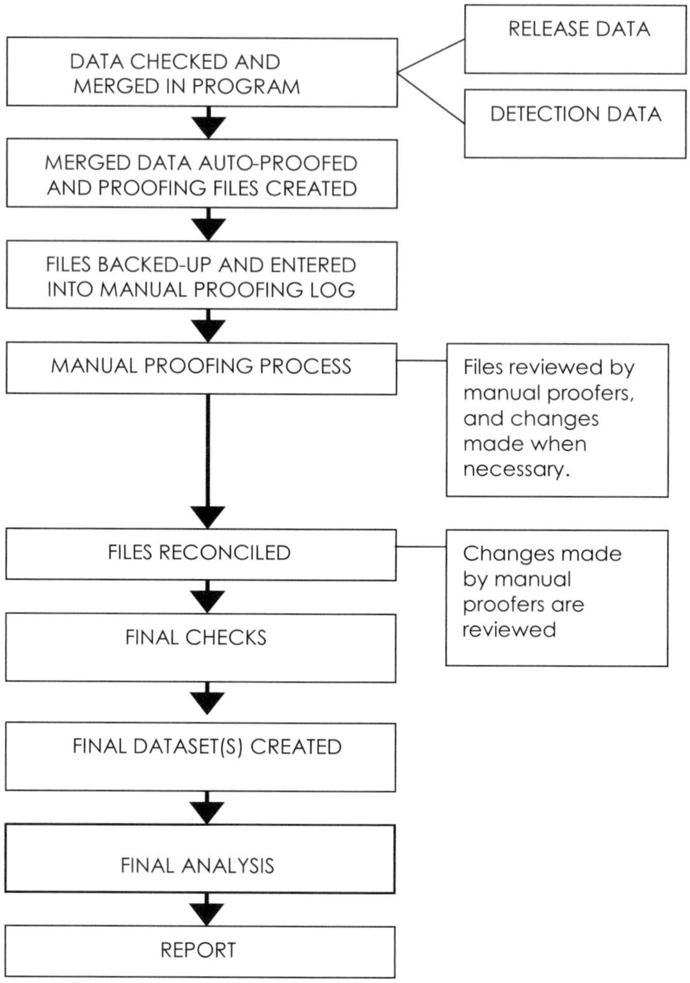

FIGURE 1. Data flow chart for data transfer from the field to the final report.

criteria. This approach allows for a consistent and repeatable procedure for obtaining the final data set without involving "professional judgment" in a manual proofing process. However, some assessment of any potential data bias resulting from the removal of identified records should be evaluated in a sensitivity analysis.

A manual proofing process begins as data records are identified and compiled by the auto-proofing program or a trained individual to be reviewed. A staff person can be designated as a manual-proofing coordinator or reconciler to manage this process. Changes made by manual proofers are logged in a spreadsheet and a hardcopy of the changes should be turned into the manual-proofing coordinator. This datasheet prevents proofers from skipping tag identification codes, serves as a medium for communicating questions or concerns, and documents all changes made by the manual proofer. A proportion of the proofed files are examined by a second manual proofer and the reconciler for a QAQC consistency check among proofers. Additionally, the reconciler is periodically checked by another person experienced with telemetry data to assure QAQC of the reconciler. Proofer performance (e.g., number of files processed, number of incorrect changes made) should be monitored and documented throughout the data processing period. The number of files checked by the reconciler will depend on proofer performance.

At the end of the proofing process described above, several final checks will be performed including an accounting of all proofing files before the dataset will be considered final. Copies of the final data set should be vetted by all appropriate user groups, and an original copy of the final database should be stored in a secure location with restricted access on a computer or network drive. Ideally, additional copies should be stored off-site. While maintaining the original versions of the database as described above, copies are then used to reduce the data according to project objectives. When the analyses are complete, steps similar to those described above should be used to archive and store the reduced processed data.

The same principles of managing the data processes using the QAP can be used to inform, manage, and monitor field activities. A necessary component of the QAP is an SOP that serves as a template for training, establishes proper implementation of the procedures in the field, and provides documentation of completed training prior to data collection. In addition to preseason training, it is important to conduct regularly scheduled site visits during the field season to ensure SOPs are being implemented correctly. Site visits should be conducted by highly trained employees who will randomly visit projects to evaluate adherence to protocols and SOPs. A site visit check sheet is used for each activity, documenting whether the selected activities are being performed according to the protocol. Protocol drift for any activity must be reported to appropriate staff and corrective actions implemented immediately. The documentation of site visits and outcomes provides a record of the actual field collection methodology against which data quality can be measured. It is recommended to conduct site visits early in project implementation to catch errors or unforeseen problems which may arise in the procedures before much of the field study has been completed. Examples of activities in which to conduct site visits from a Columbia River telemetry study included observing fish tagging and releasing procedures, telemetry receiver downloading, and water quality monitoring procedures.

In addition to the project specific QAQC procedures occurring prior to and during the field season (i.e., SOP development, training, site visits), QAQC procedures need to be developed for specific field tasks, e.g. equipment set up, calibration, and maintenance. All data collection equipment or systems needing periodic maintenance, testing, calibration, or inspection should be identified and a schedule should be developed for this. Preseason instrument testing and calibration of all data collection instruments should be established and documented, including a description of these procedures. All environmental monitoring equipment for tagged individual holding areas should have a record log documenting regular equipment calibration and proper equipment operation and maintenance checks. The precise data to be recorded and procedures for monitoring and maintaining data collection equipment should be identified in the QAP. Identification of critical spare parts, replacement equipment, and repair procedures should also be discussed. Another aspect of the QAP is to identify inspection and acceptance criteria for all critical supplies and consumables necessary for the project (e.g., telemetry tags, surgical supplies). Where appropriate, the QAP should document supply vendors; procedures for identifying, tracking, storing and retrieving supplies; identification of individuals responsible for maintaining supplies; and acceptance criteria for these supplies.

Example: QAQC procedures used in a Columbia River survival study

A critical component of any telemetry study is that tags are properly functioning for the allotted evaluation time for the study period. A tag-life study needs to be incorporated into a telemetry study in order to evaluate manufacturer tag consistency in battery life and function throughout a study period (Cowen and Schwarz 2006; Townsend et al. 2006). This is particularly important for survival studies and modeling assumptions associated with tagged individuals. Another specific example includes having environmental quality monitoring equipment in place at all animal holding areas and release sites. Monitoring environmental conditions helps ensure that tagged individuals are not being inadvertently stressed, which could then influence their behavior and bias results. Field procedures should be established to eliminate sources of stress. For example, fish can be acclimated to temperature differences gradually to avoid additional physiological stress at the time of release or collection. By regularly recording environmental quality data, study organisms that are subjected to inconsistent conditions can be removed from the data set to remove potentially biased behavior. Alternatively, the relationship between behavior and environmental conditions may be explored, but only if the environmental quality data have been recorded.

The QAQC procedures for the data management, transfer, and storage should be identified in the QAP. These procedures should be established and documented with an SOP and associated data logs that record dates and individuals who will be handling data prior to data collection. These procedures will include documentation of the data check-in, storage, entry, and proofing, as well as any final dataset revisions. Raw data will be stored, changes will be documented, and final data sets will be stored and backed up on a regular basis.

Data Processing, Review, and Storage

The QAP should include procedures used to proof, process, review, and store all intermediate and final data sets. All programs associated with data processing will also be stored on a network or file directory, including processing logs recording when the program was run and who ran it. Major components of processing telemetry data are the systematic reduction of data and distinguishing between valid and erroneous data. The assimilation of the data into a condensed, user-friendly format is critical for obtaining results in a timely manner. To speed this process, a program or procedure is developed to distinguish valid data from erroneous data and assign detection locations, time stamps, geographic order, etc., all of which will be used in the final analysis of the data. This proofing process can be done manually by individual proofers or by a program written specifically for this task. It is critical to develop consistent proofing methodology in order to avoid bias or drift in the data processing.

The criteria used to identify valid detections need to be clearly defined. Determining these criteria is a difficult task and can be very specific to a region or study. Often criteria are based on the power level of a detection signal, the frequency of detections within a specified time frame, and a pattern of either movement or residence time in a geographic area. Other examples of criteria to identify valid detection are discussed by Melynchuk, English et al. and Castro-Santos and Perry elsewhere in this book. Data processing is likely to be dynamic over time and may be tailored to a study reach or region. Thus, the data processing procedures should be well documented with supporting information. The program written to process the data should be validated by comparing the output to manually proofed data sets using a subset of the detection data from the current study. It is also a good idea to perform a sensitivity analysis to determine the implications of the data processing criteria on final results. Skalski et al. (2001a) have presented an example of such a sensitivity analysis from an in-river survival study. They found that their project survival estimates varied from 0.95 to 1.10 as they varied the criterion for the minimum number of radio signals used to determine valid detections. They had a similar result when they varied the power level criterion for a valid detection, causing their project survival estimates to vary from approximately 0.96–0.99. These authors suggest the power threshold criterion is less influential than the frequency criterion for survival estimates (English et al. 2000; Miller et al. 2000; Skalski et al. 2001b).

The data processing section of the QAP will also identify all QAQC data checks performed on the proofing programs, data sets, and proofer performance. This process was briefly described above ("Overall Data Flow"). This section describes in detail the specific checks to be done, when they are done, and who is responsible for implementing them. The frequency of each type of check should be stated, as well as the acceptable control standards and the corrective actions to be taken if control standards are not met.

The final data analysis procedures will be described in detail in the QAP or referenced from the study proposal. This will include all statistical analyses, model development, model selection, and model run completion. Final products including reports, manuscripts, and presentations will be listed as well as to whom the products will be submitted.

The data management process will be described, including all handling of data from the field, generation of data from processing programs, software and hardware associated with data processing, and the final data use and storage. A description of the final database format, including definitions of all data columns, database structure, formats, codes, etc. should be included, as well as all pertinent metadata. Also, data obtained from other sources (e.g., non-measured sources such as historical databases, or monitoring sites, such as stream gauges) that were used in project implementation should be documented. Data analysis or report writing should clearly state any limitations associated with the data. Furthermore, standard-record keeping procedures and the processes used for data storage, access rights, and retrieval of any electronic media and off-site storage should be described.

Assessment and Oversight

A periodic assessment and oversight of all procedures within the plan will be established during the field project to determine how the QAP is being implemented. A description of each assessment action and the frequency in which it is to be administered should be included. Individuals to perform independent assessment outside of the project agency should be identified and how the results of each assessment will be reported should be stated. This section also describes actions taken or to be implemented where the plan did not seem to be working and how to address such issues or procedures. All changes made to the QAP will be documented with approval as necessary.

Data Verification and Validation

Data verification and validation is the final step to determine if the data products adhere to the project's objectives and established data criteria. Data verification is the process for evaluating the completeness, correctness, and compliance of the data set against the established collection methodology and contractual or stated project objectives. This final review process ensures that the data have been collected, recorded, transferred, and processed correctly according to data criteria and QAQC procedures described in the QAP. Furthermore, it ensures that the data adhere to any necessary modeling assumptions and data quality objectives. This section can be a review of criteria previously stated for what constitutes valid data and should clearly describe any additional procedures. It will clearly state the criteria for deciding to accept, reject, or qualify all project information. Additionally, this is where all tolerances or deviations from sampling SOPs for data acceptance criteria and procedures are described for reconciliation. Any potential effects of the deviances from procedures would be addressed in the data quality assessment section. An example from fieldwork of potential deviances from data collection methodology could be prolonged holding times of study animals beyond the established holding time frame or release of study animals at alternate release sites. Data validation is the process of determining the quality of the data set relative to the final use of the data. Thus, validation is focused on the project's stated objectives or needs designed to meet the needs of the decision makers or managers and should identify departures from this. Furthermore, this section describes any limitations associated with the study results and how the results will be conveyed to data users and managers.

Reporting and Archiving Data

After all of the data have been collected, processed, and analyzed, they are then typically summarized in a technical report or publication and data and metadata are archived. The QAP should address these important aspects of the data reporting. The potential for transcription errors into tables and Figures should be addressed and appropriate proofing procedures established. Ideally, the QAP should also stipulate the disposition and format of metadata (e.g., see: http://www.pnamp.org/node/2596). Metadata can be thought of as data providing information about one or more data sets, and can include information about how, when, and by whom the data were created, the purpose of the data, and where the data are stored on computer networks, etc. After the reports or publications are peer reviewed and finalized and metadata are completed, the QAP can then be included in a packet that describes the project and can be available to decision makers and managers. By making the QAP available as an integral part of the final data products, project leaders are communicating the effectiveness of the project implementation, the quality of data collected, and how it should be interpreted and used by stakeholders.

SUMMARY

A QAP is a formal document that describes processes that ensure product reliability and aspects of a project or scientific study designed to satisfy stated objectives and expectations. The QAP can be considered a guidance document to ensure that proper planning and procedures are in place for obtaining high-quality data. A QAP should be developed prior to any data collection, and thus requires early project planning and a clear understanding of the project objectives and activities needed to achieve these. A QAP should provide guidance on how to apply quality assurance and quality control guidelines to all aspects of the project from start to finish including the reporting and archiving of data. All personnel and stakeholders involved in the project should retain or have access to a current version of the QAP. The ultimate success of any telemetry study depends on the quality of the data collected and how it addresses the project objectives. How well this is accomplished may depend on the adequacy of the QAP and its effective implementation. A QAP provides stakeholders (i.e., decision makers, reviewers, researchers) with a high level of confidence in the quality of the project data used to support potential policy and resource decisions. Although the development of a QAP may appear daunting and time consuming, the consequences of collecting inadequate or potentially unreliable data resulting from poor project implementation far outweigh the time and cost investment of a QAP. The initial time spent up front is compensated by the resulting effective and efficient studies designed to provide answers to scientific or managerial questions.

REFERENCES

Burnham, K. P., D. R. Anderson, G. C. White, C. Brownie, and K. H. Pollock. 1987. Design and analysis methods for fish survival experiments based on release-recapture. American Fisheries Society, Monograph 5, Bethesda, Maryland.

Cowen, L., and C. J. Schwarz. 2006. The Jolly-Seber model with tag loss. Biometrics 62:669–705.

Devineau, O., R. Choquet, J.-D. Lebreton. 2006. Planning Capture-Recapture Studies: Straightforward Precision, Bias, and Power Calculations. Wildlife Society Bulletin. 34(4):1028–1035.

English, K. K., Skalski, J. R., Lady, J., Koski, W. R., Nass, B. L., and Sliwinski, C. 2000. Assessment of project, pool and dam survival for run-or-river steelhead smolts at Wanapum and Priest Rapids projects using radio-telemetry techniques, 2000. Draft report prepared for Public Utility District No. 2 of Grant County, Ephrata, Washington.

Goniea, T. M., M. L. Keefer, T. C. Bjornn, C. A. Peery, D. H. Bennett, and L. C. Stuehrenberg. 2006. Behavioral thermoregulation and slowed migration by adult fall Chinook salmon in response to high Columbia River water temperatures. Transactions of the American Fisheries Society 135:408–419.

Guy, C. S., and M. L. Brown, editors. 2007. Analysis and interpretation of freshwater fisheries data. American Fisheries Society, Bethesda, Maryland.

Jones, B.E., 1999, Principles and practices for quality assurance and quality control: U.S. Geological Survey Open-File Report 98–636.

Keefer, M. L., C. A. Peery, M. A. Jepson, K. R. Tolotti, T. C. Bjornn, and L. C. Stuehrenberg. 2004. Stock-specific migration timing of adult spring Chinook salmon in the Columbia River Basin. North American Journal of Fisheries Management 24(4):1145–1162.

Miller, M., Skalski, J. R., Lady, J., Giorgi, A. E., and Stevenson, J. R. 2000. A pilot investigation to assess the feasibility of estimating juvenile steelhead survival at Wanapum and Preist Rapids projects using radiotelemetry analysis, 1999. Prepared for the Grant County Public Utility District No. 2, Ephrata Washington.

Peven, C., A. Giorgi, J. Skalski, M. Langeslay, A. Grassell, S. G. Smith, T. Counihan, R. Perry, and S. Bickford. 2005. Guidelines and Recommended Protocols for Conducting, Analyzing, and Reporting Juvenile Salmonid Survival Studies in the Columbia River Basin. Available: http://www.pnamp.org/sites/default/files/Guid__Sug_Prot_final.pdf (February 2012).

Plumb, J. M., R. W. Perry, N. S. Adams, and D. W. Rondorf. 2006. The effects of river impoundment and hatchery rearing on the migration behavior of juvenile steelhead in the lower Snake River, Washington. North American Journal of Fisheries Management 26:438–452.

Schroder, L. J., and W. J. Shampine. 1992. Guidelines for preparing a quality assurance plan for district offices of the U.S. Geological Survey: U.S. Geological Survey Open-File Report 92–136.

Skalski, J. R., J. Lady, R. Townsend, A. E. Giorgi, J. R. Stevenson, C. M. Peven, and R. D. McDonald. 2001a.

Estimating in-river survival of migrating salmonid smolts using radiotelemetry. Canadian Journal of Fisheries and Aquatic Sciences 58:1987–1997.

Skalski, J.R., J. R. Stevenson, J. Lady, R. Townsend, A. E. Giorgi, M. Miller, and K. English. 2001b. An assessment of project, pool, and dam survival for Chinook and steelhead smolts at Rocky Reach and Rock Island dams using radiotelemetry and PIT-tag techniques, 2000. Prepared for Public Utility District No. 1 of Chelan County, Wenatchee, Washington.

Skalski, J. R., R. Townsend, J. Lady, A. E. Giorgi, J. R. Stevenson, and R. D. McDonald. 2002. Estimating route-specific passage and survival probabilities at a hydroelectric project from smolt radiotelemetry studies. Canadian Journal of Fisheries and Aquatic Sciences 59:1385–1393.

Townsend, R. L., J. R. Skalski, P. Dillingham, and T. W. Steig. 2006. Correcting bias in survival estimation resulting from tag failure in acoustic and radiotelemetry studies. Journal of Agricultural, Biological, and Environmental Statistics 11(2):183–196.

U.S. Environtmental Protection Agency (U.S. EPA). 2001, EPA Requirements for Quality Assurance Plans, EPA QA/R-5.

U.S. Environtmental Protection Agency (U.S. EPA). 2002. Guidance for Quality Assurance Project Plans, EPA QA/G-5.

Section 9.4

Database Management and "Real-Time" Data Analysis Systems for Fish Telemetry Studies

Karl K. English, Dave Robichaud, and Peter Wainwright

INTRODUCTION

Radio-telemetry techniques have been used extensively over the past 25 years to study the freshwater migration of many species including salmonids (Eiler et al. 2000). In the 1980s most studies relied upon mobile survey techniques to track tagged fish (Lough 1983). In the 1990s, the use of fixed-station receivers became more common as studies expanded to cover entire watersheds (e.g., Koski et al. 1996). These fixed stations were typically deployed for several months to continuously monitor key locations along rivers (English et al. 2005) or at hydroelectric dams (Keefer et al. 2004). Advancements in tag technology (e.g., uniquely identifiable digitally coded tags; longer battery life; smaller size; lower costs) made it possible for researchers to apply more tags and to work in noisier environments, thereby generating even more data. Concurrently, fisheries managers, dam operators and research teams started requesting in-season summaries of tracking results to assist in their decision making processes. Consequently, the data management and analyses challenges grew substantially through the 1990s. The objective of this Section is to provide readers with an overview of effective database management and real-time analysis systems and present examples showing how these systems have met the typical challenges associated with telemetry studies.

Each telemetry project presents a unique set of challenges. Different study goals result in variability in numbers of tags applied, tagging/release locations, fixed-station receivers and antennas deployed (see Section 9.2). The database management methods described in this paper have been used to successfully organize and analyze fixed-station and mobile tracking data from many studies of juvenile and adult salmonid migration. In these studies, 20–4,000 fish were tagged, 5–56 fixed-station receivers were deployed and the size of the study area has ranged from a single fishway in a hydropower dam to an entire watershed. This database management approach is rigorous yet flexible, and has allowed efficient and timely analyses of ever-increasing amounts of telemetry data.

METHODS

The fundamental components of an effective database management and analysis system are:

1. rigorous data recording and verification during the tagging process;
2. on-site data verification during the data downloading process;
3. basic file management protocols;
4. a logical and simple database structure;
5. systematic and efficient data processing procedures, including:
    a. rules for assigning detections to zones;
    b. the identification and filtering of noise records;
    c. compression of large volumes of data (100–500 Mb) into rapidly accessible summary files (1–5 Mb);
    d. flexible spatial and temporal scales;
    e. customized displays for presenting results;
6. automated database updating protocols.

We elaborate on each of these components in the following sections.

1. Data recording and verification during the tagging process

Recording and verifying data during the tagging process should be a simple task. Yet, data-recording and data-entry errors can result in significant challenges during subsequent data processing, even when relatively few tags have been released. Problems are magnified for larger studies, especially when there is more than one telemetry study being conducted in the area. A systematic approach is essential for recording and entering tag codes and all the physical (e.g., release date and time, handling time, water temperature) and biological data (e.g., fish size; secondary marks; and scale, DNA, tissue, or blood samples) associated with each tagged fish (see Section 9.2).

We find that data entry personnel make fewer errors when the information comes from the taggers in a predictable manner. Taggers should provide the data in a consistent sequence (e.g., start with tag frequency and code, follow with species identification, then weight, then length, etc.), use consistent units (i.e., do not use mm for some fish, and cm for others), and give comments in standardized manner (say "split pectoral fin" instead of "ratty fins"). To effectively provide data to the recorders, it is helpful for the taggers to be familiar with the data-entry forms.

Data should be entered and stored in a structured database with built-in data-entry verification processes. Data-verification processes can range from simple to complex. Two typical simple verification processes include formatting checks (e.g., dates should be entered so that the month and day are easily identified: users are forced to enter "2 January" or "1 February" instead of "1/2" or "2/1") and value restrictions (attempts to enter a weight lower than the minimum tagable size would be rejected). An example of a complex verification process is one which uses an internal length-weight regression formula to alert the data-entry person if the weights and lengths provided do not conform to the expected values (thus prompting a double-check of data entered, or of the measurements made).

Under controlled environments, it is usually possible to use computers on site, allowing data to be entered directly into a database with built-in data-entry verification processes. For most of our studies, we have used MS Excel as the data-entry platform during tagging operations, with subsequent upload into a relational database built using Visual Fox Pro. For more rugged environments, hand-held 'palm-top' computer devices (with touch-screens, housed in clear waterproof cases) can be used, since these can run data-entry software with verification processes. In the most primitive settings, data can be initially captured on data forms (printed on water-proof paper), but it is essential that all data are entered into the structured database as promptly as possible. We prefer to archive older versions of the database, rather than overwriting it each day, and to keep multiple copies of any paper data forms. Where possible, the data related to tag releases should be entered on the same day that these data were collected. If this is not possible, the data forms should be scanned, photocopied or photographed using a digital camera to ensure none of these critical data are lost. Of course, the database should be backed up on multiple drives early and often.

2. Data verification during the data downloading process

Examining the data at the time of download can be very helpful. Often, glitches that occur during the download process are not repeated during a second download attempt. Thus overall data quality can be improved and data gaps reduced by verification of the download file prior to reinitializing the receiver and leaving the fixed-station site. For receivers equipped with easily removed memory cards, the reinitializing step is not necessary.

If the examination of the download file reveals excess noise or a lower-than-expected number of tag detections (particularly when test tags or beacon tags are used), it may indicate problems with the station equipment (broken receiver, cables, solar panel, antennas, etc.) which can be identified and corrected during the same site visit. Many receiver manufacturers have developed software that provides station performance information. For more complex studies with many fixed-station sites, beacon tags, in-line signal amplifiers, and/or multiple antennas connected to a single receiver, it is wise to develop specialized software that provides all the desired diagnostic information rapidly with each data download event. We have used Visual Fox Pro executables housed in a .NET framework to perform on-site analyses of download data.

3. Basic file management protocols

A vital part of the management of telemetry data is keeping track of all the download files. During a telemetry study, a large number of data files can be amassed, especially if multiple receivers are deployed for extended periods. Several file management methods can be employed to keep track of the download files and to help identify data gaps.

Download files should be given unique, informative names that can be easily associated with a specific receiver and the download time. For example, a file downloaded from fixed-station receiver #03 on 18 September (month "09") might be called "F030918.hex." For studies extending over multiple years, the year could be added to the file name.

Recording metadata during each download event improves the management of download files. Types of metadata include the receiver number, date, time, receiver battery voltage, the filename assigned to the downloaded data, the quantity of data downloaded, and the field technician's name. We recommend that one copy of these metadata be kept with the receiver (if possible) and another copy with the field technician. Simple software can be developed that will process a download file and output important metadata needed to assess the performance of each monitoring site (e.g., the number of noise records, data gaps, and summaries of the number of detections by signal strength).

It is useful to maintain a list of all the file names along with the metadata that summarize the contents of each file. These data can be very useful for tracking down missing data files, and they may help address potential data management errors that can occur in the downloading process. The list also allows users to predict noise and tag-detection quantities at each fixed-station site, which helps schedule periodicity of download visits, and can help identify problems with environmental noise and receiver function.

4. Database structure

The raw data for most telemetry studies include information related to each fish tagged, fixed-station tracking data, mobile tracking data and tag-recovery data. An example of the flow of tag, receiver, mobile and recovery data through the loading and processing sequence is provided in Figure 1. Structured yet open-ended database tables ensure that critical information on the date, time and location for each detection (or set of detections) of each tagged fish is available in standardized formats (see Section 9.2). Other biophysical data associated with the fish (e.g., size, age, sex, stock ID, etc.) and information on the capture and tagging process (e.g., capture method, water temperature, tagger, fish condition at release) are readily accessible for analyses. Because of the potentially large volumes of data from fixed-station receivers, automated procedures should be developed to upload these data into a rigid record format. The same format should be used for mobile tracking and tag recovery data so all tracking data can be readily extracted and examined for any fish. Database software suitable for programming these types of automated uploading and analysis procedures include MS Access, Visual Fox Pro, MySQL and SQL Server.

To link detection data to specific locations within the study area, additional database tables are required. The receiver number and antenna number data (that are included in the fixed-station receiver download files) can be linked to locations by assigning them to zone numbers. Each antenna attached to each receiver could be treated as a unique location (with a unique zone number), or detections on several different antennas can be combined into a single zone. Zones can be combined into 'superzones' for the purposes of certain types of analyses. This hierarchical organization permits flexibility in the preparation of data summaries and displays. For example, antennas deployed in individual spill-gates along the length of a dam may be treated as separate zones to determine specific problem passage locations, or combined into a superzone for questions regarding the 'whole-spillway' performance.

The database structure that we have been using includes: three source data files, six database parameter tables and five essential relational tables. The three primary source data files are:

"*Tags*": the data for each tagged fish (including a unique identifier for each fish, the frequency and code of the transmitter, the tagging and release timing, various biometrics, etc.);

"*Raw*": the complete set of data downloaded from fixed-station receivers (including date, time, receiver, antenna, frequency, code, and signal strength);

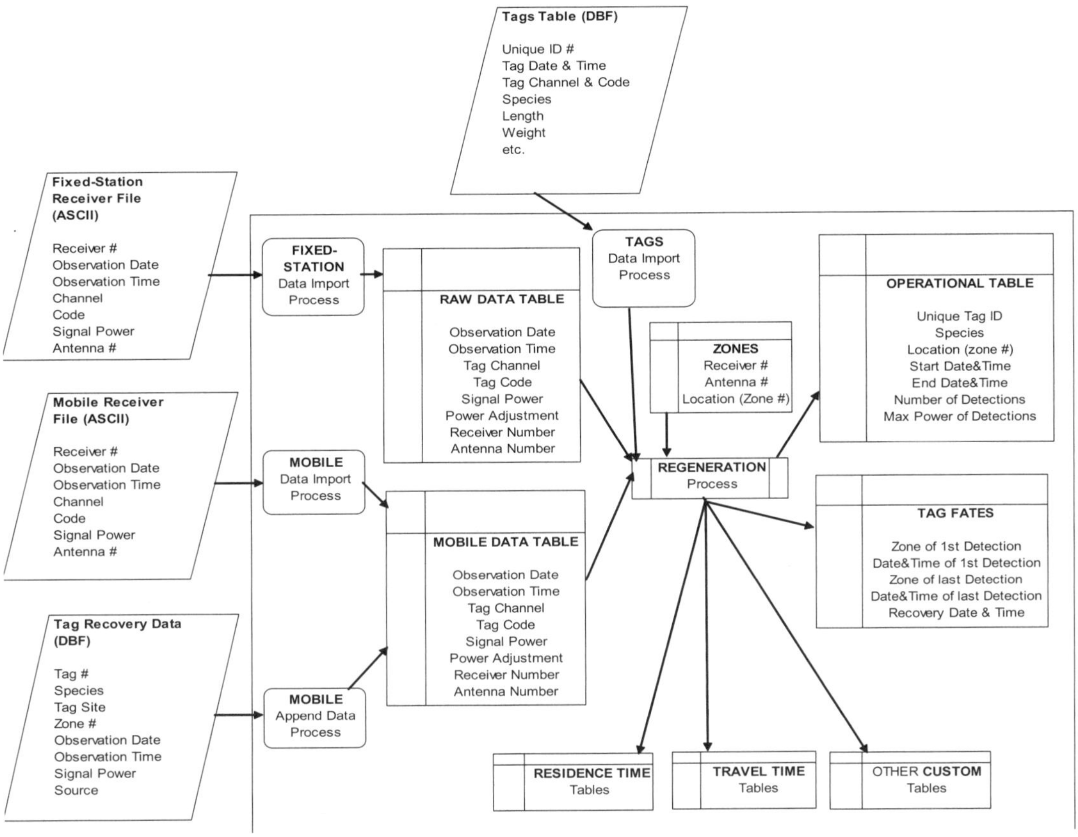

FIGURE 1. Example of the flow of receiver data through the data loading and processing sequence.

"*Mobile*": the complete set of mobile tracking data (including date, time, frequency, code, signal strength, and geolocation).

The six database parameter tables are:

"*Antennas*": a list of each unique antenna (each assigned an antenna number), along with the receiver to which it is attached, and a "low power reject" threshold (the power level below which detections are likely to be noise, and are thus ignored);

"*Zones*": a list of each detection location (each assigned a unique zone number), along with a zone-specific noise-filtering threshold (the minimum number of detections in a specified time period, below which detections are likely to be noise, and are thus ignored);

"*Super Zones*": a list of each superzone (each assigned a unique superzone number);

"*Species*": includes several-letter species codes and species names;

"*Maps*": includes a list of each map, display or aerial photo (each assigned a unique map number), along with the file location, and dimensions of the image files;

"*Tag Sites*": a list of all locations where tag/releases occurred (each assigned a unique tag site number).

The five essential relational tables are: 1) Antennas-to-Zones; 2) Zones-to-Super Zones; 3) Zones-to-Maps; 4) Super Zones-to-Maps; and 5) Tag Sites-to-Maps. These tables are used to define the spatial relationships for the input data and database displays.

A project configuration file can be added to allow the user to define the specific database and analysis parameters for each project, including: the time interval for displays from 1 min to 1 d; the temporal starting point for all displays (e.g., release date or first detection date); whether or not to include detections associated with the master antenna (combined detections from all antennas connected to the receiver); and the power threshold levels below which signals are ignored. All noise events and all signals weaker than the specified power threshold are archived, but are excluded from further analysis. The project configuration file can also be used to control the types of data summaries (tables and figures) that will be prepared or analyses that will be conducted (e.g., run reconstruction analyses).

5. Data Processing Logic

Once the project structure, noise filtering parameters and output displays are defined in the above tables, the data for multiple fixed-stations can be uploaded and processed in a single step. Processing algorithms can be designed to compress millions of raw records into a more workable-sized data table (which we call the "operational" file), including the start and end of a detection sequence, the number of detections, and the maximum power level for a set of detections of a specific tag detected at a specific station ('sequences' end on the last detection prior to a valid sequence of higher-power detection in a different zone). For studies with overlapping detection zones, the data processing algorithms are used to compare the relative signal strength for each detection so fish can be assigned to a specific spatial zone. Once the operational file has been created the tracking data can be rapidly summarized into tables, including the number of tags detected at each zone, the last zone of detection for each tag, or the travel times for each fish to move between each pairwise combination of zones. These tabulated results can be immediately plotted on the user defined maps.

By modifying certain data-filtering parameters and re-running the compression algorithm, one can easily observe the effects of parameter changes on the resulting operational file. A simple but effective approach for evaluating data-filtering parameters has been to design data displays that relate the compressed records to the raw data, using color codes (Figure 2). The sample of data provided in the upper table in Figure 2 shows the sequence of detection locations (compressed records) for one fish (tag = 632) from its release (Zone = –1) on 8 August 2006 to its last detection zone (Zone = 60). Color codes are used to show the relationship between compressed records and raw data records. The summary record, highlighted in yellow in the upper table, shows the start and end time for the five detections of tag 632 in zone 40. The five yellow records in the lower table are the individual detections in the raw data file that were compressed into a single "operational" record. The "max_power" value on the operational record is the highest power of the five detections associated with that record. The blue records are detections that were ignored because the specific antenna for these detections was not part of a defined zone. In this example, antenna 0 is the master antenna representing all the antennas attached to this receiver. Detections on the master antenna are frequently ignored when researchers need to determine the location of the fish within the antenna array for a specific receiver. The detection highlighted in brown was not included in the operational record because the number of detections on antenna 2 was below

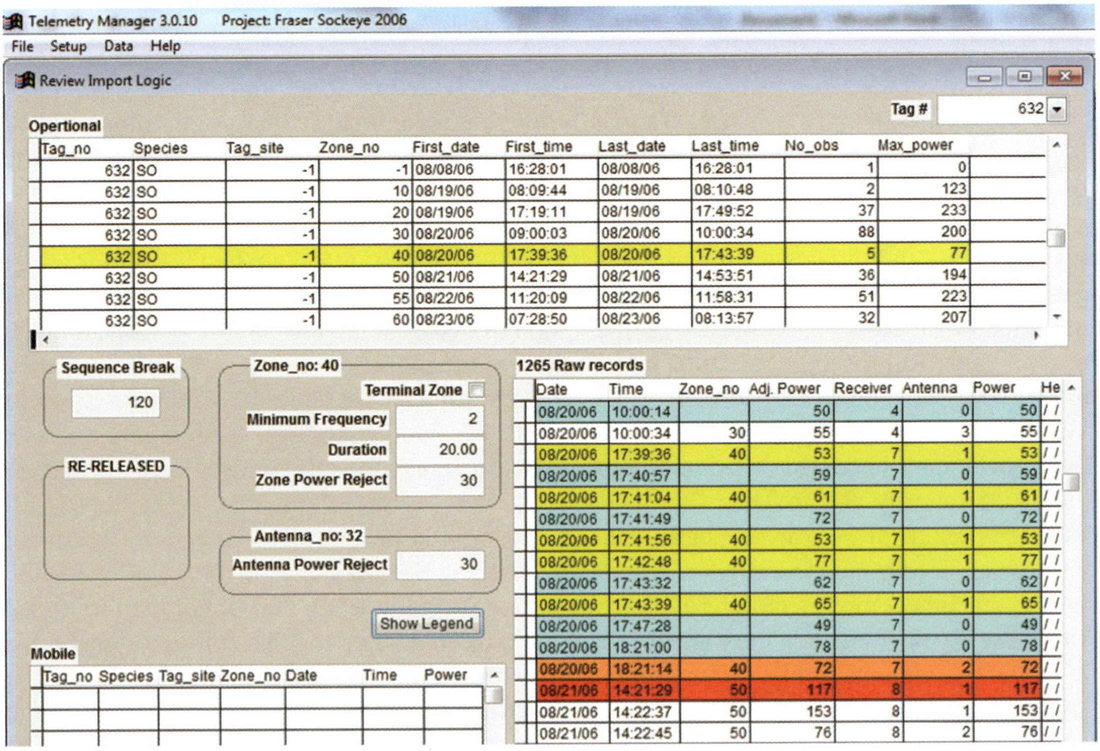

FIGURE 2. Example of an import data display showing the link between summary records and the raw data for a specific tag code. In this example, raw data records were either included in a detection sequence (yellow); or excluded from a sequence because the record was not from a defined detection zone (blue) or below the detection frequency threshold (brown). Red records indicate the start of a new detection sequence.

the detection frequency threshold set for Zone 40 (i.e., 2 hits within 20 min displayed in the Zone No: 40 box to the left of the raw data table). These detection frequency thresholds are set by the database manager to remove false positive detections that can occur due to environmental noise. The red color code is used to identify the first record of a next detection sequence following the sequence highlighted in yellow. Other color codes not shown in this sample of raw data are used to identify tag detection records that result from environmental noise (dark brown) and detections ignored because their power level was lower than the most recent detection in the current sequence (purple). The table in the lower left corner of Figure 2 is used to display all the mobile tracking records for the selected tag code. These types of relational data displays are very helpful for any study that requires experimentation with the data filtering parameters (e.g., those with overlapping detection zones, or detection sites with background environmental noise). The result of altering data processing parameters or receiver setups can be readily observed through the data displays. These displays are also helpful for designing data processing logic, evaluating the results from such logic or assessing the configuration of a specific receiver or antenna array.

A common challenge for all telemetry studies with overlapping zones is to use the detection timing and signal strength to determine the most likely location (zone) for each tagged fish at a given point in time (see Section 9.1). In watershed studies, receivers with multiple antennas are often deployed at tributary junctions to monitor fish movements in three directions: mainstem upstream, mainstem downstream, and tributary upstream (e.g., Koski et al. 1996; English et al. 2005). For dam passage studies, the detection zones for antennas deployed along the dam face or in fishways frequently overlap (see Section 9.2). A common method used to determine the location of each tagged fish is to compare the relative signal strength for a sequential series of detections and determine the "break points" when highest signal strength shifts from one antenna (i.e., zone) to another. In these types of studies, the antenna array must be carefully designed with each antenna "balanced" to ensure that the power levels accurately reflect the distance of tag from the antenna, and are not biased by antenna configuration or the losses associated with the length of coax cable between the antenna and the receiver (e.g., Robichaud et al. 2002). In studies where mobile tracking techniques are used to assess the location of tagged fish, survey crews routinely use Global Positioning Systems (GPS) to determine the precise location where the received signal was at maximum power. These techniques have been used to identify spawning areas for salmonids and determine the fate of tags shortly after release (English et al. 2006). Tag detections that are stationary for extended periods of time are generally classified as mortalities, however, a better assessment of the exact timing and location of a dead tagged fish can be obtained from tags equipped with sensors that detect motion (Eiler 1990) or tilt angle (Shardlow et al. 2007). For tags equipped with these types of sensors, the transmitting code is altered when the fish stops moving or can no longer maintain its typical orientation in the vertical plane for a specified interval.

One of the most powerful tools in the data processing tool kit is the definition of different 'analytical groups', sets of tags that are grouped together for data analysis, processing and summary displays. Once defined, the standardized analysis routines that would normally be run for each species and/or tagging location, can also be run for each analytical group. Analytical groups can be defined using release periods or locations, recovery periods or locations, or any other criteria (age, size, sex, stock, etc.). Examples of typical analytical groups include: all the fish tagged within a specific size range, time period or species. More unusual groupings include: all the fish detected at a specific location, at a specific time or with the same sequence of detections.

6. Automated Database Updating Protocols

An important feature of any modern database structure is maintaining access to historical databases. All too often historical data are lost or inaccessible because of changes in database management software or data management systems. It is especially important for large scale or multi-year projects to ensure that database updating routines are written to ensure that historical telemetry databases are upwardly compatible with newer versions of the database management software so researchers have easy access to these data and each study benefits from the latest data processing and presentation enhancements.

DISCUSSION

The above data processing procedures have been used to organize and analyze telemetry data from more than 40 radio-tag studies. These studies included multi-year assessments of Chinook *Oncorhynchus tshawytscha*, coho *O. kisutch*, steelhead (anadromous rainbow trout)

O. mykiss, and sockeye *O. nerka* for the three largest watersheds (Fraser, Skeena and Nass) in British Columbia (e.g., Koski et al. 1996; Alexander et al. 1996; English et al. 2005); multiple species assessments of adult salmon returns to the Nushagak and Copper rivers in Alaska (Daigneault et al. 2007; Wade et al. 2008); and numerous studies of adult and juvenile salmon migration through hydropower dams in the mid-Columbia River in Washington State (e.g., English et al. 1998, 2006; Nass et al. 2000; Skalski et al. 2000). Sample displays from two of these projects are provided (Figures 3–6) to demonstrate the type of summaries that have been used to communicate study results to government agencies, First Nations, scientists, managers, and the general public. These two examples were chosen because they present very different spatial and temporal challenges and clear visual presentations of study results.

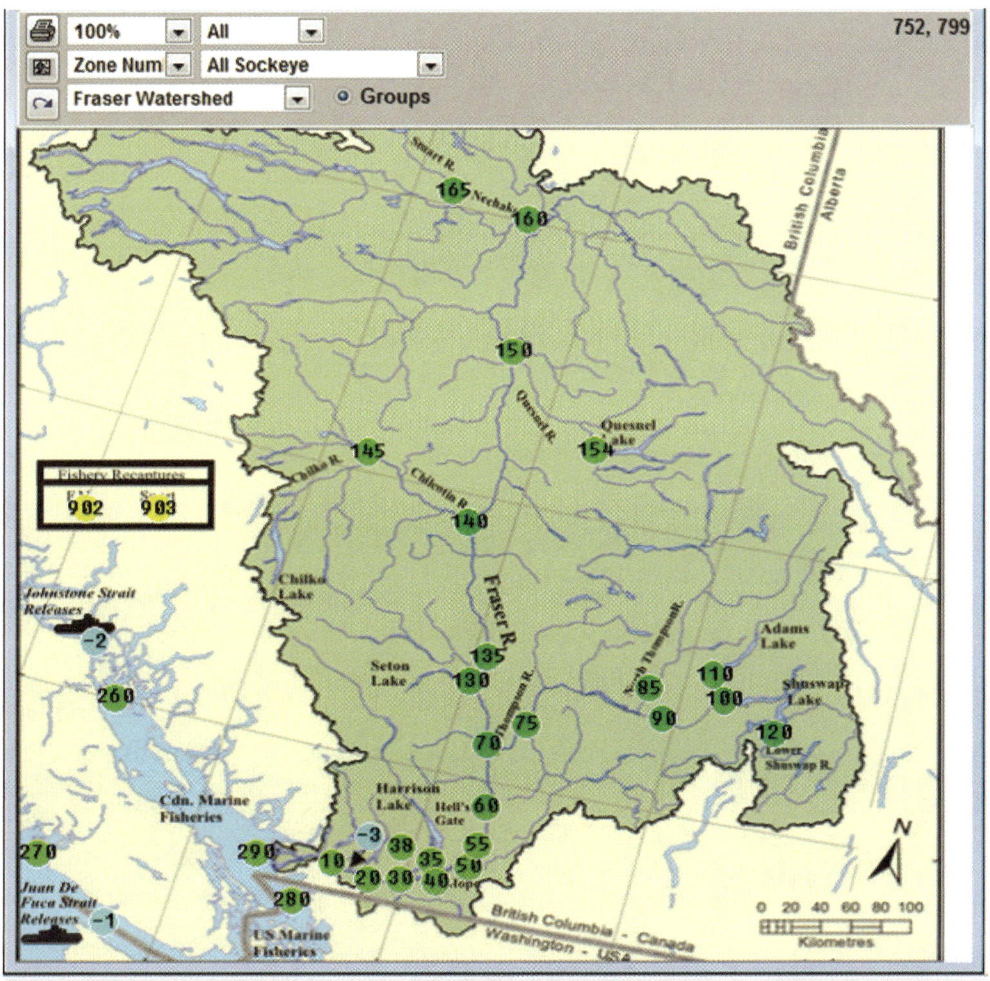

FIGURE 3. Map of the Fraser River watershed showing the locations of the release sites (blue circles), fixed-station receivers (green circles) and tag recovery bins (yellow circles). Numbers in circles are unique 'zone numbers' that have been assigned specific locations and that are linked to specific antennas and specific receivers in the study area.

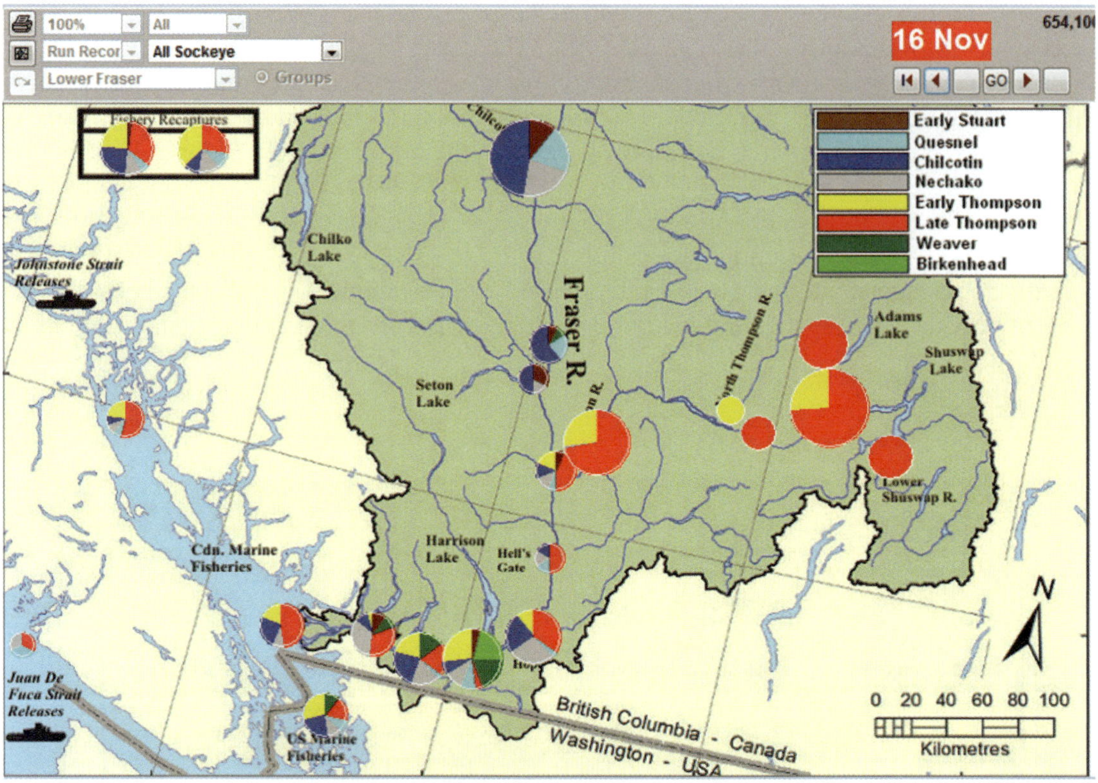

Figure 4. Map of the lower portion of the Fraser River watershed showing the distribution of radio-tagged sockeye by major stock group on 16 November 2006. The size of the circles indicates the number of fish that are assigned to each zone.

Figure 5. Aerial photograph of Wanapum Dam showing the locations of the forebay aerial antenna detection zones (white ellipses); powerhouse and spillway underwater antenna zones (green circles); and tailrace aerial antenna detection zones (yellow ellipses). Numbers in circles are unique 'zone numbers' that have been assigned specific locations and that are linked to specific antennas and specific receivers at the dam.

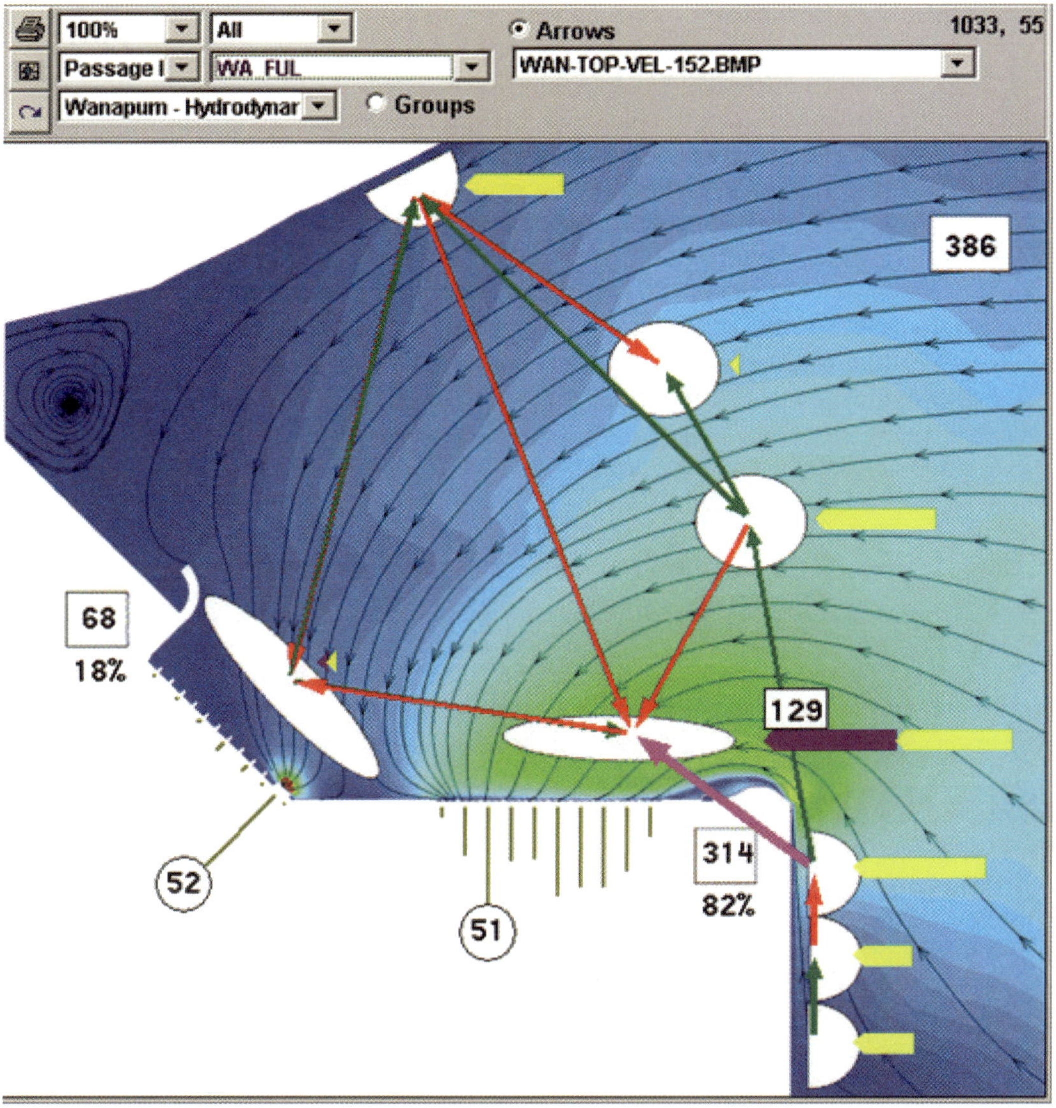

Figure 6. Example of a customized data summary display that overlays the approach, forebay movement pattern, and passage routes for radio-tagged juvenile Chinook salmon on a background showing flow patterns and color coded water velocities for a specific spillway and powerhouse configuration tested at Wanapum Dam in 2002. The large yellow and purple bars in the forebay show the distribution of first detections at the project, the arrows show the direction and relative magnitude of movements between the forebay detection zones and the thin lines on the downstream side of the dam show the relative number of fish passing via each potential turbine and spillway. The numbers indicate the total number of tagged fish that were detected at the dam during this specific period (386) and the number that passed via turbines (314) and spillways (68) and the associated passage percentages.

Fraser 2006—Adult Sockeye

In 2006, a radio-telemetry study was conducted to assess the in-river survival (successful migration from the lower Fraser to spawning areas) for all run timing groups of Fraser sockeye as well as river entry behavior for the late-run timing component (Robichaud and English 2007). Radio tags were applied to 660 adult sockeye in marine waters and to 378 adult sockeye captured in the lower Fraser River near Mission, BC. The in-river survival of these fish was assessed using tracking data from 24 fixed-station receivers deployed at strategic locations throughout the Fraser watershed (Figure 3). A DNA sample was obtained from each radio-tagged fish to determine the spawning destination prior to tracking. Tracking data confirmed that 90–95% of the fish tracked to a spawning location had been correctly identified by the DNA analyses, thus it was reasonable to use the DNA data to assign a stock-of-origin to the fish that did not survive to spawning grounds. Consequently, stock specific estimates of in-river survival could be generated and the primary locations for en-route mortalities were identified for each stock. Map displays of the Fraser watershed were used to show the day-to-day progress of the sockeye migration for each of the 8 major component stocks (e.g., Figure 4).

The Fraser watershed presented many challenges for telemetry studies including a river channel widths exceeding 500 m, substantial variability in environmental noise (100–50,000 records/d), remote locations with difficult access, and very public locations with site security issues. Frequent on-site assessment of station performance was critical for maintaining high detection efficiencies at most monitoring sites. At the Mission hydroacoustic monitoring site in the lower Fraser River, receivers were deployed on each riverbank to cover the wide river channel, receiver gains were adjusted to reduce environmental noise, and data filtering criteria set to require 3–4 sequential records for a single tag code to qualify as a valid detection. Due to these challenges, detection efficiencies at the Mission site rarely exceeded 80%. For most of the Fraser monitoring stations it was possible to select sites and configure the antennas to minimize environmental noise (generally less than 1000 records/d) and achieve detection efficiencies of 95–100%. The primary solution to maintaining high detection efficiencies over these widely disbursed monitoring sites was to conduct downloads at problem sites every 5–7 d during peak migration periods and let the more secure and reliable sites operate for 20–30 d between download visits. Efficient data loading and "real-time" data analysis allowed field technicians and project managers to assess detection efficiencies during download trips and rapidly respond to any problems encountered.

Mid-Columbia River 2002—Juvenile Chinook

In 2002, a total of 2,320 radio-tagged Chinook smolts were released into the mid-Columbia to assess their migratory behavior through the Wanapum and Priest Rapids hydroelectric projects (Robichaud et al. 2002). The detection systems consisted of 31 Lotek Wireless SRX 400/DSP units and one Grant Systems Engineering MITAS receiver monitoring over 300 individual antennas located at Wanapum and Priest Rapids dams (Figure 5). Over the study period, 99.8% and 99.4% of the Chinook that passed each dam were detected by the telemetry arrays at Wanapum and Priest Rapids, respectively.

The large number of tags being deployed created challenges during this study. In addition, we had to coordinate with other local researchers to ensure that there was no duplication of tags among the several concurrent studies in the area. As a result, we had a specific list of tags (i.e., frequency and code combinations) that were available to us, and we needed to inventory and test all purchased tags to ensure that they transmitted on the proper frequency and with the correct code. Duplicate and malfunctioning tags were sent back to the manufacturer. Once a complete list of our tags was verified, it was imported into our database. To help avoid data-entry problems during tagging sessions, we embedded additional error-checking scripts into our data-entry platform. During a tagging session, upon entry of a new frequency and code, the software would automatically alert the data-entry person if the frequency and code: 1) did not appear in the complete list of our tags; or 2) was already assigned to a previously released fish. Although not fool-proof, the program helped minimize errors during data entry, and, on occasion it helped alert us to data entry errors for fish that had been released (in some cases, the correct codes for these fish could be deduced from our detailed inventories and the tags still on hand).

Another challenging aspect of this project was the desire for in-season information about the effects of different dam operational treatments on fish survival and passage behavior. Dam operations alternated on 4 d intervals between three spill configurations. Although the tagged fish were released in groups, variable travel times meant that not all fish released together would encounter the dam during the same operational treatment. We used the date of first detection in the forebay to generate 'analytical groups' linking individual fish to the spill configuration that they encountered. These analytical groups allowed us to assess passage route efficiencies, dam passage timing, and survival (both overall and passage-route-specific) by dam operational treatment. Customized displays (e.g., Figure 6) were prepared to overlay the fish approach, forebay movement and passage locations onto the treatment-specific output of a hydrodynamic model (showing the forebay water velocities and flow lines during each spill configuration). These types of data outputs were automatically generated for all defined analytical groups, and were eventually used to optimize flow configurations for fish passage, and to evaluate alternative locations for surface fish collection and bypass structures.

SUMMARY

The development of standard database structures and flexible data processing logic facilitates the data organization, validation, and management process, thus allowing researchers to focus more of their time on the interpretation and presentation of study results. Automated data loading and processing procedures make it possible to provide reliable study results that can be used in-season to assist in decision making related to the allocation of survey effort, fisheries management, and dam operations. Customized graphics that are linked to the database allow researchers to efficiently update analysis results and compare migration behavior and survival for different groups of fish. Researchers responsible for multi-year telemetry studies should ensure that all databases are upwardly compatible with newer versions of the software so they have easy access to the data and each study benefits from the latest data processing enhancements.

REFERENCES

Alexander, R. F., K. K. English, and W. R. Koski. 1996. Distribution, timing and numbers of steelhead returning to the Skeena River watershed in 1995. Report for the B.C. Ministry of Environment, Lands and Parks, Victoria, B.C.

Daigneault, M. J., J. J. Smith, and M. R. Link. 2007. Radio telemetry monitoring of adult Chinook and sockeye salmon in the Nushagak River, Alaska, 2006. Bristol Bay Science and Research Institute, Dillingham, Alaska.

Eiler, J. H. 1990. Radio transmitters used to study salmon in glacial rivers. Pages 364–369 *in* N. C. Parker, A. E. Giorgi, R. C. Heidinger, D. B. Jester, Jr., E. D. Prince, and G. A. Winans, editors. Fish-marking techniques. American Fisheries Society, Symposium 7, Bethesda, Maryland.

Eiler, J. H., D. J. Alcorn, and M. R. Neuman. 2000. Biotelmetry 15. Proceedings of the 15th International Symposium on Biotelemetry. Auke Bay Laboratory, Juneau, Alaska.

English, K. K., D. Robichaud, C. Sliwinski, R.F. Alexander, W. R. Koski, T. Nelson, B.L. Nass, S.A. Bickford, S. Hammond, and T. R. Mosey. 2006. Comparison of adult steelhead migrations in the mid-Columbia hydrosystem and in large naturally flowing British Columbia Rivers. Transactions of the American Fisheries Society 135:739–754.

English, K. K., W. R. Koski, C. Sliwinski, A. Blakley, A. Cass, and J. C. Woodey. 2005. Migration timing and in-river survival of Late-run Fraser River sockeye using radio-telemetry techniques. Transactions of the American Fisheries Society 134:1342–1365.

English, K. K., T. C. Nelson, C. Sliwinski, J. R. Stevenson, and T. R. Mosey. 1998. Evaluation of the juvenile spring chinook, steelhead and sockeye migratory patterns at Rocky Reach Dam using radio-telemetry techniques 1998. Report for Public Utility District No. 1 of Chelan County.

Keefer, M. L., C. A. Peery, T.C. Bjornn, M.A. Jepson, and L. C. Stuehrenberg. 2004. Hydrosystem, dam and reservoir passage rates of adult Chinook salmon and steelhead in the Columbia and Snake rivers. Transactions of the American Fisheries Society 133:1413–1439.

Koski, W. R., M. R. Link, and K. K. English. 1996. Distribution, timing, fate and numbers of chinook salmon returning to the Nass River watershed in 1992. Canadian Technical Report of Fisheries and Aquatic Sciences 2129.

Lough, M. J. 1983. Radio telemetry studies of summer steelhead trout in the Cranberry, Kispiox, Kitwanga, and Zymoetz rivers and Toboggan Creek, 1980. Skeena River Fisheries Report 83–33.

Nass, B. L., K. K. English, C. Sliwinski, and S. A. Bickford. 2000. Assessment of adult chinook passage times at Wells Dam and evaluation of fishway gate alterations, 1998. Report for Public Utility District No. 1 of Douglas County.

Robichaud, D., and K. K. English. 2007. River entry timing, survival and behaviour of Fraser River sockeye salmon in 2006. Report for Pacific Salmon Commission, Vancouver, B.C.

Robichaud, D., K. K. English, C. Sliwinski, B. Nass, A. Blakley, and C. A. Grant. 2002. Telemetric assessment of Chinook smolt behavior in relation to various spill methods at Wanapum and Priest Rapids dams, 2002. Report for Public Utility District No. 2 of Grant County, Ephrata, Washington.

Shardlow, T. F., K. D. Hyatt, and S. R. Pearson. 2007. A new microcontrolled fish tag for accurately estimating life span and survey life of spawning salmon. Aquatic Living Resources 20:197–203.

Skalski, J. R., J. Jady, R. Townsend, and K. K. English. 2000. Estimates of route-specific passage and survival of run-of-river steelhead smolts at Wanapum and Priest Rapids projects using radio-telemetry techniques, 2000. Report for Public Utility District No. 2 of Grant County.

Wade, G., K. M. van den Broek, J. W. Savereide, and J. J. Smith. 2008. Spawning distribution and run timing of Copper River sockeye salmon, 2007 final report. U.S. Fish and Wildlife Service, Office of Subsistence Management, Fisheries Resource Monitoring Program (Study No. 05–501), Anchorage, Alaska.

Section 9.5

Bias from False-Positive Detections and Strategies for their Removal in Studies Using Telemetry

JOHN W. BEEMAN AND RUSSELL W. PERRY

INTRODUCTION

The use of radio and acoustic telemetry to study aquatic animals has flourished since the 1950s and 1960s (see Section 1). Electronic data-logging receivers are commonly used in both types of active telemetry to record the presence of transmitters in the detection field formed by one or more antennas or hydrophones. As described in Sections 5.1 and 7.1, the path of a transmitter signal to a telemetry receiver can be influenced by many factors and the received signal is not always detected or correctly assigned. It should be of no surprise to users of active telemetry systems that not all records in telemetry receivers are from tagged fish and not all tagged fish are recorded when present.

Four types of observations are possible in data from telemetry receiving systems based on the binary nature of presence and absence (Table 1). True positives and true negatives are what one ideally expects from telemetry systems, but in most studies they are accompanied by false negatives (not recorded when present) and false positives (recorded when absent). False negatives arise due to a variety of causes, including insufficient detection area relative to transmitter pulse rate and fish travel speed, collisions between transmitters in the detection area, interference from ambient noise, or a received signal too weak to be recorded (see Sections 3, 5, and 7). The probability of false negatives can be calculated as (1–detection probability) and can be estimated with proper study design and incorporated into estimates of fish presence (see Sections 7.2 and 9.2).

Unlike false negatives, the probability of false positives cannot be estimated from the fish data and therefore present a more serious threat to inferences made from telemetry data. The probability of false positives depends on many factors including the physical environment, the number of transmitters simultaneously present at a telemetry receiver, how the data are transmitted, and how the transmission is assigned to an individual animal. Environmental conditions creating transmissions similar to those used for animal telemetry systems increase the probability of false positives; electrical transmission lines and outboard motors are two examples (see Sections 5.1 and 7.1). Multiple transmitters simultaneously present in a detection field can create false positives when pulses from transmitters overlap to form a pattern

Table 1. The four combinations of truth and observation based on presence or absence.

Observation	Truth	
	Present	Absent
Present	True Positive	False Positive
Absent	False Negative	True Negative

of intervals identical to that of another valid transmitter, a situation often called "collisions." Collisions can also occur, primarily in the acoustic telemetry environment, when signals of one or more transmitters reach the receiver from direct and indirect (reflected) paths (often called "multipath").

Active telemetry systems for animal telemetry employ a variety of methods to encode the transmitter identity, each of which can affect the probability of false positives. The simplest method to encode a unique transmitter identity is to emit a single burst at a constant interval on a unique frequency. More complex systems use a series of short bursts within each transmission cycle that can be decoded by an electronic receiver to indicate transmitter identity; this is commonly referred to as a "coded scheme" (see Sections 5.1 and 7.1; Stuehrenberg et al. 1990). Such systems may also include an additional pulse to encode error detection data, which enables a comparison of the transmitted and received signals to reduce the probability of accepting false positives (Pincock 2011). However, one drawback to using the additional pulse is a longer transmission time, which reduces transmitter life accordingly. Other physical methods to reduce the rate of false positives due to collisions include allotting transmitters to multiple frequencies, varying burst rates among transmitters, randomly varying burst rates within a transmitter, and using more than one frequency for each transmission. Several coding types are described in Sections 5.1 and 7.1 and some manufacturers provide descriptions of the code schemes they employ on their websites.

Since false positives may bias inferences drawn from telemetry data, means to remove them should be employed. Although "data proofing" is a common step prior to analysis, few formal methods have been published for reducing false positives in telemetry data. Skalski et al. (2001) used a series of criteria to identify false positives from radio telemetry data collected using a system of narrow-band receivers scanning among several frequencies. Their criteria was implemented in an automated procedure that included 1) eliminating records from transmitter codes not used in their study or occurring prior to the time of release, 2) imposing a strength threshold of the received signal, 3) setting a minimum frequency of detections per unit of time at a given detection site, and 4) examining the spatiotemporal distribution of detections. They also examined data manually when the automated methods resulted in inconsistent results among criteria. The specific numerical criteria used in such methods are often chosen subjectively, but once the criteria are set, using an automated procedure minimizes subjectivity by ensuring uniform application across the data set. Methods similar to Skalski et al. (2001) are commonly used (English et al. 2005; Wertheimer and Evans 2005; Stutzer et al. 2006; Beeman et al. 2008). In an effort to quantify the effects of proofing criteria on the results, Skalski et al. (2001) conducted a sensitivity analysis to examine how biological parameters changed when altering the proofing criteria, but this level of scrutiny appears to be rare.

The probability of false positives in acoustic telemetry systems is theoretically greater than in radio telemetry systems, so additional means are often used to remove false positives from acoustic telemetry data. Acoustic telemetry is based on a lower frequency than radio telemetry and the transmissions therefore travel through water slower and farther than radio waves, making false positives from multipath and various sources of noise more common. Pincock (2011) described methods to reduce false positive records from an acoustic telemetry system that uses random transmission intervals and an error detection code in the transmission and an internal blanking interval following detection of a transmission to reduce multipath by preventing subsequent detections for a short period. He used output from a numerical model to estimate the frequency of occurrence of false positives and the distribution of intervals between them. From this he devised an algorithm that, among other things, rejects all single detections. However, he suggests single detections can be accepted from that telemetry system if 1) there is clear evidence of a lack of collision activity (lack of detections from one or more other transmitters at the same receiver near the same time) and 2) when the receiver is not in a location subject to false positives due to ambient noise. McMichael et al. (2010) described methods to reduce false positives from an acoustic telemetry system based on regular transmission intervals with an error detection code and a receiver with no internal blanking interval that relied on time-based criteria relative to the tag pulse rate. After accepting only records from study tags after their release time, they required a minimum of 4 records at intervals consistent with the tag pulse rate within a time period equal to 12 times the pulse rate. The time difference between the 4 detections must be integer multiples of the tag pulse rate (allowing for a small amount of variation). For example, a transmitter with a 3-s pulse rate requires a minimum of four records spaced at multiples of 3 s over a period of 36 s. The acoustic telemetry system described by McMichael et al. (2010) does not employ an internal blanking interval, so one must be applied during postprocessing prior to the criteria they describe (Mark Weiland, Battelle Pacific Northwest National Laboratory, written communication).

Clearly, a variety of methods may be used to reduce false positives in telemetry data, no single method appears to be universally applied, and few methods have been critically evaluated. Pincock (2011) concluded that no method can be guaranteed to remove all false positives from active telemetry data, but methods could be devised to minimize the probability of accepting them. Just as eliminating all false positives may be impossible, preventing removal of some true positives is difficult due to similarity in the transmission characteristics of some true- and false-positive detections. Less stringent criteria will eliminate the fewest true positives but allow a higher rate of false positives, whereas more stringent criteria may remove more false positives at the expense of losing more true positives. Thus, the criteria used for defining "valid" detections will fall along a continuum of removing false positives versus losing true positives. Our goals in this Section are to illustrate the effects of false positives in telemetry data, to understand the consequences of eliminating true positives in the quest to remove false positives, and to describe several methods for reducing the prevalence of false positives in telemetry data. First, we conduct a simulation study to quantify how biological inferences drawn from telemetry data may be influenced by: 1) the presence of false positives in telemetry data, and 2) the removal of true positives from the data set. This simulation sheds light on the tradeoff between maximizing removal of false positives and minimizing removal of true positives. Second, we apply some of the common methods described above to a real telemetry data set aimed at reducing the prevalence of false positives. Here, we vary the level of scrutiny used to remove false positives and illustrate the influence of the criteria on biological parameters estimated from the data. Although the set of criteria used will vary among studies, we hope that this Section will provide you with insights on how to develop proofing criteria well suited to your specific study.

SIMULATING THE EFFECTS OF FALSE POSITIVES AND REMOVAL OF TRUE POSITIVES

False positives in a telemetry data set may bias biological inferences drawn from telemetry data. The direction and magnitude of bias will depend on the frequency of false positives and the biological parameter of interest. To illustrate how false positives can introduce bias and to understand how removal of true positives (i.e., introducing false negatives) affects inferences from telemetry data, we simulated a simple mark–recapture study with the goal of estimating survival over a single sampling period, representing survival either between two spatial locations (e.g., migrating fish in river) or two temporal sampling occasions.

To perform such a mark–recapture study requires one release occasion and two subsequent sampling (i.e., detection) occasions to detect live tagged fish. Our primary goal is to estimate survival between release and the first sampling occasion, but the second sampling occasion is required to quantify the probability of detecting live fish during the first occasion. The model we implement is a two-occasion Cormack–Jolly–Seber (CJS) model, as described in Section 9.2 Figure 1 (Cormack 1964; Jolly 1965; Seber 1965). We set true parameter values to: $S_1 = 0.8$, $S_2 = 0.8$, $p_2 = 0.8$, and $p_3 = 0.8$, where S = survival probability, p = detection probability, and subscripts represent the survival interval or detection location.

Given these true parameter values, we then simulated two-digit detection histories for each fish, where "10," "01," "11," and "00" indicate fish detected during one, the other, both, or none of the sampling occasions respectively (we have omitted the leading '1' often associated with the initial release of the tagged animals). Individuals were simulated by first drawing a 1 or 0 randomly with probability S_1, and then if they "survived,' drawing a 1 or 0 randomly with probability p_2 to determine if they were subsequently detected (and similarly for S_2 and p_3). We simulated 500 fish, tabulated the frequency of occurrence of each detection history, and then applied a two-occasion CJS model to this data set to estimate S_1 and p_2. We conducted 1,000 simulations of the mark–recapture study, with each simulation comprised of 500 fish. When data are simulated under this model, the mean estimated survival and detection probability over the 1,000 simulations provides an unbiased estimate of the true survival and detection probabilities used to generate the data.

Next, we conducted two sets of simulations, one where false positives were added to the data and another where true positives were removed from the data to create false negatives. After simulating the data, we randomly assigned false positives at each detection site with probability F. For example, if $F = 0.1$, then each fish had a 10% probability of being assigned a false positive detection at each detection site. Thus, a "00" detection history was changed to "10" when a random number generator drew a false positive at the first occasion. Alternatively, a "10" detection history remained unchanged under the same circumstance. For the second simulation, true positives were removed with probability R at each site. For instance, a "10" detection history was changed to "00" when a random number generator determined that the first detection was to be removed. For each of the 1,000 simulated mark–recapture experiments for each scenario, we applied the CJS model, estimated survival and detection

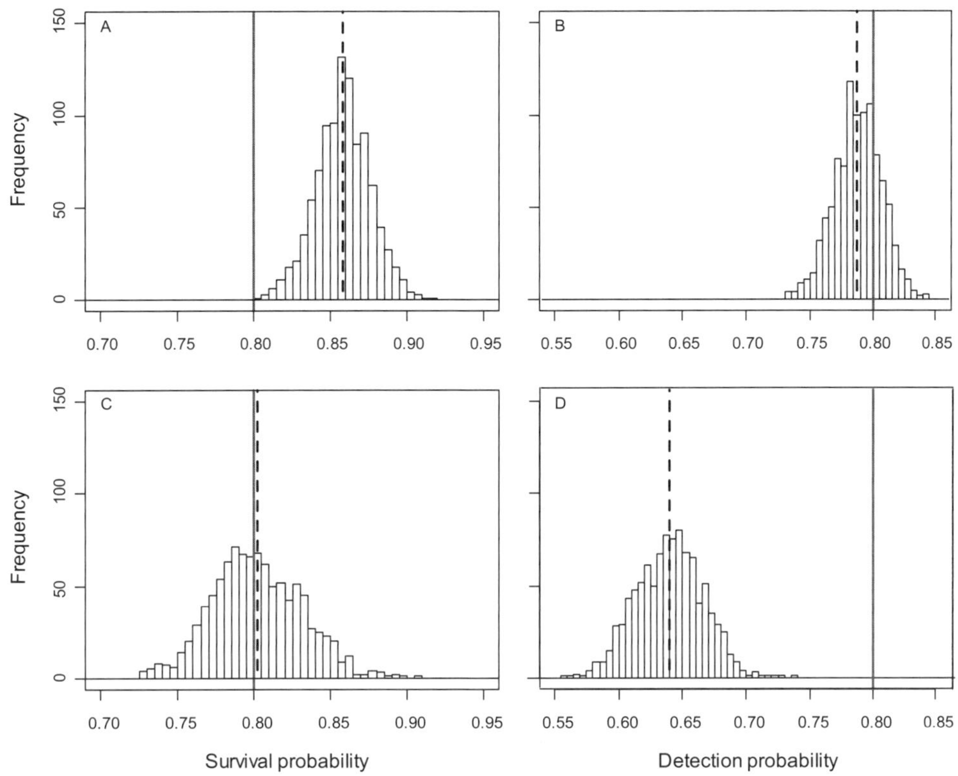

FIGURE 1. Summary of 1,000 simulations of a mark–recapture study with 500 tagged fish showing the estimated survival and detection probability. Panels A and B show estimated survival and detection probability when false positives are randomly assigned to 10% of fish at two detection sites. Panels C and D show estimated survival and detection probability when true positives are randomly eliminated from 20% of fish at two detection sites. The solid line shows the true survival or detection probability, whereas the dashed line shows mean survival or detection probability for the 1,000 simulations.

probabilities, and then calculated the mean survival and detection probabilities over the 1,000 simulations. The difference between the mean of the 1,000 simulations and the true parameter values is an estimate of the bias introduced by either adding false positives or removing true positives. We then varied the probability of adding false positives or removing true positives between 0 and 0.3 to examine how these probabilities affected bias and precision. Precision was represented using the 90% confidence interval estimated from the 5th and 95th percentiles of the 1,000 simulations.

We found that adding false positives positively biased survival estimates, but survival estimates remained unbiased when true positives were removed (Figure 1). For example, a 10% rate of false positives at each occasion resulted in a mean survival estimate of 0.86, resulting in a 6 percentage point positive bias in survival (Figure 1A). Bias was less than the rate of false positives because some fish may have already been detected on a particular occasion at which a false positive had occurred. Interestingly, detection probability was slightly negatively biased, likely because false positives resulted in a higher frequency of "01" detection histories, indicating a fish had survived but gone undetected at the first sampling occasion (Figure 1B). In contrast, when 20% of true positives were removed at each site, estimates of survival remained unbiased but detection probability was reduced from 0.80 to about 0.64 (Figure 1C and 1D).

Varying the rate of false positives showed that positive bias increased as the rate of false positives increased (Figure 2A). However, when the rate of removing true positives increased, survival estimates remained unbiased (Figure 2B) but detection probability declined (Figure 2D). Although survival remained unbiased when true positives were removed, the confidence interval widths increased as the removal rate increased (Figure 2B). Removing true positives caused precision to decrease, because precision of survival declines with detection probability. We also conducted the same simulation, but reduced the sample size from 500 to 100 tagged fish. We found nearly identical results as with 500 fish, but observed a small positive bias in survival (up to 2 percentage points) as the rate of removing true positive detections exceeded 15%. Although our simulation was simple, it illustrated the consequences of false positives in a telemetry dataset and sheds light on the tradeoff between accepting false positives versus removing true positives.

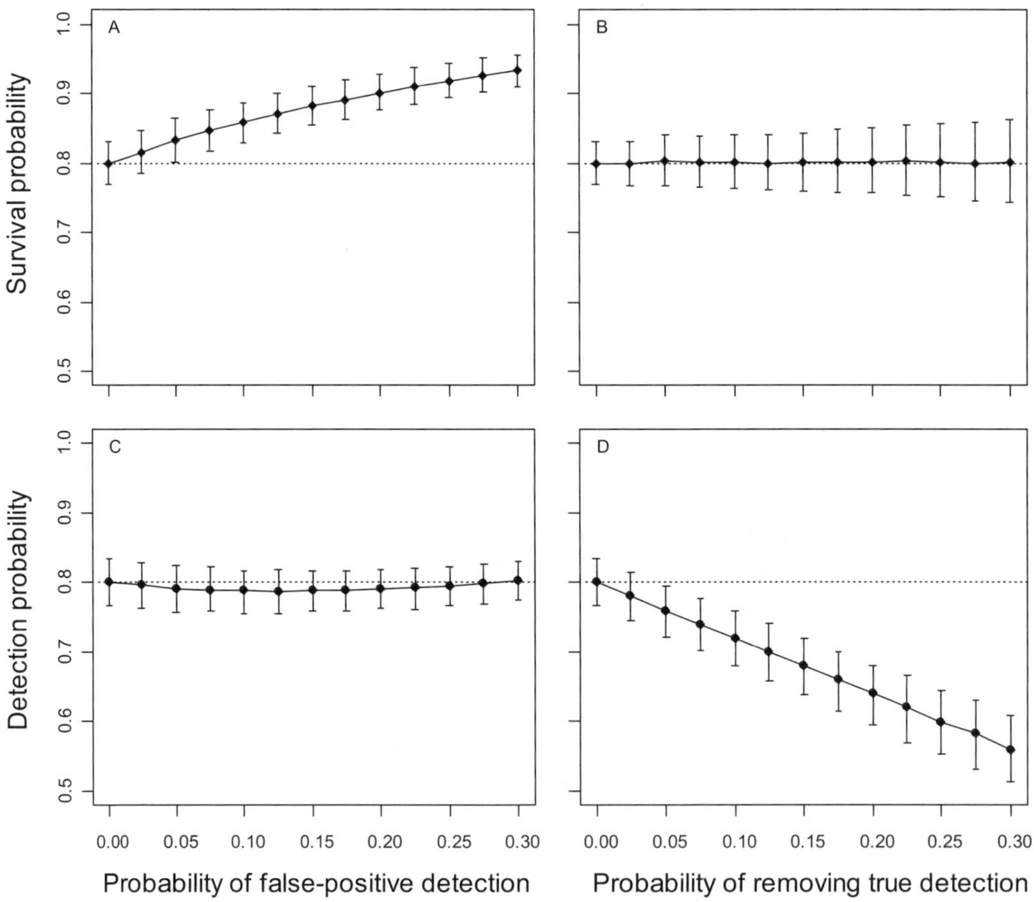

FIGURE 2. Mean survival and detection probabilities as a function adding false positives or removing true positives. Panels A and C show estimated survival and detection probability when false positives are randomly assigned to fish at two detection sites. Panels B and D show estimated survival and detection probability when true positives are randomly eliminated from fish at two detection sites. The dotted lines show the true survival and detection probability used to simulate the data. Data points show the mean of 1,000 simulations of a mark–recapture study with 500 fish and error bars show the 5th and 95th percentiles.

Clearly, we should strive for unbiased estimates that are less precise over biased estimates that are precise. Thus, our findings suggest that proofing protocols should be designed with relatively stringent criteria that seek to remove the maximum amount of false positives at the risk of possibly removing true positives. Such an approach will likely minimize bias, although the magnitude of bias will depend on the particular biological parameter being estimated in a given study. In contrast, if criteria are designed in an attempt to retain every true positive, we may unknowingly increase the rate of false positives, which would introduce bias. When conducting a field study using telemetry, we lack the luxury of knowing the true values of biological parameters we are attempting to estimate. It is for this reason that varying the proofing criteria and examining the effect on the parameters of interest can help to determine how restrictive the proofing criteria should be. In the next section we illustrate this approach by applying different proofing criteria to a real telemetry data set.

APPLYING PROOFING CRITERIA TO TELEMETRY DATA FROM THE KLAMATH RIVER, CALIFORNIA.

For this example we will use data from a study of radio-tagged juvenile coho salmon *Oncorhynchus kisutch* in the Klamath River in northern California described by Beeman et al. (2009). The purpose of the study was to estimate the effects of river discharge on apparent survival of juvenile salmon. Fish from a local hatchery were surgically implanted with a radio transmitter and released into the river at the upstream end of a 296-km study area. Detection sites were placed at eight locations downstream positioned 12–71 km apart. Each site consisted of several Yagi antennas combined into a single input which passed data to both a narrow-band receiver and a wide-band receiver. Data collected from the wide-band receiver (Orion, Sigma Eight, Inc., Newmarket, Ontario, Canada) at each site will be used in this example. The Orion wide-band receivers used in the study had a bandwidth of 1 MHz and could record radio transmissions matching the specified code set within their frequency range from several transmitters almost simultaneously. Orion receivers record the time of detection to the nearest second. The radio transmitters used a four-pulse coded transmission scheme with 512 unique codes and operated at frequencies of 164.320, 164.360, and 164.480 MHz with pulse rates of 7.8, 7.9, 8.0, 8.1, and 8.2 s nearly evenly divided among the 221 transmitters released. More than one pulse rate was used to improve the probability of detection by reducing the occurrence of collisions caused by multiple tagged fish occupying a detection field at the same time.

The parameter of interest in the study was the apparent survival between detection sites. As described in the previous section, detection at a site is the product of the probability of survival to the site and the probability of detection at the site. We used CJS mark–recapture methods within Program MARK to estimate these parameters (White and Burnham 1999).

We employed several methods of reducing false positives in data from Beeman et al. (2009; Figure 3). The first step in all methods was to retain only records from radio frequencies and tag codes used in the study occurring after their initial release. These were the sole criteria applied to the data for Method A. Four other data sets were created by applying progressively more stringent criteria. Method B was similar to the method described by Skalski et al. (2001) and in addition to the steps for Method A included retaining detections which met the following criteria: time since tag activation of less than or equal to the 91 d maximum tag life (determined empirically using a subset of transmitters), a minimum received power of

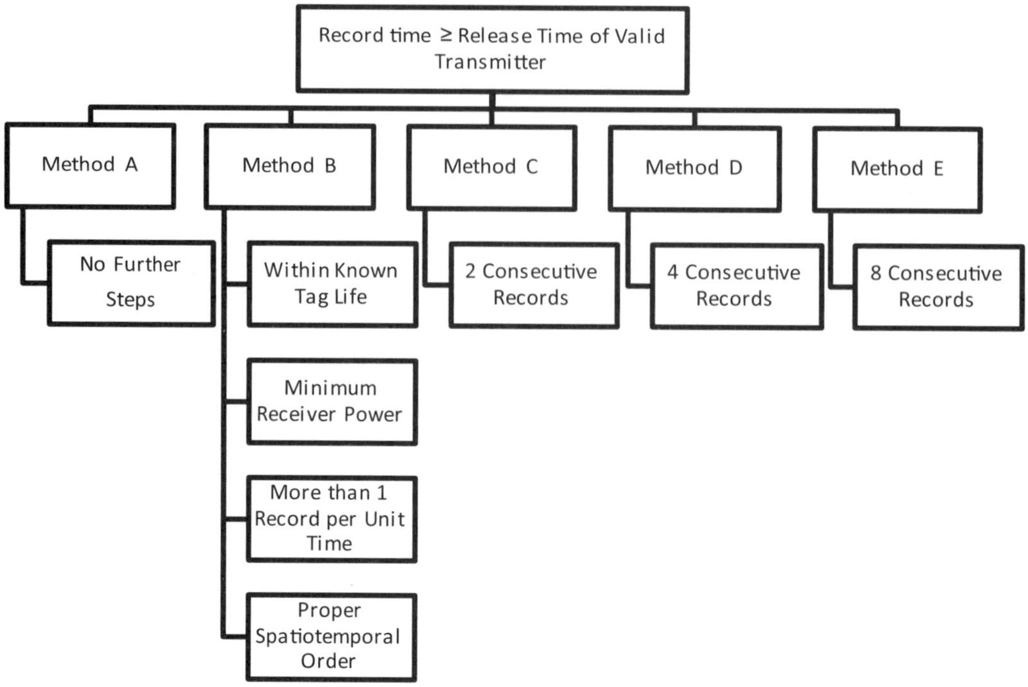

FIGURE 3. Flow chart of steps used to reduce false positives in data from radio-tagged fish in the Klamath River.

–110 decibels relative to 1 mW (dBm) at all sites except #1 and #2 for which a setting of –105 dBm was used due to an abundance of ambient noise (the receiver recorded received power in dBm), at least 2 records within 10 min, and correct spatiotemporal order among sites. Records that did not meet the spatiotemporal criterion were independently examined by two observers to 1) determine why the errors occurred and correct them, or 2) omit the entire fish history if no obvious resolution was apparent to both observers. Detections from fish with site-specific residence times greater than the 95th percentile were also manually examined to ensure the data were not erroneous, however, that criterion alone was not cause for omission. As part of the original study, detection records from a randomly-selected 10% of the tagged fish were manually examined to ensure the automated criteria performed as intended. In addition to the steps used for Method A, Methods C–E included a criterion that specified a minimum number of consecutive detections consistent with the pulse interval of the tag. Methods C, D, and E required 2, 4, or 8 consecutive records, respectively. For example, for a tag with a pulse interval of 7.8 s requiring a minimum of 4 consecutive detections (Method D), we retained the initial detection as well as detections that were between 7 and 8, 15–17, and 22–24 s after the initial detection. The broad ranges of time intervals were required because the receiver recorded detection time to the nearest second and some tags had pulse intervals that were not integers. For the example above tags with the 8-s pulse interval required detections at 8, 16, and 24 s after the initial detection.

Results from the five methods were assessed by comparing the estimated number of fish passing each site and the estimated detection and survival probabilities. As expected, the number of fish that were estimated to have passed the monitoring sites was highest when using Method A and lowest when using Method E (Figure 4). For example, estimates from Method A indicated 219 of the 221 fish released passed site 1, Method B indicated 187 passed, Method C indicated 192 passed, Method D indicated 182 passed, and Method E indicated 170 passed. The results from applying methods B, C, and D were similar at most sites.

Detection probabilities from Method A were not always higher than those of the other methods (Figure 5) because the false positives remaining after applying Method A were spread among the sites. When false positives occurred at upstream sites and not downstream sites, detection probabilities of the upstream sites were not affected, however, those present downstream and not upstream lowered detection probabilities at the sites from which detections were absent. This is similar to the effect from the simulation, in which the occurrence of '01' detection histories increased in the presence of false positives. The highest detection probabilities were often from methods B and C, which were generally similar to one another. Increasing the criterion to 8 consecutive detections (method E) reduced detection probabilities sharply at all sites other than site 4, where detection probability was high in all cases.

There were clear differences in estimates of apparent survival from the data based on Method A versus the other methods (Figure 6). The estimates of apparent survival from Method A ranged from 0.9909 (standard error, SE = 0.0064) to 1.0 (SE = 0.0, no mortality in the sample data) in the first four intervals and ranged from 0.8848 (SE = 0.0286) to 0.9444 (SE = 0.0286) in the last three intervals. Results from each of the other methods were similar to one another, but all led to considerably lower estimates of survival in the first four intervals compared to Method A, suggesting that estimates from Method A were positively biased due to presence of false positives. However, the results indicate that any of Methods B-E would likely result in unbiased estimates of survival in this study. The chief difference among Methods B-E was in the precision of apparent survival estimates. The standard errors of the estimates increased as the stringency of the proofing criteria increased and detection probabilities decreased, which is consistent with the results of the simulation.

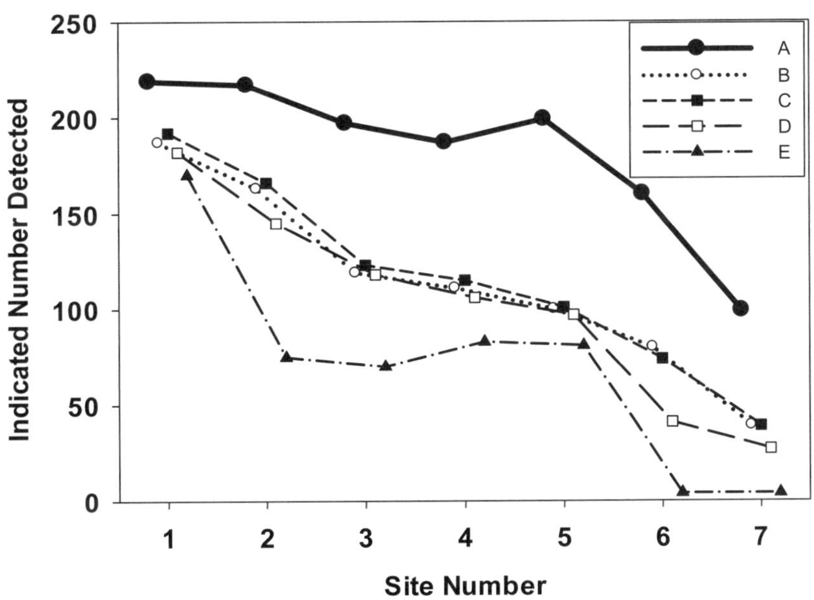

FIGURE 4. Indicated number of radio-tagged fish detected at each site after applying methods A–E to reduce false positives.

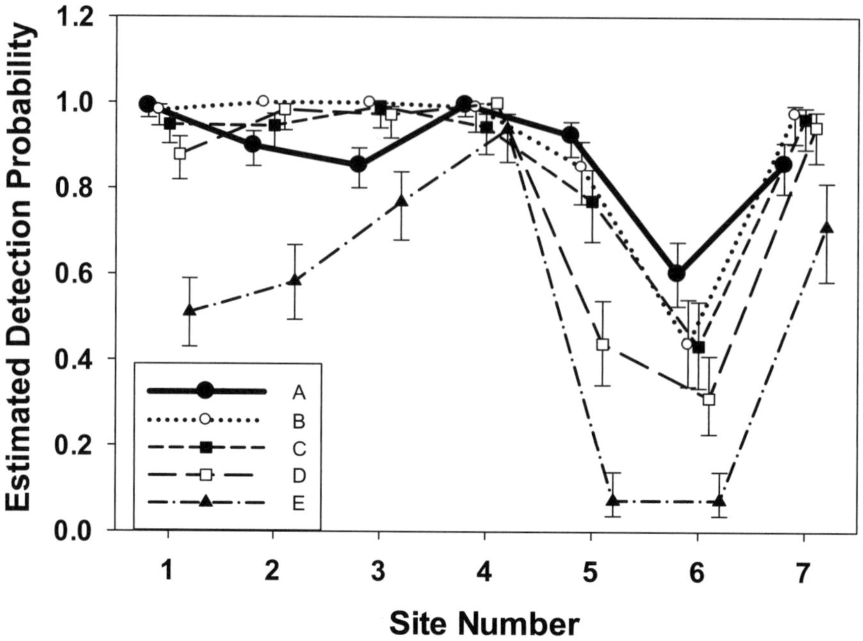

Figure 5. Recapture probabilities and 95% confidence intervals estimated from the data after applying methods A–E to reduce false positives.

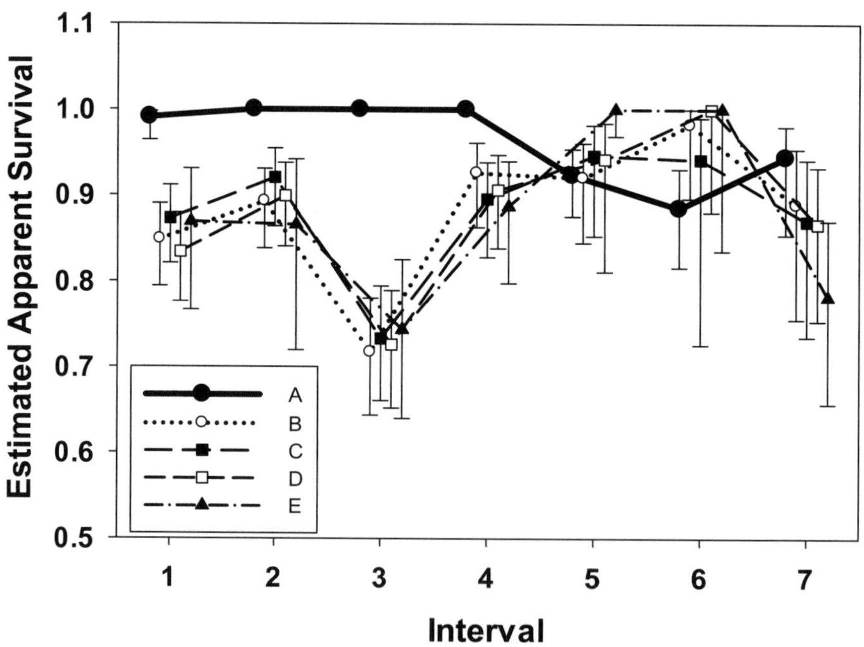

Figure 6. Apparent survival and 95% confidence intervals estimated from the data after applying methods A–E to reduce false positives.

DISCUSSION

Results from our simulation helped to explain the patterns observed when different proofing criteria were applied to real data. In the simulation, false positives in the data positively biased survival estimates, whereas with the real data, higher survival estimates under Method A relative to other methods suggested that Method A failed to remove many false positives. When we randomly removed true positives from simulated telemetry data, survival estimates remained unbiased but detection probabilities declined. With the real data, progressively more stringent criteria led to declines in detection probabilities, but survival probabilities stabilized, suggesting that most false positives had been removed and that survival estimates were unbiased for all but Method A. While some true positives were likely removed, particularly under the most restrictive criteria (Method E), survival probabilities did not change appreciably. Both approaches revealed that precision of survival estimates declined with more restrictive criteria, but most importantly, reducing the presence of false positives reduced or eliminated bias.

Our findings should be directly applicable to most studies using telemetry, but particularly those incorporating detection probabilities into estimates of fish presence. In our opinion, estimation of detection probabilities in telemetry studies appears to be the exception rather than the rule, other than studies explicitly designed to estimate certain population parameters (see Section 9). Thus, in many studies based on animal telemetry there is no feedback mechanism with which to determine when false positives have been reduced to an acceptable level without overly omitting true positives. We agree with Pincock (2011) that the goal must be to reduce the probability of false positives to an acceptable level, preferably to remove all of them (which may never be known), even if that results in some true positives being omitted in the process.

We suggest that one reasonable means to address the trade-off between removal of false positives and removal of true positives is to apply a range of criteria to the data and compare the results. The number of fish present at a detection site can be estimated by dividing the number of individuals detected by the detection probability. As the criteria becomes more restrictive, removal of false positives will result in a decline in the estimated number of fish present, whereas when true positives are removed, detection probability will decline but the estimated number of fish present should stabilize. Such an approach can be used to judge the appropriate level of scrutiny in proofing criteria. However, if detection probabilities are not explicitly incorporated into estimates of biological parameters, there is no reliable indicator to gauge the effectiveness of removing false positives or the level of deleting true positives in the process. If the purpose of a study is to estimate fish presence based on the number of fish detected in a particular habitat encompassed by the detection area of a given site, the results based on Methods A and E would likely be biased high and low, respectively. It is for this reason that incorporation of detection probability is vital to studies in which animals are not directly observed (see Section 7.2).

We applied a variety of methods to the real data to illustrate methods for removing false positives, but other methods could also be devised to accomplish this task. We suggest that at a minimum all studies using telemetry apply the criteria of Method A. Method A has a zero probability of omitting true positives and a nonzero probability of removing false positives. After that step, the most appropriate methods will vary based on the type of transmission, the environment between the transmitter and the receiver, and how the transmission is processed in the receiver. All telemetry systems have some probability of recording false positives and therefore some means of removing false positives must be employed prior to analysis (Pincock 2011). We have commonly used Method B with good success in studies based on radio telemetry. It is our opinion that methods based on the number of detections consistent with the tag pulse interval (time-based methods) whether consecutively, as in our Methods C–E, or within some reasonable time interval (McMichael et al. 2010), are less subjective than Method B and represent an improvement. Employing time-based methods must be less stringent with data from acoustic-telemetry systems than from radio-telemetry systems, because there is often a lower probability of recording consecutive transmissions from acoustic telemetry. Thus, applying criteria such as methods C–E described here may be too restrictive for acoustic telemetry systems. For one acoustic telemetry system McMichael et al. (2010) used the criterion of a number of detections at correct tag pulse intervals within a specific time period rather than a number of consecutive detections at the correct interval as in our methods C–E. In practice, the two approaches may achieve similar results. Time-based methods may be less useful when multiple frequencies are scanned by the telemetry receiver, such as in some narrow-band types used in radio-telemetry, because the time for a complete scan among frequencies can vary with the number of transmitters detected during a scan cycle and with the configuration of the receiver. Further research into the use of time-based methods with narrow-band receivers would be useful.

It is only possible to apply proofing criteria that are supported by the data. For example, requiring two consecutive detections, in our opinion a modest criterion for studies using radio telemetry, will not be fruitful if the telemetry detection sites are not designed to collect at least two detections per transmitter. We do not advocate accepting single detections as evidence of fish presence in most cases, particularly from systems that do not employ an error detection code. Thus, it is critical to design detection sites to record more than one true positive as a tagged animal passes the site (more detections would be better). Achieving this should be elementary in most studies. For example, to achieve this level of detection in the Klamath River study we designed detection field sizes based on the tag pulse interval and expected fish movement rate by combining inputs from several antennas—a cost effective solution to a simple problem. In the Klamath River study, transmitters with one of five pulse rate intervals (to reduce the probability of collisions) and no error detection code were used, and fish were expected to pass most sites relatively quickly.

In some studies researchers routinely accept single detections as evidence of animal presence. These are generally studies over very large areas, such as along the continental shelf of North America, where detection areas may be many kilometers wide (Welch et al. 2003, 2009; Section 8.3). In these studies the spatial area to monitor is often large and to reduce system costs the receivers are spaced too far apart to ensure multiple detections during all conditions. For example, Melnychuk and Walters (2010) estimated that up to 21% of their fish presence records were based on single detections of acoustic transmitters after applying criteria similar to those described by Pincock (2011). The acoustic telemetry system employed in many of the large river and ocean-deployed systems, such as the one used by Melnychuk and Walters (2010), uses a slow series of random transmission intervals including error detection code and a receiver with an internal blanking interval following detection, which aid in reducing the probability of accepting false positives. This comes at the cost of a greater probability of false negatives, which explains the desire to accept single detections. We are not stating that accepting single detections is wrong. Rather, we contend that accepting single detections increases the risk of false positives and therefore the risk of bias, and recommend against it. In our simulation we showed the consequence of accepting false positives (increase in bias) is more detrimental than that of omitting true positives (decrease in precision). However, balancing these risks is up to the researcher and the decision will depend partly on the specific telemetry system being used, because some are more prone to accepting false positives than others.

In summary, false positives can result in biased inferences and their presence should be minimized in telemetry data prior to analysis. In our examples, false positives resulted in a positive bias in estimates of survival and in the estimated number of fish present by introducing data that were not from the study transmitters. The goal of methods designed to remove false positives is to reduce or eliminate their influence on the inferences from the data, while at the same time, minimizing the omission of true positives. Our results indicate that the omission of true positives in the process of addressing false positives, (such as may occur if criteria "too stringent" are employed), can result in unbiased results if the concomitant reduction of detection probability is accounted for explicitly. We suggest that in studies which do not account for detection probability explicitly a variety of methods should be employed to remove false positive detections ranging from benign to severe (e.g., Method E). A comparison among methods can then be used to determine if a common ground has been achieved.

REFERENCES

Beeman, J. W., S. Juhnke, G. Stutzer, and N. Hetrick. 2008. Survival and migration behavior of juvenile coho salmon in the Klamath River relative to discharge at Iron Gate Dam, northern California, 2007. U.S. Geological Survey Open-File Report 2008–1332.

Beeman, J. W., S. Juhnke, H. C. Hansel. 2009. Summary of survival data from juvenile coho salmon in the Klamath River, northern California, 2008. U.S. Geological Survey Open-File Report 2009–1019.

Burnham, K. P., D. R. Anderson, G. C. White, C. Brownie, and K. H. Pollock. 1987. Design and analysis methods for fish survival experiments based on release-recapture. American Fisheries Society Monograph 5, Bethesda, Maryland.

Cormack, R. M. 1964. Estimates of survival from the sighting of marked animals. Biometrika 51:429–438.

English, K. K., W. R. Koski, C. Sliwinski, A. Blakley, A. Cass, and J. C. Woodey. 2005. Migration timing and river survival of late-run Fraser River sockeye salmon estimated using radiotelemetry techniques. Transactions of the American Fisheries Society 134(5):1342–1365.

Jolly, G. M. 1965. Explicit estimates from capture-recapture data with both death and immigration-stochastic model. Biometrika 52:225–247.

Melnychuk, M. C., and C. J. Walters. 2010. Estimating detection probabilities of tagged fish migrating past fixed receiver stations using only local information. Canadian Journal of Fisheries and Aquatic Sciences 67:641–658.

McMichael, G. A., M. B. Eppard, T. J. Carlson, J. A. Carter, B. D. Ebberts, R. S. Brown, M. Weiland, G. R. Ploskey, R. A. Harnish, and Z. D. Deng. 2010. The juvenile salmon acoustic telemetry system: a new tool. Fisheries 35(1):9–22.

Pincock, D. G. 2011. False detections: What they are and how to remove them from detection data. Document #: DOC-004691 version 02 April 13, 2011. Amirix Systems Inc. Available: http://www.vemco.com/pdf/false_detections.pdf (February 3, 2012).

Seber, G. A. F. 1965. A note on the multiple recapture census. Biometrika 52:249–259.

Skalski, J. R., J. Lady, R. Townsend, A. E. Giorgi, J. R. Stevenson, C. M. Peven, and R. D. McDonald. 2001. Estimating in-river survival of migrating salmonid smolts using radio-telemetry. Canadian Journal of Fisheries and Aquatic Sciences 58:1887–1997.

Stuehrenberg, L., A. Giorgi, and C. Bartlett. 1990. Pulse-coded radio tags for fish identification. Pages 370–374 *in* N. C. Parker, A. E. Giorgi, R. C. Heidinger, D. B. Jester, Jr., E. D. Prince, and G. A. Winans, editors. Fish-marking techniques. American Fisheries Society, Symposium 7, Bethesda, Maryland.

Stutzer, G. M., J. Ogawa, N. J. Hetrick, and T. Shaw. 2006. An initial assessment of radio telemetry for estimating juvenile coho salmon survival, migration behavior, and habitat use in response to Iron Gate Dam discharge on the Klamath River, California. U.S. Fish and Wildlife Service, Arcata Fish and Wildlife Office, Arcata Fisheries Technical Report TR2006–05, Arcata, California.

Wertheimer, R. H., and A. F. Evans. 2005. Downstream passage of steelhead kelts through hydroelectric dams on the lower Snake and Columbia Rivers. Transactions of the American Fisheries Society 134:853–865.

Welch, D. W., G. W. Boehlert, and B. R. Ward. 2003. POST–the Pacific Ocean salmon tracking project. Oceanologica Acta 25:243–253.

Welch, D. W., M. C. Melnychuk, E. R. Rechinsky, A. D. Porter, M. C. Jacobs, A. Ladouceur, R. S. McKinley, and G. D. Jackson. 2009. Freshwater and marine migration and survival of engendered Cultus Lake sockeye salmon (*Oncorhynchus nerka*) smolts using POST, a large-scale acoustic telemetry array. Canadian Journal of Fisheries and Aquatic Sciences 66:736–750.

White, G. C., and K. P. Burnham. 1999. Program MARK: survival estimation from populations of marked animals. Bird Study 46(Supplement):120–138.

Index

A

B

D

E

G

H

I

J

K

L

M

N

O

P

Q

R

S

T

U

V

W

Y

Z